Synthetic Biology of Yeasts

Farshad Darvishi Harzevili
Editor

Synthetic Biology of Yeasts

Tools and Applications

Springer

Editor
Farshad Darvishi Harzevili
Department of Microbiology
Alzahra University
Tehran, Iran

ISBN 978-3-030-89679-9 ISBN 978-3-030-89680-5 (eBook)
https://doi.org/10.1007/978-3-030-89680-5

This Springer imprint is published by the registered company Springer Nature Switzerland AG
The registered company address is: Gewerbestrasse 11, 6330 Cham, Switzerland

Preface

Yeasts are one of the most attractive microbial cell factories for the production of a wide range of valuable products, including pharmaceuticals, nutraceuticals, cosmetics, agrochemicals, biofuels, and so on. Tools of synthetic biology have been developed to improve the metabolic engineering of yeasts in a faster and more reliable manner. Nowadays, synthetic biology tools have been used to make synthetic pathways and rewiring metabolism to produce valuable products at high titer, rate, and yield.

This book covers recent advances and future trends in the synthetic biology of yeasts. The book consists of two parts. The first part opens with an introduction to rational metabolic pathway prediction and design using computational tools and their applications for yeast systems and synthetic biology (Chap. "Rational Metabolic Pathway Prediction and Design: Computational Tools and Their Applications for Yeast Systems and Synthetic Biology"). After rational metabolic pathway design, synthesis, and engineering tools for assembly of standardized biobricks, powerful genome editing and optimization of synthetic pathways are very important in metabolic engineering of yeasts. Hence, construction and assembly of standardized biobricks for synthetic pathways engineering in yeasts were summarized in Chap. "Construction and Assembly of Standardized Biobricks for Synthetic Pathways Engineering in Yeasts".

Chapter "Cellular Engineering of *Yarrowia lipolytica* for Biomanufacturing of High-Value Products from Oils and Fats" presented an example about cellular engineering of yeasts. A combined cell morphology engineering and metabolic pathway optimization of non-conventional yeast *Yarrowia lipolytica* is used to biomanufacturing of high-value products from oils and fats as hydrophobic substrates. Chapter "Whole Cell Yeast-Based Biosensors" provided an overview about engineering of yeast cells to act as whole-cell yeast-based biosensors. The highly modular nature of yeast-based biosensors suggests that they can be a useful tool for detecting a wide range of biologically relevant analytes, including pharmaceuticals, toxins, and chemically undefined or complex mixtures ranging from environmental mutagens to bioavailable phosphorous.

The second part, Chaps. "Yeast Synthetic Biology Approaches for the Production of Valuable Polyphenolic Compounds"–"Synthetic Biology in the *Candida*

(CTG) Clade", covers applications of synthetic biology to produce diverse and attractive products by some well-known yeasts. Chapter "Yeast Synthetic Biology Approaches for the Production of Valuable Polyphenolic Compounds" summarized yeast synthetic biology approaches for the production of valuable phenolic compounds including hydroxycinnamic acids, flavonoids, stilbenoids, coumarins, curcuminoids, and polyphenolic amides. Production of artemisinin as an antimalarial drug with anti-cancer, anti-inflammation, anti-viral, and anti-SARS-CoV-2 activities via yeast metabolic engineering was discussed in Chap. "Yeast Synthetic Biology for Production of Artemisinin as an Antimalarial Drug". Chapter "Yeast Synthetic Biology for the Production of Terpenoids Derived from Traditional Chinese Medicinal Plants" reviewed yeast synthetic biology applications for the production of terpenoids derived from traditional Chinese medicinal plants. The synthetic biology of yeasts for the production of fragrant compounds like citrus flavors was reviewed in Chap. Application of Yeast Synthetic Biology for the Production of Citrus Flavors. Synthesis of polyols and organic acids by metabolically engineered *Yarrowia lipolytica* strains is summarized in Chap. "Synthesis of Polyols and Organic Acids by Wild-Type and Metabolically Engineered *Yarrowia lipolytica* Strains". Chapter Recent Advances in Synthetic Biology Applications of *Pichia* Species reviewed recent advances in synthetic biology applications for *Pichia* species. *Kluyveromyces marxianus* as a platform in synthetic biology for the production of useful materials was considered in Chap. "*Kluyveromyces marxianus* as a Platform in Synthetic Biology for the Production of Useful Materials". In the last chapter, synthetic biology in the *Candida* (CTG) clade was summarized.

Overall, this book will serve as a suitable reference for students, scientists, and researchers at universities, industries, corporations, and government agencies interested in yeast synthetic biology, systems biology, and metabolic engineering as well as all disciplines related to biotechnology, microbiology, bioengineering, and chemical engineering.

Harzevil, Gilan, Iran Farshad Darvishi Harzevili

Contents

Synthetic Biology Tools and Cellular Engineering

Rational Metabolic Pathway Prediction and Design: Computational Tools and Their Applications for Yeast Systems and Synthetic Biology 3
Pedro A. Saa

Construction and Assembly of Standardized Biobricks for Synthetic Pathways Engineering in Yeasts 27
Paulina Korpys-Woźniak, Monika Kubiak, Monika Borkowska, and Ewelina Celińska

Cellular Engineering of *Yarrowia lipolytica* for Biomanufacturing of High-Value Products from Oils and Fats 63
Na Liu, Ya-Hue Valerie Soong, Andrew Olson, and Dongming Xie

Whole Cell Yeast-Based Biosensors 91
Heather A. M. Shepherd, Emilia-Maria A. Bondarenko, Katherine M. Jennings, Rachel A. Miller, and Holly V. Goodson

Applications of Yeast Synthetic Biology

Yeast Synthetic Biology Approaches for the Production of Valuable Polyphenolic Compounds 119
Daniela Gomes, João Rainha, Ligia R. Rodrigues, and Joana L. Rodrigues

Yeast Synthetic Biology for Production of Artemisinin as an Antimalarial Drug 157
Arman Beyraghdar Kashkooli, Karim Farmanpour-Kalalagh, and Alireza Babaei

Yeast Synthetic Biology for the Production of Terpenoids Derived from Traditional Chinese Medicinal Plants 181
Yongjun Wei

Application of Yeast Synthetic Biology for the Production of Citrus Flavors 207
Karim Farmanpour-Kalalagh, Arman Beyraghdar Kashkooli, and Alireza Babaei

Synthesis of Polyols and Organic Acids by Wild-Type and Metabolically Engineered *Yarrowia lipolytica* Strains 227
Chong Li, Weichao Lin, Khai Lun Ong, Jinhua Mou, Carol Sze Ki Lin, and Patrick Fickers

Recent Advances in Synthetic Biology Applications of *Pichia* Species ... 251
Wan Sun, Yimeng Zuo, Zhanyi Yao, Jucan Gao, Zengyi Shao, and Jiazhang Lian

***Kluyveromyces marxianus* as a Platform in Synthetic Biology for the Production of Useful Materials** 293
Noppon Lertwattanasakul, Mochamad Nurcholis, Nadchanok Rodrussamee, Tomoyuki Kosaka, Masayuki Murata, and Mamoru Yamada

Synthetic Biology in the *Candida* (CTG) Clade 337
Dalal Kasir, Sébastien Besseau, Marc Clastre, Audrey Oudin, Monzer Hamze, Vincent Courdavault, Marwan Osman, and Nicolas Papon

Synthetic Biology Tools and Cellular Engineering

Rational Metabolic Pathway Prediction and Design: Computational Tools and Their Applications for Yeast Systems and Synthetic Biology

Pedro A. Saa

Abstract

Continuous progress in metabolic engineering of microbial cell factories like yeast requires the support of computational tools for finding novel unintuitive biotransformations routes. In this chapter, a succinct overview is provided of the most relevant computational tools for pathway prediction by retro-biosynthesis, and pathway design through stoichiometry-based optimization methods. Illustrative case studies are also presented showcasing different strategies for pathway optimization in yeast, namely redox cofactor balancing, improved precursor supply, and heterologous expression of carbon fixation pathways. Finally, challenges and limitations hindering the broad adoption and implementation of these tools for metabolic engineering will be discussed.

Keywords

Pathway design • Pathway analysis • Metabolic engineering • Yeast

1 Introduction

Advancements in metabolic engineering and synthetic biology have enabled accelerated engineering of microbial factories for the production of valuable chemicals (Smolke and Tyo 2012; Lee and Kim 2015; Isaacs et al. 2011), realizing the

P. A. Saa (✉)
Departament of Chemical and Bioprocess Engineering, Pontificia Universidad Católica de Chile, Santiago, Chile
e-mail: pnsaa@ing.puc.cl

Institute for Mathematical and Computational Engineering, Pontificia Universidad Católica de Chile, Santiago, Chile

F. Darvishi Harzevili (ed.), *Synthetic Biology of Yeasts*,
https://doi.org/10.1007/978-3-030-89680-5_1

promise of a more sustainable (bio)economy (Voigt 2020). To keep pace with these expectations, pathway prediction and design play a crucial role in finding novel pathways for various applications like drug discovery (Galanie et al. 2015; Moura et al. 2016; Hafner et al. 2021) and value-added biochemical production (Yim et al. 2011; Tokic et al. 2018; Henry et al. 2010a, b). In this scenario, metabolic workhorses like yeast could be greatly benefited by broadening their product spectrum and improving their metabolic capabilities and performance in terms of their yields, titers, and productivities (Nielsen and Keasling 2016; Ko et al. 2020). For this task, progress in computational tools and methods capable of guiding experimental efforts is crucial for the optimization of cellular metabolism and incorporation of synthetic designs for the production of unnatural heterologous compounds.

In this chapter, an overview of the most relevant retro-biosynthesis and pathway optimization methods is provided with a focus on tools with direct application in metabolic engineering tasks. Starting with the reconstruction of a comprehensive reaction network from public databases and resources, both retro-biosynthesis tools for de novo pathway prediction and stoichiometry-based pathway optimization methods for metabolic redesign are described. Particularly in the latter case, convenient engineering objectives taking into consideration product yield, cofactor use, thermodynamic plausibility, and enzyme cost are discussed. Additionally, several relevant pathway engineering case studies in yeast are also presented, highlighting the improvement potential from the implementation of rational pathway designs. Finally, perspectives on the increasing adoption of these tools for metabolic engineering as well as limitations reducing their effectiveness are discussed.

2 In Silico Pathway Prediction and Design

Retro-biosynthesis and stoichiometry-based optimization methods have been established for pathway design and prediction, differing mostly in their scope and methodology. While both tools generate metabolic pathways producing the target metabolite, they do so by applying fundamentally different reaction network representations (i.e., graph or stoichiometric matrix) and search algorithms (i.e., optimization-based enumeration or retro-synthetic search) (the reader is referred to Wang et al. (2017) for a comprehensive review). Furthermore, their computational complexity and efficacy can vary significantly depending on the product of interest, and thus, careful selection of the appropriate tool for the case at hand is a must (Saa et al. 2019). In the following, the most relevant data- and knowledge-bases for reconstructing and parameterizing reaction networks are presented, which constitute the starting point for the application of any of these tools (Fig. 1a). Then, the most relevant retro-biosynthesis (Fig. 1b) and stoichiometry-based optimization methods for pathway prediction and design (Fig. 1c) are described.

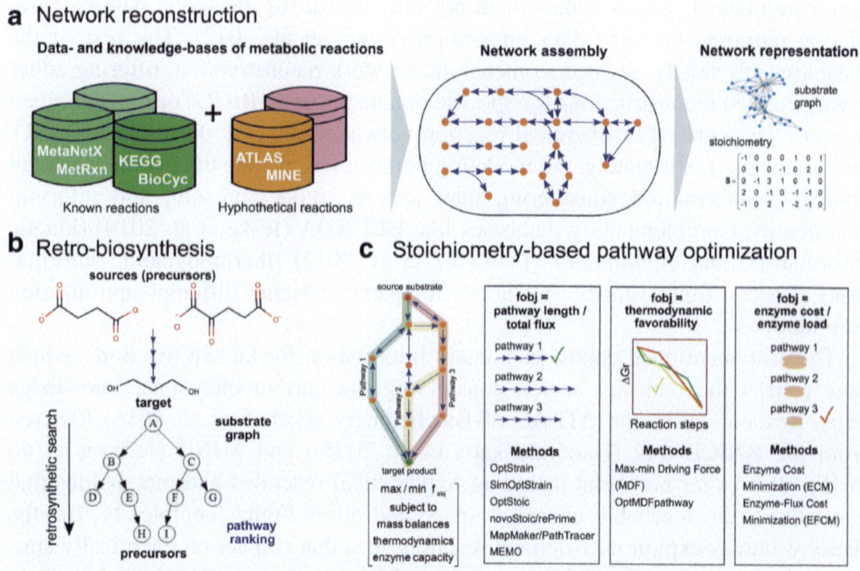

Fig. 1 Workflow for reaction network reconstruction and application of metabolic pathway prediction and design tools. **a** Assembly of accumulated metabolic reaction data into a comprehensive reaction network is a requirement for the application of the reviewed tools. Depending on the application objective, different network representations are employed for either predicting de novo pathways (i.e., retrosynthesis typically using a graph representation), or (re)designing pathways for higher metabolic performance (i.e., optimization-based pathway design using an stoichiometric representation). **b** Retro-biosynthetic tools explore a substrate graph seeking to connect the target molecule with some predefined precursors. Starting with the target molecule and moving backwards, these tools can generate several possible pathways that are typically ranked using different criteria (e.g., length, enzyme availability, thermodynamics, among others). **c** Stoichiometry-based pathway prediction methods employ a reaction network with known and fixed reactions to enumerate mass-balanced pathways that optimize a desired objective such as product yield, pathway length, thermodynamic favorability, and enzyme cost. In this case, different types of constraints can be defined to restrict the feasible solution space and narrow the search upfront

2.1 Data- and Knowledge-Bases for Metabolic Reaction Network Reconstruction

Databases for pathway search are an absolute requirement for exploring the feasible reaction space, as they contain the critical information of how metabolites are connected to others through biochemical reactions. There are numerous public data- and knowledge-bases populated with metabolic reaction data. Among the most popular, KEGG (Kanehisa et al. 2016), MetaCyc (Caspi et al. 2016), BIGG (King et al. 2016), KBase (Arkin et al. 2018), ModelSEED (Henry et al. 2010a, b), MetRxn (Kumar et al. 2012), and MetaNetX (Ganter et al. 2013; Moretti et al. 2021) stand out to name a few (for a more details refer to Wang et al. (2017)). Some of these databases (KEGG, MetaCyc, and Kbase) integrate multiple sources of biological information, e.g., genetic, molecular, physicochemical, and experimental, which makes them not only useful for metabolic pathway prediction purposes but also data integration (Lewis et al. 2012). The rest of the databases are mostly devoted to metabolic network reconstruction, offering either highly curated reconstructions for specific organisms (e.g., BIGG) or broader albeit possibly less curated biochemical reaction networks (e.g., MetRxn, ModelSEED, and MetaNetX). Ultimately, the modeling purpose will dictate the most convenient source of information considering their specific scope, breadth, and information quality. Complementary databases like BRENDA (Jeske et al. 2019) (kinetic information) and eQuilibrator (Flamholz et al. 2012) (thermodynamic information) also constitute valuable resources for parameterizing different optimization formulations.

The aforementioned databases contain information for known reactions, which may restrict the pathway search considering the current enzymatic knowledge gaps. Resources like the ATLAS of Biochemistry (Hadadi et al. 2016) (derived from the BNICE tool (Hatzimanikatis et al. 2005)) and MINE (Jeffryes et al. 2015) offer larger networks including hypothetical reactions and metabolites that can expand the reachable chemical space and allow higher complexity. Briefly, these resources exploit user-defined reaction rules that can act on chemically similar compounds, thereby yielding new hypothetical reactions. The latter reactions have recently been shown to enable filling some of the gaps in current enzyme-reaction associations (Hadadi et al. 2019). Lastly, another significant and recent tool for proposing hypothetical reactions that has been employed for pathway prediction is rePrime used by the novoStoic tool (Kumar et al. 2018) (see subsection 2.3 for more details). The former method extracts reaction rules from molecular signatures found in annotated reactions—defined by the presence of a set of chemical 'moieties'—for proposing hypothetical enzymatic transformations with a high structural encoding fidelity. Unfortunately, this tool currently lacks an associated open database for its use.

2.2 Pathway Prediction Using Retro-Biosynthesis Tools

Firstly, a distinction is made between retro-biosynthesis and classical retrosynthesis, as the latter is focused on the design of chemical reaction pathways, typically without relying on enzyme catalysis (Lin et al. 2019). Retro-biosynthesis tools seek to identify de novo biosynthetic pathways for the production of valuable compounds from inexpensive precursors using known and hypothetical enzyme activities (Wang et al. 2017; Lin et al. 2019). Another—though less explored—application of these tools involves the opposite, that is, the prediction of novel enzymatic routes for the degradation of recalcitrant compounds, e.g., for bioremediation purposes (Finley et al. 2009, 2010; Ellis et al. 2006). For pathway prediction, retro-biosynthesis tools explore the full chemical space for synthetic pathways toward the target compound. For this task, these tools typically represent the network as a (substrate) graph that can be readily traversed using known enumeration algorithms. Graph traversal is possible by connecting the substrates using various criteria based on structural (chemical) similarity, reaction promiscuity, and defined reaction rules. In the following, a relevant subset of retro-biosynthesis tools employed for metabolic engineering/synthetic biology applications is described (Table 1).

One of the most established tools for de novo pathway retro-biosynthesis is BNICE (Hatzimanikatis et al. 2005). This framework employs predefined 'generalized enzymatic reaction rules' (encoded in a bond-electron matrix) that are applied to precursor molecules on their reactive sites to yield new product molecules. BNICE uses a substrate graph representation of the chemical network, which can be traversed using graph search algorithms starting from the target compound and moving backwards until connecting with one of the defined precursors. Different pruning criteria are employed to keep the search breadth computationally tractable.

Table 1 Retro-biosynthesis tools for metabolic pathway prediction

Tool	Database	Pathway scoring	Applications and references	Source
BNICE	KEGG, ATLAS	Pruning criteria assessment (thermodynamics, pathway length, etc.)	In silico identification of novel pathways and enzyme candidates validated experimentally for the production of (S)-tetrahydropalmatine derived from intermediates of the known benzylisoquinoline alkaloid biosynthetic pathway (Hafner et al. 2021) In silico evaluation of putative feasible biodegradation pathways for 1,2,4-trichlorobenzene (Finley et al. 2010) Enzyme annotation for orphan and novel reactions from KEGG (Hadadi et al. 2019) Discovery of novel metabolic pathways for the biosynthesis of 3-hydroxypropanoate (Henry et al. 2010a)	https://lcsb-databases.epfl.ch/pathways/GraphList
PathPred	KEGG, RPAIR	Compound similarity and pathway scores	Prediction of putative biodegradation pathways for xenobiotics (e.g., 1,2,3,4-tetrachlorobenzene) as well as biosynthetic pathways for plants secondary metabolites (e.g., gentiodelphin)	https://www.genome.jp/tools/pathpred/

(continued)

Table 1 (continued)

Tool	Database	Pathway scoring	Applications and references	Source
SimPheny (BioPathway Predictor)	BiGG	Pathway length, thermodynamics, product yield, number of known metabolites/enzymes	In silico prediction and subsequent experimental validation of an unprecedented synthetic pathway for 1,4-butanediol production in *E. coli* (Yim et al. 2011)	-
RetroPath	MetaCyc, BioCyc	Pathway thermodynamics, sequence similarity, pathway length, number of putative enzyme steps, and product yield	Prediction of successfully-implemented biosynthetic pathways for 146 compounds in metabolic engineering projects. 81% of the compounds were predicted with at least one pathway (Delépine et al. 2018)	https://www.myexperiment.org/workflows/4987.html

At the end of the algorithm, pathways are ranked by features such as pathway length, thermodynamics, among others. Methodologically close to BNICE, PathPred (Moriya et al. 2010) uses instead RDM patterns consisting of reaction center atoms (R), atoms of different regions (D), and atoms of the matched region (M) for exploring the substrate graph. Pruning of the network is executed using structural similarity criteria, and pathway ranking is performed using compound similarity and pathway scores. SimPheny (Yim et al. 2011; Schilling et al. 2005) uses reactions rules from the third Enzyme Commission (EC) number level for generating reaction rules that enable reaction promiscuity for broader explorations. In this case, a retro-synthetic search is employed for enumerating feasible routes that produced intermediates of reasonable size (i.e., below a predefined size), and later they are ranked based on various criteria. Another retro-biosynthesis tool with recent important applications is RetroPath (Delépine et al. 2018). This tool uses a retro-synthetic search, albeit combined with MILP formulations for the application of various ranking criteria, such as thermodynamics, gene prediction, pathway length, number of putative steps, and product yield. In contrast to BNICE, RetroPath maintains a stoichiometric representation of the network (as opposed to a substrate graph) that enables computation of various scores. Moreover, molecular signatures are used to generate reaction rules based on a substructure of adjacent atoms, enabling the generation of substantially more and flexible reaction rules (Duigou et al. 2018). A recent implementation of reinforcement learning in RetroPath (RetroPath RL) has yielded promising results in the retro-biosynthetic prediction of biologically relevant pathways (Koch et al. 2020).

2.3 Stoichiometry-Based Optimization Methods for Pathway (Re)design

Given a universal metabolic reaction network, the 'pathway design' problem seeks to identify 'optimal' route(s) for the production of the target compound. As opposed to the retro-biosynthesis problem, possible connecting reactions are fixed and known upfront. Construction of the reaction network knowledge base is achieved by combining metabolic data from curated databases and/or from databases that also include putative reactions derived, for example, from generalized reaction rules (Hadadi et al. 2016; Hatzimanikatis et al. 2005; Jeffryes et al. 2015) or molecular signatures (Kumar et al. 2018). Regardless of the source of the data, reaction data must be charge- and mass-balanced to yield correct results, which typically is ensured in a manual curation step. Metabolite/reaction name inconsistencies are also an important source of issues that affect network connectivity and consistency, which often have to be resolved manually. While there have been attempts to standardize reaction and metabolite identifiers (King et al. 2016; Kumar et al. 2012; Alcántara et al. 2012), name reconciliation is challenging due to the incessant annotation of new metabolites and enzymatic activities, albeit important progress has been made in recent database versions (Moretti et al. 2021).

Once the reaction network has been assembled and mathematically formulated into a stoichiometric matrix, prediction of different pathway designs can be readily computed using optimization-based methods provided a convenient objective function. Among the most relevant objectives, one can name the minimization of the pathway length ensuring a minimum product yield (e.g., by fixing the overall stoichiometry) (Pharkya et al. 2004), maximization of the product yield observing thermodynamic constraints (Kumar et al. 2018; Kamp and Klamt 2020; Chowdhury and Maranas 2015), maximization of the thermodynamic favorability of the pathway (Flamholz et al. 2013; Noor et al. 2014; Hädicke et al. 2018; Yang et al. 2020; Ng et al. 2019), and minimization of the pathway´s enzymatic cost (Flamholz et al. 2013; Ng et al. 2019; Court et al. 2015; Bar-Even et al. 2010). For each of these objectives, different optimization problems must be formulated and solved, often requiring various parameters (e.g., thermodynamic and kinetic) from other sources for computing the optimal solution(s). In the following, the most relevant stoichiometry-based optimization methods for metabolic pathway prediction are presented. Further details about the methods and applications can be found in Table 2.

2.3.1 Pathways with Desired Stoichiometric Properties

Constraint-based modeling methods (Edwards and Palsson 2000) can be readily adapted for the computation of pathways exhibiting a desired stoichiometry (i.e., yield) (Kumar et al. 2018; Chowdhury and Maranas 2015; Ng et al. 2019), shortest length (Chowdhury and Maranas 2015; Ng et al. 2019), convenient precursor use (Kamp and Klamt 2020), and if using an organism's reaction network as metabolic chassis, minimum addition of exogenous reactions (Pharkya et al. 2004; Kim et al.

Table 2 Optimization-based methods for metabolic pathway design

Pathway design objective	Method	Goal(s)	Applications and references
Desired stoichiometric properties (e.g., yield, length, among others)	OptStrain	Maximization of product yield using the minimum number of heterologous reactions in a production host	Prediction of in silico production pathways for hydrogen in *E. coli*, *C. acetobutylicum*, and *M. extorquens* AM1, as well as exploration of interventions strategies for achieving growth-coupling (Pharkya et al. 2004) In silico prediction of heterologous vanillin production pathways in *E. coli* and computation of their maximum theoretical yields and number of non-native enzymes to be included (Pharkya et al. 2004)
	SimOptStrain	Maximization of growth-coupled production including heterologous reactions in a production host	Computation of optimal succinate and glycerol production pathways using *E. coli* as metabolic chassis in terms of gene deletions and non-native reaction additions (Kim et al. 2011)
	OptStoic	Optimization of overall reaction stoichiometry (e.g., max. product yield, co-factor use, etc.), and minimization of the length or total flux through the pathway capable of sustaining the desired stoichiometry	In silico verification of the non-oxidative glycolysis pathway design (Chowdhury and Maranas 2015) Computation of optimal conversions and novel pathways for co-utilization of carbon dioxide and methanol for the production of C2+ metabolites (e.g., malonate, citrate, succinate, acetate, among others) (Chowdhury and Maranas 2015) Evaluation of feasible glycolytic alternatives and optimality analysis of trade-offs in ATP yield, enzymatic cost, and maximum-minimum driving force (Ng et al. 2019)

(continued)

Table 2 (continued)

Pathway design objective	Method	Goal(s)	Applications and references
	MEMO	Minimization of the size of the regeneration pathway (module) needed for sustaining cofactor(s) requirement(s) of a production pathway	Enumeration of optimal regeneration modules capable of sustaining the operation of the synthetic crotonyl-CoA/ethylmalonyl-CoA/hydroxybutyryl-CoA (CETCH) cycle for in vitro carbon dioxide fixation in cell-free systems (Kamp and Klamt 2020)
	GEM-Path	Maximization of growth-coupled production including putative (promiscuous) enzymatic reactions in a production host	Computation of 245 unique synthetic pathways for the production of bulk chemicals and integration of these pathways into *E. coli* metabolism for yield analysis and metabolic optimization (Campodonico et al. 2014)
	novoStoic	Minimization of the pathway length Or Maximization of profit	In silico verification of synthetic pathways for 1,4-butendiol (BDO) production, identification of putative pathways for phenylephrine synthesis and benzo[a]pyrene degradation (Kumar et al. 2018)
	MapMaker/ PathTracer	Minimization of carbon transfers between reactants and products using the shortest pathway	Identification of possible conversion pathways to putrescine from glutamate, and CO_2 fixation pathways in *E. coli* (Tervo and Reed 2016)

(continued)

Table 2 (continued)

Pathway design objective	Method	Goal(s)	Applications and references
Thermodynamic favorability	OptMDFpathway	Maximization of the minimum driving force of the production pathway	Computation of all substrate-product combinations in *E. coli* that enable net CO_2 assimilation via thermodynamically feasible pathways (Hädicke et al. 2018) In silico design of multi-strain communities of a single-species (*E. coli*) that achieve maximum thermodynamic driving force for product synthesis under the assumption of pathway specialization (i.e., parts of the production pathway are performed by different community members) (Bekiaris and Klamt 2021)
Enzymatic cost or burden	ECM	Minimize enzyme cost or metabolic burden	In silico computation of putative efficient CO_2 fixation pathways into C2-C4 compounds (Bar-Even et al. 2010) Optimality analysis of enzyme cost minimization in central carbon metabolism of *E. coli* (Noor et al. 2016) Prediction of rate/yield trade-offs in metabolic pathways under oxygen-limited conditions in *E. coli* (Wortel et al. 2018)

2011). In practice, these pathway enumeration methods rely on the solution of various LP and/or MILP optimization problems that optimize some of the aforementioned objectives subject to not only stoichiometric constraints but possibly to thermodynamic and/or economic constraints.

OptStrain, SimOptStrain, and OptStoic are classical tools for pathway prediction, although they differ in their scope. While the first two seek to predict optimal pathways and metabolic interventions for the production of a target metabolite leveraging the microbial host reaction network (Pharkya et al. 2004; Kim et al. 2011), the latter aims to find complete mass-balanced conversion pathways that yield a desired stoichiometry from precursors to product(s) using metabolic databases as the input reaction network (Chowdhury and Maranas 2015). Additional constraints related to a minimum guaranteed product yield, thermodynamic plausibility of the pathway, and/or substrate costs can be readily included to obtain more convenient designs. Recently, a computational method called MEMO (Kamp and Klamt 2020) has been proposed for identifying the smallest metabolic modules with specified stoichiometric and thermodynamic properties. For instance, this approach has been employed to find small cofactor regeneration (e.g., ATP/ADP, NAD(P)H, NAD(P), among others) modules that can sustain bioconversions in the context of cell-free applications under defined thermodynamic conditions.

The aforementioned methods rely on existing annotated enzymatic reactions for metabolic conversions. However, as mentioned in the previous subsection, promiscuous enzymatic activities are characteristic features of metabolic reaction networks, likely playing an evolutionary role as a starting point in enzyme functions (Khersonsky and Tawfik 2010). Importantly, the existence of this feature suggests that there is still untapped potential for a broader chemical reaction space to be explored. By using various extraction techniques for learning putative reactions from known enzymatic reactions, it is possible to populate and assemble larger databases for pathway prediction. An example of these methods is MapMaker/PathTracer (Tervo and Reed 2016), which employs precomputed carbon transfer maps (CTMs) based on chemical and stoichiometric information (MapMaker) for the prediction of short, carbon-balanced pathways from substrates to products (PathTracer). GEM-Path (Campodonico et al. 2014) is another framework that, using a genome-scale metabolic reconstruction of *E. coli* as base reaction network, combines heterologous pathway integration (similar to OptStoic) with constraint-based growth-coupled methods for the computation of metabolic designs. Increased biochemical reaction exploration is achieved through the introduction of a chemical similarity measure to assess enzyme-catalyzed reaction promiscuity. Lastly, the novoStoic/rePrime framework (Kumar et al. 2018) enables exploration of a far greater chemical transformation space through the imposition of chemical 'moiety' conservation (refer to Sect. 2.1) that is particularly suited for the prediction of optimal pathways with maximum yield or length. Importantly, this mathematical treatment avoids chemical reaction information loss (e.g., stereoselectivity) as opposed to other approaches like MapMaker/PathTracer.

2.3.2 Pathways with Maximum Thermodynamic Favorability

Pathway thermodynamics exerts a fundamental control in metabolic flux with seemingly important consequences for microbial fitness (Du et al. 2018). While there have been different methods for combining stoichiometric-based analysis with reaction thermodynamics (Henry et al. 2007; Kummel et al. 2006), it has not been until recently that thermodynamic favorability has been mathematically formalized. For this task, the Max-min Driving Force (MDF) index (Noor et al. 2014) has been proposed for quantifying the smallest absolute Gibbs free energy (or driving force) of a given pathway under the most favorable metabolic conditions. As the latter captures the driving force of the most unfavorable conversion step (i.e., thermodynamic bottleneck), its maximization yields the most favorable operating conditions for a *given* pathway. More recently, the OptMDFpathway method (Hädicke et al. 2018) was introduced to identify the most thermodynamically favorable pathways in a *given* reaction network, thereby enabling exploration of thermodynamically plausible production pathways in the context of microbial metabolism (Hädicke et al. 2018; Yang et al. 2020), and more recently, in microbial communities (Bekiaris and Klamt 2021).

2.3.3 Pathways with Minimum Enzymatic Cost

Cellular metabolism incurs a metabolic cost when committing to the synthesis of a particular set of proteins (enzymes). As seemingly similar enzymes can still display large differences in their catalytic properties (Bar-Even et al. 2011), it is natural to seek pathways that can yield the maximum return of investment (flux) per protein (enzyme) mass synthesized. For this task, the Enzyme Cost Minimization (ECM) (Noor et al. 2016)—and later termed the Enzyme-Flux Cost Minimization (EFCM) (Wortel et al. 2018)—formulation computes the minimum enzyme load (i.e., the aggregated enzyme mass allocated) required for a metabolic pathway to yield a *given* flux (Flamholz et al. 2013; Bar-Even et al. 2010). While this formulation originally required a thermodynamically consistent, fully parameterized kinetic model for this calculation (Saa and Nielsen 2017), increasingly enzymatically-constraint GSMMs (Sánchez et al. 2017) and ME-models (metabolic and expression) (Lerman et al. 2012) are being considered and employed for these calculations under the optimistic scenario of enzymatic catalysis at capacity. Finally, the ECM/EFCM does not support performing pathway enumeration, although it can be readily employed as a ranking index when combined with the previous approaches.

3 Case Studies of Metabolic Pathway Prediction and Optimization in Yeast

In this section, selected case studies illustrate different pathway engineering aspects required for improving metabolic performance overall, and particularly, in yeast. These examples showcase strategies for redox cofactor balancing, increased precursor supply, and engineering of central pathways for carbon fixation. The impact of the latter applications is especially highlighted in the context of harnessing the metabolic potential of yeast for industrial bioproduction. Figure 2 illustrates the details of the revised strategies.

3.1 Balancing Redox Cofactor Supply for Improving Substrate Utilization and Isoprenoids Production

Regeneration of either redox and/or energy cofactors often limits the production of high-value metabolites. In order to increase the availability of the required cofactor(s), central carbon metabolism must be intervened and engineered in such a way that it favors bioproduction without extremely affecting microbial growth (Lee et al. 2013). This challenge is particularly relevant for many NAD(P)H-expensive valuable compounds that are being produced in yeast (Cataldo et al. 2020; López et al. 2020, 2019) and other microbes (Ko et al. 2020).

Increased supply of redox cofactors can be achieved by either overexpressing key enzymes involved in cofactor generation (Lee et al. 2007; San et al. 2002; Lim et al. 2002) or by increasing the expression of alternative redox partner systems. A recent application of the latter has proved effective for enhancing the unprecedented heterologous production of violaxanthin in *S. cerevisiae* by approx. two-fold (Cataldo et al. 2020). However, the success of these approaches is likely limited due to the presence of different intrinsic balancing mechanisms for maintaining homeostasis in yeast (Hou et al. 2010). An illustrative example of the latter can be found in the study of Nissen et al. (2001). Here, heterologous expression of the pyridine nucleotide transhydrogenase system (*sth* gene, absent in yeast) that transfers reducing equivalents from NADPH to NADH (and vice versa), did not improve ethanol formation in anaerobic conditions. On the contrary, ethanol production was reduced concomitantly with the increase of fermentation by-products (glycerol and 2-oxoglutarate) required for redox rebalancing. Another, less intuitive and possibly more effective, strategy for (re)balancing redox cofactors supply and demand involves cofactor swapping (Verho et al. 2003; Martínez et al. 2008). Computational studies in *S. cerevisiae* and *E. coli* support this strategy as a promising intervention for forcing higher metabolic performance (King and Feist 2014). Simply put, this approach seeks to replace native (redox-consuming) enzymes with heterologous counterparts with a different cofactor specificity (e.g., NAD(P)H—for a NA(D)H-dependent enzyme).

The first application of the latter strategy involved the optimization of D-xylose utilization for ethanol production in *S. cerevisiae* (Verho et al. 2003). This carbon source is assimilated through the pentose phosphate pathway (PPP) as D-xylulose-5-phosphate and then incorporated as glyceraldehyde 3-phosphate in glycolysis. In theory, D-xylose should produce CO_2 and ethanol in a 1:1 molar ratio under redox-neutral anaerobic conditions (Kötter and Ciriacy 1993). However, D-xylose assimilation requires extra NADPH and NAD^+ that must be regenerated by other separate processes, which are very inefficient in yeast, rendering D-xylose fermentation slow. To overcome this bottleneck and force higher

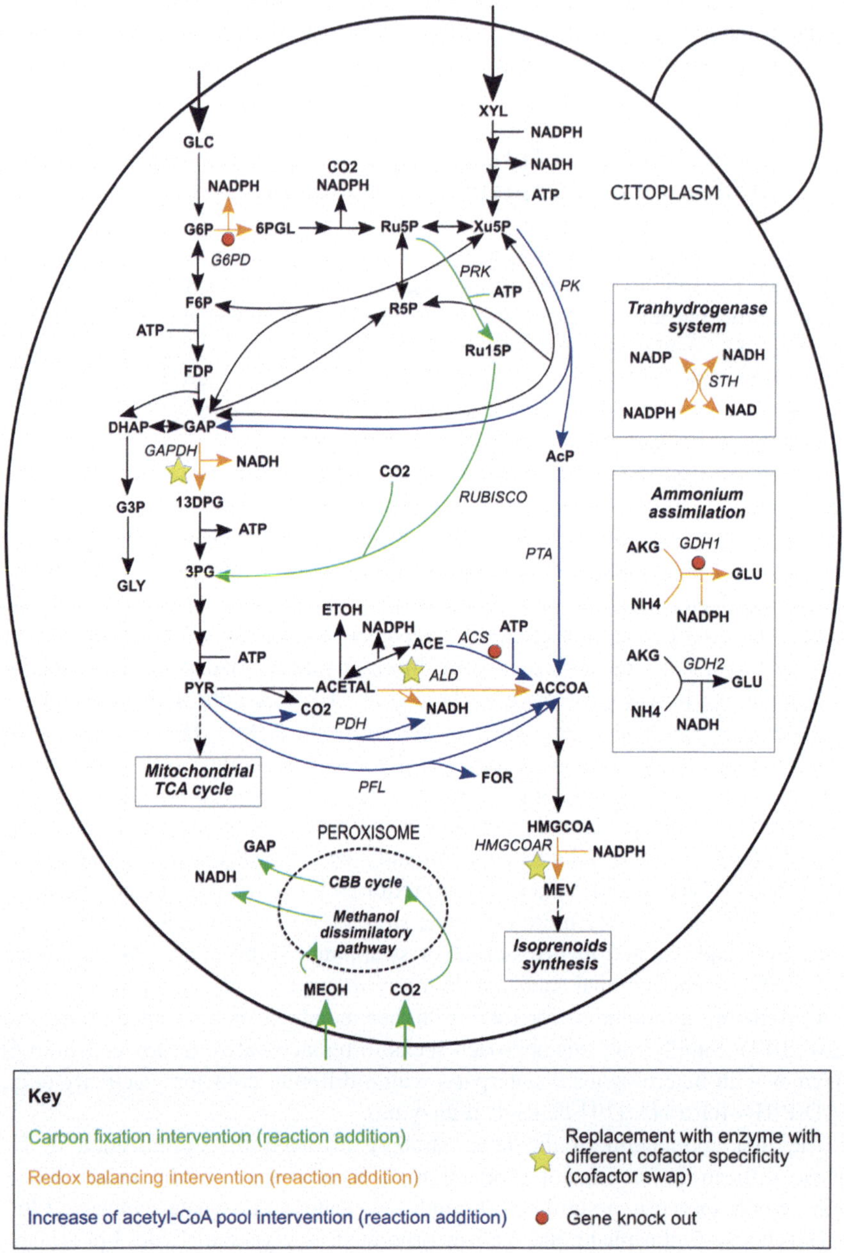
GLC
XYL
NADPH
NADH
ATP
CITOPLASM
CO2
NADPH
G6P
6PGL
Ru5P
Xu5P
G6PD
PRK
PK
F6P
R5P
ATP
Ru15P
FDP
DHAP
GAP
GAPDH
NADH
13DPG
G3P
GLY
CO2
RUBISCO
AcP
3PG
PTA
ETOH
NADPH
ACE
ACS
ATP
ALD
PYR
ACETAL
ACCOA
CO2
PDH
NADH
FOR
PFL
Mitochondrial TCA cycle
Tranhydrogenase system
NADP
NADH
STH
NADPH
NAD
Ammonium assimilation
AKG
GDH1
GLU
NH4
NADPH
AKG
GDH2
GLU
NH4
NADH
PEROXISOME
GAP
NADH
CBB cycle
Methanol dissimilatory pathway
MEOH
CO2
HMGCOA
HMGCOAR
NADPH
MEV
Isoprenoids synthesis
Key
Carbon fixation intervention (reaction addition)
Redox balancing intervention (reaction addition)
Increase of acetyl-CoA pool intervention (reaction addition)
Replacement with enzyme with different cofactor specificity (cofactor swap)
Gene knock out

◂**Fig. 2** Illustration of selected reported strategies for achieving improved cofactor balancing, increased acetyl-CoA supply and engineering CO_2 fixation in yeast. The details of each strategy are discussed in Sect. 3. Relevant metabolite names are represented by uppercase bold fonts, whereas enzyme names are represented by uppercase italics fonts. Abbreviations: 13DPG, 1,3-diphosphoglycerate; 3PG, 3-phosphoglycerate; 6PGL, 6-phospho-D-glucono-1,5-lactone; ACCOA, acetyl-CoA; ACE, acetate; ACETAL, acetaldehyde; AcP, acetyl phosphate; ACS, acetyl-CoA synthetase; AKG, alpha-ketoglutarate; ALD, aldehyde dehydrogenase; DHAP, dihydroxyacetone phosphate; ETOH, ethanol; F6P, D-fructose 6-phosphate; FDP, D-fructose 1,6-disphosphate; FOR, formate; G3P, glycerol 3-phospate; G6P, D-glucose 6-phosphate; G6PD, glucose 6-phosphate dehydrogenase; GAP, glyceraldehyde 3-phosphate; GAPDH, glyceraldehyde 3-phosphate dehydrogenase; GDH, glutamate dehydrogenase; GLC, D-glucose; GLU, glutamate; GLY, glycerol; HMGCOA, 3-hydroxy-3-methyl-glutaryl-CoA; HMGCOAR, HMG-CoA reductase; MEOH, methanol; MEV, mevalonate; NH4, ammonia; PDH, pyruvate dehydrogenase; PFL, pyruvate formate lyase; PK, phosphoketolase; PRK, phosphoribulokinase; PTA, phosphotransacetylase; PYR, pyruvate; R5P, D-ribose 5-phosphate; Ru15P, ribulose 1,5-disphosphate; Ru5P, D-ribulose 5-phosphate; STH, transhydrogenase; Xu5P, D-xylulose 5-phosphate; XYL, D-xylose

NADPH supply and flux through lower glycolysis, the native NAD-dependent GAP dehydrogenase (GAPDH) was replaced by an NADP-dependent GAPDH and the NADPH-dependent glucose-6-phosphate dehydrogenase (G6PD) was knocked out, which also prevented carbon loss as CO_2 (Verho et al. 2003). This strategy almost doubled the ethanol yield on D-xylose (from 18 to 41%) and reduced the CO_2/ethanol molar ratio close to the theoretical 1:1 (from 2.5 to 1.3). Later, expression of the heterologous phosphotransacetylase (PTA) and phosphoketolase (PK) for improving NADH reoxidation in the D-xylose utilization pathway generated an increase in ethanol yield (25% higher) without affecting the growth rate (Sonderegger et al. 2004).

Cofactor rebalancing and swapping strategies for the synthesis of NADPH-expensive isoprenoid-derived compounds have shown to be particularly effective in yeast. For instance, α-santalene production yields a net production of NADH and consumption of NADPH, which calls for the rebalancing of the cofactor supply (Scalcinati et al. 2012). By deleting known reactions involved in glutamate metabolism (ammonium assimilation) that consume NADPH (GDH1) and activating NAD-dependent counterparts (GDH2) (Nissen et al. 2000), the production of α-santalene was substantially improved. Similarly in a different study of protopanaxadiol production—another isoprenoid-derived compound—the availability of NADPH was enhanced by replacing the native NADH-generating acetaldehyde dehydrogenase (ALD2) with a functionally equivalent NADPH-generating enzyme (ALD6), resulting in a 11-fold increase in titer (Kim et al. 2018). Lastly, swapping of the native NADP-dependent 3-hydroxy-3-methyl-glutaryl-CoA reductase (HMG-CoA reductase)—third committed step of the mevalonate pathway responsible for the production of isoprenoid precursors—has also shown to increase the overall pathway flux in *E. coli* (Ma et al. 2011). This result was leveraged by Meadows et al. (2016) whereby an NADH-consuming HMG-CoA reductase from *Silicibacter pomeroyi* was employed for the overproduction of the sesquiterpene farnesene. Implementation of other computationally predicted major metabolic cofactor swaps like the alcohol dehydrogenase (ALCD2) and GAPD for the improved production of isoprenoids remains to be tested experimentally (King et al. 2016), as they could be potentially beneficial for boosting production as shown in other microorganisms (Martínez et al. 2008).

3.2 Increasing Cytosolic Acetyl-CoA Availability for Metabolic Production

Cytosolic acetyl-CoA is a key metabolite for the production of a range of valuable compounds in yeast (Rossum et al. 2016). Native production of this compound requires 2 mol of ATP and yields 2 mol of acetyl-CoA and 4 mol of NAD(P)H per mol of glucose (Rossum et al. 2016). To improve the availability of this precursor and lower the ATP cost, different heterologous enzymes have been introduced to either bypass the native aldehyde dehydrogenase (ALD) and acetyl-CoA synthetase (ACS) system using bacterial counterparts, i.e., A-ALD and PFL, that do not incur such high cost (Kozak et al. 2014a, b), or to enable acetyl-CoA biosynthesis in situ by expressing the pyruvate dehydrogenase (PDH) complex in the cytoplasm (Kozak et al. 2014a, b). While the former application showed mixed results in terms of growth and yield (mainly due to by-product accumulation), the second approach along with a knock-out of the native ACS reaction exhibited similar metabolic performance to the control, but at a lower ATP cost.

A different approach for improving acetyl-CoA availability relies on increasing its yield. For this task, the phosphoketolase pathway (PKP) was early suggested for the conversion of 1 mol of F6P into 3 mol of acetyl-P without carbon loss (Schramm and Racker 1957). Conversion of acetyl-P to acetyl-CoA can be later achieved by the reversible phosphotransacetylase (PTA) reaction (Rossum et al. 2016). This was initially implemented in yeast for improving D-xylose fermentation (Sonderegger et al. 2004) (refer to previous section). More recently, Bogorad et al. (2013) implemented the full PKP in *E. coli* and demonstrated almost stoichiometric conversion of C5 and C6 sugars into acetate under anaerobic conditions. A similar approach was replicated in yeast accompanied by several genetic interventions to increase acetyl-CoA-derived farnesene (Meadows et al. 2016). This non-native pathway increased carbon utilization by 25%, decreased oxygen consumption by 75%, and reached 15% v/v of farnesene. As illustrated here, increasing acetyl-CoA availability may be critical not only for maximizing production but also for overall improving metabolic performance.

3.3 Engineering a Heterologous CBB Cycle for CO_2 Fixation

There is a growing interest in the field for engineering carbon assimilation pathways in heterotrophs for improving product yields—e.g., by reducing carbon loss as CO_2 –, and most notably, for implementing one-carbon (C1) compounds (e.g., CO_2) fixation pathways to develop more sustainable fermentation bioprocesses.

In an early effort from Guadalupe-Medina et al. (2013), a heterologous Calvin–Benson–Bassham (CBB) cycle was implemented in *S. cerevisiae* seeking to improve ethanol yield by reducing carbon loss under anaerobic conditions. The authors noted that by expressing the CBB enzymes phosphoribulokinase (PRK) and ribulose-1,5-bisphosphate carboxylase-oxygenase (RuBisCO), a working pathway could be realized where CO_2 is effectively used as an electron acceptor for NADH oxidation, thereby coupling CO_2 fixation by RuBisCO with the fermentation redox balance. Importantly, this mechanism rendered NADH reoxidation through the native glycerol formation pathway unnecessary (90% reduction in glycerol titer), increasing ethanol yield by 14% (Guadalupe-Medina et al. 2013).

While the latter strategy was successful in increasing product yield, it did so by reducing carbon loss as glycerol and not by significantly increasing CO_2 assimilation (Guadalupe-Medina et al. 2013). A more radical approach is to engineer a CO_2 assimilation pathway capable of sustaining growth and production. In a pioneer work from Antonovsky et al. (2016), *E. coli* was transformed and evolved to grow solely on CO_2 as a carbon source and pyruvate as an electron source. Again, expression of the missing CBB enzymes PRK and RuBisCO, and knock-out of the phosphoglycerate mutase (PGM)—revealed by Flux Balance Analysis (Lewis et al. 2012)—forced CBB operation by decoupling carbon fixation from energy production. This metabolic phenotype was termed hemi-autotrophic growth, and it has since been implemented in other bacteria like the methanol-consuming bacteria *Methylobacterium extorquens* AM1 through the expression of the previous CBB enzymes and deletion of essential genes for methanol assimilation (Borzyskowski et al. 2018). Building on these strategies, a recent study reported the conversion of the yeast *P. pastoris* into an autotroph that grows on CO_2 as the sole carbon source and methanol as an energy source (Gassler et al. 2020). Briefly, *P. pastoris* can use methanol as both energy and carbon sources. By decoupling the formaldehyde—the assimilated product of methanol oxidation—dissimilatory (carbon-fixating) and assimilatory (energy-producing) pathway branches, one can force CO_2 assimilation by blocking the dissimilatory branch through the deletion of the dihydroxyacetone synthase (DAS1 and DAS2) and alcohol oxidase 1 (AOX1). Then, complementation of the native peroxisomal xylose monophosphate (XuMP) cycle with six enzymatic steps enables operation of the CBB cycle allowing growth on CO_2. In stoichiometric terms, 1 mol of oxidized methanol produces 2 mol of NADH, which can be used to fuel the CBB cycle though not in stoichiometric proportions with CO_2 (3 mol of ATP and 2 mol of NADH are needed to fix 1 mol of CO_2). The resulting mutant strain reached a maximum specific growth rate of 0.018 h^{-1} (Gassler et al. 2020) and constitutes an unprecedented advance for compartmentalized C1 carbon fixation in yeast differing from seemingly similar efforts in bacteria (Antonovsky et al. 2016; Bang and Lee 2018).

4 Challenges and Outlook

During the past decade, yeast metabolic engineering has shown great progress and promise (Smolke and Tyo 2012; Nielsen and Keasling 2016), quickly becoming one of the preferred microbial factories for realizing the bioproduction of new chemicals or improving the production of traditional ones. This success has been largely driven by the continuous advances in the development of genetic and molecular tools (Smolke and Tyo 2012), as well as novel computational frameworks for pathway discovery and optimization (Wang et al. 2017; Saa et al. 2019). The latter has brought not only new possibilities for the evaluation of novel biochemical synthesis routes but also has provided more rational methods for designing metabolic pathways with superior performance by rewiring metabolism at a whole-cell scale (Saa et al. 2019). In time, such capabilities will become increasingly essential for arriving at designs that scale industrially and meet commercial expectations.

Pathway discovery is supported by the use of retro-synthesis tools that generate putative routes connecting substrates to products. A comprehensive exploration of the chemical space typically rests on the availability of reaction rules, which fills the gaps between the metabolic precursors and target chemical(s). Generation and application of such rules must be carefully performed, as they may provide infeasible pathways that may obscure results interpretation (Wang et al. 2017). Atom mapping information can be of great aid for validating the application and generation of certain reaction rules, see e.g., RouteSearch (Latendresse et al. 2014) and ReTrace (Pitkänen et al. 2009), which can be further completed with enzyme promiscuity knowledge if available (Mazurenko et al. 2020). Another incipient alternative for learning novel chemical reaction routes rests on machine learning techniques (Koch et al. 2020), which can potentially increase exponentially the size of the reachable chemical space as shown elsewhere (Coley et al. 2019; Mikulak-Klucznik et al. 2020). Efficient navigation of such vast space would necessarily have to rely on the introduction of pathway scores and rankings to focus the attention on the most promising and realizable designs. For this task, evaluation of the objectives reviewed here along with others—e.g., use of enzymes with known promiscuous activity or cofactor specificity—constitutes a natural way for prioritizing and selecting desired pathways. Rational integration of the various objectives can be achieved by leveraging mature multi-decision multi-criteria techniques (Bonissone et al. 2009), which remains largely unexplored in the field. Notably, the latter techniques are also transferable to optimization-based methods for pathway prediction, which could enable a more holistic evaluation of pathway performance and robustness.

While the revised computational methods and tools for pathway prediction have provided unintuitive and useful insights, their experimental application and validation remain still limited. Although there have been recent applications in yeast (Hafner et al. 2021) and other model organisms (Yim et al. 2011) where some of the tools have proven to be critical for finding effective in vivo metabolic designs, there is still resistance to their broad adoption. Indeed, in vivo implementation of complex in silico metabolic designs is not trivial, typically demanding great amounts of experimentation time before arriving at a working pathway (Antonovsky et al. 2016; Schwander et al. 2016; Savile et al. 2010). Such efforts could gain from recent computational frameworks for kinetic model construction (Saa and Nielsen 2017) that could help to predict a priori the expected performance of the pathway (see, for example, (Theisen et al. 2016)), greatly reducing the time and resources needed. As the metabolic prediction capabilities of current models increase (Foster et al. 2021), it is expected that the use of these tools for rational pathway engineering in yeast and other microbial factories will progressively become part of the basic toolbox for metabolic engineering.

Acknowledgements The author acknowledges the financial support of ANID Fondecyt Iniciacion Grant No 11190871 and FONDAP Grant No 15090007 of the Center for Genome Regulation (CGR).

References

Alcántara R, Axelsen KB, Morgat A, Belda E et al (2012) Rhea-a manually curated resource of biochemical reactions. Nucleic Acids Res 40:D754–D760

Antonovsky N, Gleizer S, Noor E, Zohar Y et al (2016) Sugar Synthesis from CO_2 in *Escherichia coli*. Cell 166:115–125

Arkin AP, Cottingham RW, Henry CS, Harris NL et al (2018) KBase: the united states department of energy systems biology knowledgebase. Nat Biotechnol 36:566–569

Bang J, Lee SY (2018) Assimilation of formic acid and CO_2 by engineered Escherichia coli equipped with reconstructed one-carbon assimilation pathways. Proc Natl Acad Sci 115:E9271

Bar-Even A, Noor E, Lewis NE, Milo R (2010) Design and analysis of synthetic carbon fixation pathways. Proc Natl Acad Sci 107:8889

Bar-Even A, Noor E, Savir Y, Liebermeister W et al (2011) The moderately efficient enzyme: evolutionary and physicochemical trends shaping enzyme parameters. Biochemistry-Us 50:4402–4410

Bekiaris PS, Klamt S (2021) Designing microbial communities to maximize the thermodynamic driving force for the production of chemicals. Plos Comput Biol 17:e1009093

Bogorad IW, Lin TS, Liao JC (2013) Synthetic non-oxidative glycolysis enables complete carbon conservation. Nature 502:693–697

Bonissone PP, Subbu R, Lizzi J (2009) Multicriteria decision making (mcdm): a framework for research and applications. IEEE Comput Intell Mag 4:48–61

Campodonico MA, Andrews BA, Asenjo JA, Palsson BO, Feist AM (2014) Generation of an atlas for commodity chemical production in *Escherichia coli* and a novel pathway prediction algorithm GEM-Path. Metab Eng 25C:140–158

Caspi R, Billington R, Ferrer L, Foerster H et al (2016) The MetaCyc database of metabolic pathways and enzymes and the BioCyc collection of pathway/genome databases. Nucleic Acids Res 44:D471-480

Cataldo VF, Arenas N, Salgado V, Camilo C et al (2020) Heterologous production of the epoxycarotenoid violaxanthin in *Saccharomyces cerevisiae*. Metab Eng 59:53–63

Chowdhury A, Maranas CD (2015) Designing overall stoichiometric conversions and intervening metabolic reactions. Sci Rep 5

Coley CW, Thomas DA, Lummiss JAM, Jaworski JN et al (2019). A robotic platform for flow synthesis of organic compounds informed by AI planning. Science 365:eaax1566

Court SJ, Waclaw B, Allen RJ (2015) Lower glycolysis carries a higher flux than any biochemically possible alternative. Nat Commun 6:8427

Delépine B, Duigou T, Carbonell P, Faulon J-L (2018) RetroPath2.0: a retrosynthesis workflow for metabolic engineers. Metab Eng 45:158–170
Du B, Zielinski DC, Monk JM, Palsson BO (2018) Thermodynamic favorability and pathway yield as evolutionary tradeoffs in biosynthetic pathway choice. Proc Natl Acad Sci 115:11339
Duigou T, du Lac M, Carbonell P, Faulon J-L (2018) RetroRules: a database of reaction rules for engineering biology. Nucleic Acids Res 47:D1229–D1235
Edwards JS, Palsson BO (2000) The *Escherichia coli* MG1655 in silico metabolic genotype: Its definition, characteristics, and capabilities. P Natl Acad Sci USA 97:5528–5533
Ellis LBM, Roe D, Wackett LP (2006) The University of Minnesota biocatalysis/biodegradation database: the first decade. Nucleic Acids Res 34:D517–D521
Finley SD, Broadbelt LJ, Hatzimanikatis V (2009) Computational framework for predictive biodegradation. Biotechnol Bioeng 104:1086–1097
Finley SD, Broadbelt LJ, Hatzimanikatis V (2010) In silico feasibility of novel biodegradation pathways for 1,2,4-trichlorobenzene. Bmc Syst Biol 4:7
Flamholz A, Noor E, Bar-Even A, Milo R (2012) eQuilibrator-the biochemical thermodynamics calculator. Nucleic Acids Res 40:770–775
Flamholz A, Noor E, Bar-Even A, Liebermeister W, Milo R (2013) Glycolytic strategy as a tradeoff between energy yield and protein cost. Proc Natl Acad Sci 110:10039
Foster CJ, Wang L, Dinh HV, Suthers PF, Maranas CD (2021) Building kinetic models for metabolic engineering. Curr Opin Biotech 67:35–41
Galanie S, Thodey K, Trenchard IJ, Filsinger Interrante M, Smolke CD (2015) Complete biosynthesis of opioids in yeast. Science 349:1095–1100
Ganter M, Bernard T, Moretti S, Stelling J, Pagni M (2013) MetaNetX.org: a website and repository for accessing, analysing and manipulating metabolic networks. Bioinformatics 29:815–816
Gassler T, Sauer M, Gasser B, Egermeier M et al (2020) The industrial yeast Pichia pastoris is converted from a heterotroph into an autotroph capable of growth on CO_2. Nat Biotechnol 38:210–216
Guadalupe-Medina V, Wisselink HW, Luttik MA, de Hulster E et al (2013) Carbon dioxide fixation by Calvin-cycle enzymes improves ethanol yield in yeast. Biotechnol Biofuels 6:125
Hadadi N, Hafner J, Shajkofci A, Zisaki A, Hatzimanikatis V (2016) ATLAS of Biochemistry: a repository of all possible biochemical reactions for synthetic biology and metabolic engineering studies. ACS Synth Biol 5:1155–1166
Hadadi N, MohammadiPeyhani H, Miskovic L, Seijo M, Hatzimanikatis V (2019) Enzyme annotation for orphan and novel reactions using knowledge of substrate reactive sites. Proc Natl Acad Sci 116:7298
Hädicke O, von Kamp A, Aydogan T, Klamt S (2018) OptMDFpathway: Identification of metabolic pathways with maximal thermodynamic driving force and its application for analyzing the endogenous CO_2 fixation potential of *Escherichia coli*. Plos Comput Biol 14:e1006492
Hafner J, Payne J, MohammadiPeyhani H, Hatzimanikatis V, Smolke C (2021) A computational workflow for the expansion of heterologous biosynthetic pathways to natural product derivatives. Nat Commun 12:1760
Hatzimanikatis V, Li C, Ionita JA, Henry CS et al (2005) Exploring the diversity of complex metabolic networks. Bioinformatics 21:1603–1609
Henry CS, Broadbelt LJ, Hatzimanikatis V (2007) Thermodynamics-based metabolic flux analysis. Biophys J 92:1792–1805
Henry CS, Broadbelt LJ, Hatzimanikatis V (2010a) Discovery and analysis of novel metabolic pathways for the biosynthesis of industrial chemicals: 3-hydroxypropanoate. Biotechnol Bioeng 106:462–473
Henry CS, DeJongh M, Best AA, Frybarger PM et al (2010) High-throughput generation, optimization and analysis of genome-scale metabolic models. Nat Biotechnol 28:977–982
Hou J, Scalcinati G, Oldiges M, Vemuri GN (2010) Metabolic impact of increased NADH availability in *Saccharomyces cerevisiae*. Appl Environ Microb 76:851–859

Isaacs FJ, Carr PA, Wang HH, Lajoie MJ et al (2011) Precise manipulation of chromosomes in vivo enables genome-wide codon replacement. Science 333:348

Jeffryes JG, Colastani RL, Elbadawi-Sidhu M, Kind T et al (2015) MINEs: open access databases of computationally predicted enzyme promiscuity products for untargeted metabolomics. Journal of Cheminformatics 7:44

Jeske L, Placzek S, Schomburg I, Chang A, Schomburg D (2018) BRENDA in 2019: a European ELIXIR core data resource. Nucleic Acids Res gky1048-gky1048

Kamp AV, Klamt S (2020) MEMO: A Method for computing metabolic modules for cell-free production systems. ACS synthetic biology 9:556–566

Kanehisa M, Sato Y, Kawashima M, Furumichi M, Tanabe M (2016) KEGG as a reference resource for gene and protein annotation. Nucleic Acids Res 44:D457–D462

Khersonsky O, Tawfik DS (2010) Enzyme promiscuity: a mechanistic and evolutionary perspective. Annu Rev Biochem 79:471–505

Kim J, Reed JL, Maravelias CT (2011) Large-scale bi-level strain design approaches and mixed-integer programming solution techniques. Plos One 6:e24162

Kim J-E, Jang I-S, Sung BH, Kim SC, Lee JY (2018) Rerouting of NADPH synthetic pathways for increased protopanaxadiol production in *Saccharomyces cerevisiae.* Sci Rep 8:15820

King ZA, Feist AM (2014) Optimal cofactor swapping can increase the theoretical yield for chemical production in *Escherichia coli* and *Saccharomyces cerevisiae.* Metab Eng 24:117–128

King ZA, Lu J, Dräger A, Miller P et al (2016) BiGG Models: a platform for integrating, standardizing and sharing genome-scale models. Nucleic Acids Res 44:D515–D522

Ko Y-S, Kim JW, Lee JA, Han T et al (2020) Tools and strategies of systems metabolic engineering for the development of microbial cell factories for chemical production. Chem Soc Rev 49:4615–4636

Koch M, Duigou T, Faulon J-L (2020) Reinforcement learning for bioretrosynthesis. ACS Synth Biol 9:157–168

Kötter P, Ciriacy M (1993) Xylose fermentation by *Saccharomyces cerevisiae.* Appl Microbiol Biot 38:776–783

Kozak BU, van Rossum HM, Benjamin KR, Wu L et al (2014a) Replacement of the *Saccharomyces cerevisiae* acetyl-CoA synthetases by alternative pathways for cytosolic acetyl-CoA synthesis. Metab Eng 21:46–59

Kozak BU, van Rossum HM, Luttik MAH, Akeroyd M et al (2014) Engineering acetyl coenzyme a supply: Functional expression of a bacterial pyruvate dehydrogenase complex in the cytosol of *Saccharomyces cerevisiae.* mBio 5:e01696–01614

Kumar A, Suthers PF, Maranas CD (2012) MetRxn: a knowledgebase of metabolites and reactions spanning metabolic models and databases. BMC Bioinformatics 13:6

Kumar A, Wang L, Ng CY, Maranas CD (2018) Pathway design using de novo steps through uncharted biochemical spaces. Nat Commun 9:184

Kummel A, Panke S, Heinemann M (2006) Putative regulatory sites unraveled by network-embedded thermodynamic analysis of metabolome data. Mol Syst Biol 2006(2):0034

Latendresse M, Krummenacker M, Karp PD (2014) Optimal metabolic route search based on atom mappings. Bioinformatics 30:2043–2050

Lee SY, Kim HU (2015) Systems strategies for developing industrial microbial strains. Nat Biotechnol 33:1061–1072

Lee WH, Park JB, Park K, Kim MD, Seo JH (2007) Enhanced production of epsilon-caprolactone by overexpression of NADPH-regenerating glucose 6-phosphate dehydrogenase in recombinant *Escherichia coli* harboring cyclohexanone monooxygenase gene. Appl Microbiol Biotechnol 76:329–338

Lee WH, Kim MD, Jin YS, Seo JH (2013) Engineering of NADPH regenerators in *Escherichia coli* for enhanced biotransformation. Appl Microbiol Biotechnol 97:2761–2772

Lerman JA, Hyduke DR, Latif H, Portnoy VA et al (2012) In silico method for modelling metabolism and gene product expression at genome scale. Nat Commun 3

Lewis NE, Nagarajan H, Palsson BO (2012) Constraining the metabolic genotype-phenotype relationship using a phylogeny of in silico methods. Nat Rev Microbiol 10:291–305

Lim SJ, Jung YM, Shin HD, Lee YH (2002) Amplification of the NADPH-related genes zwf and gnd for the oddball biosynthesis of PHB in an *E. coli* transformant harboring a cloned phbCAB operon. J Biosci Bioeng 93:543–549

Lin G-M, Warden-Rothman R, Voigt CA (2019) Retrosynthetic design of metabolic pathways to chemicals not found in nature. Curr Opin Syst Biol 14:82–107

López J, Bustos D, Camilo C, Arenas N, Saa PA (2020) Engineering *Saccharomyces cerevisiae* for the overproduction of β -ionone and its precursor β -carotene. Front Bioeng Biotechnol 8:1–13

López J, Cataldo VF, Peña M, Saa PA et al (2019) Build your bioprocess on a solid strain—β-carotene production in recombinant *Saccharomyces cerevisiae*. Front Bioeng Biotechnol 7

Ma SM, Garcia DE, Redding-Johanson AM, Friedland GD et al (2011) Optimization of a heterologous mevalonate pathway through the use of variant HMG-CoA reductases. Metab Eng 13:588–597

Martínez I, Zhu J, Lin H, Bennett GN, San K-Y (2008) Replacing *Escherichia coli* NAD-dependent glyceraldehyde 3-phosphate dehydrogenase (GAPDH) with a NADP-dependent enzyme from *Clostridium acetobutylicum* facilitates NADPH dependent pathways. Metab Eng 10:352–359

Mazurenko S, Prokop Z, Damborsky J (2020) Machine learning in enzyme engineering. ACS Catal 10:1210–1223

Meadows AL, Hawkins KM, Tsegaye Y, Antipov E et al (2016) Rewriting yeast central carbon metabolism for industrial isoprenoid production. Nature 537:694–697

Mikulak-Klucznik B, Gołębiowska P, Bayly AA, Popik O et al (2020) Computational planning of the synthesis of complex natural products. Nature 588:83–88

Moretti S, Tran, Van Du T, Mehl F, Ibberson M, Pagni M (2021) MetaNetX/MNXref: unified namespace for metabolites and biochemical reactions in the context of metabolic models. Nucleic Acids Res 49:D570–D574

Moriya Y, Shigemizu D, Hattori M, Tokimatsu T et al (2010) PathPred: an enzyme-catalyzed metabolic pathway prediction server. Nucleic Acids Res 38:W138–W143

Moura M, Finkle J, Stainbrook S, Greene J et al (2016) Evaluating enzymatic synthesis of small molecule drugs. Metab Eng 33:138–147

Ng CY, Wang L, Chowdhury A, Maranas CD (2019) Pareto optimality explanation of the glycolytic alternatives in nature. Sci Rep 9:2633

Nielsen J, Keasling JD (2016) Engineering cellular metabolism. Cell 164:1185–1197

Nissen TL, Kielland-Brandt MC, Nielsen J, Villadsen J (2000) Optimization of ethanol production in *Saccharomyces cerevisiae* by metabolic engineering of the ammonium assimilation. Metab Eng 2:69–77

Nissen TL, Anderlund M, Nielsen J, Villadsen J, Kielland-Brandt MC (2001) Expression of a cytoplasmic transhydrogenase in *Saccharomyces cerevisiae* results in formation of 2-oxoglutarate due to depletion of the NADPH pool. Yeast 18:19–32

Noor E, Bar-Even A, Flamholz A, Reznik E et al (2014) Pathway thermodynamics highlights kinetic obstacles in central metabolism. Plos Comput Biol 10:e1003483

Noor E, Flamholz A, Bar-Even A, Davidi D et al (2016) The protein cost of metabolic fluxes: prediction from enzymatic rate laws and cost minimization. Plos Comput Biol 12: e1005167

Pharkya P, Burgard AP, Maranas CD (2004) OptStrain: a computational framework for redesign of microbial production systems. Genome Res 14:2367–2376

Pitkänen E, Jouhten P, Rousu J (2009) Inferring branching pathways in genome-scale metabolic networks. Bmc Syst Biol 3:103

Saa PA, Nielsen LK (2017) Formulation, construction and analysis of kinetic models of metabolism: a review of modelling frameworks. Biotechnol Adv 35:981–1003

Saa PA, Cortés MP, López J, Bustos D et al (2019) Expanding metabolic capabilities using novel pathway designs: computational tools and case studies. Biotechnol J 14:1800734

San KY, Bennett GN, Berrios-Rivera SJ, Vadali RV et al (2002) Metabolic engineering through cofactor manipulation and its effects on metabolic flux redistribution in *Escherichia coli*. Metab Eng 4:182–192

Sánchez BJ, Zhang C, Nilsson A, Lahtvee P-J et al (2017) Improving the phenotype predictions of a yeast genome-scale metabolic model by incorporating enzymatic constraints. Mol Syst Biol 13:935

Savile CK, Janey JM, Mundorff EC, Moore JC et al (2010) Biocatalytic asymmetric synthesis of chiral amines from ketones applied to sitagliptin manufacture. Science 329:305

Scalcinati G, Partow S, Siewers V, Schalk M et al (2012) Combined metabolic engineering of precursor and co-factor supply to increase α-santalene production by *Saccharomyces cerevisiae*. Microb Cell Fact 11:117

Schilling C, Thakar R, Travnik E, Dien S, Wiback S (2005) SimPheny: a Computational Infrastructure for Systems Biology. In: US Department of Energy, Genomic Science Program publications

Schramm M, Racker E (1957) Formation of erythrose-4-phosphate and acetyl phosphate by a phosphorolytic cleavage of fructose-6-phosphate. Nature 179:1349

Schwander T, von Schada Borzyskowski L, Burgener S, Cortina NS, Erb TJ (2016) A synthetic pathway for the fixation of carbon dioxide in vitro. Science 354: 900–904

Smolke CD, Tyo KEJ (2012) Synthetic biology: Emerging methodologies to catalyze the metabolic engineering design cycle. Metab Eng 14:187–188

Sonderegger M, Schumperli M, Sauer U (2004) Metabolic engineering of a phosphoketolase pathway for pentose catabolism in *Saccharomyces cerevisiae*. Appl Environ Microbiol 70:2892–2897

Tervo CJ, Reed JL (2016) MapMaker and PathTracer for tracking carbon in genome-scale metabolic models. Biotechnol J 11:648–661

Theisen MK, Lafontaine Rivera JG, Liao JC (2016) Stability of ensemble models predicts productivity of enzymatic systems. Plos Comput Biol 12:e1004800

Tokic M, Hadadi N, Ataman M, Neves D et al (2018) Discovery and evaluation of biosynthetic pathways for the production of five methyl ethyl ketone precursors. ACS Synth Biol 7:1858–1873

van Rossum HM, Kozak BU, Pronk JT, van Maris AJA (2016) Engineering cytosolic acetyl-coenzyme A supply in *Saccharomyces cerevisiae*: pathway stoichiometry, free-energy conservation and redox-cofactor balancing. Metab Eng 36:99–115

Verho R, Londesborough J, Penttilä M, Richard P (2003) engineering redox cofactor regeneration for improved pentose fermentation in *Saccharomyces cerevisiae*. Appl Environ Microb 69:5892–5897

Voigt CA (2020) Synthetic biology 2020–2030: six commercially-available products that are changing our world. Nat Commun 11:6379

von Schada Borzyskowski L, Carrillo M, Leupold S, Glatter T et al (2018) An engineered Calvin-Benson-Bassham cycle for carbon dioxide fixation in Methylobacterium extorquens AM1. Metab Eng 47:423–433

Wang L, Dash S, Ng CY, Maranas CD (2017) A review of computational tools for design and reconstruction of metabolic pathways. Synth Syst Biotechnol 2:243–252

Wortel MT, Noor E, Ferris M, Bruggeman FJ, Liebermeister W (2018) Metabolic enzyme cost explains variable trade-offs between microbial growth rate and yield. Plos Comput Biol 14:e1006010

Yang X, Mao Z, Zhao X, Wang R et al (2020) Integrating thermodynamic and enzymatic constraints into genome-scale metabolic models. bioRxiv 2020.2011.2030.403519

Yim H, Haselbeck R, Niu W, Pujol-Baxley C et al (2011) Metabolic engineering of *Escherichia coli* for direct production of 1,4-butanediol. Nat Chem Biol 7:445–452

Construction and Assembly of Standardized Biobricks for Synthetic Pathways Engineering in Yeasts

Paulina Korpys-Woźniak, Monika Kubiak, Monika Borkowska, and Ewelina Celińska

Abstract

Over the past years, a tremendous variety of modular cloning and DNA assembly strategies has been developed. By definition, they enable one-step fusion of multiple DNA modules, arranged in a defined position and orientation. They differ in terms of flexibility, robustness and precision, but they all overcome limitations of time-consuming traditional cloning. Irrespectively of technical details, modular cloning and standardized bioparts assembly strategies contribute to great progress within the field of genetic engineering of yeast. Their exploitation allows the researchers to focus on addressing the scientific questions, rather than on execution of tedious laboratory procedures. Modularity and standardization facilitate wide-spread collaboration within the yeast community and speed-up accomplishing of set goals. This chapter provides an overview of synthetic biology tools for assembly of large DNA constructs (in many cases—from predesigned modules) to construct and fine-tune complex DNA assemblies for yeasts. The principles of the techniques are presented, and then followed by examples of yeast species-specific implementations for *Saccharomyces cerevisiae, Komagataella phaffii* and *Yarrowia lipolytica.* A general division into restriction digestion/recombination-based techniques and overlap extension assembly strategies were adopted.

Paulina Korpys-Woźniak and Monika Kubiak—These Authors contributed equally to the work and should be considered as the two first Authors.

P. Korpys-Woźniak · M. Kubiak · M. Borkowska · E. Celińska (✉)
Department of Biotechnology and Food Microbiology, Poznan University of Life Sciences, ul. Wojska Polskiego 48, 60-637 Poznań, Poland
e-mail: ewelina.celinska@up.poznan.pl

F. Darvishi Harzevili (ed.), *Synthetic Biology of Yeasts*,
https://doi.org/10.1007/978-3-030-89680-5_2

1 General Concept of Modularity

Gene cloning has gone a long way since its origin in '70 of the XX century. For many years, the DNA fragments to be cloned were processed with specific and unique *R*estriction *E*nzymes (RE or ER—*E*ndonuclease *R*estriction) and individually ligated with a vector that enabled maintenance of the fragment in the host cell, and in some cases—its expression. The procedure had to be individually tailored to a given sequence's specificity, i.e., its nucleotide sequence and the presence of ER *R*ecognition *S*ites (RS) at a specific location. Along with time and complexity of generated DNA constructions, the need for streamlined cloning procedures has been growing. Traditional *R*estriction *D*igestion (RD) with specific RE, and cloning of the fragments one-by-one did not meet growing demands. Until the concept of modularity has been transplanted to molecular biology from engineering and programming fields.

Modularity dissects a system (= complex DNA construction) into individual modules (= biobricks/bioparts), each having a specific position and task to execute, in order to develop the desired functionality of the whole system. A given module (e.g., promoter/ORF) may have different variants (e.g., strong, weak, inducible/fluorescent, enzymatic) that can be interchangeably shuffled. Importantly, in a specific design of the system, each module has a strictly defined location and function that cannot be exchanged with another module (only its variants can be exchanged at the same position). A given module's location (and function) is predefined by a scaffold, which is a system of linking and docking adjacent modules with each other. By organizing the modules within the scaffold, they are correctly ordered and oriented, and cannot be translocated/shifted to another location (only to another variant of the same module). Modularity gives the advantage of an open upgrading option, provided that the new module/variant will fit into the scaffold. This quality is not available for cloning through the traditional, sequence-specific RE-based strategies.

To put the modularity concept into a molecular biology perspective, we need to translate a term "module" into "a function at a specific location" within a multipart DNA construction (Fig. 1). This can be, for example, a promoter element at the second *T*ranscription *U*nit (TU; TU is a system of promoter-ORF-terminator) within a two TU-bearing DNA construction (*P2* in Fig. 1).

This (*P2*) element is defined by specific linking/docking elements, flanking the module. These may be specific ER RS, Recombinase (Rec) RS, OverLapping protruding elements (OL; overlaps) generated by PCR or in vitro DNA synthesis—anything that facilitates correct assembly (hybridization and ligation) of the subsequent DNA fragments. They are specific for a given element (as *P2* in our example), and unique within the whole DNA construction. These elements guarantee correct positioning and orientation of the (*P2*) element within the multipart DNA assembly. Depending on the linking/docking system adapted for a particular technique, the final assembly may bear scars or be scarless.

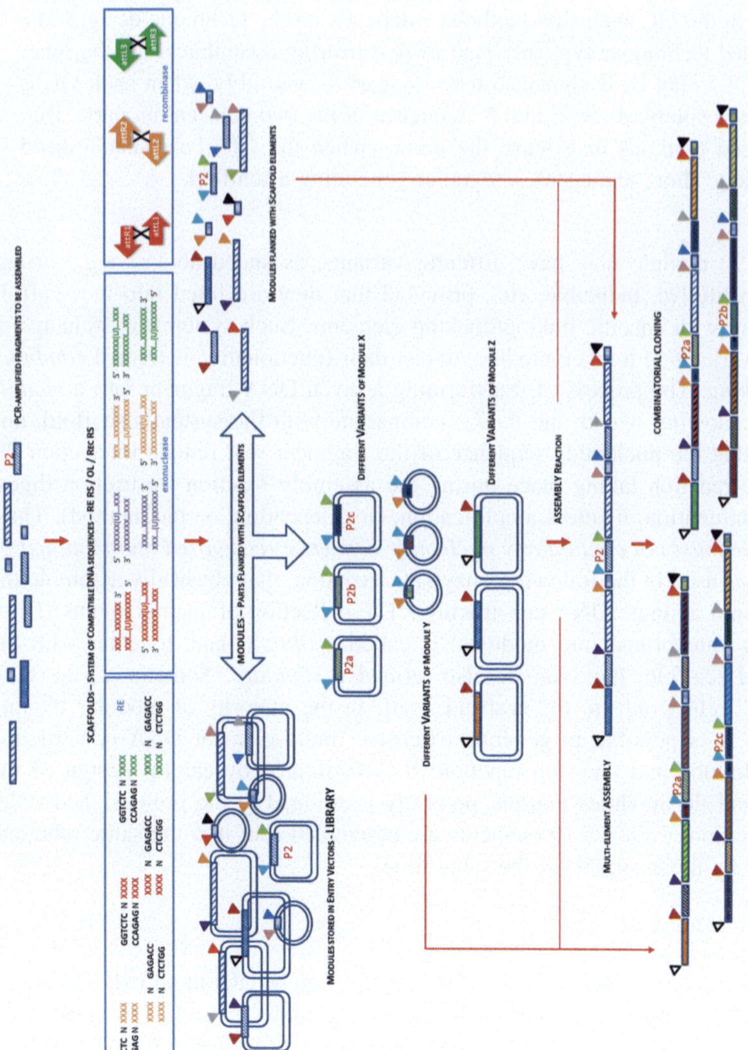

Fig. 1 General outline of bioparts assembly and combinatorial cloning

Box 1

A scar is a short nucleotide sequence, present in the final assembly, that is not a functional part of any of the subsequent bioparts, but was necessary at the assembly stage for correct hybridization of the modules. Typical scars, in traditional and in BioBrick clonings, are ER RS, in GoldenGate—4 nt overhangs, in the OL extension methods—depends on the technique design. The ER-based techniques typically lead to scar-bearing assemblies. On the other hand, OLs may be designed to leave a scarless assembly, when each OL is composed solely of the 3' and 5' sequence of the two adjacent bioparts. But, OL-based methods may leave the scars—when the OLs bear standardized but unique short sequences, altogether generating a scaffold.

The (*P2*) module may have different variants, as stated above—e.g., strong, weak, constitutive, inducible, etc., provided that they are fitted into the scaffold by flanking with specific linking/docking elements. Such systematic exchange of different variants of a given module, to test their functionality, is termed *combinatorial cloning*. The process of transforming a given DNA fragment into a *module* requires endowing it with the *flanks*, compatible with the system's scaffold, and ensuring that the nucleotide sequence of the fragment will remain inert upon the enzymatic reaction taking place during the assembly reaction (restriction digestion, recombination, ligation, amplification, etc., depending on the method). Thus, the *module must not contain any nucleotide sequence recognized and modified by the enzymes* used in the following enzymatic reaction, that physically assemble the modules into a single DNA construction. The collection of such elements (DNA fragments transformed into modules) is called a *library*; and, together with the predefined scaffold, they can be also termed a *standard*. Sometimes, the term "standard" refers only to the scaffold itself. In the majority of modular cloning strategies, it is possible to generate extensive, multi-element DNA construction in a single-tube and one-step reaction. It is facilitated by careful design of the scaffold and the modules that are precisely positioned in the pre-designed order and orientation; even if >10 elements are assembled (put into the same tube and processed by the enzymes) at the same time.

Box 2

Once assigned to a given module functionality can be later modified. Considering our example, if a gene (ORF; G2) cloned under control of P2 must be transcriptionally fused to an element (signal peptide or fusion partner) at its 5' terminus, and this element will be shuffled in the following cloning (different variants will be tested), two different strategies could be followed: (i) expanding the scaffold by modifying the linking/docking elements of P2 at 3' terminus and G2 at 5' terminus (adding additional fusion site) so that another module can be inserted; (ii) functionality of all the elements upstream of the

G2 can be changed, but the linking/docking sites remain unchanged—this approach would require the generation of a modified library of all elements upstream from the modified site.

By definition, entering a modular cloning approach is a long-term strategy of DNA construction assembling. To take full advantage of benefits it offers, one must first make an effort to design the standard, the system of nomenclature and cataloging, and to expand the library. Once entered, working within a given system offers a multitude of profits. Laboratories working in the same standard may easily exchange modules, expanding their libraries. Additionally, standardization of the methods greatly simplifies the design of cloning strategies and minimizes the number of steps required to obtain a desired construct. Most techniques enable combinatorial construction of complex DNA constructions (cassettes) using DNA parts from libraries, as well as from open registries of modular components, available for the synthetic biology community. These standards facilitate the reuse of fragments between experiments and their exchange between research groups. Apart from constructing expression cassettes of a defined number of TUs, DNA assembly is useful to build for example whole chromosomes.

Box 3

Modular cloning may have different formats in DNA construction assembly. The number of shuffled modules in a single reaction may differ significantly—from a single biopart (when all the remaining elements necessary for the gene expression in yeast cell are already present in the destination/acceptor vector), up to all the "yeast elements" (when only the "bacterial part" in a shuttle vector* remains unchanged). Technically, the approach depends on the scaffold's structure. An extensive and highly specific scaffold (with a large number of unique and specific linking/docking elements) accepts more elements to be shuffled in a single reaction. Limited scaffold, allow shuffling one or two bioparts at a time. In the latter case, all the remaining elements, necessary for full functionality of the expression cassette must be already present in the acceptor vector. (* typically the DNA bioparts assembly is conducted in a shuttle vector—bearing a "bacterial part", to be assembled and amplified in E. coli, and a "yeast part"—composed of yeast-specific regulatory elements and the target genes, to be transformed and expressed in a yeast host cell).

Over the past years, a tremendous variety of modular cloning and DNA assembly strategies has been developed. They differ in terms of flexibility, robustness and precision of cloning, but they all overcome limitations of time-consuming traditional cloning using RD and DNA ligation, typically joining only two, "single-use" DNA parts simultaneously. DNA assembly (incl. modular cloning) typically enables one-step fusion of multiple DNA modules, arranged in a defined position and orientation. This chapter provides an overview of "state-of-the-art" synthetic biology tools for assembly of large DNA constructs (in many cases—from pre-designed modules) to construct and fine-tune complex DNA assemblies for yeasts. For the sake of clarity and comprehensiveness, the principles of the techniques are presented and then followed by yeast species-specific examples. *Saccharomyces cerevisiae, Komagataella phaffii* and *Yarrowia lipolytica* were chosen as the case studies, due to their fundamental role or high interest in the yeast community. Selection of methods was an arbitrary choice of the authors, based on their popularity in research by the yeast community. General division into RE-dependent and OL assembly methods was adopted.

2 Historical Outline

The evolution of modular cloning and DNA bioparts assembly strategies spans the last 30 years. General trend of the techniques' evolution leads from founding the basic principles of a method, through demonstration of its applicability in a given yeast species, up to development of the so-called, *toolkit* (or *toolbox*), built of species-specific standardized modules and/or DNA constructions and recipient strains. In order to put the techniques into a research community perspective and illustrate dynamics of the methods' evolution, in this section we summarize timeline of development of the most popular techniques, according to that key—establishing the technical concept, through "the first mention" for a given yeast species, up to a toolkit development (where applicable; Fig. 2a, b). A more detailed description of the techniques with deeper insight into current state of the art is given in the following sections of this chapter.

Technical implementation of the DNA bioparts assembling concept was at first addressed by Overlap Extension PCR (OE-PCR). The method was initially adopted for assembly of relatively short elements (300–400 bp), incl. gene's exons, or parts located upstream and downstream of the mutation site (two parts) (Ho et al. 1989; Horton et al. 1989). The technique was later adopted for creating chimeric genes or gene alleles in *S. cerevisiae* (Anthony and Liebman 1995), *K. phaffii* (Eldin et al. 1997) and *Y. lipolytica* (Zhang et al. 2010). In many cases, due to simplicity of the method, lack of spectacular technical background behind it and low efficiency when compared to the other assembly method, the authors tend to use it as a routine laboratory technique for simple assemblies, without special mention. Nevertheless, later elaborated variants of OE-PCR, combined with "*in yeasto*" assembly, demonstrated the great potential of that simple method, which ultimately enabled assembly of extensive DNA constructions (Shao et al. 2009; Gao et al. 2014).

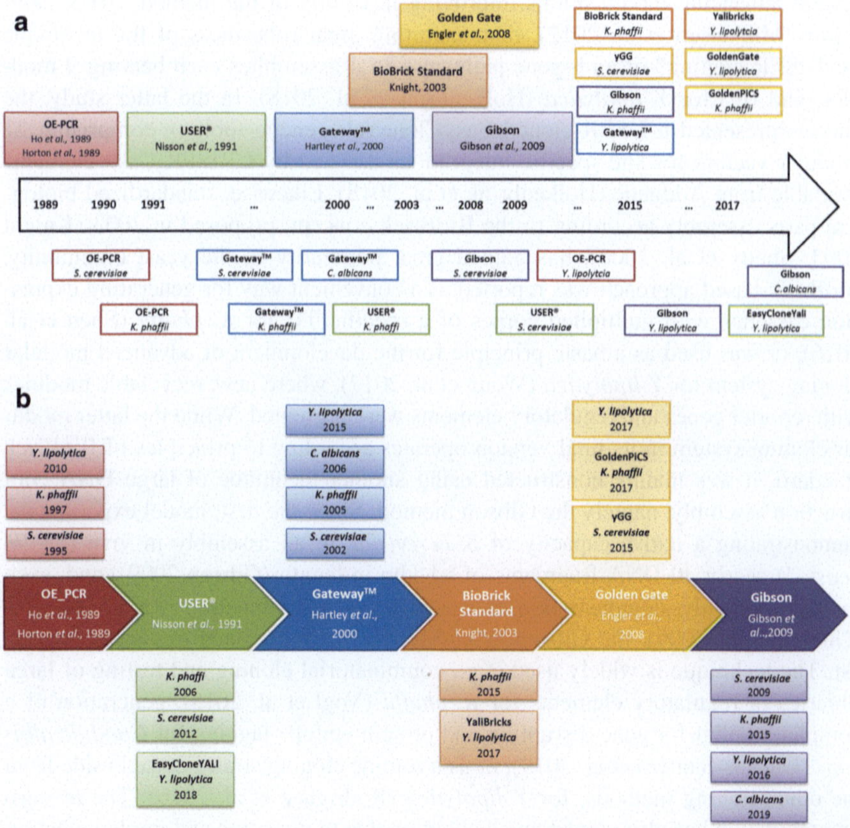

Fig. 2 Historical outline of DNA bioparts assembly and modular cloning strategies invention and implementation. **a** Inventions and implementations are given chronologically. **b** Inventions and implementations are sorted by the principal technique

> **Box 4**
> The term "in yeasto" refers to phenomena or reactions taking place within the living yeast cell. It may be considered as an abbreviation of stating "in vivo in the yeast cell". It can be sometimes found in the research papers issued by/dedicated to the yeast community.

In contrast, an assembly technique developed only 2 years later, USER® (*U*racil-*S*pecific *E*xcision *R*eaction; Nisson et al. 1991; Smith et al. 1993) has gained immediate, great attention by the yeast research community. The principles were developed and commercialized by New England Biolabs in 2003, and the toolkit is commercially available. The USER standard was introduced to a molecular biology tools portfolio for (i) *K. phaffii* (Nour-Eldin et al. 2006), with several advancing modifications improving flexibility of the method, (ii) *S. cerevisiae* (Mikkelsen et al. 2012), demonstrating great robustness of the technique by "USER cloning" of an 8-gene pathway—in 4 assemblies each bearing 9 modules, and (iii) for *Y. lipolytica* (Holkenbrink et al. 2018). In the latter study, the authors presented a USER-cloned EasyCloneYali genetic toolbox comprising 27 modular vectors for site-specific integrations assisted by CRISPR-Cas9 elements available from Addgene (Holkenbrink et al. 2018). Likewise, standardized biological parts assembly according to the BioBrick concept, proposed in 2003 (Knight 2003; Shetty et al. 2008), has gained great popularity in the yeast community. BioBrick-based approach was reported as a convenient way for generating expression cassettes with multiplied copies of a specific TU for *K. phaffii* (Shen et al. 2016), or was used as a basic principle for the development of advanced modular cloning system for *Y. lipolytica* (Wong et al. 2017), where new, recyclable modules with reporter genes and regulatory elements were presented. While the latter modular cloning system in its final version operates according to principles of BioBrick standard, it was mainly constructed using another technique of large DNA construction assembly, namely the Gibson method. Since the first, model experiments, demonstrating a native capacity of *S. cerevisiae* to an assembly in vivo (or "*in yeasto*") nearly 40 DNA fragments of >1 kbp in length (Gibson 2009), and, even more importantly, demonstration of mastering of recombination reaction in vitro (Gibson et al. 2009), the Gibson assembly method has gained tremendous interest. The technique is widely used, e.g., combinatorial cloning and testing of large libraries of regulatory elements for *K. phaffii* (Vogl et al. 2015), generation of a complete toolkit for gene disruption and protein epitope tagging for *Candida albicans* (Dueñas-Santero et al. 2019), or as a routine cloning strategy, used aside from the other cloning methods, for *Y. lipolytica* (Rodriguez et al. 2016). The in vitro recombination has also gained much attention due to invention and implementation into laboratory practice of Gateway™ cloning technique, first proposed in 2000 (Hartley et al. 2000). Only two years later, a rich library of Gateway™ bioparts and destination vectors was developed for constitutive or inducible expression of (various) epitope-tagged proteins in *S. cerevisiae* (Funk et al. 2002). Gateway™ standard was successfully adopted to *Y. lipolytica* by developing a broad-use destination vector, a platform strain and several modules with reporter proteins (Leplat et al. 2015). Earlier, Gateway™—assembled constructions, merging modules bearing signal peptides and secretory proteins, were tested in *K. phaffii* (Esposito et al. 2005) with positive outcome. Gateway™ combinatorial cloning was also used for efficient generation of a panel of deletion cassettes dedicated for *C. albicans* (Lebel et al. 2006).

Extensive and widely available toolkits for a generation of multiple, highly complex DNA constructions were developed using another modular cloning method, namely GoldenGate. The principles of modular cloning according to GoldenGate strategy were first reported in 2008 (Engler et al. 2008; Weber et al. 2011), and it took 7 years to develop a first comprehensive toolkit for GoldenGate cloning in yeast (*S. cerevisiae*) (Agmon et al. 2015). That first toolkit set the standards and paved the way for further species-specific libraries of bioparts and destination vectors. In 2017, customized collections of standardized bioparts for modular cloning in *Y. lipolytica* (Celińska et al. 2017) and *K. phaffii* (Prielhofer et al. 2017) were reported. The new standards were further developed and introduced to open repositories of bioparts (Prielhofer et al. 2017; Larroude et al. 2019) and are now widely available for the yeast community.

3 Restriction Digestion/Recombination-Based Assembly Strategies

Restriction digestion/recombination-assisted bioparts assembly relies on flanking the modules with short sequences, specifically recognized by the enzymes used in the next step. By definition, these methods generate small scars at the fusion sites. Cloning according to the ER/Rec-based strategies requires elimination of unintended ER/Rec RS from the module's sequence, to maintain it intact during the assembly reaction. Moreover, a scaffold of linking/docking elements must be developed in advance. On the other hand, exploitation of the predesigned scaffolds facilitates combinatorial cloning, which is usually limited in OL-based techniques. Additionally, these techniques are considered more precise when compared to OL-dependent methodologies (discussed hereafter).

3.1 BioBricks

BioBrick assembly (BB) (Knight 2003) of standardized bioparts, otherwise known as BBF RFC 9, was first proposed by Tom Knight in 2003 (version RFC 10 in 2007). The principle of this method relies on the use of two pairs of standard Res—(i) *U*nique *RE*s located outside of a biopart (URE1/2) and (ii) *Iso*C*audo*M*ers (IsoCM1/2)—located in direct proximity of the element (Fig. 3). As in any method relying on RE—their RS must not be present in the bioparts to be assembled. It might be thus necessary to remove the RS by site-directed mutagenesis before the assembly reaction.

Box 5

Isocaudomers are pairs of RE that recognize slightly different RS but upon cleavage, they generate compatible overhangs. After ligation of such protruding elements, an asymmetrical sequence is generated and thus cannot be recognized by any of the ERs—it is lost, but in this way, the sequence is stable during the next cycle of assembly. Examples of IsoCMs: 1: BamHI/BglII, 2: NdeI/MseI, 3: XbaI/AvrII/NheI/SpeI, 4: XhoI/SalI.

There is no specific scaffold of linking/docking in BB technique. While it saves the initial design effort, only two parts can be assembled in a single reaction. To get a multipart assembly, the cycles of the two-part-joining must be repeated. The

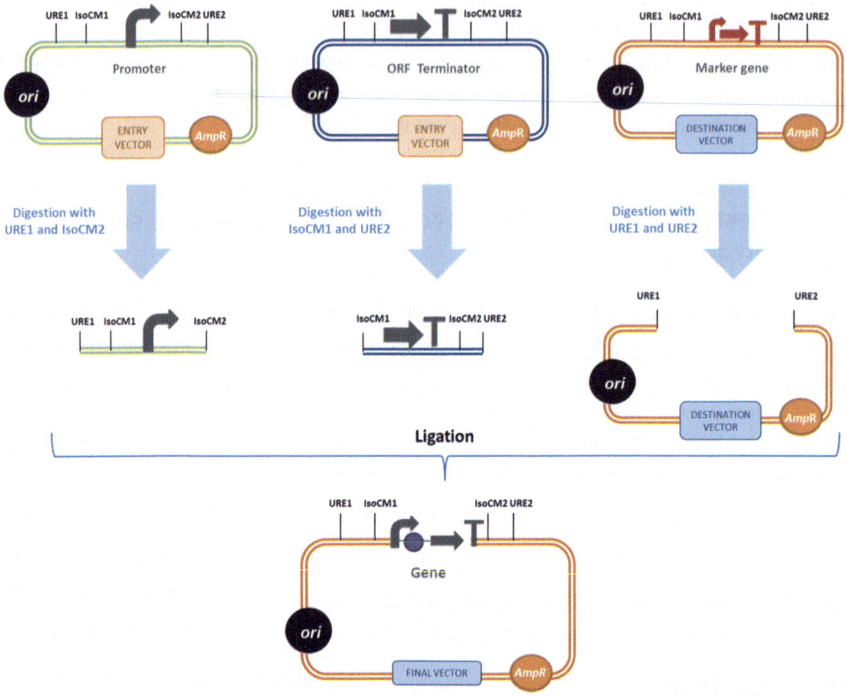

Fig. 3 General strategy for BioBrick parts assembly (e.g.: promoter, ORF and terminator). Each of the modules is first equipped with prefix and suffix bearing RS for a single URE (unique restriction endonuclease recognition site) and a single IsoCM (isocaudomer restriction endonuclease recognition site). Full description is given in the text. "Scar" site is shown as a purple circle in the final assembly. Circular objects indicate the bacterial ori of replication (black) and ampicillin resistance gene AmpR (orange) contained in the "bacterial" part of the assembly, which is typically discarded before yeast cell transformation

modules are prepared by adding specific prefixes and suffixes in the PCR to a biopart, each bearing RS for URE at 5' and IsoCM at 3'. The two modules to be assembled are then cleaved in a specific pattern each with URE and IsoCM to generate compatible overhangs at the linking/docking site. The overhangs at the fusion site are generated by different IsoCM for each of the modules. RS of URE flank the assembly, and compatible RS are present in the destination vector. As a result of the following ligation reaction, the sequence between the fused DNA fragments bears a 6 bp scar, but the URE-bearing prefix and suffix sequences remain unchanged, hence the other BB parts can be assembled in the next step. In this technique, the destination vector bearing two assembled bioparts becomes an entry vector in the next cycle of assembly.

The reaction conditions are standard for ER and ligation (typically 1 h at 37 °C and 1 h 16–22 °C). Most of compatible BBs for yeasts are listed on the website of the Registry of Standard Biological Parts (http://partsregistry.org/ Main_Page), and the bioparts are available from Addgene and iGEM.

As compared with the traditional cloning, several advantages of the BB standard assembly can be highlighted, such as utilization of only four RE, "endless" number of cycles of pairwise assembly, practically limited only by the plasmid size, and no necessity of recloning or even re-PCR amplification. Despite facilitating automation and reuse of parts, the method is not truly high-throughput, as the fusion is always only pairwise, and the scar left at the fusion site can affect the coding sequence or mRNA's secondary structure.

The BB standard has been subjected to specific adjustments, relying on the same principle (including the BglBrick standard (Anderson et al. 2010) and 2ab assembly (Leguia et al. 2013)) but with improved flexibility and efficiency. A collection of several thousands of *S. cerevisiae*-specific modules, compatible with BB standard, is available from the iGEM Registry of Standard Biological Parts.

Specific implementations of BB parts assembly in the selected yeast species cover several relatively recent reports. In 2012, Schneider et al. (Schneider et al. 2012) presented at annually held iGEM competition (International Genetically Engineered Machine) a Yeast BioBrick Assembly toolkit for *S. cerevisiae*, compatible with BB Standard. Presented cloning system facilitated cloning of seven modules in an acceptor vector, bearing already auxotrophic selection marker, one promoter and one terminator, flanking the cloning site. Hence, altogether three complete TUs could be cloned. A mevalonate pathway, composed of three TUs served as a proof-of-concept. BB standard was implemented for multicopy cloning of expression cassettes encoding enzymes: protease from *Trichoderma koningii* (Shu et al. 2016), organophosphorus hydrolase from *Pseudomonas pseudoalcaligenes* (Shen et al. 2016) and two L-amino acid oxidases (Rao et al. 2020) in *K. phaffii*. In the two former reports, the BB cycle was repeated four times, to generate four TU-bearing assemblies of ~15 kbp. A single module was built by a gene flanked with regulatory elements (three elements already subcloned through traditional methods). In the latter report, the authors repeated the BB cycle up to three times, obtaining up to three copies of TUs. This time, each TU was built of five fragments: a promoter, a gene-encoding enzyme flagged at N- and C-termini, and a terminator. The generated vector was more than 18 kbp. Altogether, the principal aim of the multicopy cloning in *K. phaffii* was to increase gene dosage and produce more of the heterologous proteins. As another implementation of the BB cloning standard, (Wong et al. 2017) proposed a complete toolkit for modular cloning in *Y. lipolytica*. The authors generated reusable and ready-to-use modules of 12 native promoters of 1 kbp, intronic sequences (as they proposed and validated operon-type cloning system in *Y. lipolytica*, instead of traditional TUs) and terminators. Additionally, four modules bearing reporter genes (fluorescent and chromogenic) were constructed in accordance with the BB standard. Robustness of the toolkit was validated by a generation of 5-gene violacein biosynthetic pathway of 12 kbp. Thanks to BB modular cloning assisted by rapid DNA assembly technique (Gibson assembly was adopted), the extensive pathway was constructed in one week. Thanks to the modularity of the system, combinatorial cloning of the pathway was executed, and positional effects and different genes' configurations could be studied. The system of reusable modules compatible with the BB standard is available from Addgene.

3.2 GoldenGate

Introduction of GoldenGate (GG) cloning strategy to molecular biology portfolio is dated back to 2008 (Engler et al. 2008). The technique relies on the use of type IIS RE (e.g., *Bsa*I, *Bsm*BI) that recognizes a 6 bp sequence but cleaves 5 bp downstream of the RS, leaving a 4 bp overhang. But the key element is that *the cleaved region may be of any nucleotide sequence.*

Box 6
Type IIS RE belongs to a specific group of RE that recognize asymmetric DNA sequences (BsaI: GGTCTC) and cleave at a defined distance outside of their RS, (1–20 bp) but the combination of nucleotides at the cleavage site is not defined. Since the sequence at the cutting site is indifferent to the RE, it can be modified/designed. It is possible to generate a system of unique sequences, that will form overlaps after the cleavage with type IIS RE, which is the essence of designing a scaffold for GG cloning.

In this way, the user may design overhangs that will be generated upon cleavage. A set of such unique overhangs, serving as a linking/docking system, constitutes a *scaffold of the GG* standard defining position and orientation of assembled parts (Fig. 4). To assure compatibility of the modules, the scaffold must remain unchanged. Adjacent bioparts share the same overhang at the junction points. Modules are prepared by PCR amplification with primers adding ER IIS RS at 5', followed by the 4 bp overhang in the middle of the primer, and a sequence compatible with the biopart at its 3' end. Such construction allows to "loose" the type IIS RE RS after cleavage, leaving only the 4 bp overhang after digestion and a scar after fusion. To generate a library of GG modules, the PCR-amplified bioparts are cloned in the entry vectors. The entry vectors with the modules are pooled together with a destination vector in the assembly reaction. The destination vector typically bears some chromophore gene that serves as a negative control after assembly and cloning in *E. coli*. The chromophore region in the destination vector is flanked with type IIS RE RS cloned inwardly, so the cleavage leaves 4 bp overhangs at the vector's termini, and the chromophore cassette with the type IIS RE RS is discarded. The reaction is conducted in a cycling profile of subsequent incubations at 37 °C for ER action and 16 °C for the action of T4 ligase. With each ER/ligation cycle, new fusions are generated, devoid of type IIS RE RS—disallowing any further modifications. The number of elements that can be fused in a single cycling reaction in a single tube is limited by the scaffold design. More than ten modules are

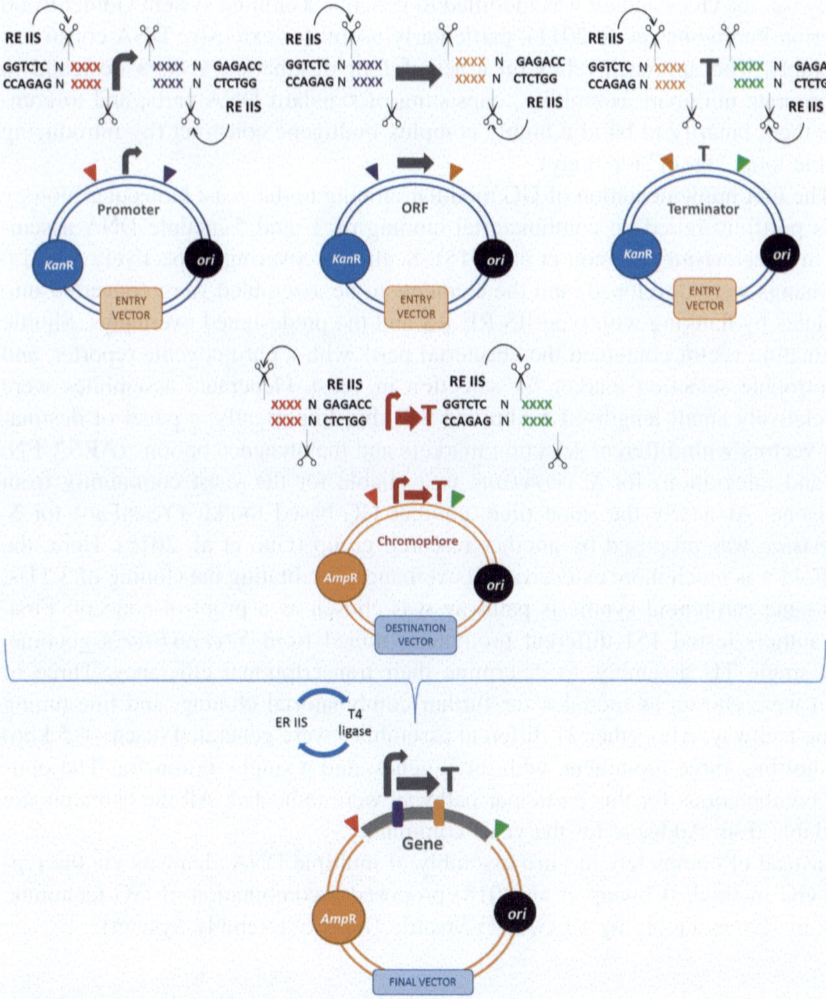

Fig. 4 General outline of GoldenGate parts assembly (e.g., promoter, ORF and terminator). Each of the modules is first equipped with RS for type IIS ER and 4 bp overhang sequences (indicated as red, purple, orange and green XXXX), according to a predesigned scaffold. Full description is given in the text. 4-bp scar sites are shown as purple and orange rectangles in the final assembly. Circular objects indicate the bacterial ori of replication (black) and bacterial resistance gene: KanR (blue) and AmpR (orange) contained in the "bacterial" part of the assembly, which is typically discarded before yeast cell transformation

regularly fused (examples are given hereafter) with a cloning efficiency of about 90% (Engler et al. 2008).

The GG concept-based variation, called MoClo (Weber et al. 2011), allows generating large multigene construction (up to 33 kb) in three cloning steps, from basic modules, through pre-assembled TUs, finally merged into an expression construct.

Likewise, the GG standard was modified to generate a cloning system GoldenBraid (Sarrion-Perdigones et al. 2011), particularly useful for extensive DNA constructs assembly. The approach relies on a set of four destination vectors designed to incorporate multipart assemblies, consisting of standard DNA parts, and to combine them binarily to build a highly complex multigene construct (by introducing double loop, "braid" topology).

The first implementation of GG modular cloning to the yeast molecular biology tools portfolio relied on combinatorial cloning of 3- and 5-module DNA assembly in *S. cerevisiae* (Agmon et al. 2015). Scaffolds covering respectively 4 and 6 overhangs were developed, and the elements to be assembled were converted into modules by flanking with type IIS RE RS and the predesigned overhangs. Shuttle destination vector contained the "bacterial part" with a chromogenic reporter, and auxotrophic selection marker for selection in yeast. Generated assemblies were of relatively small length—1,5 kbp and 2,2 kbp. Importantly, a panel of destination vectors with different selection markers and maintenance options (ARS/CEN, 2μ and integration) for *S. cerevisiae* is available for the yeast community from Addgene. At nearly the same time, another GG-based toolkit (YeastFab) for *S. cerevisiae* was proposed by another research group (Guo et al. 2015). Here, the scaffold was much more extensive (10 overhangs) facilitating the cloning of 3 TUs. A 3-gene carotenoid synthesis pathway was chosen as a proof-of-concept. First, the authors tested 151 different promoters, mined from *S. cerevisiae*'s genome, in a single TU assembly, to determine their transcriptional efficiency. Three of them were chosen as modules for further combinatorial cloning, and fine-tuning of the pathway. Altogether 27 different assemblies were generated (each ~6,5 kbp) by shuffling three promoters, with three genes, and a single terminator. The optimal combinations for this particular pathway were indicated. All the elements are available from Addgene for the yeast community.

Instead of completely in vitro assembly of multiple DNA elements via the typical GG method, (Chuang et al. 2018) proposed a combination of GG technique with in vivo assembly by VEGAS (*VE*rsatile *G*enetic *A*ssembly *S*ystem).

Box 7
VEGAS is a molecular biology technique exploiting the innate capacity of a microbial cell to combine DNA fragments with terminal complementarity by homologous recombination using orthogonal adapter sequences (VAs). The adapters are introduced to the DNA fragments as terminal modules flanking the TUs in the core. Their location and orientation direct desired order of the core pathway genes.

First, five individual DNA assemblies (violacein synthesis pathway), each composed of five modules in a 6-overhang scaffold, were constructed in vitro in a typical GG reaction. In each assembly, central TU (promoter, gene, terminator) was flanked with terminal adapters, facilitating in vivo homologous recombination. The five DNA assemblies were transformed into *E. coli* and correctly fused in a complete construction of ~13 kbp by the native recombination machinery of a cell.

Another interesting example of GG cloning usage in *S. cerevisiae*'s genetic engineering is assembling of CRISPR-Cas9-sgRNA cassette, for multiplexed genome editing (Zhang et al. 2019). The authors exploited modular cloning to assemble a CRISPR-Cas9 cassette made of 4 up to 11 modules in a scaffold composed of 5 to 12 predesigned overhangs. All the generated assemblies bear a gene for Cas9, auxotrophic marker, a promoter and a different number of sgRNAs, to facilitate simultaneous disruption of up to 6 genes at 96% efficiency.

Several comprehensive GG-based systems were proposed for *K. phaffii*. One of the most extensive toolkits was developed and provided to the yeast community by (Prielhofer et al. 2017). The system GoldenPiCS (available from Addgene) covers ready-to-use, multi-use modules bearing 20 different promoters, 10 terminators, 4 integration *loci* and 4 dominant selection markers. The modules can be cloned in a scaffold composed of 13 fusion sites. The authors developed a hierarchical system of three destination vectors (BB1-3; 3 different variants of BB1, 8—BB2, 21—BB3), each dedicated for specific cloning—BB1 serves as the entry vectors with single, individually cloned modules, BB2 is for cloning of individual TUs, and BB3—for merging up to 8 TUs in a single DNA assembly. Each of the vectors' type is prepared to be assembled via a specific type IIS RE, and bears some specific traits, like incompatible RS between subsequent levels of hierarchical cloning, or elements targeting genomic integration (BB3). Based on provided information, the largest DNA assembly generated via hierarchical cloning from BB1 to BB3 bears 27 modules of total length ~23 kbp. The GoldenPiCS technique was recently used for construction of a 10-modules-bearing, very convenient CRISPRi system for genome editing in *K. phaffii* (Gassler et al. 2019). The system requires only one-step cloning of sgRNA sequence in a ready-to-use BB3 vector to a get fully operable construction (also available from Addgene).

MoClo *Pichia* Toolkit (termed Yeast Toolkit; YTK) for hierarchical, modular cloning was extended in a work on standardized regulatory elements and secretory tags for fine-tuning of heterologous secretory proteins in *K. phaffii* (Obst et al. 2017). Twenty-one new bioparts, compatible with the standard, were added to the existing library; these were 4 promoters, 10 secretion tags, 1 terminator and 2 ori. All the DNA assemblies were composed of 8 modules in a 9-fusion sites scaffold. With the new modules, added to existing elements, it was possible to overexpress a given gene in 264 different ways. The authors built and characterized the expression and secretion efficiency of 124 constructs that combined different regulatory elements with two fluorescent reporter proteins, of approximate length 4 kbp. The authors pointed that it was possible to cross-use *S. cerevisiae*'s YTK bioparts in *K. phaffii*, which, when combined, give >4000 different possibilities of a given gene cloning.

Schreiber et al. (2017) developed a 7-fusion sites GG scaffold to perform combinatorial cloning of chimeric proteins in *K. phaffii*. The study comprised construction of 11 new modules, incl. 2 promoters, 3 pre-sequences, 1 pro-sequence, 1 protease cleavage site, 3 genes of interest and a single terminator. Shuffling of all the module variants allowed to rapid construction and test of 18 DNA assemblies. A very similar approach was undertaken in a very recent study by (Püllmann et al. 2020) aiming at overexpression of eight enzyme-encoding genes. In that study, the authors prepared destination vectors with a pre-cloned dominant selection marker, promoter and terminator for the gene of interest transcription, and ARS for episomal maintenance. Thus, only a 4-fusion sites GG scaffold was used for the modules shuffling. The three shuffled modules were: (i) signal peptide (one of 17 variants), (ii) target gene (out of 8), (iii) protein tag for purification (7 variants available). The authors deposited all the modules in Addgene for the yeast community.

A remarkable variety of GG modular cloning implementations were reported for *Y. lipolytica*, with the first mention published in 2017 (Celińska et al. 2017). In that report, a scaffold composed of 13 fusion sites was developed for modular cloning of 12 bioparts, equipped with specific 4 bp overhangs. The authors proposed 48 new modules, to be fitted into the scaffold, covering different promoters, terminators, selection markers and elements directing targeted genomic integration. Soon after that first mention, a report on biotechnologically relevant exploitation of that scaffold and the modules was published (Larroude et al. 2018). The authors shuffled promoters of different strengths to optimize expression of a 3-gene carotenoids synthesis pathway in *Y. lipolytica*. The shuffling process was conducted in a single-tube, single-step reaction to assembly 12-module pathway of >12 kb. The initial scaffold was later on modified by the addition of a new overhang into the scaffold (Celińska et al. 2018). The additional fusion site was used to enable combinatorial cloning of different signal peptides, to be tested with different heterologous secretory proteins. The initial scaffold was designed to enable the rapid fusion of 12 bioparts—3 TUs + selection marker +2 integration targeting sites. Introduction of this additional fusion site enabled the reuse of the previously developed modules (the problem is addressed in Box 1) and insertion of an additional one for a signal peptide (between promoter and target gene). The site was carefully designed to maintain the reading frame unchanged and to minimize amino acid changes within the encoded secretory proteins. The authors shuffled ten novel signal peptides with two reporter proteins to indicate optimal fusions. In the following study (Celińska et al. 2020), the best fusions (signal peptide + gene) were amplified as single modules and could be cloned in the former scaffold, as multiple TUs bearing DNA assembly. In that study, the authors investigated positional effect on the genes' expression, by combinatorial cloning of optimized fusions (signal peptide and the genes) in either the first or the second TU. Thanks to the modularity of the GG method, it was possible to rapidly generate the necessary set of DNA constructions. An improved scaffold with an enriched library of modules (altogether 64 modules, validated with three different fluorescent reporters) was finally made available to the yeast community through Addgene (Larroude et al. 2019). The same scaffold, either in its full length or limited to a single/double TUs was later exploited in different applications, like rapid single genes cloning under strong promoter (Korpys-Woźniak et al. 2020) or testing the strength of newly described promoters (Park et al. 2019).

Recently, (Liu et al. 2019) exploited the GG assembly method to construct chimeric pathways of lycopene synthesis in *Y. lipolytica.* The authors selected nine pathways from *S. cerevisiae*, generating a panel of 100 heterologous genes, to be transformed into the modules and fit into GG scaffolds of up to 5 TUs. The heterologous genes, cloned in GG assemblies, operated together with native lipophilic terpene synthesis genes in *Y. lipolytica* generating a variety of different phenotypes. Stochastic, combinatorial approach, facilitated through efficient modular cloning, enabled rapid optimization of the pathway expression and >150-fold enhancement in lycopene synthesis. Additionally, new knowledge on molecular background behind this biotechnologically relevant trait was gained.

Finally, the GG assembly technique was exploited to generate a CRISPR-Cas9 acceptor vector for genome editing in *Y. lipolytica* (Larroude et al. 2020). The acceptor vector bears a typical bacterial part, ARS for episomal maintenance in *Y. lipolytica*, an excisable selection marker gene, an optimized Cas9 gene under a strong hybrid promoter, and a *Bsm*BI RS flanking RFP chromophore, for convenient selection of positive (white) clones. The *Bsm*BI RS site was designed for cloning of the targeting sgRNA. A collection of CRISPR-Cas9 vectors with different selection markers is now available from Addgene.

3.3 Gateway

Gateway™ cloning technology (GW) relies on a system of short *att* sequences recognized by a viral DNA recombinase. The mechanism was discovered in bacteriophage lambda where it facilitates the integration of the phage's DNA into *E. coli*'s genome (Hartley et al. 2000). The cloning system (vectors and enzymes mixes) was developed and commercialized by Invitrogen (http://www.invitrogen.com/; now ThermoFisher Scientific). The technology is routinely used to transfer DNA fragments between vectors containing compatible *att* recombination sites.

Box 8

There are four different variants of att sites (B, P, L, R). All of them contain a core 25-bp "recognition region" (black box) whereas L, R/P contain "arms" on either/both sides of the core, acting as interaction sites for the recombination enzymes (left arm—white box; right arm—lined box). The core region bears a central 7-bp "asymmetric overlap" (yellow box) that is recognized by clonases and thus determines where DNA is cut and rejoined.

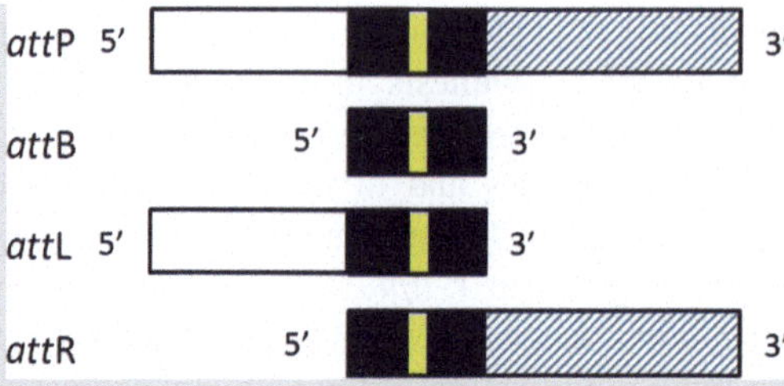

In its basic format, GW is used for single-fragment subcloning, but can be expanded to a multipart format for simultaneous assembly of multiple fragments in desired sequence and orientation (Fig. 5). For the latter, a scaffold of specific *att* sites is required. The modules are thus generated by flanking the bioparts with specific *attB* sequences (only 25-bp core). In this system, the modules must be first cloned into the donor vectors (which could be de facto omitted for BB and GG). Cloning in the donor vectors is executed by BP clonase that recombines *att*B sites from PCR-amplified modules with *att*P sites, present in the vectors. Upon recombination, the BP clonase generates *att*L (or *att*R—depends on the scaffold design) sites, which are further processed in the next step, conducted by LR clonase. In the multipart format, it is necessary to carefully design *subtypes* of the flanking regions (e.g., *att*B1 compatible with *att*P1, but not with *att*P2 or *att*P3; Fig. 5), so that only the border modules (the most 5' and the most 3' in the multipart assembly) bear *att*L(1) sites compatible with *att*R(1) sites present in the destination vector (Fig. 5).

Box 9
Definition of att sites' subtype is executed by making nucleotide changes within the 25-bp region. Setting different subtypes of att sites facilitates cloning of each DNA fragment in only one position and orientation.

In the final cloning stage, the LR clonase executes the reverse reaction to BP clonase, generating *att*B sites at the fusion sites. The destination vector bears *att*R sequences flanking ccdB toxic cassette, which, when maintained, disable clones growth. The reaction is conducted in vitro, over short incubation (1 h) at 25 °C. The multipart format allows for one-step cloning of up to five modules. As the other RD-/Rec-based assembly techniques, GW leaves (relatively long—25-bp) scars at the fusion sites. One of the biggest advantages of GW techniques is a wide array of destination vectors available, which can be either purchased or obtained from Addgene/iGEM collections.

Apart from the GW assembly system, some other recombination-based cloning systems have been issued, like Univector (Liu et al. 1998) and Creator (Siegel et al. 2004) relying on the use of Cre recombinase and *lox*P sites, or In-Fusion

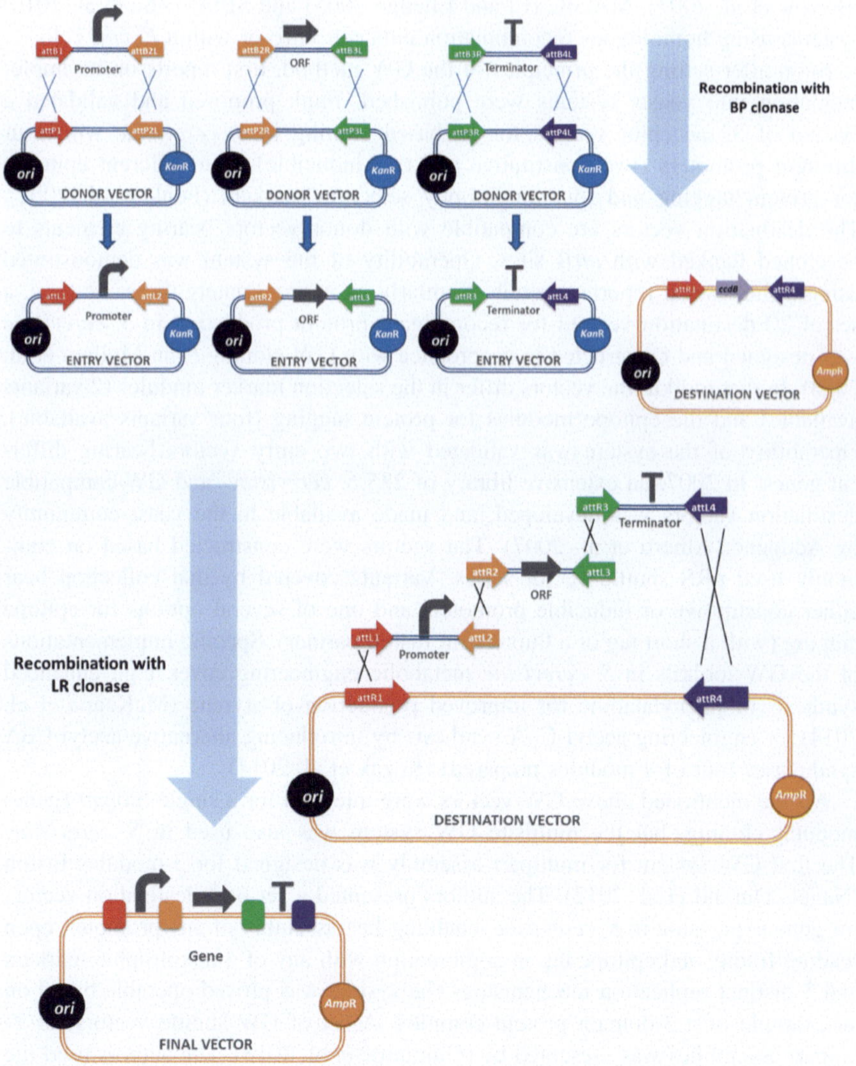

Fig. 5 General outline of Gateway cloning strategy (e.g., promoter, ORF and terminator). Each of the modules is first equipped with RecRS by the introduction of attB sites (indicated as red, purple, orange and green arrows), according to a predesigned scaffold. The reaction is two-stage as described in detail in the text. 25-bp scar sites are shown as red, green, purple and orange rectangles in the final assembly. Circular objects indicate the bacterial ori of replication (black) and bacterial resistance gene: KanR (blue) and AmpR (orange) contained in the "bacterial" part of the assembly, which is typically discarded before yeast cell transformation

(Berrow et al. 2007), MAGIC (Li and Elledge 2005) and SEFC (Zhu et al. 2010) systems using homologous recombination either in vitro or within *E. coli* cells.

Soon after setting the principles of the GW method, first reports on its implementation into yeasts systems were published. Funk proposed and validated a system of 32 acceptor vectors for *att*-based cloning in *S. cerevisiae* with four different promoters (two constitutive and two inducible), four different epitopes for protein tagging and four auxotrophy selection markers (Funk et al. 2002). The destination vectors are compatible with donor vectors bearing elements to be cloned flanked with *att*B sites. Operability of the system was demonstrated using a fluorescent reporter protein. Similarly, at approximately the same time, a set of 20 destination vectors for recombinant protein production in *S. cerevisiae* was designed and constructed in accordance with GW cloning (Van Mullem et al. 2003). In that toolkit, the vectors differ in the selection marker modules (2 variants available) and the epitope modules for protein tagging (four variants available). Operability of the system was validated with two entry vectors bearing different genes. In 2007, an extensive library of 285 *S. cerevisiae*- and GW-compatible destination vectors was developed, and made available to the yeast community by Addgene (Alberti et al. 2007). The vectors were constructed based on commonly used pRS shuttle vector series. Variants covered by that collection bear either constitutive or inducible promoters and one of several options for epitope tagging (with a short tag or a fluorescent fusion partner). Specific implementations of the GW toolkits in *S. cerevisiae* metabolic engineering cover, e.g., enhanced synthesis of phenylalanine for improved production of styrene (McKenna et al. 2014), or engineering acetyl-CoA synthesis by introducing alternative acetyl-CoA synthetases (out of 4 modules prepared) (Kozak et al. 2014).

All the mentioned above GW vectors were intended for a single biopart (gene) modular cloning, but the multisite GW system was also used in *S. cerevisiae*. The first GW system for multipart assembly was designed for 3-modules fusion (Nagels Durand et al. 2012). The authors presented a set of 3 destination vectors for gene expression in *S. cerevisiae*, enabling fast assembly of any promoter, open reading frame, and epitope tag in combination with any of 4 auxotrophic markers and 3 distinct replication mechanisms. The system was proved operable based on an example of a 3-domain protein complex. A set of GW shuttle vectors for 2-biopart assemblies was presented by (Giuraniuc et al. 2013). The authors used the system to develop 2-parts fusions to make a new, synthetic regulatory element, and to generate combinatorial fusions between five different promoters and 12 ORFs.

The first implementation of GW standard to *K. phaffii* system was reported by (Esposito et al. 2005). The authors generated a GW destination vector by modification of a popular pPICZ vector for secretory overproduction of proteins in a methanol-inducible manner. Interestingly, the authors used modified *att*B sites, which were reported to give 4-times more clones after transformation, than the original *att*Bs. Of key importance was the fact, the authors evidenced a lack of negative impact of the scar (*att* site) left between N-terminal fusion and the mature polypeptide-encoding sequence. With the same aim and highly corresponding approach, two GW-/*K. phaffii*-compatible destination vectors were developed based on pPICZ and pBGP1 (inducible/constitutive promoter) (Sasagawa et al. 2011). The vectors were used for single biopart cloning of secretory proteins tagged with an epitope and an additional 6xHis tag. Operability of the new vectors was tested with different hydrolase-encoding proteins.

A comprehensive toolkit for high-throughput gene cloning according to the GW principle was proposed for *Y. lipolytica* (Leplat et al. 2015). The authors developed a destination vector with a species-specific promoter, selection marker and terminator, enabling *att*-recombination cloning under control of a strong constitutive promoter. Moreover, a docking platform within *Y. lipolytica* strain's genome was constructed. The platform was located within *ura3 locus* (generating auxotrophy for the following GW-cassettes cloning), bearing a fluorescent reporter and zeta-docking sites for homologous recombination with the GW-generated cassette. The system was validated with a series of fluorescent and enzymatic reporters. In the following study by the same Research Team, the GW system was used for optimization of heterologous reporter genes overproduction by adopting promoter shuffling approach (Dulermo et al. 2017). The authors took advantage of modularity offered by GW cloning and prepared modules for 7 promoters of different strengths and 3 reporter proteins (fluorescent and enzymatic). Random shuffling of different promoters was conducted for each of the reporters individually, which allowed for interesting observations. In the following study, the authors cloned over 150 genes in the entry vectors and, using a newly developed high-throughput cloning protocol, subcloned them to a pre-developed destination vector, and ultimately—transformed the assemblies to a compatible *Y. lipolytica* host strain (Leplat et al. 2018). It was possible to conduct large-scale screens for the desired trait. That study is an excellent demonstration of GW cloning's high-throughput character and its pronounced capacity. The proposed system and developed protocols are very convenient and efficient in terms of high-throughput testing of different genes overexpression in *Y. lipolytica.*

4 Overlap Assembly Strategies

Overlap assembly strategies rely on exploitation of *binary primers*, partly complementary to DNA sequences of the two fused bioparts. The primers can be designed to leave no scar between the two modules if they are composed solely of the adjacent bioparts sequences. No scaffold per se is needed, as the primers contain only the adjacent modules' sequences. On the other hand, all the elements must be specifically designed, and no generic elements are typically used. When used without a scaffold, the system lacks a modularity option. Depending on the specific approach, the bioparts *termini* are processed by different enzymes to generate single-strand DNA overhangs, that are ligated in the next step. In terms of the assembly reaction precision, OL assembly techniques are considered to be less efficient and precise when compared to RE / Rec-based techniques. For the classic experiment published by (Gibson 2009) on "*in yeasto*" assembly of short overlapping oligomers, frequently reported efficiency is 64.3%. However, it relates to the number of correctly assembled full-length products, but—bearing mutations. The actual number of full-length clones, with an identical sequence as designed was approx. 10%. Showing that OL-based techniques tend to be more susceptible to errors. On the other hand, the modules preparation stage is much simplified, as no elimination of internal RE RS step is needed.

4.1 Gibson

The principle of Gibson (GB) assembly technique relies on exploitation of specific binary primers covering sequences of adjacent modules to be fused and a mix of three enzymatic activities processing *termini* of the modules. So, by definition, the modularity of this technique is impaired as no scaffold is used. On the other hand, GB does not leave any scars at the fusion sites. During amplification reaction, the binary primers introduce 15–40-bp OLs between the adjacent elements and it is the only step required for the preparation of the modules (Fig. 6). Complementary

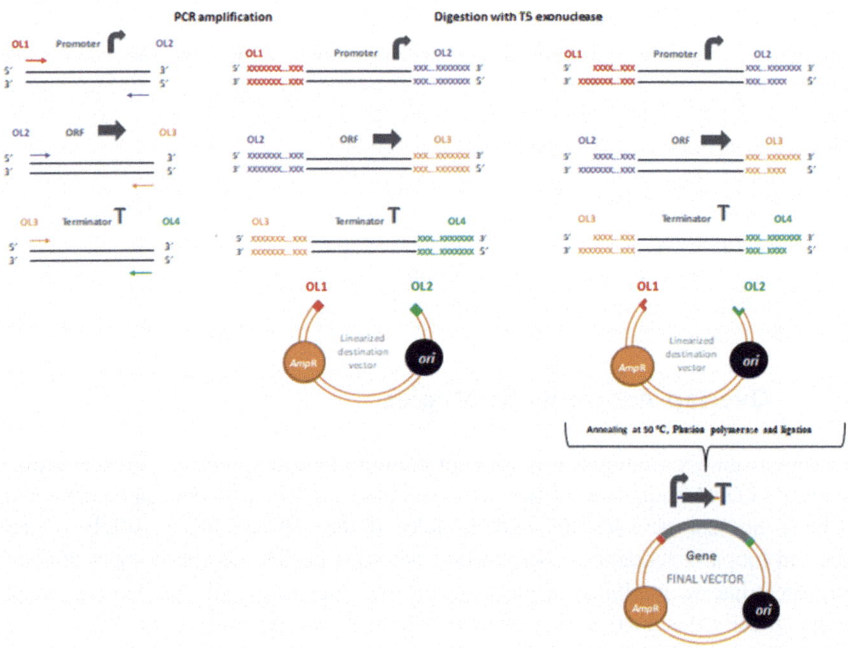

Fig. 6 General outline of Gibson cloning (e.g., promoter, ORF and terminator). Each of the modules is first amplified using binary primers covering sequences of adjacent elements (indicated as red, purple, orange and green OL XXX). The OLs are typically 40 nucleotides in length and define the final arrangement of the construct. The reaction is two-stage as described in detail in the text. No scars are left at the fusion sites. Circular objects indicate the bacterial ori of replication (black) and ampicillin resistance gene AmpR (orange) contained in the "bacterial" part of the assembly, which is typically discarded before yeast cell transformation

OLs must be also present in the destination vector (OL compatible with 5' of the first module and 3' of the last module within the assembly). Overlapping bioparts are mixed in a single tube with the destination vector and subjected to the action of three different enzymes. The 5'-exonuclease generates single-stranded DNA OLs by acting from the 5'-end so that the complementary regions are exposed and can be annealed. High-fidelity DNA polymerase fills the gaps and *Taq* DNA ligase covalently joins fragments. The reaction is a single step and isothermal (50 °C, over 15–60 min). Typically, with the use of a commercial mix of enzymes, up to 5 modules can be assembled simultaneously in a single-tube reaction using a one-step master mix of enzymes, whereas a two-step reaction allows combining as many as 15 bioparts. The size limit for this in vitro DNA assembly is unknown. Typically, efficient cloning of the cassettes reaching 20-kbp is reported, but the largest demonstrated construct has reached 900-kbp (Gibson 2009; Gibson et al. 2009). The GB method was a basis for the development of other techniques, like MODAL (Casini et al. 2013) and UNS (*U*nique *N*ucleotide *S*equences)-guided isothermal assembly (Torella et al. 2014).

The GB method's principles were first described in a paper on the enzymatic assembly of DNA bioparts in vitro (Gibson et al. 2009), where an isothermal, single-step reaction for assembling multiple overlapping DNA molecules by the concerted action of the three enzymes was described. As in the other OL-based techniques, no *sensu stricte* scaffold is needed (as it is for the GG method). Recently, (Cataldo et al. 2020) constructed an expression cassette for *S. cerevisiae* via the so-called "full in vitro Gibson assembly." The authors observed a toxic effect of the cloned genes on *E. coli* (the subcloning host). The in vitro-assembled GB cassette was composed of 8 bioparts (2 genes, bidirectional promoter, 2 terminators, up and down elements for integration, and a selection marker) bearing 40 bp OLs. To bypass the bottleneck of subcloning in *E. coli*, the full-length assembly was used as a template in a PCR. The amplified cassette was then transformed into the final host—*S. cerevisiae*.

Several interesting implementations of the GB method to complex DNA assemblies construction for *K. phaffii* were described. Combination of OE-PCR and GB assembly methods was adopted to construct an extensive, 12-bioparts bearing DNA assembly in 12 different variants (Vogl et al. 2016). A 4-gene pathway for carotenoids synthesis was used as a model. At first, the authors generated a library of 12 methanol-dependent promoters of different strengths and a different mechanism of induction; and an additional collection of 4 terminators. They assembled a complete expression cassette in a step-wise manner. First, 3 individual elements building individual TUs (promoter, gene, terminator) were fused using OE-PCR. Subsequently, 4 such TUs were assembled via the GB method to generate a complete DNA construction. A combinatorial approach was followed to construct different variants of the complete DNA assembly, each bearing 4 TUs. This way, in the initial OE-PCR, 3 bioparts were assembled, and in the following GB reaction, 4 modules (each bearing individual TU) were shuffled to build a complete carotenoid pathway, but controlled by different regulatory elements. The strategy enabled fine-tuning of the model pathway.

GB assembly was used to optimize vectors for CRISPR-Cas9-based genome editing in *K. phaffii* (Weninger et al. 2016). In that study, full advantage of the GB method's efficiency and modularity was taken. The authors shuffled multiple elements in the CRISPR-Cas9 vectors (constructed on a pPpT4_GAP shuttle vector's backbone), including gRNA, ribozymes and structural components of sgRNA, promoters for Cas9 and sgRNA expression, as well as different variants of codon-optimized Cas9 with different fusions. From among 95 DNA assemblies, only 5 were operable. The optimized system was proved highly efficient in multiplexed gene deletions and genomic integration of DNA expression cassettes.

Based on conducted literature search, it seems, that the yeast community working with *K. phaffii* widely uses GB technology for both: more complex cloning strategies (Royle and Polizzi 2017) and routine single-gene assemblies (Torres et al. 2019). Likewise, GB is broadly used for the generation of various DNA constructions dedicated to cloning in *Y. lipolytica.* By the simple introduction of specific OLs in the primers used during DNA fragments amplification and the following GB reaction, multiple different constructions were generated in work on glycogen synthase modification in *Y. lipolytica* (Bhutada et al. 2017) or xylose utilization pathway engineering (Rodriguez et al. 2016); where it enabled rapid incorporation of sgRNA into modular pCRISPRyl, or assembly of multipart overexpression cassettes.

4.2 USER®

Uracil-Specific Excision Reaction (USER®; NEB, New England Biolabs)-based cloning is a RE- and ligase-independent technique that enables assembly of multiple DNA fragments by the means of short OLs of 7–12 bp in length (Nour-Eldin et al. 2006). The basic principle is to use binary primers (as in GB), complementary to both adjacent elements, to generate the bioparts. But the key distinguishing element is to *replace* a single deoxythymidine (*dT*) nucleotide with one deoxyuridine (*dU*) nucleotide at the 3' terminus of the primers (Fig. 7). The primers are used at the stage of the preparation of the modules, so the DNA polymerase used here must be tolerant to dU (e.g. PfuX7 (Nørholm 2010)). The presence of dU at the borders of the modules precisely marks the site for an uracil DNA glycosylase and endonuclease VIII, contained in the Uracil-specific excision reagent. By the action of the two enzymes, the uracil residue is excised together with upstream sequence, releasing single-stranded 3'OL of 7–12-bp, permitting the directional assembly of several modules (up to 7). The modules are assembled in a single-tube reaction incubated at 37 °C for 20 min, and 25 °C for 20 min, reaching up to 90% efficiency (Lund et al. 2014). While no scar is present between the two fused modules, a 13-bp fragment of the USER cassette is present between the vector backbone and the assembly. A panel of yeast species-specific modules compatible with USER cloning is available from Addgene.

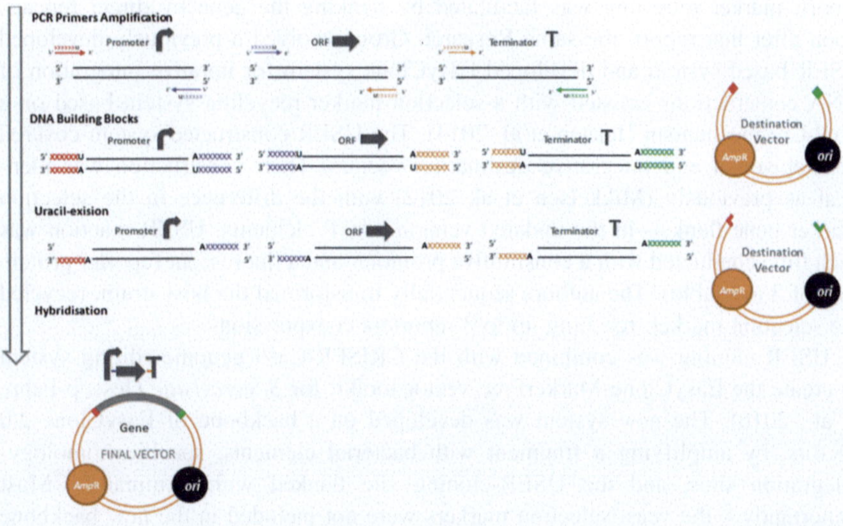

Fig. 7 General outline of USER® cloning (e.g., promoter, ORF and terminator). Each of the modules is first amplified using binary primers covering sequences of adjacent elements (indicated as red, purple, orange and green XXX). The primers include a single deoxyuracil residue (dU) flanking the 3' ends of the homology region. The OLs are typically 7–12 nucleotides in length and define the final arrangement of the construct. The reaction details are given in the text. No scars are left at the fusion sites. Circular objects indicate the bacterial ori of replication (black) and ampicillin resistance gene AmpR (orange) contained in the "bacterial" part of the assembly, which is typically discarded before yeast cells transformation

Several interesting implementations of the USER cloning strategy to the selected yeast species genetic engineering have been published. In 2012, the technique was exploited for the construction of a comprehensive cloning system for complex pathway assembly and their site-specific genomic integration in *S. cerevisiae* (Mikkelsen et al. 2012). The authors generated a chassis for cloning up to 22 genes: 2 genes under bidirectional promoter per DNA assembly and 11 validated, individual integration sites within *S. cerevisiae*'s genome. The integration sites in the yeast genome were validated by tracking growth rate and reporter enzyme synthesis after a reporter cassette integration. The operability of the system was demonstrated by the heterologous expression of an 8-gene pathway for indolylglucosinolate synthesis. The generated destination vector comprised genomic integration targeting sites (up and down), 2 terminators, auxotrophic selection marker, and the "bacterial part." Three modules were fused in each reaction to obtain operable cassettes—two genes flanking a bidirectional promoter. The 8-gene pathway was thus generated by construction of 4 expression cassettes, each bearing 2 genes under a bidirectional promoter. Each time, 3 modules were fused at 4 fusion sites, resulting in expression cassettes of 3–4-kbp (without the backbone vector). The whole pathway was assembled in the yeast host by 4 consecutive transformations with the selection marker recycling in-between. In that report, marker recycling was facilitated by flanking the gene by direct repeats. Soon after that report, the same Research Group evolved a previously developed USER-based system and developed EasyClone system for iterative integration of DNA constructions assisted with a selection marker recycling system based on a *lox*P-Cre mechanism (Jensen et al. 2014). The USER-constructed system covered both episomal and integrative destination vectors. Their construction was identical as previously (Mikkelsen et al. 2012) with the difference in the selection marker gene flanks—in the updated version—*lox*P elements. USER-reaction was each time conducted with a constitutive promoter and a fluorescent reporter protein (out of 3 available). The authors sequentially transformed the host strain, recycled the selection marker, reaching up to 3 reporters coexpression.

USER cloning was combined with the CRISPR-Cas9 genome-editing system to create the EasyClone-MarkerFree vector toolkit for *S. cerevisiae* (Jessop-Fabre et al., 2016). The new system was developed on a backbone of EasyClone 2.0 vectors, by amplifying a fragment with bacterial elements, specific homology-integration sites, and the USER-cloning site flanked with terminators. Most importantly – the yeast selection markers were not included in the new backbone vector. The expression cassettes (two genes flanking a bidirectional promoter) to be cloned in such a new backbone vector were first assembled (via USER) and then fused with the marker-less vector. To precisely direct genomic integration of these cassettes, a gRNA-expressing vector was prepared. In the proof-of-concept example, the authors prepared three USER-cloned assemblies, each bearing two genes under bidirectional promoter and specific homology-integration regions, and a vector bearing 3 gRNAs, compatible with the homology regions. The three expression cassettes were then cotransformed with the triple gRNA vector into a *S. cerevisiae* strain already synthetizing Cas9 from an episomal vector. The cassettes are then integrated into *loci* specified by gRNA and the 500-bp homology regions. The vector expressing gRNAs is then lost by subculturing the strains in a non-selective medium. Similarly, Cas9-bearing helper vector could be lost, if no further engineering was planned. The developed system enables simultaneous introduction of up to three integration cassettes into the genome of *S. cerevisiae* without the use of any yeast selection markers. Using standardized USER-bioparts, the integration cassettes can be constructed for overexpression of one or two genes per integration site. The authors suggested that for better stability, the selected integration sites should not be in direct proximity to each other. The system was validated not only in laboratory strain (CEN.PK) but also in an industrial diploid (EthanolRed), which is of great interest. The EasyClone-MarkerFree vector toolkit is available to the yeast community from Addgene.

The USER® assembly system was used in multiple following studies by that Research Team. One of the interesting examples is a paper presenting a method enabling simultaneous disruption of multiple genetic targets, where a combination of CRISPR/Cas9, in vivo recombination, USER assembly, and RNAi was developed (Kildegaard et al. 2019). USER cloning was used there for assembly of multipart expression cassettes. Interestingly, the authors improved the modularity of USER by introducing standardized spacers/overhangs of 60-bp at the fusion sites, flanking individual TUs. These were used in combination with standard overhangs adopted for USER cloning, for fusions inside the TUs. Such specific design allowed reusing a standard set of primers for amplification of different TUs, so the genes and regulatory elements could be cloned in a combinatorial manner. By adopting that sophisticated system of cloning, the authors constructed an extensive assembly, covering 3 TUs, with a selection marker element, and 2 genomic integration regions flanking the whole assembly.

The USER-based EasyClone system was also adopted for *Y. lipolytica* in a form of a comprehensive toolbox (Holkenbrink et al. 2018). The system comprises a broad set of integrative expression vectors, compatible with USER cloning, which allows cloning up to 2 genes. As in the case of the EasyClone toolkit for *S. cerevisiae*, the "Yali" system was extensively studied in terms of the selection of intergenic and highly transcribed integration *loci*. As for the former system, stringent requirements were set to propose candidate integration sequences, and the presumptions were carefully verified using a reporter protein system. Within the EasyClone Yali toolbox, the Authors proposed 5 integrative, marker-less vectors, precisely directed for integration by CRISPR/Cas9 system, and 11 vectors to be integrated and maintained using auxotrophic/resistance selection markers. Additionally, the vectors and protocols for marker-less deletion of up to 2 genes simultaneously using the CRISPR/Cas9 system were provided. Using 90-bp double-stranded oligos as the DNA repair templates, the authors obtained efficiencies of up to 90% and 66% for individual and double knockouts, respectively. The expression/integration cassette construction in EasyClone Yali is facilitated by USER cloning and a rich collection of standardized and reusable bioparts. The EasyCloneYali vectors can be obtained via Addgene. Operability and robustness of the system were evidenced in another paper by that research Team on cloning and optimization of astaxanthin synthesis pathway in *Y. lipolytica* (Kildegaard et al. 2017). In that study, a great variety of diverse combinations of 6 genes of different origins was shuffled and cloned in both laboratory auxotrophs and native *Y. lipolytica* strains. Thanks to intensive engineering of several side-branches of the terpenes synthesis pathway, the flux through the target astaxanthin was enhanced. Among the other, efficient cloning techniques were the enabling factor for that study.

4.3 OE-PCR

*O*verlap *E*xtension *P*olymerase *C*hain *R*eaction (OE-PCR) is a method that enables extending DNA fragments of interest to obtain assembled DNA constructions. In its standard form, OE-PCR is not a typical module assembly strategy, as no scaffold and no specific fusions system is needed. But the technique was proven to be useful for site-directed mutagenesis (Ho et al. 1989) and gene splicing (Horton et al. 1989). In principle, the assembly takes place solely by PCR (DNA amplification by DNA polymerase). First, the bioparts are individually amplified using binary primers partly complementary to the adjacent elements. 20–50-bp of complementarity was proved sufficient (Anthony and Liebman 1995; Nelson and Fitch 2011). In the next step, the bioparts are pooled in a single tube for PCR. The denaturation step renders single-stranded DNA and the complementary regions of the adjacent fragments act as priming sites for DNA polymerase at the annealing step (so-called "*auto-priming*"). Hence, the fragments can be fused in a PCR without additional, flanking oligonucleotide primers (Fig. 8). Typical primers, flanking a complete assembly from 5' and 3' *termini*, are added to the reaction after several rounds of auto-priming. As reported, the OE-PCR allows assembling up to 8 elements of different lengths in a single in vitro reaction (Kadkhodaei et al. 2016). The reaction is conducted under a standard PCR cycling profile, adjusted to the properties of the assembled elements. Due to its simplicity and wide availability, this method has gained popularity, but rather for assembling a limited number of small elements. The biggest drawback in terms of the complex assemblies generation is low efficiency of fusion, and that typical OE-PCR is not amenable neither for standardization nor modularity, as no specific scaffold is generated and no destination vectors are necessary. It was demonstrated that even upon using 40-bp OLs

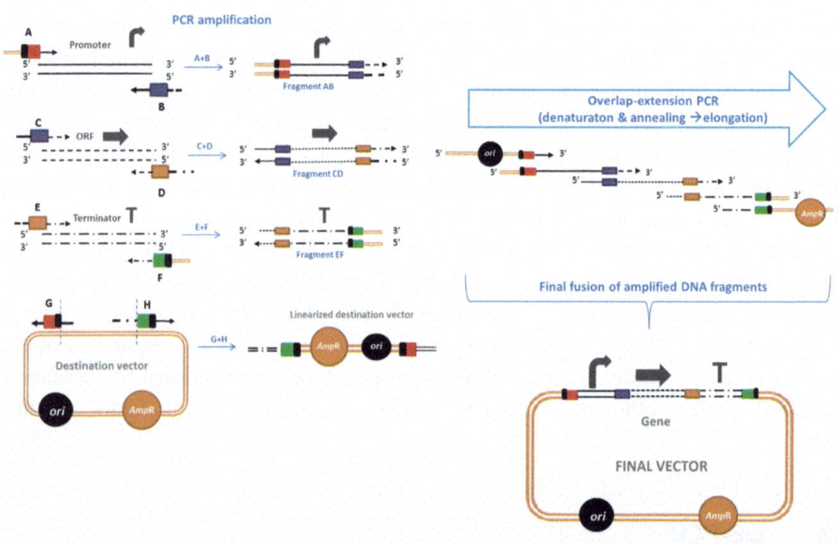

Fig. 8 General outline of OE_PCR assembly strategy. Each of the modules is first amplified using binary primers (represented as letters A-H) covering sequences of adjacent elements: promoter (solid lines), open reading frame (ORF, dotted lines), terminator (dashes and dots) and destination vector. Complementary regions to the adjacent parts are marked in colors (red, purple, orange and green squares). Product AB is generated using "A" and "B" primers on the promoter template. Parts "CD" "EF" and "GH" are amplified in the same way. Restriction enzyme sites are introduced in primers (black boxes). The reaction details are given in the text. No scars are left at the fusion sites. Circular objects indicate the bacterial ori of replication (black) and ampicillin resistance gene AmpR (orange) contained in the "bacterial" part of the assembly, which is typically discarded before yeast cells transformation

between adjacent fragments, the fusion product may not be correctly assembled (Cha-aim et al. 2009). However, several specific improvements and modifications were incorporated to the techniques, to address these issues (Xiao and Pei 2011; Guo et al. 2019). For example in 2007, (Dong et al. 2007) improved the OE-PCR technique by introducing into the protocol several elements, like primers design by DNA Works software instead of manual design, provided the Melting Temperatures™ of the OLs to ensure their specificity, reduced the length of OLs down to < 5 bp, and promoted the use of high fidelity DNA polymerase, which was specifically designed for OE-PCR conditions. The technique in the new format was validated with several genes of ~1–1.5 kbp.

While in its basic format OE-PCR technique may seem inferior to the other modular cloning and DNA assembly techniques, several interesting modifications advanced the technique and scope of its exploitation. To more easily apply OE-PCR to the generation of DNA assemblies and improve its specificity, a system of short, GC-rich OLs at the fusion sites was proposed (Cha-aim et al. 2009). The authors demonstrated that introduction of short G/C stretches can significantly improve specific annealing of compatible DNA fragments in a fusion PCR. Precisely, $C_{(15)}$ OL (but not GC repeat) turned out to be highly efficient in directing specific hybridization between the fusion elements under all examined temperatures. Such stretch cannot be used for fusion of functionally related parts (signal peptide and mature polypeptide or promoter and ORF) but could be used for fusion of independent bioparts—integration sites, selection markers, TUs. The authors demonstrated operability of their method on 3 bioparts fusion in a single reaction.

In 2009 (Shao et al. 2009) reported on "DNA Assembler" strategy for "*in yeasto*" assembly of complex, multi-element DNA constructions. In fact, the method highly corresponds to the initial, model experiment described by (Gibson 2009), when a 1170-bp DNA fragment was dissected into 38 short oligomers (60-bp long) and flanked with 20-bp complementarity regions to adjacent fragments, that were assembled "*in yeasto.*" The key difference between the two reports is the size of the assembled bioparts. Gibson assembled ~1 kb fragment from 38 synthetic oligos, and (Shao et al. 2009) executed complex pathway assembly of >10-kpb. Indeed, both techniques should be classified in the same sort of cloning strategies, as they rely on identical principles. Basically, the method relies on designing short OLs flanking adjacent fragments to be assembled. These can be either short oligos (Gibson 2009), or modules bearing regulatory elements and heterologous genes, as in the case of (Shao et al. 2009). For the latter, all proof-of-concept cassettes were assembled step-wise. The cassettes were as follows: (i) xylose utilization pathway (∼9 kb DNA consisting of 3 genes), (ii) zeaxanthin biosynthesis pathway (∼11 kb DNA consisting of 5 genes) and (iii) combined xylose utilization and zeaxanthin biosynthesis pathway (∼19 kb consisting of 8 genes). Basically, all TUs were assembled in vitro by standard OE-PCR. Adjacent TUs were already flanked with specific OLs, enabling correct assembly in desired orientation upon transformation into the yeast cell. The length of OLs was individually adjusted to the complexity of the DNA construction—less complex assembly (3–5 genes, ~10 kb) could be correctly fused with only 50-bp OLs reaching very high efficiency of 80–100%, while for more complex DNA constructions (8 genes, ~19 kb) the efficiency of 40–70% could be achieved with longer OLs (~125–430 bp). It was also demonstrated that by increasing the amount (in ng) of the modules per the same amount of the backbone vector, the efficiency of assembly can be increased. The authors stressed simplicity and rapid character of the DNA assembly method, which actually requires only appropriate design of PCR and exploitation of native recombination system of the yeast cell.

Many other, aim-specific implementations of OE-PCR to *S. cerevisiae* genetic engineering have been described. Just to give several different examples, OE-PCR was exploited for high-throughput production of linear DNA templates, that were later processed by cell-free transcription-translation system from *S. cerevisiae* (Gan and Jewett 2014), or for generation of chimeric genes involved in phenylpropanoid pathway (Jiang et al. 2005) or assembly of resveratrol synthesis pathway (Wang et al. 2011).

Likewise, multiple literature reports demonstrate wide applicability of OE-PCR for the assembly of synthetic DNA constructions that are later used for *K. phaffii* modifications. OE-PCR was used to adjust standard pGAPZα vector for removing unnecessary elements, and high-throughput mutagenesis of *Candida rugosa* lipase to be overproduced in *K. phaffii* (Chang et al. 2005). The scientific aim of that study was to exchange a rare codon for a frequent codon, which improved enzyme production. OE-PCR was also used in *K. phaffii*-related applications to conduct in vitro splicing (Jeya et al. 2009), or generate chimeric DNA templates encoding a gene with a fusion partner and cell-surface anchor (Yang et al. 2017).

The "DNA Assembler" technique reported by (Shao et al. 2009) was later "transplanted" to *Y. lipolytica* system to rapidly assemble multipart DNA cassettes (Gao et al. 2014). While the authors termed their cloning strategy as one-step assembly, in fact the approach was executed in a step-wise manner. First, the TUs and the flanking regions homologous to rDNA sequences in *Y. lipolytica* genome were assembled by OE-PCR in groups of 3–4 bioparts. Terminal elements of individual OE-PCR-assembled TUs were overlapping adjacent bioparts according to a design. The OLs between the adjacent parts were 45–60-bp, depending on the constructed variant. Such pre-assembled TUs were then mixed and transformed into the recipient strains which assembled them. The efficiency of the assembly reaction of a 4-gene pathway ranged between 11 and 21%, but the most important advancement relates to a decrease in the time necessary for the procedure completion. According to the authors calculations, the total procedure could be completed in less than one week, as compared to a previously reported sequential gene integration method that required n weeks for n genes (Celińska and Grajek 2013).

In 2020, (Guo et al. 2020) used OE-PCR to assemble 9 elements, altogether forming the first completely synthetic yeast artificial chromosome for *Y. lipolytica* (Yl-YAC), with centrally located ARS and flanked with telomeres. As the authors mainly aimed at optimization of Yl-YAC construction, they adopted combinatorial approach to testing different combinations of chromosomal elements (ARS and

TEL) and reporter gene-encoding modules, which were transformed into *Y. lipolytica* cells as free DNA elements. Each module contained 50-bp flanking regions compatible with adjacent sequence, according to a design. Importantly, as in the case of previous works with *S. cerevisiae*, *Y. lipolytica* native recombination mechanisms correctly assembled pre-designed DNA construction. After examining the replicative character and transferability of the Yl-YAC, the authors set at transferring both xylose and cellobiose utilization pathways to *Y. lipolytica* within the newly constructed Yl-YAC. A complete construction comprised three key genes for xylose (*XYL1*, *XYL2* and *XKS1*) and three genes for cellobiose utilization (*CBP1*, *CDT1* and *scPGM2*), of total length 23 kb.

5 Summary

A great variety of other modular cloning and DNA parts assembly methods have been developed, that were not covered by the main part of this chapter. For specific reasons, those methods did not gain popularity in the yeast research community. One of such methods is ligase cycling reaction (LCR) that employs single-stranded bridging oligomers (partly complementary to adjacent bioparts) that guide correct hybridization of DNA fragments. The bioparts to be fused must be phosphorylated at 5'ends (either by kinase or by amplification with 5'phosphorylated primers). The method is amenable to modularity by possible use of a 40-bp scaffold of oligonucleotide connectors (complementary to adjacent elements) which can be reused in alternative assembly design (Kok et al. 2014). The LCR is executed by hot-start thermostable ligase and polymerase under cycling profile of denaturation, annealing of the fragments and ligation at 66 °C. As reported, up to 20 bioparts of total length 20 kbp can be assembled in a single reaction without leaving scars at the fusion sites. The other such technique is circular polymerase extension cloning (CPEC) conducted with DNA polymerase solely (Quan and Tian 2009). The fragments must bear 15–25 bp OLs to anneal correctly. CPEC was reported to be highly efficient, but only up to 4 bioparts.

Irrespectively of technical details, modular cloning and standardized bioparts assembly strategies contribute to great progress within the field of genetic engineering of yeast. Their exploitation allows the researchers to focus on addressing the scientific questions, and to expand the scope of experiments to accurately test hypotheses, rather than execution of tedious laboratory procedures. Modularity and standardization facilitate wide-spread collaboration within the yeast community and speed-up accomplishing of new goals.

Acknowledgements PKW was financially supported by the Ministry of Sciences and Higher Education in Poland project: DI 2017 000947. MK was financially supported by the Ministry of Sciences and Higher Education in Poland project: DI2017 001047.

References

Agmon N et al (2015) Yeast Golden Gate (yGG) for the efficient assembly of *S. cerevisiae* transcription units. ACS Synthetic Biology. American Chemical Society 4(7):853–859. https://doi.org/10.1021/sb500372z

Alberti S, Gitler AD, Lindquist S (2007) A suite of Gateway® cloning vectors for high-throughput genetic analysis in *Saccharomyces cerevisiae*. Yeast 24(10):913–919. https://doi.org/10.1002/yea.1502

Anderson JC et al (2010) BglBricks: a flexible standard for biological part assembly. J Biol Eng 4:1–12. https://doi.org/10.1186/1754-1611-4-1

Anthony RA, Liebman SW (1995) Alterations in ribosomal protein RPS28 can diversely affect translational accuracy in *Saccharomyces cerevisiae*. Genetics 140(4):1247–1258

Berrow NS et al (2007) A versatile ligation-independent cloning method suitable for high-throughput expression screening applications. Nucleic Acids Res 35(6):e45. https://doi.org/10.1093/nar/gkm047

Bhutada G et al (2017) Sugar versus fat: elimination of glycogen storage improves lipid accumulation in *Yarrowia lipolytica*. FEMS Yeast Res 17(3):1–10. https://doi.org/10.1093/femsyr/fox020

Casini A et al (2013) One-pot DNA construction for synthetic biology: the modular overlap-directed assembly with linkers (MODAL) strategy. Nucleic Acids Res 42(1):1–13. https://doi.org/10.1093/nar/gkt915

Cataldo VF et al (2020) Genomic integration of unclonable gene expression cassettes in *Saccharomyces cerevisiae* using rapid cloning-free workflows. MicrobiologyOpen 9(3):1–10. https://doi.org/10.1002/mbo3.978

Celińska E et al (2017) Golden gate assembly system dedicated to complex pathway manipulation in *Yarrowia lipolytica*. Microbial Biotechnology. Wiley 10(2):450–455. https://doi.org/10.1111/1751-7915.12605

Celińska E et al (2018) Robust signal peptides for protein secretion in *Yarrowia lipolytica*: identification and characterization of novel secretory tags. Applied Microbiology and Biotechnology. Springer Verlag 102(12):5221–5233. https://doi.org/10.1007/s00253-018-8966-9

Celińska E et al (2020) Optimization of *Yarrowia lipolytica*-based consolidated biocatalyst through synthetic biology approach: transcription units and signal peptides shuffling. Appl Microbiol Biotechnol 104(13):5845–5859. https://doi.org/10.1007/s00253-020-10644-6

Celińska E, Grajek W (2013) A novel multigene expression construct for modification of glycerol metabolism in *Yarrowia lipolytica*. Microb Cell Fact 12(1):1–16. https://doi.org/10.1186/1475-2859-12-102

Cha-aim K et al (2009) Reliable fusion PCR mediated by GC-rich overlap sequences. Gene. Elsevier B.V. 434(1–2):43–49. https://doi.org/10.1016/j.gene.2008.12.014

Chang SW et al (2005) Multiple mutagenesis of the *Candida rugosa* LIP1 gene and optimum production of recombinant LIP1 expressed in *Pichia pastoris*. Appl Microbiol Biotechnol 67(2):215–224. https://doi.org/10.1007/s00253-004-1815-z

Chuang J, Boeke JD, Mitchell LA (2018) Coupling yeast golden gate and VEGAS for efficient assembly of the violacein pathway in *Saccharomyces cerevisiae*. Methods Mol Biol 1671:211–225. https://doi.org/10.1007/978-1-4939-7295-1_14

Dong B et al (2007) An improved method of gene synthesis based on DNA works software and overlap extension PCR. Mol Biotechnol 37(3):195–200. https://doi.org/10.1007/s12033-007-0039-8

Dueñas-Santero E et al (2019) A new toolkit for gene tagging in *Candida albicans* containing recyclable markers. PLoS ONE 14(7):1–17. https://doi.org/10.1371/journal.pone.0219715

Dulermo R et al (2017) Using a vector pool containing variable-strength promoters to optimize protein production in *Yarrowia lipolytica*. Microb Cell Fact. BioMed Central Ltd. 16(1). https://doi.org/10.1186/s12934-017-0647-3

Eldin P et al (1997) High-level secretion of two antibody single chain Fv fragments by *Pichia pastoris*. J Immunol Methods 201(1):67–75. https://doi.org/10.1016/S0022-1759(96)00213-X

Engler C, Kandzia R, Marillonnet S (2008) A one pot, one step, precision cloning method with high throughput capability. PLoS ONE 3(11). https://doi.org/10.1371/journal.pone.0003647

Esposito D et al (2005) Gateway cloning is compatible with protein secretion from *Pichia pastoris*. Protein Expr Purif 40(2):424–428. https://doi.org/10.1016/j.pep.2004.12.006

Funk M et al (2002) Vector systems for heterologous expression of proteins in *Saccharomyces cerevisiae*. Methods Enzymol 350(1994):248–257. https://doi.org/10.1016/S0076-6879(02)50967-8

Gan R, Jewett MC (2014) A combined cell-free transcription-translation system from *Saccharomyces cerevisiae* for rapid and robust protein synthesis. Biotechnol J 9(5):641–651. https://doi.org/10.1002/biot.201300545

Gao S et al (2014) One-step integration of multiple genes into the oleaginous yeast *Yarrowia lipolytica*. Biotechnol Lett. Kluwer Academic Publishers 36(12):2523–2528. https://doi.org/10.1007/s10529-014-1634-y

Gassler T et al (2019) CRISPR/Cas9-mediated homology-directed genome editing in *Pichia pastoris*. Methods Mol Biol (Clifton, N.J.). United States, 1923 211–225. https://doi.org/10.1007/978-1-4939-9024-5_9

Gibson DG (2009) Synthesis of DNA fragments in yeast by one-step assembly of overlapping oligonucleotides. Nucleic Acids Res 37(20):6984–6990. https://doi.org/10.1093/nar/gkp687

Gibson DG et al (2009) Enzymatic assembly of DNA molecules up to several hundred kilobases. Nat Methods 6(5):343–345. https://doi.org/10.1038/nmeth.1318

Giuraniuc CV, MacPherson M, Saka Y (2013) Gateway vectors for efficient artificial gene assembly in vitro and expression in yeast *Saccharomyces cerevisiae*. PLoS ONE 8(5). https://doi.org/10.1371/journal.pone.0064419

Guo W et al (2019) An improved overlap extension PCR for simultaneous multiple sites large fragments insertion, deletion and substitution. Sci Rep. Springer US 9(1):1–6. https://doi.org/10.1038/s41598-019-52122-8

Guo Y et al (2015) YeastFab: The design and construction of standard biological parts for metabolic engineering in *Saccharomyces cerevisiae*. Nucleic Acids Res 43(13):e88. https://doi.org/10.1093/nar/gkv464

Guo Z-P et al (2020) An artificial chromosome ylAC enables efficient assembly of multiple genes in *Yarrowia lipolytica* for biomanufacturing. Commun Biol. Springer US 3(1). https://doi.org/10.1038/s42003-020-0936-y

Hartley JL, Temple GF, Brasch MA (2000) DNA cloning using in vitro site-specific recombination. Genome Res 10(11):1788–1795. https://doi.org/10.1101/gr.143000

Ho SN et al (1989) Site-directed mutagenesis by overlap extension using the polymerase chain reaction. Gene 77(1):51–59. https://doi.org/10.1016/0378-1119(89)90358-2

Holkenbrink C et al (2018) EasyCloneYALI: CRISPR/Cas9-based synthetic toolbox for engineering of the yeast *Yarrowia lipolytica*. Biotechnol J. Germany 1–23. https://doi.org/10.1002/biot.201700543

Horton RM et al (1989) Engineering hybrid genes without the use of restriction enzymes: gene splicing by overlap extension. Gene 77(1):61–68. https://doi.org/10.1016/0378-1119(89)90359-4

Jensen NB et al (2014) EasyClone: Method for iterative chromosomal integration of multiple genes in *Saccharomyces cerevisiae*. FEMS Yeast Res. Blackwell Publishing Ltd, 14(2):238–248. https://doi.org/10.1111/1567-1364.12118

Jessop-Fabre MM et al (2016) EasyClone-MarkerFree: a vector toolkit for marker-less integration of genes into *Saccharomyces cerevisiae* via CRISPR-Cas9. Biotechnol J. Wiley-VCH Verlag 11(8):1110–1117. https://doi.org/10.1002/biot.201600147

Jeya M et al (2009) Cloning and expression of a GH11 xylanase gene from *Aspergillus fumigatus* MKU1 in *Pichia pastoris*. J Biosci Bioeng. The Society for Biotechnology, Japan 108(1):24–29. https://doi.org/10.1016/j.jbiosc.2009.02.003

Jiang H, Wood KV, Morgan JA (2005) Metabolic engineering of the phenylpropanoid pathway in *Saccharomyces cerevisiae*. Appl Environ Microbiol 71(6):2962–2969. https://doi.org/10.1128/AEM.71.6.2962-2969.2005

Kadkhodaei S et al (2016) Multiple overlap extension PCR (MOE-PCR): an effective technical shortcut to high throughput synthetic biology. RSC Adv 6(71):66682–66694. https://doi.org/10.1039/c6ra13172g

Kildegaard KR et al (2017) Engineering of *Yarrowia lipolytica* for production of astaxanthin. Synth Syst Biotechnol. KeAi Communications Co. 2(4):287–294. https://doi.org/10.1016/j.synbio.2017.10.002

Kildegaard KR et al (2019) CRISPR/Cas9-RNA interference system for combinatorial metabolic engineering of *Saccharomyces cerevisiae*. Yeast 36(5):237–247. https://doi.org/10.1002/yea.3390

Knight T (2003) Idempotent vector design for standard assembly of Biobricks, MIT Libraries. http://hdl.handle.net/1721.1/21168

Kok SD et al (2014) Rapid and reliable DNA assembly via ligase cycling reaction. ACS Synth Biol 3(2):97–106. https://doi.org/10.1021/sb4001992

Korpys-Woźniak P et al (2020) Impact of overproduced heterologous protein characteristics on physiological response in *Yarrowia lipolytica* steady-state-maintained continuous cultures. Appl Microbiol Biotechnol. https://doi.org/10.1007/s00253-020-10937-w

Kozak BU et al (2014) Replacement of the *Saccharomyces cerevisiae* acetyl-CoA synthetases by alternative pathways for cytosolic acetyl-CoA synthesis. Metab Eng. Elsevier 21:46–59. https://doi.org/10.1016/j.ymben.2013.11.005

Larroude M et al (2018) A synthetic biology approach to transform *Yarrowia lipolytica* into a competitive biotechnological producer of β-carotene. Biotechnol Bioeng 115(2):464–472. https://doi.org/10.1002/bit.26473

Larroude M et al (2019) A modular Golden Gate toolkit for *Yarrowia lipolytica* synthetic biology. Microb Biotechnol 12(720824):1249–1259. https://doi.org/10.1111/1751-7915.13427

Larroude M et al (2020) A set of *Yarrowia lipolytica* CRISPR/Cas9 vectors for exploiting wild-type strain diversity. Biotechnol Lett. Springer Netherlands 42(5):773–785. https://doi.org/10.1007/s10529-020-02805-4

Lebel K, MacPherson S, Turcotte B (2006) New tools for phenotypic analysis in *Candida albicans*: the WAR1 gene confers resistance to sorbate. Yeast (Chichester, England) 23(4):249–259. https://doi.org/10.1002/yea.1346

Leguia M et al (2013) 2ab assembly: A methodology for automatable, high-throughput assembly of standard biological parts. J Biol Eng 7(1):1. https://doi.org/10.1186/1754-1611-7-2

Leplat C, Nicaud J-MM, Rossignol T (2018) Overexpression screen reveals transcription factors involved in lipid accumulation in *Yarrowia lipolytica*. FEMS Yeast Res. England 18(5):1–9. https://doi.org/10.1093/femsyr/foy037

Leplat C, Nicaud JM, Rossignol T (2015) High-throughput transformation method for *Yarrowia lipolytica* mutant library screening. FEMS Yeast Res. Oxford University Press 15(6):1–9. https://doi.org/10.1093/femsyr/fov052

Li MZ, Elledge SJ (2005) 'MAGIC, an in vivo genetic method for the rapid construction of recombinant DNA molecules. Nature genetics. United States, 37(3):311–319. https://doi.org/10.1038/ng1505g

Liu D et al (2019) Constructing yeast chimeric pathways to boost lipophilic terpene synthesis. ACS Synth Biol. United States 8(4):724–733. https://doi.org/10.1021/acssynbio.8b00360

Liu Q et al (1998) The univector plasmid-fusion system, a method for rapid construction of recombinant DNA without restriction enzymes. Curr Biol. England 8(24):1300–1309. https://doi.org/10.1016/s0960-9822(07)00560-x

Lund AM et al (2014) A versatile system for USER cloning-based assembly of expression vectors for mammalian cell engineering. PLoS ONE 9(5). https://doi.org/10.1371/journal.pone.0096693

McKenna R et al (2014) Rational and combinatorial approaches to engineering styrene production by *Saccharomyces cerevisiae.* Microb Cell Fact 13(1):1–12. https://doi.org/10.1186/s12934-014-0123-2

Mikkelsen MD et al (2012) Microbial production of indolylglucosinolate through engineering of a multi-gene pathway in a versatile yeast expression platform. Metab Eng. Elsevier, 14(2):104–111. https://doi.org/10.1016/j.ymben.2012.01.006

Nagels Durand AN et al (2012) A MultiSite Gateway TM vector set for the functional analysis of genes in the model *Saccharomyces cerevisiae.* BMC Mol Biol 13. https://doi.org/10.1186/1471-2199-13-30

Nelson MD, Fitch DHA (2011) Overlap extension PCR: an efficient method. Methods Mol Biol 772:459–470. https://doi.org/10.1007/978-1-61779-228-1

Nisson PE, Rashtchian A, Watkins PC (1991) Rapid and efficient cloning of Alu-PCR products using uracil DNA glycosylase. Genome Res 1(2):120–123. https://doi.org/10.1101/gr.1.2.120

Nørholm MHH (2010) A mutant Pfu DNA polymerase designed for advanced uracil-excision DNA engineering. BMC Biotechnol 10. https://doi.org/10.1186/1472-6750-10-21

Nour-Eldin HH et al (2006) Advancing uracil-excision based cloning towards an ideal technique for cloning PCR fragments. Nucleic Acids Res 34(18). https://doi.org/10.1093/nar/gkl635

Obst U, Lu TK, Sieber V (2017) A modular toolkit for generating *Pichia pastoris* secretion libraries. ACS Synth Biol. American Chemical Society 6(6):1016–1025. https://doi.org/10.1021/acssynbio.6b00337

Park Y-K et al (2019) Engineering the architecture of erythritol-inducible promoters for regulated and enhanced gene expression in *Yarrowia lipolytica.* FEMS Yeast Res. England, 19(1):1. https://doi.org/10.1093/femsyr/foy105

Prielhofer R et al (2017) GoldenPiCS: a golden gate-derived modular cloning system for applied synthetic biology in the yeast *Pichia pastoris.* BMC Syst Biol. 11(1):1–14. https://doi.org/10.1186/s12918-017-0492-3

Püllmann P et al (2020) A modular two yeast species secretion system for the production and preparative application of fungal peroxygenases. bioRxiv 2020.07.22.216432. https://doi.org/10.1101/2020.07.22.216432

Quan J, Tian J (2009) Circular polymerase extension cloning of complex gene libraries and pathways. PLoS ONE 4(7). https://doi.org/10.1371/journal.pone.0006441

Rao B et al (2020) Efficient surface display of L-glutamate oxidase and L-amino acid oxidase on *Pichia pastoris* using multi-copy expression strains. Biotechnol Bioprocess Eng 25(4):571–579. https://doi.org/10.1007/s12257-019-0370-5

Rodriguez GM et al (2016) Engineering xylose utilization in *Yarrowia lipolytica* by understanding its cryptic xylose pathway. Biotechnol Biofuels. BioMed Central 9(1). https://doi.org/10.1186/s13068-016-0562-6

Royle KE, Polizzi K (2017) A streamlined cloning workflow minimising the time-to-strain pipeline for *Pichia pastoris.* Sci Rep. Springer US 7(1):1–10. https://doi.org/10.1038/s41598-017-16172-0

Sarrion-Perdigones A et al (2011) GoldenBraid: An iterative cloning system for standardized assembly of reusable genetic modules. PLoS ONE 6(7). https://doi.org/10.1371/journal.pone.0021622

Sasagawa T et al (2011) High-throughput recombinant gene expression systems in *Pichia pastoris* using newly developed plasmid vectors. Plasmid. Elsevier Inc. 65(1):65–69. https://doi.org/10.1016/j.plasmid.2010.08.004

Schneider M, Fresenborg L, Schadeweg V (2012) Yeast BioBrick assembly (YBA) standardized method for vector assembly of BioBrick devices via homologous recombination in *Saccharomyces cerevisiae.* iGEM report, Frankfurt, Germany

Schreiber C et al (2017) A high-throughput expression screening platform to optimize the production of antimicrobial peptides. Microb Cell Fact. BioMed Central 16(1):1–13. https://doi.org/10.1186/s12934-017-0637-5

Shao Z, Zhao H, Zhao H (2009) DNA assembler, an in vivo genetic method for rapid construction of biochemical pathways. Nucleic Acids Res 37(2):1–10. https://doi.org/10.1093/nar/gkn991

Shen W et al (2016) High level expression of organophosphorus hydrolase in *Pichia pastoris* by multicopy ophcM assembly. Protein Expr Purif. Elsevier Ltd, 119:110–116. https://doi.org/10.1016/j.pep.2015.11.012

Shetty RP, Endy D, Knight TF (2008) Engineering BioBrick vectors from BioBrick parts. J Biol Eng 2:1–12. https://doi.org/10.1186/1754-1611-2-5

Shu M et al (2016) High-level expression and characterization of a novel serine protease in *Pichia pastoris* by multi-copy integration. Enzyme Microb Technol. Elsevier Inc., 92:56–66. https://doi.org/10.1016/j.enzmictec.2016.06.007

Siegel RW et al (2004) Recombinatorial cloning using heterologous lox sites. Genome Res 14(6):1119–1129. https://doi.org/10.1101/gr.1821804

Smith C, Day PJR, Walker MR (1993) Generation of cohesive ends on PCR products by UDG-mediated excision of dU, and application for cloning into restriction digest-linearized vectors. Genome Res 2(4):328–332. https://doi.org/10.1101/gr.2.4.328

Torella JP et al (2014) Rapid construction of insulated genetic circuits via synthetic sequence-guided isothermal assembly. Nucleic Acids Res 42(1):681–689. https://doi.org/10.1093/nar/gkt860

Torres P et al (2019) Contextualized genome-scale model unveils high-order metabolic effects of the specific growth rate and oxygenation level in recombinant *Pichia pastoris*. Metab Eng Commun. Elsevier Ltd 9(July):e00103. https://doi.org/10.1016/j.mec.2019.e00103

Van Mullem V et al (2003) Construction of a set of *Saccharomyces cerevisiae* vectors designed for recombinational cloning. Yeast 20(8):739–746. https://doi.org/10.1002/yea.999

Vogl T et al (2015) Restriction site free cloning (RSFC) plasmid family for seamless, sequence independent cloning in *Pichia pastoris*. Microb Cell Fact. BioMed Central 14(1):1–15. https://doi.org/10.1186/s12934-015-0293-6

Vogl T et al (2016) A toolbox of diverse promoters related to methanol utilization: functionally verified parts for heterologous pathway expression in *Pichia pastoris*. ACS Synth Biol 5(2):172–186. https://doi.org/10.1021/acssynbio.5b00199

Wang Y et al (2011) Stepwise increase of resveratrol biosynthesis in yeast *Saccharomyces cerevisiae* by metabolic engineering. Metab Eng. Elsevier 13(5):455–463. https://doi.org/10.1016/j.ymben.2011.04.005

Weber E et al (2011) A modular cloning system for standardized assembly of multigene constructs. PLoS ONE 6(2). https://doi.org/10.1371/journal.pone.0016765

Weninger A et al (2016) Combinatorial optimization of CRISPR/Cas9 expression enables precision genome engineering in the methylotrophic yeast *Pichia pastoris*'. J Biotechnol. Elsevier B.V., 235:139–149. https://doi.org/10.1016/j.jbiotec.2016.03.027

Wong L et al (2017) YaliBricks, a versatile genetic toolkit for streamlined and rapid pathway engineering in *Yarrowia lipolytica*. Metab Eng Commun. Elsevier B.V., 5(August):68–77. https://doi.org/10.1016/j.meteno.2017.09.001

Xiao Y-H, Pei Y (2011) Asymmetric overlap extension PCR method for site-directed mutagenesis. Methods Mol Biol (Clifton, N.J.). United States 687:277–282. https://doi.org/10.1007/978-1-60761-944-4_20

Yang S et al (2017) Cell-surface displayed expression of trehalose synthase from *Pseudomonas putida* ATCC 47054 in *Pichia pastoris* using Pir1P as an anchor protein. Front Microbiol 8(DEC):1–9. https://doi.org/10.3389/fmicb.2017.02583

Zhang L et al (2010) Gene cloning, codon optimization and functional expression of *Yarrawia lipolytica* lipase Lip1]. Wei sheng wu xue bao = Acta microbiologica Sinica. China 50(7):969–974

Zhang Y et al (2019) A gRNA-tRNA array for CRISPR-Cas9 based rapid multiplexed genome editing in *Saccharomyces cerevisiae*. Nat Commun. Springer US 10(1):1–10. https://doi.org/10.1038/s41467-019-09005-3

Zhu D et al (2010) High-throughput cloning of human liver complete open reading frames using homologous recombination in *Escherichia coli*. Anal Biochem. United States 397(2):162–167. https://doi.org/10.1016/j.ab.2009.10.018

Cellular Engineering of *Yarrowia lipolytica* for Biomanufacturing of High-Value Products from Oils and Fats

Na Liu, Ya-Hue Valerie Soong, Andrew Olson, and Dongming Xie

Abstract

Yarrowia lipolytica is a safe and robust yeast to efficiently use lipid as the sole carbon source, which provides us opportunities for biomanufacturing of a series of high-value products from cost-effective agriculture feedstocks such as plant oils and animal fats. *Y. lipolytica* has a unique propensity for high flux through tricarboxylic acid (TCA) cycle intermediates and biological precursors such as acetyl-CoA and malonyl-CoA that can be diverted into a variety of heterologous value-added bioproducts. With recent advances in metabolic engineering and synthetic biology tools, the potential of using *Y. lipolytica* for biomanufacturing of high-value products has been expanded. Examples include industrial enzymes, extracellular proteins, fatty alcohols, wax esters, long-chain diacids, omega-3 fatty acids, and carotenoids. For large-scale biomanufacturing using oils/fats as substrate, the poor mixing and mass transfer caused by the insolubility of substrates in an aqueous medium is one of the major challenges that have to be addressed in addition to the pathway engineering and optimization for both fatty acid biosynthesis and conversion. The multi-phase computational fluid dynamics (CFD) simulation can be used as a powerful tool for analysis of mixing and mass transfer behaviors in bioreactors and further guide the bioreactor design and optimization of operating conditions. Cell morphology has a profound effect on cell growth, oil substrate uptake, and product formation. Both PKA and cAMP-dependent signaling pathways are involved in the dimorphic transition in *Y. lipolytica*. Maintaining the dimorphic yeast shape via

Na Liu and Ya-Hue Valerie Soong are contribute equally to the paper.

N. Liu · Y.-H. V. Soong · A. Olson · D. Xie (✉)
Department of Chemical Engineering, University of Massachusetts Lowell, Lowell, MA 01854, USA
e-mail: Dongming_Xie@uml.edu

F. Darvishi Harzevili (ed.), *Synthetic Biology of Yeasts*,
https://doi.org/10.1007/978-3-030-89680-5_3

morphology engineering strategies has been explored. This chapter also introduced several examples of how we combined cell morphology engineering, metabolic pathway optimization, and bioreaction engineering to significantly improve the production of intracellular lipids, citric acid, and wax esters from plant oils. Potential strategies for further improving the biosynthesis efficiency via transporter engineering in yeast were also introduced.

Keywords

Biomanufacturing • Metabolic engineering • Cell morphology • CFD simulation • Oils and fats • *Yarrowia lipolytica*

1 Introduction

Yarrowia lipolytica is one of the most intensively studied "non-conventional" yeast which possesses the potential to act as a biotechnological workhorse. It has a wide range of biotechnological applications that include degradation of hydrophobic substrates such as fatty acids (FAs), oils, fats, n-alkanes, production of organic acids, and secretion of homologous and heterologous proteins (Morales-Vargas et al. 2012). *Y. lipolytica* has a unique propensity for high flux through tricarboxylic acid (TCA) cycle intermediates and biological precursors such as acetyl-CoA and malonyl-CoA that can be diverted into a variety of heterologous products. *Y. lipolytica* is generally regarded as safe (GRAS) by the American Food and Drug Administration (FDA) (Soong et al. 2019). This safety approval is particularly important for broadening the range of possible applications of the products derived from the fermentation of *Y. lipolytica.*

The dimorphic yeast *Y. lipolytica* has been considered an adequate model for dimorphism studies, as it can grow as yeast-like (single oval cells) or in the form of a mycelium (pseudohypha and hypha) depending on the environmental conditions, and to be reversible between each one (Kawasse et al. 2003). Under nutrient-rich conditions, the organism grows as a mixture of yeast-like and short mycelial cells (Ruiz-Herrera and Sentandreu 2002). It is believed that yeast dimorphism is related to a defense mechanism to adverse conditions (Kawasse et al. 2003). Different effectors have been reported to be implicated in the dimorphic transition of *Y. lipolytica.* There are a number of genes that have been identified as apparent regulators of the dimorphic transition of *Y. lipolytica*, the genes *YlMHY1, YlHOY1, YlBEM1, YlCLA4, YlRPK1, YlTPK1, YlRAS2,* and *YlSTE11* have been reported responsible for yeast-to-hypha transition in *Y. lipolytica* while the genes *YlTPK1* and *YlZNC1* have been reported as negative regulators of *Y. lipolytica* filamentation (Bankar et al. 2018; Martinez-Vazquez et al. 2013; Pomraning et al. 2018; Soong et al. 2019).

Y. lipolytica is often found naturally in environments rich in hydrophobic substrates, such as oils, fats, fatty acids, or n-alkanes. The unique physiological features that *Y. lipolytica* has developed to grow in oils and fats have made this yeast an important biotechnological yeast, which has been proposed for the treatment of petroleum oil-polluted soil or water (Ledesma-Amaro and Nicaud 2016b). In addition, *Y. lipolytica* has been efficiently cultivated on oils or fats for producing many intra- or extracellular metabolites of industrial significance (Fickers et al. 2005a, b; Papanikolaou et al. 2007). Vegetable oils and animal fats were reported as promising substrates for biosurfactant production, as the oily or fatty carbon sources are consumed by microorganisms could work as a building block for biosurfactant synthesis (Goncalves et al. 2014). Various studies have indicated that the production and secretion of lipases in *Y. lipolytica* strains are stimulated by the presence of long-chain fatty acids (LCFAs) (Fickers et al. 2003), plant oils (Braga et al. 2012; Deive et al. 2010; Kebabci and Cihangir 2012; Najjar, et al. 2011), and animal fats (Kamzolova et al. 2005) in the culture medium. Conversely, glucose in the medium might repress the production of lipases (Liu et al. 2021). The oleaginous yeast *Y. lipolytica* has been reported capable of accumulating a significant amount of intracellular lipid, stored in lipid bodies, during growth on vegetable oil (Najjar et al. 2011) and animal fats or their industrial derivatives (Papanikolaou et al. 2002, 2007). With recent improvements in the synthetic biology tools, the industrial potential of *Y. lipolytica* has been expanded to include organic acids such as citric acid, α-ketoglutaric acid, and itaconic acid, enzymes such as lipases, RNase, and esterase, lipids and lipid-derived compounds such as biodiesel, dicarboxylic acids, and biosurfactants (Liu et al. 2021).

The great potential for industrial applications of *Y. lipolytica* has driven the development of metabolic engineering tools for it as well as the basic research to understand its physiological features. In recent years, different metabolic engineering tools and strategies have been created and applied in *Y. lipolytica*, which has further expanded the application of this yeast.

This chapter summarizes the degradation of oil and fat substrates by *Y. lipolytica*, fermentation and cellular engineering efforts of the *Y. lipolytica* metabolic engineering program to improve the utilization efficiency and bioconversion of oils and fats. The distribution of oil–water mixing in stirred bioreactor was studied by computational fluid dynamics (CFD) simulation. The CFD results were further used to guide the optimization of the bioreactor design and operation conditions during bioreactor fermentation experiments. The effect of cell morphology on the fermentation with lipid substrates was examined by creating the hyphal strain and the yeast-like strain through overexpression and disruption of the gene *YlMHY1* in the wild-type *Y. lipolytica* ATCC20362, respectively. The strategies of bioreactor design and operation conditions and cell morphology engineering can also be used for the production of other high-value products (e.g., fatty alcohol, wax esters). The metabolic engineering for the biosynthesis of wax esters by *Y. lipolytica* from oils and fats is introduced as a case study. This chapter also summarizes the engineering of transporters in yeast that are used to enhance bioproduction.

2 Oils- or Fats-Substrate Utilization

Y. lipolytica is often found in environments with the presence of plant oils or animal fats, being able to assimilate FAs, oils, and fats efficiently. The route of oils or fats substrates into the cells induces several modifications of the substrates to improve their accessibility. The initial challenge is a contact between the poorly water-miscible substrate, and the cell surface. *Y. lipolytica* can produce surfactants, amphiphilic compounds consist of hydrophilic and hydrophobic moieties, which can reduce surface and interfacial tensions in aqueous media and hydrophobic substrates and also reduce the size of the oils/fats droplets, thus increasing the contact between substrates and cell surface (Fickers et al. 2005a, b). In bioreactor fermentation, this step could be additionally facilitated by powerful agitation. Moreover, surfactants can also facilitate cell adhesion to oils/fats droplets (Beopoulos et al. 2009).

The main component of oils and fats is triglycerides (TAGs) which cannot be directly uptake by *Y. lipolytica*, as no TAGs transporters have been identified (Liu et al. 2021). *Y. lipolytica* secretes extracellular lipases (*LIP2, LIP7* and *LIP8*), which help hydrolyze TAGs to form glycerol and the respective free fatty acids (FFAs). Grown in oils/fats leads to structural changes on the cells surface of *Y. lipolytica* and results in the formation of protrusions that enable cells to take up FFAs from the medium (Mlickova et al. 2004). The action of biosurfactant and lipase occurs progressively, the formation of numerous small-sized droplets facilitating the surface-mediated substrate transport (Beopoulos et al. 2009). Once in the cytoplasm, FFA becomes activated by conversion into a fatty acyl-CoA by the enzyme acyl-CoA synthetase FAA1. Then the fatty acyl-CoA can enter the Kennedy pathway to be stored into the lipid body as TAGs (formed by DGA1 and DGA2) (Abghari and Chen 2014), as precursors for the synthesis of fatty-acid-derived products, or be transported into the peroxisome to carry out β-oxidation (*POX1-6*, *MFE1,* and *POT1*). FFAs can be released from the TAGs through intracellular lipases (*TGL3* and *TGL4*), and thereafter, they can be activated to form fatty acyl-CoA then transported into the peroxisome (Ledesma-Amaro et al. 2016). The β-oxidation pathway, the main pathway for the breakdown of these fatty acyl-CoA esters, is a four-reaction cycle. After each cycle, the CoA ester of FA gets two carbons shorter and one molecule of acetyl-CoA is released. In *Y. lipolytica,* the first and most important step of β-oxidation is carried out by six acyl-CoA oxidases (*POX1-6*). The strain with *POX1-6* genes knock-out is unable to degrade FFAs resulting in a lipid accumulation (Wang et al. 1999). The second and third steps in β-oxidation are catalyzed by the multi-functional enzyme (MFE). In contrast to acyl-CoA oxidases, which are encoded by six genes, MFE is encoded by a single gene *MFE1*. The deletion of the *MFE1* gene has been extensively studied in *Y. lipolytica* for lipid production due to its technical simplicity (Blazeck et al. 2014). The fourth and last step is carried out by a peroxisomal thiolase POT1. As β-oxidation takes place in the peroxisome, this pathway has been blocked by disrupting the genes, *PEX3*, *PEX10,* and *PEX11*, involved in peroxisome biogenesis

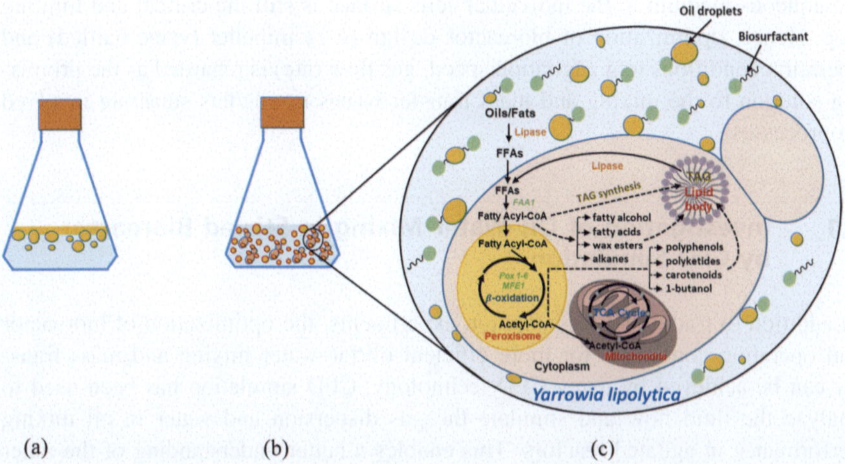

Fig. 1 **a** Culture medium (blue) and oil (yellow) in the flask, oil stay on the top of the aqueous medium due to hydrophobic nature and light density. **b** Flask culture of *Y. lipolytica* with oil substrate. **c** The assimilation of oils by *Y. lipolytica*: (i) Oil–water mixing was facilitated by surfactant from *Y. lipolytica*; (ii) TAGs (oils/fats) are cleaved by extracellular lipases to give FFAs; (iii) oils/FAAs droplets bind onto the cell surface; (iv) FAAs enter into cell interior via transport/export mechanisms; (v) FAAs in the cytoplasm can be activated by cytoplasmic fatty acyl-CoA synthase (FAA1); (vi) the activated fatty acid (fatty acyl-CoA) directly transported into the peroxisome for β-oxidation degradation, (vii) converted into fatty acids derived products, or (viii) store into lipid bodies as TAGs; (ix) lipid bodies could work as storage pools for fatty acids derived products, TAG in lipid bodies could be hydrolyzed by lipases to release FFAs. Dash lines: Putative route, not confirmed

(Ledesma-Amaro and Nicaud 2016b). The cycle is repeated several times, theoretically until the fatty acyl-CoA been completely breakdown. The acetyl-CoA formed in β-oxidation is a key intracellular metabolite, which plays a major role in various metabolic pathways that link catabolism and anabolism such as TCA cycle, biosynthesis of acetyl-CoA-derived products (e.g., polyphenol, carotenoids) (Fig. 1) (Chen et al. 2012).

3 CFD Modeling and Fermentation Engineering for Improving Biosynthesis from Oils and Fats

In biomanufacturing processes, the growth of microorganism requires contact between the substrate and cell surface. However, due to their relative immiscibility in water and the lighter density, oils/fats substrates tend to stay as a single layer on the top of aqueous media. *Y. lipolytica* can produce some surfactants to help emulsify oils/fats in the aqueous phase which is helpful in flask-scale culture. However, in bioreactors fermentation, the transport of oils/fats from the organic phase to the oils-aqueous medium and the delivery of the oils/fats droplets from the aqueous medium to the individual cells surface is still the critical and limiting step. Hence, optimization of bioreactor design (e.g., impeller types, baffled) and operation conditions (e.g., agitation speed, gas flow rate) is regarded as the promising solution to the mixing and mass transfer issues in oils/fats substrate involved bioprocesses.

3.1 Investigation of Oil–Water Mixing in Stirred Bioreactor by CFD Simulation

In addition to traditional fermentation experiments, the optimization of bioreactor and operation conditions for more efficient oil/fat–water mixing and mass transfer can be achieved by using CFD technology. CFD simulation has been used to analyze the fluid flow and simulate the gas dispersion and water in oil mixing performance in agitated reactors. This enables a better understanding of the inner mechanisms of the multi-phase flow and the factors that modulate the hydrodynamics (Hutmacher and Singh 2008; Wang et al. 2014). CFD modeling is a relatively low cost and less time-consuming methodology that allows more efficient prediction and possibly optimization of bioreactor design and operation conditions in fermentation experiments.

In our research group, three-dimensional CFD models have been developed to simulate the distribution of oil droplets in an aqueous medium in agitated bioreactors with two standard Rushton turbines. The bioreactors were filled with water with 5% (v/v) of vegetable oil as the dispersed phase. Simulations were performed at the agitation speed of 500 rpm. The entire vessel was considered as the computational domain. The simulation results showed that most larger oil droplets gathered in the upper and central region of the vessel. This result suggests that pitched-blade impellers with the blades set at a certain angle (e.g., 45°) should be applied in the fermentation experiment, as they could more efficiently facilitate the mixing of the fluid from between the upper and bottom regions of the vessel as well as from between the center to and the side regions (Fig. 2b), thus being able to overcome the oil droplets' buoyance forces to improve dispersion in an aqueous medium. The Rushton turbines used in the CFD model usually produce unidirectional radial flow (Fig. 2a). In addition, in the CFD simulation, oil droplets around the impeller blades were broken into smaller droplets, of which some were pushed into side regions of the bioreactor. From this result, it can be postulated that increasing the agitation speed increases the shearing forces on the oil droplets, which would lead to the formation of smaller oil droplets from larger ones and an improvement in the distribution of the droplets to other regions of the vessel. The CFD results and predictions were validated in 1-L fermentation experiments.

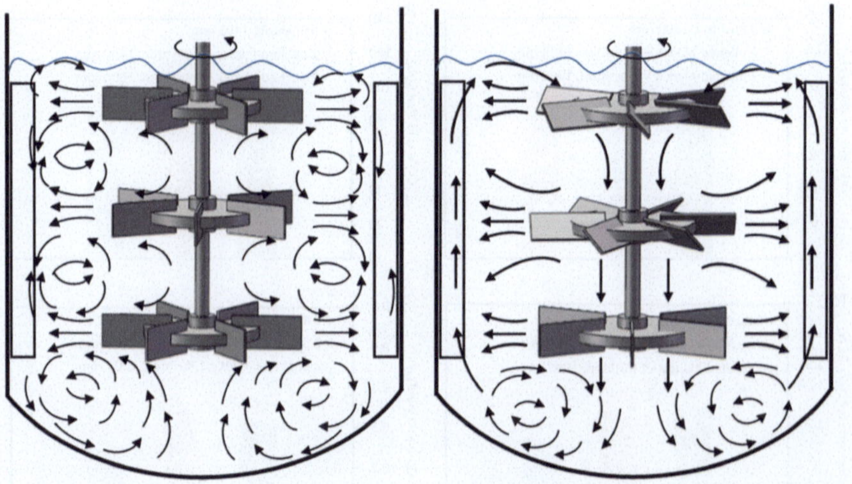

Fig. 2 Two typical impellers used for fermentation. **a** The blades on Rushton impellers are flat and set vertically along an agitation shaft, which produces a unidirectional radial flow. **b** The blades on pitched-blade impellers (top and medium) are flat and set at ~45° angles, which produces a simultaneous axial and radial flow. The combination of pitched-blades and Rushton impellers provides better overall mixing and creates a higher mass transfer rate

3.2 Fermentation Engineering for Improved Biosynthesis from Oils and Fats

In our recent bioreactor fermentation experiments, the cell growth and citric acid production of wild-type *Y. lipolytica* ATCC20362 on soybean oil substrate were studied as a representation of the overall mixing and mass transfer, since better mixing and mass transfer usually led to faster cell growth and/or higher products formation. To compare the effect of impellers on mixing and mass transfer, two sets of impellers were tested in 1-L bioreactors, with one set of impellers consisted of three evenly-spaced Rushton impellers (Fig. 2a), while the other consisted of two pitched-blade impellers on top and one Rushton impeller at the bottom (Fig. 2b). The agitation speed through fermentation was controlled between 500 and 1400 rpm. The bioreactor is equipped with three impellers, which has one more impeller as compared to the bioreactor geometry in CFD simulation. The third impeller was added to the top region of the bioreactor to break up and disperse the oil droplets accumulated in the upper region. Furthermore, two maximal agitation speeds, 1000 rpm and 1400 rpm, were tested in the bioreactor equipped with two pitched-blade impellers on top and one Rushton impeller at the bottom to examine if more powerful input provides better mixing and improves the

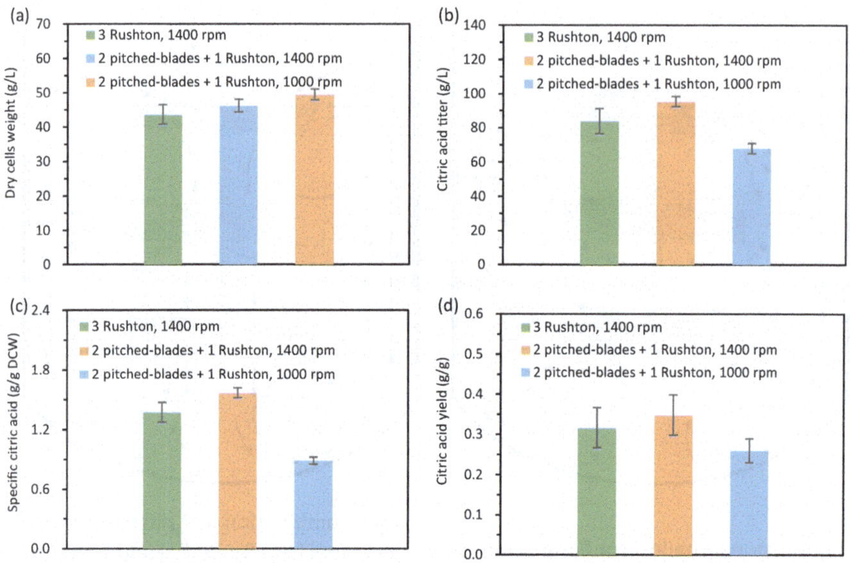

Fig. 3 Fed-batch fermentation of *Y. lipolytica* ATCC20362 with soybean oil substrate in bioreactors equipped with different impellers and under different maximal agitation speed. **a** Dry cells weight (DCW), **b** citric acid titer, **c** specific citric acid, **d** citric acid yield at 144 h

fermentation performance, as suggested by the CFD model. The agitation speed through fermentation was controlled within 500–1000 rpm and 500–1400 rpm, respectively. Other than the impeller setup or agitation speed, conditions for each fermentation were the same.

As shown in Fig. 3a, the dry cell weight (DCW) from the fermentation with different impellers and maximal agitation speeds were similar, indicating that the oil–water mixing condition with any set of the impellers and agitation speed met the minimal requirements for cell growth. However, citric acid production from soybean oil was significantly impacted by impeller type and agitation speed (Fig. 3b, c, and d). With three Rushton impellers and the maximal agitation speed of 1400 rpm, a citric acid titer of 84 g/L, a specific citric acid titer of 1.4 g/g DCW, and a citric acid conversion yield of 0.32 g/g from soybean oil was obtained at 144 h. By replacing the two top Rushton impellers with two pitched-blade ones, the titer, the specific titer and the yield of citric acid at 144 h increased to 95 g/L, 1.6 g/g DCW, and 0.35 g/g soybean oil, respectively. This confirmed the hypothesis that the pitched-blade impellers are more beneficial for overall mixing and mass transfer rate so that the oil droplets were dispersed into the whole bioreactor to be uptake by the yeast cells to convert to citric acid. In addition, when the maximal agitation speed decreased from 1400 to 1000 rpm, the titer of citric acid at 144 h decreased from 95 g/L to 68 g/L, which also validated what the CFD simulation results suggested, i.e., more powerful agitation could break down the oil droplets into smaller ones and disperse them into the regions farther from the center so that the cells were able to assimilate them and convert them into citric acid.

4 Cell Morphology Engineering in Dimorphic Yeast *Y. lipolytica*

Understanding microbial dynamics is crucial in response to the interaction of the gene expression and environmental conditions during cell growth course and product formation (Guan et al. 2017; Timoumi, et al. 2018). The impacts of cell microenvironment on metabolism, such as the supplementation of nutrients, variations in temperature (Arsène, et al. 2000; Barria et al. 2013) and pH, are known factors to affect the cellular behavior (Arsène et al. 2000; Barria et al. 2013; Ruiz-Herrera and Sentandreu 2002; Yuzbashev et al. 2010). Among the changes in biotic responses of microorganisms, the morphological change provides important information about the cell differentiation process, cell function, and signal responses. Microorganisms are capable of altering their cellular structure dynamically, including the shape and size, to adapt to a new environment and optimize the interactions physicochemically and biologically (Okano et al. 1995). However, most previous studies have focused on the construction of novel biosynthetic pathways and their metabolic regulatory processes. Increasing attention has recently been paid to morphology engineering as a new strategy for constructing efficient microbial cell factories, the purpose of which is to control cell shape and cell growth pattern by manipulating cell morphology-related genes (Huo et al. 2020). Therefore, improving tolerance of microbial species to various environmental stresses in a predictable manner has been proposed to achieve optimal product titer, rate, and yield at a large scale.

Among the microorganisms of biotechnological interest in the production of metabolites with high values, the non-conventional oleaginous yeast *Y. lipolytica* has been successfully used in recent years as microbial cell factories due to its unique biochemical characteristics suitable for large-scale biomanufacturing and a wide range of potential applications (Liu et al. 2015; Papanikolaou et al. 2020). *Y. lipolytica* is considered as one of the most attractive hosts capable of assimilating an extremely broad range of raw substrates, including both hydrophilic (e.g., glucose, fructose, mannose, glycerol, ethanol, and organic acids) and hydrophobic (e.g., fatty acids, alkanes, triacylglycerols, plants oils, and animal fats) carbon sources (Fickers et al. 2005a, b; Ledesma-Amaro and Nicaud 2016a; Soong et al. 2019; Spagnuolo et al. 2018).

4.1 Regulation of Yeast-To-Hyphal Transition in *Y. lipolytica*

Y. lipolytica is emerging as a model organism for dimorphism studies. It can grow in two distinct phenotypes, oval-shaped yeast single cells or filamentous form (i.e., pseudohyphae or hyphae) (Berman and Sudbery 2002; Dominguez et al. 2000; Palecek et al. 2002; van der Walt and von Arx 1980). Morphologic switching is dependent on the environmental and physiological conditions together with genetic characteristics. Induction of yeast-to-hyphal transition is triggered by the signal transduction from the cell surface sensors to signal transducer and then launches a series of intracellular signaling pathways, which activate the expression of morphology-related genes and result in cellular response (Berman and Sudbery 2002; Lengeler et al. 2000; Su et al. 2018). The external signals in the environmental conditions, such as physicochemical (i.e., temperature, pH, and dissolved oxygen levels), mechanical (i.e., pressure and mixing), or nutritional (i.e., carbon/nitrogen sources and concentration of metal ions and salts) can initiate the dimorphic transition in *Y. lipolytica* (González-López et al. 2006; Ruiz-Herrera and Sentandreu 2002). For example, in the presence of N-acetylglucosamine and blood serum, organic sources of nitrogen (i.e., amino acid and peptone), thermal shock and hypoxia could contribute to the filamentous phenotype, whereas in the presence of glutamine and glutamate, exposure of the osmotic stress (i.e., 0.2 M NaCl) and sufficient aeration have been reported to suppress the hypha formation (Guevara-Olvera et al. 1993; Kim et al. 2000; Pérez-Campo and Domínguez 2001; Szabo and Štofaníková 2002; Zinjarde et al. 2008).

4.2 The MAPK Pathway in *Y. lipolytica*

Two major signal transduction pathways are involved in the regulation of the yeast-to-hyphal transition in dimorphic yeast. One is the mitogen-activated protein kinase (MAPK) pathway (Fig. 4). MAPK signaling cascades are multi-functional pathways that are evolutionarily conserved in all eukaryotic cells (Ligterink and Hirt 2001; Xu et al. 2017). The MAPK pathway is initiated by specific extracellular cues, amplifies, and integrates the signals through the activation of a particular MAPK following the consecutive of a MAPK kinase kinase (MAPKKK) and a MAPK kinase (MAPKK), and consequently elicits an appropriate physiological response. Typically, the MAPKKK is activated by interactions with a small GTPase and/or phosphorylation by protein kinases downstream from cell surface receptors transducing signal downstream (Cuevas et al. 2007; Zhang and Liu 2002). Several types of protein kinases (i.e., YlCLA4p, YlSTE11p, YlSTE7p) contributed to the MAPK pathway in *Y. lipolytica* have been characterized (Cervantes-Chávez and Ruiz-Herrera 2006; Martínez-Soto and Ruiz-Herrera 2017; Martinez-Vazquez et al. 2013; Szabo 2001). YlCLA4p is highly homologous to CLA4 protein kinase of *Candida albicans* and *Saccharomyces cerevisiae*, which are members of the p21-activated kinase (PAK) family containing conserved internal Cdc42p-binding regions (Bartholomew and Hardy 2009; Lai et al. 2012). Deletion of *YlCLA4* in *Y. lipolytica* is not lethal but completely loses the ability to form filaments or invade agar (Szabo 2001). Disruption of the *YlSTE11*, which exhibits high homology to fungal MAPKKK, loses the capacity to mate and grows constitutively in the yeast-like form (Cervantes-Chávez and Ruiz-Herrera 2006). The expression level of *YlSTE7* encoding MAPKK is increased during yeast-to-hyphal transition (Chaleff and Tatchell 1985; Leberer et al. 1996; Martinez-Vazquez et al. 2013).

Transcription factor is a downstream effector of upstream signaling pathway. YlZNC1p contains a $Zn(II)_2C_6$ fungal-type zinc finger DNA-binding domain and a leucine zipper domain that acts as a transcription factor repressing hyphal formation. YlZNC1p is involved in the MAPK pathway via the regulation of *YlCLA4*,

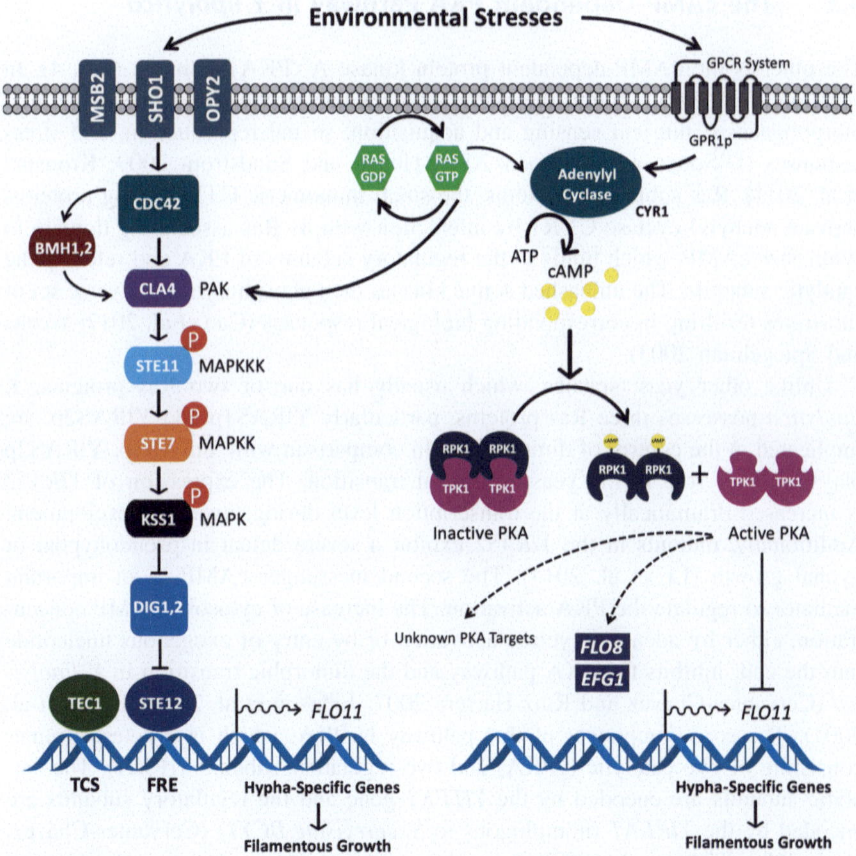

Fig. 4 Simplified scheme describing the dimorphic transition in *Y. lipolytica*. Both MAPK and cAMP-dependent signaling pathways are involved in the yeast-to-hyphal transition but have opposing actions. The activation of the MAPK pathway induces filamentation. The activation of the cAMP-dependent pathway represses the filamentous growth. Abbreviations: PAK, p21-activated kinase; MAPKKK, MAP kinase kinase kinase; MAPKK, MAP kinase kinase; MAPK, MAP kinase; TCS, TEC1 consensus binding sequence; FRE, filamentation response element; GPCR, G-protein coupled receptor; PKA, protein kinase A

YlSTE11, and *YlSTE7* gene expressions (Martinez-Vazquez et al. 2013). Furthermore, both YlBMH1p and YlBMH2p are closely related to the *S. cerevisiae* proteins ScBMH1p and ScBMH2p, which have been found to associate with ScSTE20p (PAK family kinase) required for the MAPK signaling cascade that regulate the hypha-specific genes (Costa et al. 2007; Hurtado and Rachubinski 2002). Therefore, by regulating MAPK cascades, cells can respond to a range of stresses and lead to change in gene expression, and eventually adapt to changing environmental conditions.

4.3 The cAMP-Dependent PKA Pathway in *Y. lipolytica*

The other is the cAMP-dependent protein kinase A (PKA) pathway (Fig. 4). In yeast and filamentous fungi, this pathway takes part in the pathogenesis, cellular morphogenesis, nutrient sensing and acquisition, sexual reproduction, and stress responses (D'Souza and Heitman 2001; Hogan and Sundstrom 2009; Kronstad et al. 2011). Ras subfamily proteins, the small monomeric GTP-binding proteins, activate adenylyl cyclase CYR1 by interaction with its Ras associating domain to synthesize cAMP, which binds to the regulatory subunits of PKA and releases the catalytic subunits. The unleashed active kinases then phosphorylate a diverse set of substrates resulting in corresponding biological responses (Cao et al. 2017; Weeks and Spiegelman 2003).

Unlike other yeast species, which usually has one or two Ras proteins, *Y. lipolytica* possesses three Ras proteins, particularly YlRAS1p and YlRAS2p, are implicated in the control of dimorphism. In comparison with YlRAS1p, YlRAS2p plays a major role in the yeast-to-hyphal transition. The expression of *YlRAS2* is increased dramatically at the transcription level during mycelial development. Additionally, mutants in the *YlRAS2* exhibit a severe defeat in pseudohyphal or hyphal growth (Li, et al. 2014). The second messenger cAMP is an important mediator to regulate the PKA activation. The increase of cytosolic cAMP concentration, either by adenylyl cyclase activation or by entry of exogenous nucleotide into the cell, inhibits the PKA pathway and the dimorphic transition in *Y. lipolytica* (Cervantes-Chávez and Ruiz-Herrera 2007; Johnson et al. 2001; Taylor et al. 2004). The core component of this pathway is PKA, which is a heterotetramer consisting of two catalytic (cPKA) and two regulatory subunits (rPKA). The catalytic subunits are encoded by the *YlTPK1* gene and the regulatory subunits are encoded by the *YlRKA1* (homologous to *S. cerevisiae BCY1*) (Cervantes-Chávez, et al. 2009; Toda et al. 1987). It has been previously reported that the *YlRKA1* gene was up-regulated at the transcriptional level under conditions that induce mycelial morphology, indicating that dimorphic transition is regulated by the PKA pathway via YlPKA1p in *Y. lipolytica* (Cervantes-Chávez and Ruiz-Herrera 2007). Expression of YlPKA1p in PKA cascade was also mediated by YlZNC1p. Hence, YlZNC1p acting through both MAPK and PKA pathways represses the yeast-to-hyphal transition. Mutant strains deleted in the *YlTPK1* grew constitutively in the mycelial form, whereas *YlSTE11* and *YlTPK1* double mutants grew constitutively in the yeast form, suggesting that the default growth pattern of the *Y. lipolytica* is a yeast-like form (Cervantes-Chávez et al. 2009). Surprisingly, all these data provide evidence that the opposite actions of the MAPK and PKA pathways in regulating *Y. lipolytica* dimorphism, in contrast to *S. cerevisiae* and *C. albicans* where both pathways act together to regulate the dimorphic switching (Biswas et al. 2007).

Multiple genes, including *YlHOY1* and *YlMHY1*, are also responsible for the filamentous growth of *Y. lipolytica*. The *YlHOY1* encoding a putative nuclear protein with a homeodomain function as transcriptional regulatory protein. Disruption of *YlHOY1* results in a defect in filamentation (Torres-Guzman and Domínguez

1997). Like the *YlHOY1* gene, YlMHY1p is localized to the nucleus during filamentous growth and acts as a transcription factor. *YlMHY1* encoding a C_2H_2-type zinc finger protein, YlMHY1p, exhibits strong homology to the *S. cerevisiae* stress response factors MSN2p and MSN4p, which is involved in the regulation of morphogenesis (Kobayashi and McENTEE 1993; Marchler et al. 1993). Like these factors, YlMHY1p specifically recognizes and binds to putative *cis*-acting DNA stress response elements (STREs) located on the upstream of a number of genes conferring tolerance to a variety of stress conditions, such as heat shock, carbon source starvation, osmotic stress, and oxidative shock (Treger et al. 1998). Transcription of *YlMHY1* is dramatically increased during the dimorphic transition in *Y. lipolytica.* Deletion of *YlMHY1* is unable to undergo mycelial growth, indicating that YlMHY1p promotes hyphal development. Interestingly, overexpression of *YlMHY1* in *YlRAS2* mutant cells form abundant pseudohyphae or hyphae, while overexpression of *YlRAS2* in *YlMHY1* mutant cells fail to induce filamentous growth (Li et al. 2014). *YlMHY1* is a gene whose overexpression could restore hyphal growth and is required for *Y. lipolytica* YlRAS2p function, like *S. cerevisiae* Ras protein does to MSN2p and MSN4p. Therefore, the transcription factor YlMHY1p may function as a signal transducer downstream of YlRAS2p in the control of the dimorphic transition.

Besides, the transcriptome analysis reveals the downstream target regulated by YlMHY1p. These downstream genes encode proteins that are similar to the flocculin FLO11, cell wall mannoprotein TIR3 and agglutinin AGA1 in *S. cerevisiae* and HYR1 in *C. albicans* (Bailey et al. 1996; Rupp et al. 1999). Therefore, YlMHY1p mediates the expression of a large group of cell wall proteins and enzymes involved in cell wall maintenance. In addition to cell wall-related genes, YlMHY1p also regulates genes involved in nutrient uptake, protein processing, and lipid metabolism (Wu et al. 2020). YlMHY1p has multiple cellular functions.

4.4 Cell Morphology Engineering for Enhanced Utilization of Oils and Fats

Morphology engineering strategies have been previously explored for improving bacterial growth rate, enlarging cell size, and simplifying downstream separation. It has also become an effective method to maintain the dimorphic yeast shape (Jiang and Chen 2016; Zakhartsev and Reuss 2018). The manipulation of morphology-related genes allows changing the cell shape from a sphere to a hyphal form (Hurtado et al. 2000; Hurtado and Rachubinski 2002; Jiménez-Bremont et al. 2012; Ruiz-Herrera and Sentandreu 2002). Improving stress tolerance in a predictable manner in yeast cell factories should facilitate their widespread utilization in the biobased economy and extend the range of products successfully produced on large scale in a sustainable and economically profitable way.

In the dimorphic yeast *Y. lipolytica*, the MSN2p/MSN4p-like protein YlMHY1p is a key positive regulator of yeast-to-hyphal transition. Both *YlMHY1* knock-out and overexpression affect filamentation (Konzock and Norbeck 2020; Wu et al.

2020). Unlike the *Y. lipolytica* ATCC20362 wild-type strain, cells of the *YlMHY1* deletion are unable to form filaments (Fig. 5a). Cells with *YlMHY1* overexpression form filaments longer than those of control cells (Fig. 5a), higher dry cell weight (DCW) (Fig. 5b), decreased citric acid titer (Fig. 5c), and low lipid accumulation (Fig. 5d) in liquid medium using glucose, soybean oil or waste cooking oil as main carbon source. Citric acid is defined as a key precursor in lipid production. Deletion of *YlMHY1* in *Y. lipolytica* exhibits larger lipid bodies (LB), higher citric acid titer (Fig. 5c), and increased lipid/DCW (Fig. 5d) in 1-L fed-batch fermentation, suggesting that the morphology arresting in yeast-like form has a beneficial effect on the lipid-derived product formation and accumulation (Liu et al. 2021). In comparison with glucose, lipid-derived feedstock (soybean oil and waste cooking oil) benefits both *Y. lipolytica* cells' growth and lipid production. Mutations in *YlMHY1* are shown to be more efficient in lipid feedstock utilization. Consistent with this finding, Wang et al. reported that YlMHY1p regulates lipid biosynthesis since the *Y. lipolytica* with *YlMHY1* deletion increased carbon flux through lipid biosynthesis and accumulated more intracellular oil than the wild-type strain dose (Wang et al. 2018). Therefore, in combination with morphology engineering and metabolic engineering strategies together with lipid-derived feedstock utilization is expected to become an efficient and economical approach for constructing *Y. lipolytica* cell factories.

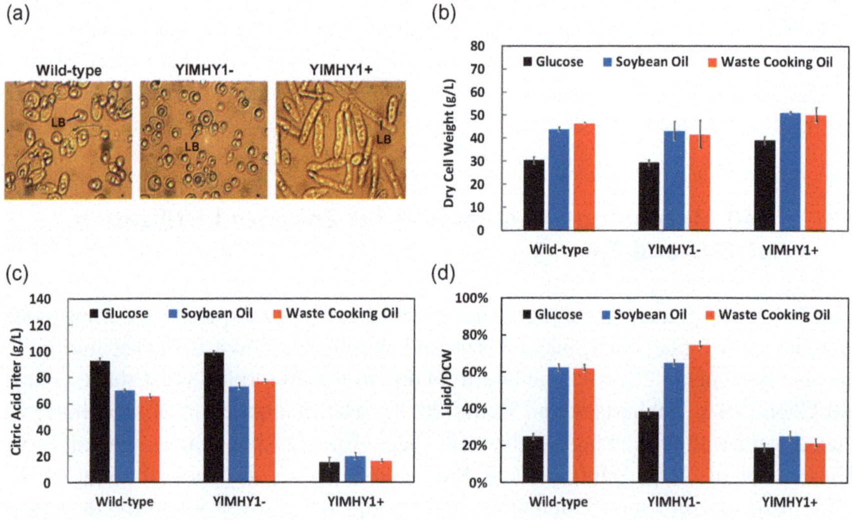

Fig. 5 Cellular responses to different carbon sources for *Y. lipolytica* strains with *YlMHY1* deletion and overexpression. **a** Cell morphology under microscope: Cells of strains ATCC20362 (wild-type), *YlMHY1* knock-out (YlMHY1−), and *YlMHY1* overexpression (YlMHY1+) were grown at 30 °C for 120 h in liquid media containing glucose as the main carbon source. **b** dry cell weight (DCW), **c** citric acid titer, and **d** lipid/DCW of *Y. lipolytica* wild-type, YlMHY1- and YlMHY1+ strains after 144 h cultivation in 1-L fed-batch fermentation using glucose, soybean oil, or waste cooking oil as the main carbon sources

5 Biosynthesis of Wax Esters from Oils and Fats

Wax esters are widely distributed nature compounds that are found in high evolution plants, algae, microorganisms, and even insects and mammals. Natural occurring waxes, consisting of fatty acids esterified to long-chain alcohols, are a group of highly hydrophobic neutral lipids but structurally diverse. Physical properties and applications of wax esters are varied due to different chain lengths of the fatty acid and the fatty alcohol components as well as the degree of unsaturation that affect melting temperature, oxidation stability, and pressure stability. Wax esters have a variety of biological functions that provide the protective coating on the surface so that they are resistant to dehydration, ultraviolet rays, and pathogens (Jetter and Kunst 2008; Wältermann et al. 2005). Wax esters are used commercially to serve as a wide range of applications, such as cosmetics, printing inks, lubricants, coatings, pharmaceuticals, and the food industries (Doan et al. 2017; Fiume et al. 2015; Petersson et al. 2005). Although the wax esters are found in nature universally, the abundance of wax ester source is still low because only a few organisms such as Jojoba plant (*Simmondsia chinensis*) and sperm whale (*Physeter macrocephalus*) are capable of accumulating a considerable amount of intracellular wax esters. Currently, wax esters are in a short supply due to the hunting ban for sperm whale, the high extraction cost, and the harsh requirement of agriculture system for Jojoba (Miwa 1971). Therefore, establishment of an efficient expression platform will be expected to advance the economic feasibility of wax esters from low-cost substrates as carbon sources, especially from waste cooking oil.

5.1 Characterization of the Wax Ester Biosynthesis Pathway

The biosynthesis and accumulation of TAG and wax esters (WEs) have been reported in some soil and marine bacteria including *Acinetobacter* (Ishige et al. 2002; Santala et al. 2014), *Marinobacter* (Willis et al. 2011), *Mycobacterium* (Sirakova et al. 2012), *Streptomyces* (Röttig et al. 2016), *Euglena* (Tomiyama et al. 2017), and *Rhodococcus* (Round et al. 2019) genera under nitrogen-limited conditions. In prokaryotes, mechanism of WE synthesis proceeds via sequential reactions (Fig. 6a): First, the fatty acyl-CoA or fatty acyl-ACP substrate is reduced to a respective long-chain fatty aldehyde by fatty acyl-CoA reductase (FAR) in an NADPH-dependent manner; the aldehyde is further reduced to corresponding fatty alcohol by uncharacterized fatty aldehyde reductase (Alvarez 2016; Mcdaniel et al. 2011). Second, the fatty alcohol is esterified with acyl-CoA through the existence of a CoA-dependent acyltransferase enzyme known as wax ester synthase/diacylglycerol acyltransferase (WS/DGAT).

In *Y. lipolytica*, biosynthesis of fatty alcohols by FAR has been intensively studied (Madzak 2018; Zeng et al. 2018). Heterologous expression of FAR from *Marinobacter hydrocarbonoclasticus* strain VT8 produced 5.75 g/L fatty alcohols in *Y. lipo*lytica when grown on modified YPD medium containing 91 g/L glucose

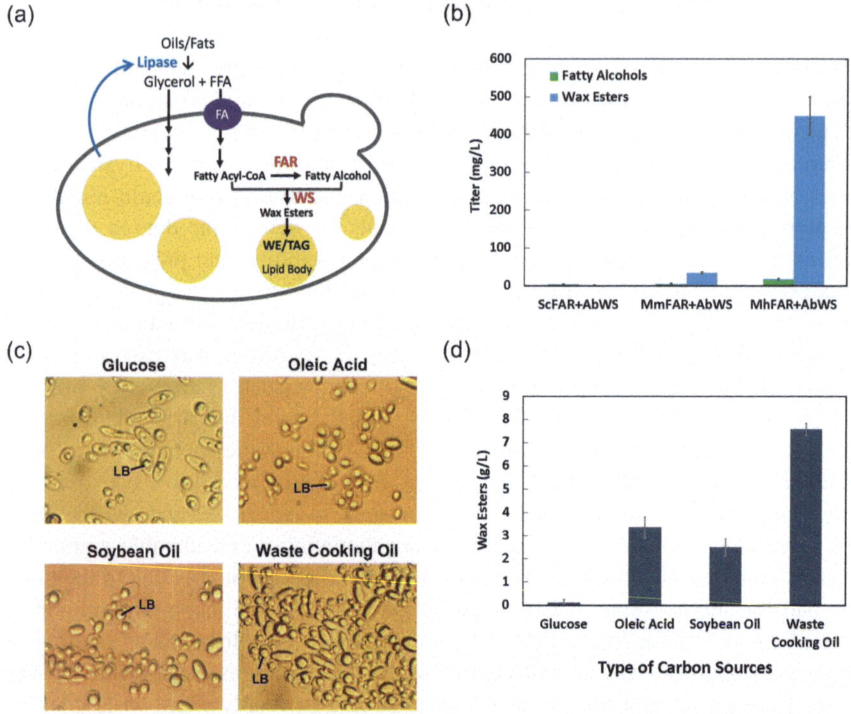

Fig. 6 Introducing wax ester biosynthesis pathway in *Y. lipolytica*. **a** Biosynthesis pathway of wax esters in *Y. lipolytica* during the growth on oils/fats as carbon source. **b** Fatty alcohol and wax ester titers in the engineered *Y. lipolytica* Po1f strain, which contained the fatty acyl CoA reductase (*ScFAR, MmFAR or MhFAR*) and wax ester synthase (*AbWS*) expression via plasmid transformation. **c** Phenotype and **d** wax ester titer of engineered *Y. lipolytica* ATCC20362 strains with co-expression of *MhFAR* and *AbWS* via chromosomal integration grown on glucose, oleic acid, soybean oil or waste cooking oil as a carbon source for 120 h in a shaking flask. *ScFAR* from *S. chinensis*; *MmFAR* from *Mus musculus*; *MhFAR* from *M. hydrocarbonoclasticus* strain VT8; *AbWS* from *A. baylyi* ADP1; LB, Lipid bodies

in shaking flask scale (Zhang et al. 2019). A recent report also showed that *Y. lipolytica* possessing two genes coding for MhFAR achieved at titers of 5.8 g/L fatty alcohol production in a minimal glucose media under fed-batch fermentation (Cordova et al. 2020). Also, *Y. lipolytica* was engineered to produce fatty acid methyl esters (FAME) and fatty acid ethyl esters (FAEE) by introducing wax ester synthase (Gao et al. 2018; Xu et al. 2016).

5.2 Introducing Wax Ester Biosynthesis Pathway into *Yarrowia lipolytica*

Recently, several studies have been conducted to achieve significantly greater cell growth rate and higher production yield of microbial-derived biodiesel (Blazeck et al. 2014; Darvishi et al. 2017). Other research efforts include metabolic engineering of microorganisms to enhance the conversion of hydrophobic substrates into value-added, lipid-derived products (Sabirova et al. 2011; Xie 2017; Xue et al. 2013; Yang et al. 2019). In *Y. lipolytica*, co-expression of *MhFAR* (*M. hydrocarbonoclasticus* strain VT8) and the *AbWS* (*A. baylyi* ADP1) via plasmid transformation produced the most fatty alcohols and wax esters as compared to the strains co-expressing the *ScFAR* (*S. chinensis*) or the *MmFAR* (*Mus musculus*) together with the *AbWS* (Fig. 6b). Oil-based substrates have been employed for the improving production of lipid-derived products by *Y. lipolytica* (Enshaeieh et al. 2014; Li et al. 2008). In particular, oleic acid (C18:1) is the major fatty acid present in *Y. lipolytica*, whereas palmitic (C16:0) and linoleic (C18:2) acids have also been detected in high content in the cells (Dobrowolski et al. 2016; Magdouli, Guedri et al. 2020).

In our recent study (Soong et al. 2021), as shown in Fig. 6c, the enlarged lipid bodies were produced when the engineered *Y. lipolytica* (integrating the *MhFAR* and *AbWS* into the chromosome) cells grown on the oleic acid or oil-containing media, indicating that the accumulation of lipid-derived products could be improved. In addition, by switching the main carbon source from glucose to waste cooking oil, the wax ester titer was increased by 70-fold, reaching a maximum of 7.58 g/L after 120 h of cultivation in a shaking flask (Fig. 6d). Thus, in the presence of exogenous oleic acid or oil-based substrates (soybean oil or waste cooking oil) to the culture medium, the engineered *Y. lipolytica* strain tends to rely on an external source of fatty acids rather than endogenous de novo fatty acid biosynthesis for wax ester production. This work opens a door toward the economical production of lipids, lipid-derived, and lipid-assisted products for applications as fuels, chemicals, nutraceuticals, and pharmaceuticals.

6 Transporter Engineering in Yeasts

Transporters are key components for efficient import and export activities. The transport of the substrates, such as sugars and fatty acids, across the cytoplasmic membrane into the cell, which links extracellular substrates utilization and intracellular metabolic pathways, is a critical step for consolidated yeast bioprocessing. Furthermore, the transport of the produced products out of the cell is also critical for maintaining a high production rate (Hara et al. 2017). This section summarizes the engineering of specific transporters that are used in yeast to enhance bioproduction.

6.1 Sugar Transport

Yeast can utilize varieties of sugars including monosaccharides and disaccharides, but sugars are not able to freely transport across the cell membrane. Transporter proteins exist at the interface of the cell membrane that can selectively transport sugars across cell membranes into the cytoplasm (Spagnuolo et al. 2018). Monosaccharides (e.g., hexose, pentose) are widely used as substrates in biotechnology since they are readily assimilated by microorganisms to convert them into desired products (Ledesma-Amaro and Nicaud 2016a).

Hexose (C_6) sugars (e.g., glucose, galactose, fructose, mannose) are the favored carbon and energy sources for most biotechnological microorganisms. There are 24 sugar transporters have been identified in *Y. lipolytica* (e.g., W29, H222), among them at least 6 proteins (e.g., YALI0C06424p, YALI0F19184p, YALI0E23287p, YALI0B06391p, YALI0C08943p, and YALI0C08943p) can function as hexose transporters in a study of the heterologous host *S. cerevisiae* (Lazar et al. 2017). Generally, one transporter is capable of transporting more than one hexose but with a preference for one (Spagnuolo et al. 2018). Hexose transporters YALI0C06424p, YALI0F19184p, and YALI0E23287p appear to transport a wider range of sugars, efficiently importing four hexoses (e.g., glucose, galactose, fructose, mannose) into *S. cerevisiae,* demonstrated with the drop-test assays. The transporter YALI0B06391p is seemingly dedicated only to the transport of galactose and mannose. YALI0C08943p did not transport glucose, and YALI0B01342p did not enable fructose uptake at any concentration (Hara et al. 2017; Spagnuolo et al. 2018). Sugar transporters have been well-studied in the conventional yeast *S. cerevisiae* (Luyten et al. 2002; Rintala et al. 2008). Transport of hexoses in *S. cerevisiae* is through facilitated diffusion mediated by several transporters (Leandro et al. 2009). There are 17 hexose transporters, HXT1p-HXT17p, and one galactose permease, GAL2p, that have been identified in *S. cerevisiae* (Leandro et al. 2009). Wieczorke et al. (1999) demonstrated that all 18 proteins, except HXT12, can transport hexoses. HXT1p-HXT7p are the main hexose transporters (Reifenberger et al. 1997). Kim et al. (2015) have enhanced glucose uptake by overexpression of 5 hexose transporters including HXT1p, HXT2p, HXT3p, HXT4p, and HXT7p, and overexpression of HXT7p was most effective in increasing the glucose uptake rate. Wieczorke et al. (1999) restored growth of the *HXT1-17* and *GAL2* deletion *S. cerevisiae* on glucose, fructose, and mannose but not on galactose by overexpressing the individual HXP5p, HXP8p, HXP13p, HXP15p, HXP16p, and HXP17p, while HXT9p, HXP10p, and HXP11p were also identified to transport galactose.

Pentose (C_5) sugars (e.g., xylose, arabinose) are the second most prevalent carbon and energy sources for biotechnological microorganisms (Ryu and Trinh 2018). The uptake of pentose sugars in *Y. lipolytica* has been debated. The wild *Y. lipolytica* has been known to be very inefficient at utilizing pentose sugars. Xylose transporters are an important rate-limiting step that causes poor xylose assimilation (Ryu and Trinh 2018). Xylose could be weakly transported into *Y. lipolytica* by endogenous hexose transporters YALI0C06424p and YALIB06391p (Young et al. 2014). More recently, two pentose-specific transporters, YALI0C04730p and

YALI0B00396p, have been identified in *Y. lipolytica* (Ryu and Trinh 2018). The yeast *S. cerevisiae* lacks pentose-specific transporters; it is known to uptake D-xylose and L-arabinose via endogenous hexose/glucose transporters. The efficiency of hexose transporters for D-xylose uptake was as follows: HXT7 > HXT5 > GAL2 ≫ HXT1 > HXT4 (Sedlak and Ho 2004). Overexpression of endogenous hexose transporters and/or heterologous xylose transporters have effectively facilitated D-xylose uptake into engineered *S. cerevisiae* strains (Hector et al. 2008). Because D-xylose uptake is competitively inhibited by D-glucose, cofermentation of D-glucose and D-xylose, such as in cellulosic and hemicellulose hydrolysates, is not cost-efficient (Farwick et al. 2014). Many studies have focused on effective D-xylose fermentation from mixed sugars by developing more xylose-specific transporters. Farwick and coworkers (Farwick et al. 2014) engineered hexose transporters HXT7p and GAL2p and have achieved glucose-insensitive xylose transporters. These engineering efforts have facilitated the economic production of value-added products from renewable feedstocks.

6.2 Fatty Acid Transport

Cells cannot directly uptake oils or fats in the format of triacylglycerides (TAGs), as no known TAG transporters have been identified (Beopoulos et al. 2009). After hydrolysis of TAGs catalyzed by extracellular lipases, the released free fatty acids (FFAs), usually are long-chain fatty acids (LCFAs), can be transported into cells by membrane-bound FA-transporters. In *S. cerevisiae*, the transport system of LCFAs has been well characterized. In this yeast, exogenous LCFAs traverse the membrane via the FA-transporter ScFAT1p. Once transported across the membrane, LCFAs are activated by conversion into a fatty acyl-CoA by the fatty acyl-CoA synthetases ScFAA1p and ScFAA4p (DiRusso and Black 1999). *Y. lipolytica* has LCFAs transport and activation proteins similar to those of *S. cerevisiae*, but the transporter YlFAT1p had a different function than ScFat1p, and it is also involved in the export of FAs from lipid bodies after the TAGs hydrolysis under the catalyzation of intracellular lipases (Dulermo et al. 2014). Once FAs have been exported from intracellular lipid bodies and activated, they can enter the peroxisomes and be degraded as a result of β-oxidation cycle. In *S. cerevisiae*, the membrane-bound transport heterodimer ScPXA1p/ScPXA2p is responsible for transporting fatty acyl-CoA into peroxisomes (Shani and Valle 1996). Disruption of the peroxisomal fatty acyl-CoA transporter ScPXA1p increased the intracellular TAGs accumulation by 14% (Ferreira et al. 2018). In *Y. lipolytica*, there is a membrane-bound transport protein pair YlPXA1p (YALI0A06655) and YlPXA2p (YALI0D04246) corresponding to transporters in *S. cerevisiae*, which were responsible for the transport of fatty acyl-CoA into peroxisome (Dulermo et al. 2015). In *Y. lipolytica*, the utilization of FAs is largely dependent on transporters YlFAT1p, YlPXA1p, and YlPXA2p. The deletion of *YlPXA1*and *YlPXA2* in *Y. lipolytica* has increased the accumulation of FAs when cells were grown in an oleate (C18:1) medium (Dulermo et al. 2015).

6.3 Product Export

In bioproduction, the accumulation of intracellular products is often toxic to the host cells and impairing cell growth and the production of target materials. Transporting the products to the extracellular space is an effective strategy to release the toxicity. In addition, the export of target products also could avoid the negative feedback regulation, thus resulting in more efficient fermentation processes. So far, transporter engineering has been employed in a few cases for improving yeast bioproduction. In yeasts, the transporter MAE1p from fission yeast *S. pombe* has been applied to increase the export of various organic acids, including malate, maleic, oxaloacetic, and succinic (Darbani et al. 2019). The tolerance limit of *S. cerevisiae* against alkanes has been improved by about 80-fold through harnessing *Y. lipolytica* ATP-binding cassette (ABC) transporters pump alkanes out of the cell (Chen et al. 2013). The expression of human transporter FATP1p, which has been known as a fatty acid importer, resulted in a higher biomass yield and a 4.5-fold higher fatty alcohol production in *S. cerevisiae* (Hu et al. 2018).

Engineering transporters to facilitate substrate uptake and reduce products cytotoxicity then increase the production is important for yeast bioproduction. However, the transporters and their mechanisms are still unclear, and further studies on developing efficient expression methods for membrane transporters and improving specific functional properties of the transporter are required.

7 Conclusions

The GRAS status of *Y. lipolytica* and its metabolic traits, such as the ability to utilize diverse hydrophobic substrates, high flux through acetyl-CoA, has made the oleaginous yeast an important host for the production of fuels, commodity chemicals, nutraceuticals, and pharmaceuticals that can be derived from acetyl-CoA, FAs, and lipids. Plant oils and animal fats, especially the low-cost waste oils and fats, can be the preferred substrates for the biomanufacturing of these products at low cost and high yield. However, due to the insolubility of oils and fats in water, poor oil–water mixing and mass transfer is one of the major challenges to be addressed for the fermentation processes using plant oils or animal fats as substrates. CFD model and simulation were applied to analyze the distribution of oil droplets in water in a stirred bioreactor, which further guided the bioreactor design and optimization of operating conditions to significantly improve the fermentation performance. Controlling cell morphology in a yeast-like form via morphology engineering enhanced the cell growth on oils/fats and improved the production of lipid-derived products. However, compared to the hydrophilic substrates, more work in both cellular engineering and bioreaction engineering of *Y. lipolytica* should be conducted in the future to facilitate the fatty acid transport for more efficient extracellular substrate uptake and intracellular bioconversion. In a summary, with advances in metabolic engineering in various promising microorganisms, oils or fats can be used as a great substitute for sugars for biomanufacturing a series of high-value products.

Acknowledgements This research work presented in this chapter was supported by NSF (#1911480) and UML-WPI seed grant (2019–2020). The authors would also like to thank Dr. Carl Lawton and Massachusetts Biomanufacturing Center for providing experimental facilities and technical support.

References

Abghari A, Chen S (2014) *Yarrowia lipolytica* as an oleaginous cell factory platform for production of fatty acid-based biofuel and bioproducts. Front Energy Res 2. https://doi.org/10.3389/fenrg.2014.00021

Alvarez HM (2016) Triacylglycerol and wax ester-accumulating machinery in prokaryotes. Biochimie 120:28–39

Arsène F, Tomoyasu T, Bukau B (2000) The heat shock response of *Escherichia coli*. Int J Food Microbiol 55(1–3):3–9

Bailey DA, Feldmann P, Bovey M, Gow N, Brown A (1996) The *Candida albicans HYR1* gene, which is activated in response to hyphal development, belongs to a gene family encoding yeast cell wall proteins. J Bacteriol 178(18):5353–5360

Bankar A, Zinjarde S, Telmore A, Walke A, Kumar AR (2018) Morphological response of *Yarrowia lipolytica* under stress of heavy metals. Can J Microbiol 64(8):559–566. https://doi.org/10.1139/cjm-2018-0050

Barria C, Malecki M, Arraiano C (2013) Bacterial adaptation to cold. Microbiology 159(Pt_12):2437–2443

Bartholomew CR, Hardy CF (2009) p21-activated kinases Cla4 and Ste20 regulate vacuole inheritance in *Saccharomyces cerevisiae*. Eukaryot Cell 8(4):560–572

Beopoulos A, Chardot T, Nicaud JM (2009) *Yarrowia lipolytica*: A model and a tool to understand the mechanisms implicated in lipid accumulation. Biochimie 91(6):692–696. https://doi.org/10.1016/j.biochi.2009.02.004

Berman J, Sudbery PE (2002) *Candida Albicans*: a molecular revolution built on lessons from budding yeast. Nat Rev Genet 3(12):918–930. https://doi.org/10.1038/nrg948

Biswas S, Van Dijck P, Datta A (2007) Environmental sensing and signal transduction pathways regulating morphopathogenic determinants of *Candida albicans*. Microbiol Mol Biol Rev 71(2):348–376

Blazeck J, Hill A, Liu L, Knight R, Miller J, Pan A, Alper HS (2014) Harnessing *Yarrowia lipolytica* lipogenesis to create a platform for lipid and biofuel production. Nat Commun 5:3131. https://doi.org/10.1038/ncomms4131

Braga A, Gomes N, Belo I (2012) Lipase induction in *Yarrowia lipolytica* for castor oil hydrolysis and its effect on γ-decalactone production. J Am Oil Chem Soc 89(6):1041–1047. https://doi.org/10.1007/s11746-011-1987-5

Cao C, Wu M, Bing J, Tao L, Ding X, Liu X, Huang G (2017) Global regulatory roles of the cAMP/PKA pathway revealed by phenotypic, transcriptomic and phosphoproteomic analyses in a null mutant of the PKA catalytic subunit in *Candida albicans*. Mol Microbiol 105(1):46–64

Cervantes-Chávez JA, Ruiz-Herrera J (2006) *STE11* disruption reveals the central role of a MAPK pathway in dimorphism and mating in *Yarrowia lipolytica*. FEMS Yeast Res 6(5):801–815

Cervantes-Chávez JA, Ruiz-Herrera J (2007) The regulatory subunit of protein kinase a promotes hyphal growth and plays an essential role in *Yarrowia lipolytica*. FEMS Yeast Res 7(6):929–940

Cervantes-Chávez JA, Kronberg F, Passeron S, Ruiz-Herrera J (2009) Regulatory role of the PKA pathway in dimorphism and mating in *Yarrowia lipolytica*. Fungal Genet Biol 46(5):390–399

Chaleff DT, Tatchell K (1985) Molecular cloning and characterization of the *STE7* and *STE11* genes of *Saccharomyces cerevisiae*. Mol Cell Biol 5(8):1878–1886

Chen B, Ling H, Chang MW (2013) Transporter engineering for improved tolerance against alkane biofuels in *Saccharomyces cerevisiae*. Biotechnol Biofuels 6(1):21–21. https://doi.org/10.1186/1754-6834-6-21

Chen Y, Siewers V, Nielsen J (2012) Profiling of cytosolic and peroxisomal acetyl-CoA metabolism in *Saccharomyces cerevisiae*. PLoS One 7(8):e42475. https://doi.org/10.1371/journal.pone.0042475

Cordova LT, Butler J, Alper HS (2020) Direct production of fatty alcohols from glucose using engineered strains of *Yarrowia lipolytica*. Metabol Eng Commun 10:e00105

Costa M, Borges CL, Bailao AM, Meirelles GV, Mendonça YA, Dantas SF, Mendes-Giannini MJ (2007) Transcriptome profiling of *Paracoccidioides brasiliensis* yeast-phase cells recovered from infected mice brings new insights into fungal response upon host interaction. Microbiology 153(12):4194–4207

Cuevas B, Abell A, Johnson G (2007) Role of mitogen-activated protein kinase kinase kinases in signal integration. Oncogene 26(22):3159–3171

Darbani B, Stovicek V, van der Hoek SA, Borodina I (2019) Engineering energetically efficient transport of dicarboxylic acids in yeast *Saccharomyces cerevisiae*. Proc Natl Acad Sci 116(39):19415. https://doi.org/10.1073/pnas.1900287116

Darvishi F, Fathi Z, Ariana M, Moradi H (2017) *Yarrowia lipolytica* as a workhorse for biofuel production. Biochem Eng J 127:87–96

Deive FJ, Sanromán MA, Longo MA (2010) A comprehensive study of lipase production by *Yarrowia lipolytica* CECT 1240 (ATCC 18942): from shake flask to continuous bioreactor. J Chem Technol Biotechnol 85(2):258–266. https://doi.org/10.1002/jctb.2301

DiRusso CC, Black PN (1999) Long-chain fatty acid transport in bacteria and yeast. Paradigms for defining the mechanism underlying this protein-mediated process. Mol Cell Biochem 192(1):41–52. https://doi.org/10.1023/A:1006823831984

Doan CD, To CM, De Vrieze M, Lynen F, Danthine S, Brown A, Patel AR (2017) Chemical profiling of the major components in natural waxes to elucidate their role in liquid oil structuring. Food Chem 214:717–725

Dobrowolski A, Mituła P, Rymowicz W, Mirończuk AM (2016) Efficient conversion of crude glycerol from various industrial wastes into single cell oil by yeast *Yarrowia lipolytica*. Biores Technol 207:237–243

Dominguez A, Ferminan E, Gaillardin C (2000) *Yarrowia lipolytica*: an organism amenable to genetic manipulation as a model for analyzing dimorphism in fungi. Contrib Microbiol 5:151–172. Retrieved from https://www.ncbi.nlm.nih.gov/pubmed/10863671

D'Souza CA, Heitman J (2001) Conserved cAMP signaling cascades regulate fungal development and virulence. FEMS Microbiol Rev 25(3):349–364

Dulermo R, Gamboa-Melendez H, Dulermo T, Thevenieau F, Nicaud JM (2014) The fatty acid transport protein Fat1p is involved in the export of fatty acids from lipid bodies in *Yarrowia lipolytica*. FEMS Yeast Res 14(6):883–896. https://doi.org/10.1111/1567-1364.12177

Dulermo R, Gamboa-Meléndez H, Ledesma-Amaro R, Thévenieau F, Nicaud JM (2015) Unraveling fatty acid transport and activation mechanisms in *Yarrowia lipolytica*. Biochim Biophys Acta 1851(9):1202–1217. https://doi.org/10.1016/j.bbalip.2015.04.004

Enshaeieh M, Nahvi I, Madani M (2014) Improving microbial oil production with standard and native oleaginous yeasts by using Taguchi design. Int J Environ Sci Technol 11(3):597–604

Farwick A, Bruder S, Schadeweg V, Oreb M, Boles E (2014) Engineering of yeast hexose transporters to transport D-xylose without inhibition by D-glucose. Proc Natl Acad Sci USA 111(14):5159–5164. https://doi.org/10.1073/pnas.1323464111

Ferreira R, Teixeira PG, Gossing M, David F, Siewers V, Nielsen J (2018) Metabolic engineering of *Saccharomyces cerevisiae* for overproduction of triacylglycerols. Metabol Eng Commun 6:22–27. https://doi.org/10.1016/j.meteno.2018.01.002

Fickers P, Nicaud JM, Destain J, Thonart P (2003) Overproduction of lipase by *Yarrowia lipolytica* mutants. Appl Microbiol Biotechnol 63(2):136–142. https://doi.org/10.1007/s00253-003-1342-3

Fickers P, Benetti PH, Wache Y, Marty A, Mauersberger S, Smit MS, Nicaud JM (2005a) Hydrophobic substrate utilisation by the yeast *Yarrowia lipolytica*, and its potential applications. FEMS Yeast Res 5(6–7):527–543. https://doi.org/10.1016/j.femsyr.2004.09.004

Fickers P, Fudalej F, Nicaud JM, Destain J, Thonart P (2005b) Selection of new over-producing derivatives for the improvement of extracellular lipase production by the non-conventional yeast *Yarrowia lipolytica*. J Biotechnol 115(4):379–386. https://doi.org/10.1016/j.jbiotec.2004.09.014

Fiume MM, Heldreth BA, Bergfeld WF, Belsito DV, Hill RA, Klaassen CD, Slaga TJ (2015) Safety Assessment of alkyl esters as used in cosmetics. Int J Toxicol 34(2_suppl):5S-69S

Gao Q, Cao X, Huang Y-Y, Yang J-L, Chen J, Wei LJ, Hua Q (2018) Overproduction of fatty acid ethyl esters by the oleaginous yeast *Yarrowia lipolytica* through metabolic engineering and process optimization. ACS Synth Biol 7(5):1371–1380

Goncalves FA, Colen G, Takahashi JA (2014) *Yarrowia lipolytica* and its multiple applications in the biotechnological industry. Sci World J 476207. https://doi.org/10.1155/2014/476207

González-López CI, Ortiz-Castellanos L, Ruiz-Herrera J (2006) The ambient pH response Rim pathway in *Yarrowia lipolytica*: identification of YlRIM9 and characterization of its role in dimorphism. Curr Microbiol 53(1):8–12

Guan N, Li J, Shin H-D, Du G, Chen J, Liu L (2017) Microbial response to environmental stresses: from fundamental mechanisms to practical applications. Appl Microbiol Biotechnol 101(10):3991–4008

Guevara-Olvera L, Calvo-Mendez C, Ruiz-Herrera J (1993) The role of polyamine metabolism in dimorphism of *Yarrowia lipolytica*. Microbiology 139(3):485–493

Hara KY, Kobayashi J, Yamada R, Sasaki D, Kuriya Y, Hirono-Hara Y, Kondo A (2017) Transporter engineering in biomass utilization by yeast. FEMS Yeast Res 17(7).https://doi.org/10.1093/femsyr/fox061

Hector RE, Qureshi N, Hughes SR, Cotta MA (2008) Expression of a heterologous xylose transporter in a *Saccharomyces cerevisiae* strain engineered to utilize xylose improves aerobic xylose consumption. Appl Microbiol Biotechnol 80(4):675–684. https://doi.org/10.1007/s00253-008-1583-2

Hogan DA, Sundstrom P (2009) The Ras/cAMP/PKA signaling pathway and virulence in *Candida albicans*. Future Microbiol 4(10):1263–1270

Hu Y, Zhu Z, Nielsen J, Siewers V (2018) Heterologous transporter expression for improved fatty alcohol secretion in yeast. Metab Eng 45:51–58. https://doi.org/10.1016/j.ymben.2017.11.008

Huo K, Zhao F, Zhang F, Liu R, Yang C (2020) Morphology engineering: a new strategy to construct microbial cell factories. World J Microbiol Biotechnol 36(9):1–15

Hurtado CA, Rachubinski RA (2002) *YlBMH*1 encodes a 14-3-3 protein that promotes filamentous growth in the dimorphic yeast *Yarrowia lipolytica*. Microbiology 148(11):3725–3735

Hurtado CA, Beckerich J-M, Gaillardin C, Rachubinski RA (2000) A rac homolog is required for induction of hyphal growth in the dimorphic yeast *Yarrowia lipolytica*. J Bacteriol 182(9):2376–2386

Hutmacher DW, Singh H (2008) Computational fluid dynamics for improved bioreactor design and 3D culture. Trends Biotechnol 26(4):166–172. https://doi.org/10.1016/j.tibtech.2007.11.012

Ishige T, Tani A, Takabe K, Kawasaki K, Sakai Y, Kato N (2002) Wax ester production from n-alkanes by *Acinetobacter sp.* strain M-1: ultrastructure of cellular inclusions and role of acyl coenzyme a reductase. Appl Environ Microbiol 68(3):1192–1195

Jetter R, Kunst L (2008) Plant surface lipid biosynthetic pathways and their utility for metabolic engineering of waxes and hydrocarbon biofuels. Plant J 54(4):670–683

Jiang X-R, Chen G-Q (2016) Morphology engineering of bacteria for bio-production. Biotechnol Adv 34(4):435–440

Jiménez-Bremont JF, Rodriguez-Hernandez AA, Rodriguez-Kessler M, Ruiz-Herrera J (2012) Development and dimorphism of the yeast *Yarrowia lipolytica*. In: *ER-H. José, Dimorphic fungi. their importance as models for differentiation and fungal pathogenesis*, pp 58–66

Johnson DA, Akamine P, Radzio-Andzelm E, Madhusudan A, Taylor SS (2001) Dynamics of cAMP-dependent protein kinase. Chem Rev 101(8):2243–2270

Kamzolova S, Morgunov I, Aurich A, Perevoznikova O, Shishkanova N, Stottmeister U, Finogenova T (2005) Lipase secretion and citric acid production in *Yarrowia lipolytica* yeast grown on animal and vegetable fat. Food Technol Biotechnol 1524771:663–612

Kawasse FM, Amaral PF, Rocha-Leao MH, Amaral AL, Ferreira EC, Coelho MA (2003) Morphological analysis of *Yarrowia lipolytica* under stress conditions through image processing. Bioprocess Biosyst Eng 25(6):371–375. https://doi.org/10.1007/s00449-003-0319-z

Kebabci O, Cihangir N (2012) Comparison of three *Yarrowia lipolytica* strains for lipase production_ NBRC 1658, IFO 1195, and a local strain. Turk J Biol 36:15–24. https://doi.org/10.3906/biy-1102-10

Kim J, Cheon SA, Park S, Song Y, Kim J-Y (2000) Serum-induced hypha formation in the dimorphic yeast *Yarrowia lipolytica*. FEMS Microbiol Lett 190(1):9–12. https://doi.org/10.1111/j.1574-6968.2000.tb09254.x

Kim D, Song J-Y, Hahn J-S (2015) Improvement of glucose uptake rate and production of target chemicals by overexpressing hexose transporters and transcriptional activator Gcr1 in *Saccharomyces cerevisiae*. Appl Environ Microbiol 81(24):8392–8401. https://doi.org/10.1128/AEM.02056-15

Kobayashi N, McENTEE K (1993) Identification of cis and trans components of a novel heat shock stress regulatory pathway in *Saccharomyces cerevisiae*. Mol Cell Biol 13(1):248–256

Konzock O, Norbeck J (2020) Deletion of *MHY1* abolishes hyphae formation in *Yarrowia lipolytica* without negative effects on stress tolerance. PLoS One 15(4):e0231161

Kronstad JW, Hu G, Choi J (2011) The cAMP/protein kinase a pathway and virulence in *Cryptococcus neoformans*. Mycobiology 39(3):143–150

Lai C-C, Lee M-R, Hsiao C-H, Tan C-K, Lin S-H, Liao C-H, Hsueh P-R (2012) Infections caused by *Candida lipolytica*. J Infect 65(4):372–374

Lazar Z, Neuvéglise C, Rossignol T, Devillers H, Morin N, Robak M, Crutz-Le Coq AM (2017) Characterization of hexose transporters in *Yarrowia lipolytica* reveals new groups of sugar porters involved in yeast growth. Fungal Genet Biol 100:1–12. https://doi.org/10.1016/j.fgb.2017.01.001

Leandro MJ, Fonseca C, Gonçalves P (2009) Hexose and pentose transport in ascomycetous yeasts: an overview. FEMS Yeast Res 9(4):511–525. https://doi.org/10.1111/j.1567-1364.2009.00509.x

Leberer E, Harcus D, Broadbent ID, Clark KL, Dignard D, Ziegelbauer K, Thomas DY (1996) Signal transduction through homologs of the Ste20p and Ste7p protein kinases can trigger hyphal formation in the pathogenic fungus *Candida albicans*. Proc Natl Acad Sci 93(23):13217–13222

Ledesma-Amaro R, Nicaud JM (2016a) Metabolic Engineering for Expanding the Substrate Range of *Yarrowia lipolytica*. Trends Biotechnol 34(10):798–809. https://doi.org/10.1016/j.tibtech.2016.04.010

Ledesma-Amaro R, Nicaud JM (2016b) *Yarrowia lipolytica* as a biotechnological chassis to produce usual and unusual fatty acids. Prog Lipid Res 61:40–50. https://doi.org/10.1016/j.plipres.2015.12.001

Ledesma-Amaro R, Dulermo R, Niehus X, Nicaud JM (2016) Combining metabolic engineering and process optimization to improve production and secretion of fatty acids. Metab Eng 38:38–46. https://doi.org/10.1016/j.ymben.2016.06.004

Lengeler KB, Davidson RC, D'souza C, Harashima T, Shen W-C, Wang P, Heitman J (2000) Signal transduction cascades regulating fungal development and virulence. Microbiol Mol Biol Rev 64(4):746–785

Li Q, Du W, Liu D (2008) Perspectives of microbial oils for biodiesel production. Appl Microbiol Biotechnol 80(5):749–756

Li M, Li Y-Q, Zhao X-F, Gao XD (2014) Roles of the three Ras proteins in the regulation of dimorphic transition in the yeast *Yarrowia lipolytica*. FEMS Yeast Res 14(3):451–463

Ligterink W, Hirt H (2001) Mitogen-activated protein (MAP) kinase pathways in plants: versatile signaling tools. Int Rev Cytol 209–258

Liu H-H, Ji X-J, Huang H (2015) Biotechnological applications of *Yarrowia lipolytica*: past, present and future. Biotechnol Adv 33(8):1522–1546

Liu N, Soong Y-HV, Mirzaee I et al (2021) Biomanufacturing of value-added products from oils or fats: a case study on cellular and fermentation engineering of *Yarrowia lipolytica*. Biotechnol Bioeng 1–16

Luyten K, Riou C, Blondin B (2002) The hexose transporters of *Saccharomyces cerevisiae* play different roles during enological fermentation. 19(8):713–726

Madzak C (2018) Engineering *Yarrowia lipolytica* for use in biotechnological applications: a review of major achievements and recent innovations. Mol Biotechnol 60(8):621–635

Magdouli S, Guedri T, Rouissi T, Brar SK, Blais J-F (2020) Sync between leucine, biotin and citric acid to improve lipid production by *Yarrowia lipolytica* on crude glycerol-based media. Biomass Bioenergy 142:105764

Marchler G, Schüller C, Adam G, Ruis H (1993) A *saccharomyces cerevisiae* UAS element controlled by protein kinase a activates transcription in response to a variety of stress conditions. EMBO J 12(5):1997–2003

Martínez-Soto D, Ruiz-Herrera J (2017) Functional analysis of the MAPK pathways in fungi. Rev Iberoamericana De Mycologia 34(4):192–202

Martinez-Vazquez A, Gonzalez-Hernandez A, Domínguez Á, Rachubinski R, Riquelme M, Cuellar-Mata P, Guzman JCT (2013) Identification of the transcription factor Znc1p, which regulates the yeast-to-hypha transition in the dimorphic yeast *Yarrowia lipolytica*. PLoS One 8(6):e66790

Mcdaniel R, Behrouzian B, Clark L, Hattendorf D, Valle F (2011) Production of fatty alcohols with fatty alcohol forming acyl-coa reductases (far). In: Google patents

Miwa TK (1971) Jojoba oil wax esters and derived fatty acids and alcohols: gas chromatographic analyses. J Am Oil Chem Soc 48(6):259–264

Mlickova K, Roux E, Athenstaedt K, d'Andrea S, Daum G, Chardot T, Nicaud JM (2004) Lipid accumulation, lipid body formation, and acyl coenzyme a oxidases of the yeast *Yarrowia lipolytica*. Appl Environ Microbiol 70(7):3918–3924. https://doi.org/10.1128/AEM.70.7.3918-3924.2004

Morales-Vargas AT, Dominguez A, Ruiz-Herrera J (2012) Identification of dimorphism-involved genes of *Yarrowia lipolytica* by means of microarray analysis. Res Microbiol 163(5):378–387. https://doi.org/10.1016/j.resmic.2012.03.002

Najjar A, Robert S, Guerin C, Violet-Asther M, Carriere F (2011) Quantitative study of lipase secretion, extracellular lipolysis, and lipid storage in the yeast *Yarrowia lipolytica* grown in the presence of olive oil: analogies with lipolysis in humans. Appl Microbiol Biotechnol 89(6):1947–1962. https://doi.org/10.1007/s00253-010-2993-5

Okano T, Yamada N, Okuhara M, Sakai H, Sakurai Y (1995) Mechanism of cell detachment from temperature-modulated, hydrophilic–hydrophobic polymer surfaces. In: The biomaterials: silver jubilee compendium, Elsevier, pp 109–115

Palecek SP, Parikh AS, Kron SJ (2002) Sensing, signalling and integrating physical processes during *Saccharomyces cerevisiae* invasive and filamentous growth. Microbiology 148(Pt 4):893–907. https://doi.org/10.1099/00221287-148-4-893

Papanikolaou S, Diamantopoulou P, Blanchard F, Lambrinea E, Chevalot I, Stoforos NG, Rondags E (2020) Physiological characterization of a novel wild-type *Yarrowia lipolytica* strain grown on glycerol: effects of cultivation conditions and mode on polyols and citric acid production. Appl Sci 10(20):7373

Papanikolaou S, Chevalot I, Komaitis M, Marc I, Aggelis G (2002) Single cell oil production by *Yarrowia lipolytica* growing on an industrial derivative of animal fat in batch cultures. Appl Microbiol Biotechnol 58:308-312. lipid-derived metabolites

Papanikolaou S, Chevalot I, Galiotou-Panayotou M, Komaitis M, Marc I, Aggelis G (2007) Industrial derivative of tallow: a promising renewable substrate for microbial lipid, single-cell protein and lipase production by *Yarrowia lipolytica*. Electron J Biotechnol 10(3):425–435.https://doi.org/10.2225/vol10-issue3-fulltext-8

Pérez-Campo FM, Domínguez A (2001) Factors affecting the morphogenetic switch in *Yarrowia lipolytica*. Curr Microbiol 43(6):429–433

Petersson AE, Gustafsson LM, Nordblad M, Börjesson P, Mattiasson B, Adlercreutz P (2005) Wax esters produced by solvent-free energy-efficient enzymatic synthesis and their applicability as wood coatings. Green Chem 7(12):837–843

Pomraning KR, Bredeweg EL, Kerkhoven EJ, Barry K, Haridas S, Hundley H, Baker SE (2018) Regulation of yeast-to-hyphae transition in *Yarrowia lipolytica.* mSphere 3(6):e00541–00518. https://doi.org/10.1128/mSphere.00541-18

Reifenberger E, Boles E, Ciriacy M (1997) Kinetic characterization of individual hexose transporters of *Saccharomyces cerevisiae* and their relation to the triggering mechanisms of glucose repression. Eur J Biochem 245(2):324–333. https://doi.org/10.1111/j.1432-1033.1997.00324.x

Rintala E, Wiebe MG, Tamminen A, Ruohonen L, Penttilä M (2008) Transcription of hexose transporters of *Saccharomyces cerevisiae* affected by change in oxygen provision. BMC Microbiol 8(1):53. https://doi.org/10.1186/1471-2180-8-53

Röttig A, Strittmatter CS, Schauer J, Hiessl S, Poehlein A, Daniel R, Steinbüchel A (2016) Role of wax ester synthase/acyl coenzyme a: diacylglycerol acyltransferase in oleaginous *Streptomyces sp.* strain G25. Appl Environ Microbiol 82(19):5969–5981

Round JW, Roccor R, Eltis LD (2019) A biocatalyst for sustainable wax ester production: rewiring lipid accumulation in *Rhodococcus* to yield high-value oleochemicals. Green Chem 21(23):6468–6482

Ruiz-Herrera J, Sentandreu R (2002) Different effectors of dimorphism in *Yarrowia lipolytica.* Arch Microbiol 178(6):477–483

Rupp S, Summers E, Lo HJ, Madhani H, Fink G (1999) MAP kinase and cAMP filamentation signaling pathways converge on the unusually large promoter of the yeast *FLO11* gene. EMBO J 18(5):1257–1269

Ryu S, Trinh CT (2018) Understanding functional roles of native pentose-specific transporters for activating dormant pentose metabolism in *Yarrowia lipolytica.* Appl Environ Microbiol 84(3). https://doi.org/10.1128/aem.02146-17

Sabirova JS, Haddouche R, Van Bogaert I, Mulaa F, Verstraete W, Timmis K, Soetaert W (2011) The 'LipoYeasts' project: using the oleaginous yeast *Yarrowia lipolytica* in combination with specific bacterial genes for the bioconversion of lipids, fats and oils into high-value products. Microb Biotechnol 4(1):47–54

Santala S, Efimova E, Koskinen P, Karp MT, Santala V (2014) Rewiring the wax ester production pathway of *Acinetobacter baylyi* ADP1. ACS Synth Biol 3(3):145–151

Sedlak M, Ho NW (2004) Characterization of the effectiveness of hexose transporters for transporting xylose during glucose and xylose co-fermentation by a recombinant *Saccharomyces* yeast. Yeast 21(8):671–684. https://doi.org/10.1002/yea.1060

Shani N, Valle D (1996) A *Saccharomyces cerevisiae* homolog of the human adrenoleukodystrophy transporter is a heterodimer of two half ATP-binding cassette transporters. Proc Natl Acad Sci USA 93(21):11901–11906. https://doi.org/10.1073/pnas.93.21.11901

Sirakova TD, Deb C, Daniel J, Singh HD, Maamar H, Dubey VS, Kolattukudy PE (2012) Wax ester synthesis is required for *Mycobacterium tuberculosis* to enter *in vitro* dormancy. PLoS One 7(12)

Soong YHV, Liu N, Yoon S, Lawton C, Xie D (2019) Cellular and metabolic engineering of oleaginous yeast *Yarrowia lipolytica* for bioconversion of hydrophobic substrates into high-value products. Eng Life Sci 19(6):423–443

Soong Y, Zhao L, Liu N, Yu P, Lopez C, Olson A, Wong H, Shao Z, Xie D (2021) Microbial Synthesis of Wax Esters. Metab Eng 67:428–442. https://doi.org/10.1016/j.ymben.2021.08.002

Spagnuolo M, Shabbir Hussain M, Gambill L, Blenner M (2018) Alternative substrate metabolism in *Yarrowia lipolytica.* Front Microbiol 9:1077. https://doi.org/10.3389/fmicb.2018.01077

Su C, Yu J, Lu Y (2018) Hyphal development in *Candida albicans* from different cell states. Curr Genet 64(6):1239–1243

Szabo R (2001) Cla4 protein kinase is essential for filament formation and invasive growth of *Yarrowia lipolytica.* Mol Genet Genomics 265(1):172–179

Szabo R, Štofaníková V (2002) Presence of organic sources of nitrogen is critical for filament formation and pH-dependent morphogenesis in *Yarrowia lipolytica*. FEMS Microbiol Lett 206(1):45–50

Taylor S, Yang J, Wu J, Haste N, Radzio-Andzelm E, Anand G (2004) PKA: a portrait of protein kinase dynamics. Biochim Biophys Acta (BBA)-Proteins Proteomics 1697(1–2):259–269

Timoumi A, Guillouet SE, Molina-Jouve C, Fillaudeau L, Gorret N (2018) Impacts of environmental conditions on product formation and morphology of *Yarrowia lipolytica*. Appl Microbiol Biotechnol 102(9):3831–3848

Toda T, Cameron S, Sass P, Zoller M, Wigler M (1987) Three different genes in *S. cerevisiae* encode the catalytic subunits of the cAMP-dependent protein kinase. Cell 50(2):277–287

Tomiyama T, Kurihara K, Ogawa T, Maruta T, Ogawa T, Ohta D, Ishikawa T (2017) Wax ester synthase/diacylglycerol acyltransferase isoenzymes play a pivotal role in wax ester biosynthesis in *Euglena gracilis*. Sci Rep 7(1):1–13

Torres-Guzman JC, Domínguez A (1997) *HOY1*, a homeo gene required for hyphal formation in *Yarrowia lipolytica*. Mol Cell Biol 17(11):6283–6293

Treger JM, Magee TR, McEntee K (1998) Functional analysis of the stress response element and its role in the multistress response of *Saccharomyces cerevisiae*. Biochem Biophys Res Commun 243(1):13–19

van der Walt JP, von Arx JA (1980) The yeast genus *Yarrowia* gen. nov. Antonie Van Leeuwenhoek 46(6):517–521. Retrieved from https://www.ncbi.nlm.nih.gov/pubmed/7195185

Wältermann M, Hinz A, Robenek H, Troyer D, Reichelt R, Malkus U, Von Landenberg P (2005) Mechanism of lipid-body formation in prokaryotes: how bacteria fatten up. Mol Microbiol 55(3):750–763

Wang HJ, Le Dall MT, Wach Y, Laroche C, Belin JM, Gaillardin C, Nicaud JM (1999) Evaluation of acyl coenzyme A oxidase (Aox) isozyme function in the n-alkane-assimilating yeast Yarrowia lipolytica. J Bacteriol 181(17):5140–5148. https://doi.org/10.1128/jb.181.17.5140-5148.1999

Wang H, Jia X, Wang X, Zhou Z, Wen J, Zhang J (2014) CFD modeling of hydrodynamic characteristics of a gas–liquid two-phase stirred tank. Appl Math Model 38(1):63–92. https://doi.org/10.1016/j.apm.2013.05.032

Wang G, Li D, Miao Z, Zhang S, Liang W, Liu L (2018) Comparative transcriptome analysis reveals multiple functions for Mhy1p in lipid biosynthesis in the oleaginous yeast *Yarrowia lipolytica*. Biochim Biophys Acta (BBA)-Mol Cell Biol Lipids 1863(1):81–90

Weeks G, Spiegelman GB (2003) Roles played by Ras subfamily proteins in the cell and developmental biology of microorganisms. Cell Signal 15(10):901–909

Wieczorke R, Krampe S, Weierstall T, Freidel K, Hollenberg CP, Boles E (1999) Concurrent knock-out of at least 20 transporter genes is required to block uptake of hexoses in *Saccharomyces cerevisiae*. FEBS Lett 464(3):123–128. https://doi.org/10.1016/S0014-5793(99)01698-1

Willis RM, Wahlen BD, Seefeldt LC, Barney BM (2011) Characterization of a fatty acyl-CoA reductase from *Marinobacter aquaeolei* VT8: a bacterial enzyme catalyzing the reduction of fatty acyl-CoA to fatty alcohol. Biochemistry 50(48):10550–10558

Wu H, Shu T, Mao Y-S, Gao XD (2020) Characterization of the promoter, downstream target genes and recognition DNA sequence of *Mhy1*, a key filamentation-promoting transcription factor in the dimorphic yeast *Yarrowia lipolytica*. Curr Genet 66(1):245–261

Xie D (2017) Integrating cellular and bioprocess engineering in the non-conventional yeast *Yarrowia lipolytica* for biodiesel production: a review. Front Bioeng Biotechnol 5:65

Xu P, Qiao K, Ahn WS, Stephanopoulos G (2016) Engineering *Yarrowia lipolytica* as a platform for synthesis of drop-in transportation fuels and oleochemicals. Proc Natl Acad Sci 113(39):10848–10853

Xu C, Liu R, Zhang Q, Chen X, Qian Y, Fang W (2017) The diversification of evolutionarily conserved MAPK cascades correlates with the evolution of fungal species and development of lifestyles. Genome Biol Evol 9(2):311–322

Xue Z, Sharpe PL, Hong S-P, Yadav NS, Xie D, Short DR, Wang J (2013) Production of omega-3 eicosapentaenoic acid by metabolic engineering of *Yarrowia lipolytica*. Nat Biotechnol 31(8):734–740

Yang K, Qiao Y, Li F, Xu Y, Yan Y, Madzak C, Yan J (2019) Subcellular engineering of lipase dependent pathways directed towards lipid related organelles for highly effectively compartmentalized biosynthesis of triacylglycerol derived products in *Yarrowia lipolytica*. Metab Eng 55:231–238

Young EM, Tong A, Bui H, Spofford C, Alper HS (2014) Rewiring yeast sugar transporter preference through modifying a conserved protein motif. Proc Natl Acad Sci 111(1):131. https://doi.org/10.1073/pnas.1311970111

Yuzbashev TV, Yuzbasheva EY, Sobolevskaya TI, Laptev IA, Vybornaya TV, Larina AS, Sineoky SP (2010) Production of succinic acid at low pH by a recombinant strain of the aerobic yeast *Yarrowia lipolytica*. Biotechnol Bioeng 107(4):673–682

Zakhartsev M, Reuss M (2018) Cell size and morphological properties of yeast *Saccharomyces cerevisiae* in relation to growth temperature. FEMS yeast Res 18(6):foy052

Zeng SY, Liu HH, Shi TQ, Song P, Ren LJ, Huang H, Ji XJ (2018) Recent advances in metabolic engineering of *Yarrowia lipolytica* for lipid overproduction. Eur J Lipid Sci Technol 120(3):1700352

Zhang W, Liu HT (2002) MAPK signal pathways in the regulation of cell proliferation in mammalian cells. Cell Res 12(1):9–18

Zhang J-L, Cao Y-X, Peng Y-Z, Jin C-C, Bai Q-Y, Zhang R-S, Yuan Y-J (2019) High production of fatty alcohols in *Yarrowia lipolytica* by coordination with glycolysis. Sci China Chem 62(8):1007–1016

Zinjarde SS, Kale BV, Vishwasrao PV, Kumar AR (2008) Morphogenetic behavior of tropical marine yeast *Yarrowia lipolytica* in response to hydrophobic substrates. J Microbiol Biotechnol 18(9):1522e1528

Whole Cell Yeast-Based Biosensors

Heather A. M. Shepherd, Emilia-Maria A. Bondarenko, Katherine M. Jennings, Rachel A. Miller, and Holly V. Goodson

Abstract

Analyte detection is a major component of fieldwork, environmental surveillance, and health protection, but lack of resources or access to instruments can create challenges in technology-limited environments such as remote research sites or low- and middle-income countries (LMICs). Whole-cell yeast-based biosensors provide potential solutions to these barriers. Here, we discuss the components of whole-cell yeast-based biosensors, emphasizing the process by which they are designed for their analyte of interest, approaches for harnessing them to function in particular research environments, and their advantages and disadvantages relative to other analytical tools. We provide examples of the various purposes for which existing whole-cell yeast-based biosensors have been used, focusing most on those appropriate for detecting externally-generated analytes in technology-limited settings. The further development of field-friendly whole-cell yeast-based biosensors still faces challenges, including the need to reduce the time from contact with the analyte to signal readout of the biosensor. Regardless, the inexpensive, robust, portable, environmentally friendly, and highly modular nature of yeast-based biosensors suggests that they could become useful tools for a range of analytical tasks.

Heather A.M. Shepherd and Emilia-Maria A. Bondarenko are Co-first authors.

H. A. M. Shepherd · E.-M. A. Bondarenko · K. M. Jennings · H. V. Goodson (✉)
Department of Chemistry and Biochemistry, University of Notre Dame, Notre Dame, IN 46556, USA
e-mail: hgoodson@nd.edu

H. V. Goodson
Department of Biological Sciences, University of Notre Dame, Notre Dame, IN 46556, USA

R. A. Miller
Department of Biology and Chemistry, Bethel University, Mishawaka, IN 46545, USA

F. Darvishi Harzevili (ed.), *Synthetic Biology of Yeasts*,
https://doi.org/10.1007/978-3-030-89680-5_4

1 Introduction

Sensors are tools used for analyte detection, analyte quantification, and/or gathering information relating to biological activity. One subset of sensors is biosensors. Though this term has varied meanings, here we define a biosensor as an analytical tool that utilizes biological components (e.g., proteins, nucleic acids, whole cells or even animals) to detect the analyte(s) of interest (Ostrov et al. 2017). Whole-cell biosensors are a subset of biosensors that utilize living cells for sensing tasks. One advantage of whole-cell biosensors relative to many other types of sensors is that whole-cell biosensors can provide information on the biological relevance of the analyte. This information is particularly important in cases where the sensing task is performed in a complex environment (e.g., wastewater) or where compounds of interest have varied composition (such as with endocrine disruptors) or unknown identity (e.g., as might occur with environmental mutagens). Additionally, whole-cell biosensors have the potential to provide detection with both high sensitivity and high specificity without need for expensive equipment or analytical-grade standards. These attributes suggest that whole-cell biosensor detection systems could be a useful alternative to spectral and mass-based detection systems in areas that lack sufficient equipment or resources (Martin-Yken 2020; Miller et al. 2020a, b).

Like most other sensors (biological and not), whole-cell biosensors are modular devices that can be viewed as being composed of standard components. As typically described, the fundamental parts of a whole-cell biosensor are the receptor (an element that interacts in a specific way with the analyte), the reporter (an element that produces a detectable signal in response to the receptor-analyte interaction), and the transducer (an element or process that connects the state of the receptor to the state of the reporter, sometimes with signal amplification) (Conroy et al. 2009). However, in considering the full spectrum of whole-cell biosensors and thinking about how they could be designed through synthetic biology, we suggest that it is useful to view these systems as being composed of *five* components: an *analyte*, a *detection process* (which often but not always involves a specific receptor), a *transduction process* (often with signal amplification), a *response process* (which often but not always involves production of an identifiable reporter), and a *readout process* (which involves quantifying or otherwise detecting the output of the response process) (Fig. 1, Table 1). In addition, most whole-cell biosensors also require a sixth component: some type of external *system housing* to contain the cells both during exposure to the analyte and periods of extended storage (Fig. 1). The reason for breaking the full process down in this way is that it enables a better understanding of how different modules can be rearranged, especially for cases that don't fit neatly into the "receptor-transducer-reporter" paradigm.

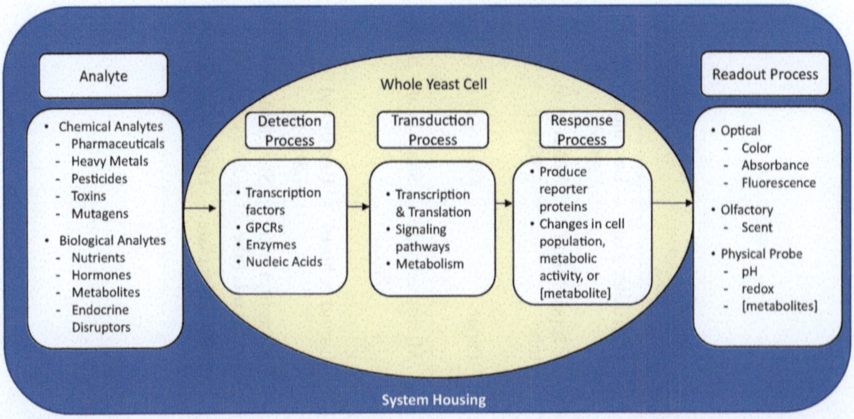

Fig. 1 Elements of whole-cell yeast-based biosensors. Whole cell biosensors are made up of five key components: *analyte*, *detection process*, *transduction process*, *response process*, and *readout process*. Here we have provided examples of each of these components featured in this review. These five components are then contained within the *system housing*, which is selected based on the environment in which the biosensor is intended to be operated. In technology-limited settings, paper-based assays or tubes read by a portable reader or the human senses (eyes, nose) would be ideal. In some settings, preloaded microtiter plates could work depending on the availability of suitable plate-reading technology

In a whole-cell biosensor, the detection, transduction, and response processes occur within or on the cell. Though the detection process often involves biological components (particularly proteins) imported from other biological systems, the transduction/amplification and response typically utilize endogenous machinery. Thus, a major challenge to designing and building effective whole-cell biosensors is figuring out how to properly integrate imported and endogenous components.

To illustrate these six components, consider a whole-cell yeast-based biosensor recently developed in our lab, one which produces a detectable smell in response to the hormone estrogen (Miller et al. 2020b):

Table 1 Examples of specific whole-cell yeast-based biosensors for exogenously-generated analytes

Analyte	Detection process (Receptor)	Transduction	Response	Readout	Housing	References
Estrogen	Human ER	Transcription/translation	Luciferase	Light level	Microtiter plate	Leskinen et al. (2003)
Estrogen	Chimeric GEV	Transcription/translation	Banana smell (isoamyl acetate)	Ordorant detection via human nose or gas chromatography	Capped 50 mL tube	Miller et al. (2020b)
Fungal pathogens	*C. albicans* GPCR Ste2	Signalling → transcription/translation	mCherry	Red lycopene pigment	Paper	Ostrov et al. (2017)
Melatonin	human GPCR MTNR1A in heavily modified yeast	Signalling → transcription/translation	GFP	Fluorescence	Microtiter plate	Shaw et al. (2019)
DNA damage	Yeast chromosome	DNA damage → transcription/translation	YFP, RFP	Fluorescence	Liquid cultures	Burrill and Silver (2011)
Bioavailable phosphorus	Multiple proteins	Metabolism	Cell growth	Absorbance	Microtiter plate or 50 mL tube	Shepherd et al. (2021)
Chemical sensitivity	Varied; potentially unknown	Metabolism	Reduced rate of cell growth	Absorbance as a function of time	Microtiter plate	Hung et al. (2018)

For a more comprehensive list see Martin-Yken (2020)

- The analyte is estrogen.
- The detection process is accomplished by binding estrogen to the cytosolic protein estrogen receptor alpha, imported from humans. More specifically, the yeast cells are exposed to a test sample in liquid media, and the estrogen (if present) diffuses across the membrane and binds to the estrogen receptor in the yeast cytoplasm.
- The transduction process involves several steps. In a simplified proposed mechanism, the estrogen/estrogen receptor complex translocates to the nucleus to bind to the DNA, activating transcription and translation of a reporter gene that encodes an enzyme which catalyzes production of an odorant detectable by the human nose.
- The response process consists of accumulating the enzyme that produces the odorant and the odorant itself.
- The readout process consists of a human using their nose to provide a yes/no answer about the presence of odorant, or a machine assessing its quantity.
- The housing device is a simple capped 50 mL tube containing the biosensor yeast and liquid growth media, to which test water or pharmaceutical samples have been added.

While the list above seems straightforward, people who are familiar with this or similar systems will realize that there are some additional steps implicit in the description above, and it is important to recognize these additional complications in considering how to design or improve particular whole-cell biosensors. For example, the transduction process for induction of transcription in response to estrogen is not as simple as implied above: it appears to involve ligand-mediated release from the estrogen receptor of the cellular component HSP90, with attendant translocation to the nucleus, though aspects of this process remain uncertain (Fuentes and Silveyra 2019). Moreover, there are implicit amplification steps in the process above. These amplification steps are important because there is no way that a human nose would be able to detect the signal if there were a 1:1 relationship between analyte (estrogen) and response (produced odorant).

The involvement of additional (and potentially unknown) eukaryote-specific cellular components and/or processes during transduction (as illustrated by the estrogen sensor above) is a challenging aspect of harnessing exogenous receptors and is a major reason why eukaryotic whole-cell biosensors are preferable to analogous prokaryotic systems for many applications. An additional issue with using prokaryotes for whole-cell biosensors is that they are often poorly suited to express eukaryotic receptors including the aforementioned estrogen receptor because they lack many protein folding factors, post-translational modification enzymes, and internal organelles specific to eukaryotic cells (Zhang et al. 2000; Vieira Gomes et al. 2018). Biosensors based on mammalian cells provide a potential solution to these problems, but they are expensive to maintain and require specialized equipment to do so, as they are sensitive to environmental conditions (Jarque et al. 2016a).

In considering these challenges, it becomes apparent that the baker's yeast *Saccharomyces cerevisiae* offers some significant advantages for construction of whole-cell biosensors relative to these systems. For example, yeast is recognized by the public as safe, and they are easy to work with due to their robustness, short doubling time, and extensive genetic toolbox (Redden et al. 2015). Whole-cell yeast-based biosensors have already been used to detect a wide range of chemical and biological analytes in the lab, and some have even been commercialized (e.g., the yeast estrogen (YES) and yeast androgen (YAS) tests as provided by Xenometrix.ch). The rapid development of synthetic biology tools and approaches in yeast (see other articles in this book) makes yeast an even more attractive organism for whole-cell biosensor development.

This review focuses on basic principles of whole-cell yeast-based biosensors, and it gives particular attention to those that would be appropriate for detecting externally-generated analytes in technology-limited environments. To improve access for readers less familiar with yeast molecular biology and avoid redundancy with other chapters in this book, we will avoid detailed discussions of the experimental technicalities of biosensor construction, though we do point out some challenges and pitfalls in biosensor development. For a review that goes more into depth in the technical side of whole-cell yeast-based biosensors, see (Adeniran et al. 2015). For more comprehensive reviews including discussions of biosensors for internally generated analytes and recent high-tech biosensor developments, see (Adeniran et al. 2015; Jarque et al. 2016a; Martin-Yken 2020).

Below we discuss in turn each of the components of a whole-cell yeast-based biosensor system: the analyte, a detection process, a transduction process (perhaps with signal amplification), a response process, a readout process, and a device in which to house the yeast and expose them to analyte (Fig. 1). As will become apparent, each of these components can be considered as a separate module that can potentially be mixed and matched to create a range of sensor functionalities. Though this text focuses specifically on yeast whole-cell biosensors, many of the principles discussed could apply to any whole-cell biosensor.

2 Initial Technical Considerations

In the discussion below, there is frequent reference to utilizing biological components (proteins, nucleic acids) imported from other organisms, enabling the yeast to detect analytes or produce reporters for which they have no natural machinery. Most often, genes encoding these components are provided on one or more plasmids, but protein-coding genes or other genetic elements can also be inserted directly into a yeast chromosome (see e.g., (Vopálenská et al. 2015; Yu et al. 2018)). The discussion below assumes a basic understanding of these tools and approaches (as well as yeast cell biology more generally), but it should still be understandable to those new to working with yeast. For information on working with yeast plasmids, see (Gnügge and Rudolf 2017). For a discussion of standard approaches to engineering genes onto yeast chromosomes and other aspects of yeast molecular biology, see (Gardner and Jaspersen 2014). For information about recent advances in yeast genetic engineering that may be relevant to biosensor engineering, see reviews including (Malcı et al. 2020; Schindler 2020) and the other chapters in this book.

3 Analytes Suitable for Whole-Cell Yeast-Based Biosensors

In theory, yeast can be harnessed to make a biosensor suitable for any analyte to which yeast naturally respond (e.g., nutrients, chemical toxins, metals, mutagens, temperature, generic stress) or anything for which an exogenous detection system can be imported from another organism (e.g., medicines, human hormones, light). In order to maintain a living sensor (and thus maintain the cell machinery), the analyte of interest must be within a concentration range compatible with yeast viability. For the purposes of detecting general hazards, one could use a sensor based on cell death. Whole-cell yeast-based biosensors can be particularly appropriate for situations where a biologically-active analyte (e.g., a drug or toxin) is present in low concentrations under conditions where detection by typical analytical techniques (e.g., mass spectrometry) is not feasible; they can also be particularly useful when the goal is to detect a biological activity (e.g., mutagenicity) instead of a particular chemical, or when the analyte is of unknown or of mixed composition. Note that these last two applications stretch the typical definition of "analyte".

In trying to design a whole-cell yeast-based biosensor, a key consideration is where the analyte is expected to be when it is detected: on the cell surface, in the cytoplasm, or in some other cellular compartment? It is important to recognize that some analytes can cross the membrane spontaneously (e.g., hydrophobic molecules such as steroid hormones), while others cross by dedicated transporters (for example, nutrients such as sugars), and still others become spontaneously concentrated in organelles such as the mitochondria (e.g., hydrophobic cationic dye molecules). However, many remain outside the cell or at best sequestered in internal vesicles because they can't cross membranes spontaneously. Considering the expected localization of the analyte is essential for proper design of any whole-cell biosensor because one needs to make sure that the necessary detection system components are in the same compartment as the analyte.

In our discussion of detection systems, we provide some examples of specific analytes for which whole-cell yeast-based biosensors have been developed. For a more comprehensive list, see (Adeniran et al. 2015; Jarque et al. 2016a; Martin-Yken 2020).

4 Detection Systems for Whole-Cell Yeast-Based Biosensors

4.1 General Considerations

The detection system of many familiar whole-cell biosensors involves a specific receptor (e.g., a locally-produced protein) that binds directly and specifically to the analyte and then undergoes a conformational change in response to the binding event. As discussed more below, this conformational change then activates, by one of a number of possible mechanisms, a downstream sequence of events (the transduction/amplification process) that eventually leads to a detected signal.

In theory, one might be able to design a receptor for a given analyte, but in practice, a much more effective approach (at least thus far) has been to find a natural receptor for the analyte of interest. Because the range of chemicals/environmental components detected by the diversity of the living world is enormous, there is a reasonable chance that a receptor exists for the analyte of interest, or if not, something similar to it. If the desired receptor is unavailable, similar natural receptors can then be used as a starting point for developing improved (e.g., more sensitive or altered specificity) receptors through approaches such as genetic engineering (see e.g., (Urlinger et al. 2000; Roney et al. 2016)) or directed evolution (see (Zhou et al. 2006; Stainbrook et al. 2017; Adeniran et al. 2018)).

To harness an exogenous receptor for a yeast whole-cell biosensor application, there are several issues to consider. First, one must understand the receptor's structure and its function in the context of the source organism. Just as whole-cell biosensors are modular, so are the receptor proteins themselves. However, the specific functionalities of a given receptor's modules depend on the natural signal transduction process it connects to. For example, some receptors work by binding DNA; others change the concentration of a second messenger, interface with the signal transduction machinery, or produce a metabolite. Knowing how the receptor works in its natural setting is essential for determining how straightforward it will be to integrate it with the endogenous yeast transduction machinery. This need to communicate properly with endogenous systems makes it challenging to utilize exogenous receptors. Another key point to consider in harnessing receptors is where the receptor is expected to interact with the analyte: on the cell surface, in the cytoplasm, or in some other cellular compartment. Making sure that the receptor has access both to the analyte and the downstream signal transduction machinery is essential. This can involve modifying the receptor to localize it to the relevant part of the cell (e.g., the nucleus), though it is important to recognize that such targeting does not always work as expected (e.g., (Emr et al. 1984)). Finally, it is important to recognize that some additional endogenous cellular processes (e.g., amplification) are important for the proper function of a receptor-based signaling process, while others such as sensory adaptation can interfere.

While the detection modules of whole-cell biosensors often utilize particular receptors that bind in a specific way to individual analyte molecules, this is not always the case. For example, as discussed more below, some detect *activities* such as the ability to induce DNA damage (Paetkau et al. 1994; Billinton et al. 1998; Burrill and Silver 2011; Lu et al. 2015; Bui et al. 2015), cause cell stress (Hollis et al. 2000; Hung et al. 2018; Gong et al. 2020), or serve as nutrients (Shepherd et al. 2021; Trentman et al. 2021).

4.2 Detectors for Cytoplasmic Analytes

Most whole-cell biosensors developed thus far detect analytes that can make their way into the cytoplasm. Many of these utilize specific receptors that interface with the transcription/translation system for the transduction part of the sensing process. Thus, these receptors typically have an analyte binding domain and a DNA binding domain; those that work as activators (not repressors) generally also have a transcription activator domain. Generally speaking, binding of analytes to these receptors changes their ability to bind DNA, thus altering the amount of transcription of a reporter gene and ultimately the amount of reporter protein.

An example of an endogenous DNA-binding receptor used in biosensors is provided by the copper-response system; the yeast *CUP1* promoter is activated in response to copper through the action of endogenous machinery including the DNA-binding Ace1p protein (Dameron et al. 1991; Smith et al. 2017). One particularly well-characterized exogenous DNA-binding intracellular receptor is the bacterial tetracycline receptor (TetR), which can either activate or repress transcription in yeast in response to binding tetracycline; the specific effect depends on the details of how the system is set up (Baron and Bujard 2000). TetR is a very "portable" type of gene response element (i.e., it can work in diverse organisms) because it needs nothing other than itself and a target sequence placed near a promoter to cause production of a reporter in response to TetR.

Another example of an imported DNA-binding receptor is provided by the human estrogen receptor (ER), mentioned earlier. However, while the DNA binding ability of Ace1 and TetR are directly regulated by binding to their ligands (copper and tetracycline), the effect of estrogen on the DNA-binding activity of ER is more indirect and involves the participation of some additional cellular factors. Ultimately, binding of estrogen to ER causes increased transcription from promoters controlled by "estrogen-response elements" (EREs), similar to what happens with TetR (Gruber et al. 2004). However, the change in transcription in response to estrogen cannot be explained simply by differential binding to EREs in the presence of estrogen. Instead, it appears to involve regulation of ER nuclear localization through differential binding to chaperone HSP90 and also involves the participation of co-receptors (Smith and Toft 1993). When such additional elements are involved in the downstream transduction process, the detection element is much less portable, and use of compatible eukaryotic systems that contain these downstream components (either naturally or by engineering) can be essential. Indeed, the compatibility of yeast with the estrogen receptor (as well as other steroid receptors, see (Ito-Harashima et al. 2020)) is an advantage of yeast over bacteria for whole-cell biosensor applications (Routledge and Sumpter 1996).

Interestingly, the modularity of these DNA-binding intracellular receptors suggests that one might be able to “mix and match” analyte-binding and DNA-binding domains (Aranda-Díaz et al. 2017). Indeed, this logic has led to generation of functional chimeric receptors where the presence of estrogen turns on a galactose-responsive promoter (Louvion et al. 1993; Gao and Pinkham 2000; Miller et al. 2020b). However, this strategy generally depends on utilizing protein domains that fold and act independently (i.e., that do not require interactions with distant components of the original protein for activity).

One type of intracellular analyte receptor that has been used in bacteria but much less so in yeast is riboswitches, which are self-regulating mRNA sequences. The binding of an analyte to a riboswitch changes the shape of the riboswitch, affecting whether the sequence is translated into a protein (Zhang et al. 2015). Most riboswitches are found in prokaryotes in nature and will not work if directly imported into yeast because of differences in translation between prokaryotes and eukaryotes. Nevertheless, harnessing of riboswitches remains an interesting idea that might have utility for yeast-based biosensors. As a related idea, ribozyme-based switches controlling gene expression have been utilized in yeast as sensors for the detection of aminoglycosides (Klauser et al. 2015).

While many detection systems for intracellular analytes use specific receptors for detection and the transcription/translation system for transduction, other designs exist. In particular, some specific receptors for cytoplasmic analytes utilize metabolism for their transduction process, with the response being increased (or decreased) cell division. Note that increased transcription may occur as part of these responses, but the effect is indirect. Often these metabolism-based sensors detect nutrients. For example, one could imagine a whole-cell yeast-based biosensor designed to detect a specific sugar; in this biosensor, the receptor would be an enzyme that starts the process of metabolizing that sugar, and the response would be cell growth when the sugar is present (assuming that carbon is otherwise limiting). In some examples of metabolism-based transduction systems, there is no specific receptor: the detector is the cell as a whole because multiple proteins are involved in parallel in the detection process. An excellent example of this type of biosensor is provided by the use of yeast for "BOD" (biological oxygen demand) assays, which are effectively assays for bioavailable carbon in all its guises (Renneberg et al. 2004; Seo et al. 2009; Yudina et al. 2015). Similarly, a yeast assay for bioavailable phosphorous has recently been developed (Shepherd et al. 2021; Trentman et al. 2021). The response for these nutrient-detection assays is typically increased cell growth, and readout of these systems can be accomplished through measurement of parameters ranging from turbidity as measured by a spectrophotometer (Shepherd et al. 2021; Trentman et al. 2021) to oxygen utilization as measured by potentiometry (Yudina et al. 2015). Using similar logic, researchers have developed assays for toxins, where the toxin is detected through decreased cell growth (Paetkau et al. 1994; Billinton et al. 1998; Hung et al. 2018). However, it is important to note that systems utilizing cell division (increased or decreased) as an output can be less specific than other typical reporting systems, so it is essential to include appropriate controls.

Using a related set of ideas, researchers have developed yeast-based assays for stress-inducing agents. Though these agents can overlap chemically and biologically with toxins, the transduction process in these assays occurs through induction of cellular stress-response pathways. A typical response for these stress-detecting sensors would be an accumulation of reporters placed under control of stress response promoters (Hartner and Glieder 2006; Adeniran et al. 2015; Zhao et al. 2019). Similarly, there are yeast-based sensors for mutagens in which the yeast genome is the detector, the transduction process is provided by the machinery that senses and responds to DNA damage, and the response is increased transcription/translation of reporters placed under control of DNA-damage-induced promoters (Burrill and Silver 2011). An even more sensitive example of a mutagen detector is provided by a biosensor where the detector is a yeast *CAN1* gene. Mutations in *CAN1* make yeast resistant to the toxin canavanine; the response to mutagen is thus an increase in the number of yeast cells that are resistant to canavanine, and the readout process is to count the number of canavanine-resistant colonies (Ong et al. 2021).

4.3 Detectors for Analytes that Remain Outside of Cells

Extracellular analytes are typically detected by specific receptors. In contrast to the intracellular-analyte-detecting receptors discussed above, a receptor that directly detects an extracellular analyte needs to have an extracellular domain, one or more transmembrane regions, and an intracellular domain (Fig. 2). The purpose of the extracellular domain is to bind to the analyte, while the transmembrane domain fixes the protein in the membrane, and the intracellular domain interfaces with one of various possible transduction systems to convey to the rest of the cell that the analyte is present. In most cases, the intracellular domain communicates with the endogenous signal transduction machinery to trigger a downstream response. For example, the intracellular domain could be an enzyme that catalyzes production of a second messenger (a small molecule that diffuses away and turns on other proteins), or a kinase that phosphorylates other proteins (which then diffuse away) to turn them on or off (Fig. 2a).

Thus far, the field of biosensors based on heterologous transmembrane receptors in yeast is under-developed, likely because of the challenges presented by the need to integrate the heterologous receptors with the endogenous yeast downstream transduction machinery and hook this machinery up to a useful response system. As yet, much of the whole-cell biosensor work done in yeast with transmembrane receptors has focused on G-protein coupled receptors (GPCRs, Fig. 2b), with a few exceptions including antibody display for point of care devices (Venkatesh et al. 2015). GPCRs are found throughout the eukaryotic world and are involved in detecting the presence of compounds ranging from peptide hormones to tastants, and even light (Hilger et al. 2018; Ahmad and Dalziel 2020). Yeast cells have two endogenous classes of GPCRs: one type consists of two proteins that respond to yeast mating factors, while the other responds to glucose (Versele et al. 2001).

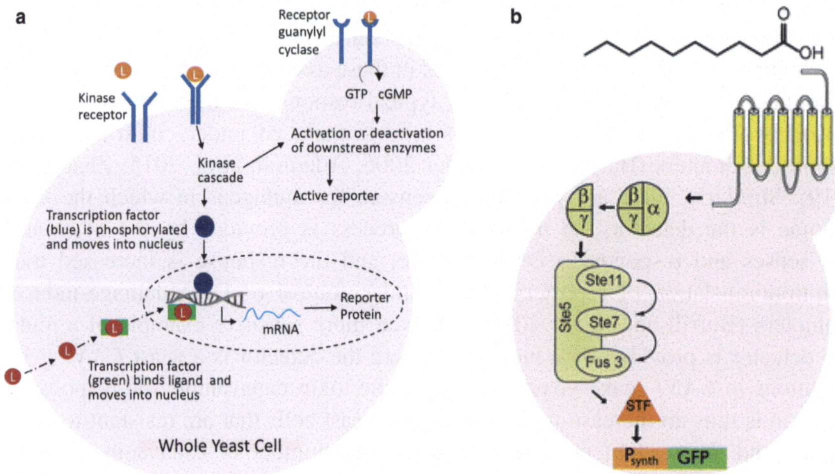

Fig. 2 Examples of biosensor pathways for extracellular analytes, where endogenous signaling pathways are typically used as part of the transduction process. The response can involve activation of transcription of a reporter gene or activation of other enzymes. **a** Analytes can either diffuse across the membrane (red) or remain outside and bind to extracellular receptors (orange). Ligands that can freely diffuse across the membrane typically bind to a receptor protein in the cytoplasm, enabling the complex to enter the nucleus and alter transcription. For analytes that remain extracellular, potential detection mechanisms include kinase receptors and second-messenger systems; these bind ligands (orange) and then can either activate transcription, alter the activity of other enzymes, or both. **b** Specific example of a medium-chain fatty acid biosensor detected through a GPCR. A heterologous GPCR (yellow) detects a medium-chain fatty acid in the culture medium, transmits this chemical signal to the yeast mating-pathway (mustard), which relays it to a synthetic transcription factor (STF, orange). The STF activates transcription of GFP. The image for **b** was reprinted with permission of the American Chemical Society (ACS). The original file can be found at https://doi.org/10.1021/sb500365m. Further permissions regarding this image should be directed to the ACS

Significantly, the mating factor-response GPCRs activate a signal transduction pathway that leads to activation of mating-factor-responsive genes. Thus, it seems logical that one could plug human GPCRs into this pathway to induce transcription of a reporter. Unfortunately, researchers have found that most medically relevant human GPCRs require significant optimization as to expression level, membrane transport, and coupling to the downstream transduction machinery in order to work usefully in yeast (reviewed by (Shaw et al. 2019)). However, progress is being made through approaches such as genome engineering to remove interfering pathways (Shaw et al. 2019) and humanization of the yeast membrane to promote proper folding, processing, and allosteric behaviors (Routledge et al. 2016). The result of all this work is that yeast expressing human GPCRs are becoming strong systems for biotechnological and medical applications (Lengger and Jensen 2020).

One question that must be addressed for extracellular analyte receptors is how the presence of the bound analyte is conveyed from the extracellular part of the receptor across the membrane to the intracellular part. The typical answer is that the information crosses the membrane in the form of a conformation change. While that is a reasonable answer for GPCRs (which cross the membrane seven times, providing a 3-D structure by which to convey allosteric change), the mechanism of this information transfer is less clear for proteins like receptor kinases, which generally cross the membrane only one time (Fig. 2). One plausible explanation is that binding of the receptor to analyte induces dimerization, and it is the dimerization event that turns on downstream signalling (Nakamura et al. 2016). Thus, this dimerization response could be considered the first step in the transduction process. This point suggests that regulation of the level of expression of the receptor itself can be very important for functionality of a sensing system: if the level is too low or too high, performance of a system that relies on regulated dimerization could be severely compromised.

5 Transduction and Amplification Pathways

5.1 Typical Transduction Pathways Used by Whole-Cell Yeast-Based Biosensors

As discussed above, while receptors can either be endogenous or exogenous, the transduction process typically occurs via cellular machinery: the transcription/translation system (as exemplified by the copper or tetracycline response systems), signaling machinery (as seen with GPCR-mediated responses), or metabolism (e.g., sensors for nutrients such as phosphate). We won't repeat these discussions, except to stress that in the case of imported detection or response systems, it is essential to identify all necessary components, ensure they are present, and determine that these components interact properly with the cellular transduction machinery. In addition, it is important to remember that many endogenous biological information processing elements integrate inputs from multiple sources. For example, metal-response promoters can sometimes respond to other metals or even other stresses, meaning that responses may not be as specific as expected. For example, the copper-responsive promoter *CUP1* is induced not only by copper (through Ace1p) (Jeyaprakash et al. 1991) but also heat shock, glucose starvation, and oxidation stress (through Hsf1p) (Tamai et al. 1994). A related issue is that the biosensor elements themselves may have potentially confounding effects caused either by simple overexpression (e.g., through selfish recruitment of cellular machinery) or protein-specific toxicity (which can cause off-target cell stress responses that are aggravated by overexpression). Thus, the strength of the promoters and/or types of plasmids can affect the overall success of the biosensor. Finally, cellular sensing pathways often include mechanisms for adaptation and desensitization, similar to those experienced by human eyes upon exposure to light. Engineering out these additional inputs/influences can be helpful or even necessary (Versele et al. 2001; Mukherjee et al. 2015; Hilger et al. 2018; Shaw et al. 2019).

5.2 Other Potentially Useful Transduction Pathways

While most extracellular reporters interface with endogenous signal transduction systems, other possibilities exist, and they may provide inspiration for synthetic biologists designing new types of whole-cell biosensors. One intriguing example is provided by the Notch pathway of animal development, where a transcription factor is released upon ligand binding by an induced proteolytic cleavage (Hori et al. 2013). This and related systems offer the possibility of utilizing tension (not just conformational change) as part of the transduction process. Harnessing the Notch system would likely be challenging because of the number of components involved and the need to introduce them to yeast, but this type of approach could potentially be useful for connecting an extracellular sensor to a transcription-based transduction system.

5.3 Amplification

A key element common to most of the transduction systems discussed above is that the signal is amplified as part of the transduction process. For example, a single analyte binding to a single DNA-binding receptor can result in production of many mRNA molecules and many protein reporter proteins. Similarly, endogenous signal transduction pathways typically incorporate amplification at multiple steps. For example, activation of an initial receptor enzyme activates multiple additional enzymes, creating a "signaling cascade" (Nelson and Cox 2017). Amplification can also occur through cell division or the use of a population of yeast as the detecting unit instead of single cells (e.g., (Ong et al. 2021)). Amplification at some point in the detection process is important to high-sensitivity sensing because without it there would be a 1:1 correspondence between analyte-binding and reporter signal, meaning that only high-concentration analytes could be detected. Amplification could occur either during the biological part of the transduction process or during readout (e.g., as part of a fluorescence detection process). However, if the goal is to harness the biosensor for a technology-limited environment, it can be advantageous to utilize biological amplification.

The need for amplification for high-sensitivity sensing tasks reduces the utility of engineered molecules such as aptamers for whole-cell sensing applications. Briefly, aptamers are short single-stranded DNA or RNA molecules that can be selected from a library to bind specifically to a target (Adachi and Nakamura 2019; Wang et al. 2020). These and similar tools (e.g., phage display antibodies (Peltomaa et al. 2019)) are attractive as sensing tools because in theory one can use selection (perhaps as optimized by directed evolution) to select molecules that bind specifically to essentially anything. However, the challenge to using these molecules for whole-cell biosensing applications is figuring out how to couple the molecule to endogenous transduction/amplification systems or design novel systems with sensitivity sufficient to detect exogenous analytes.

For example, one response system commonly used in conjunction with aptamers is FRET (Fluorescence/Förster Resonance Energy Transfer). With FRET-based aptamer-sensing molecules, the probes are placed such that FRET increases (or decreases) upon binding to analyte (reviewed by (Yoon et al. 2012)). FRET reporters can indeed be very effective. However, the ~ 1:1 ratio between detection (binding event) and response means that the concentration of analytes needs to be relatively high for this method to be effective, or that the detection system needs to be very sensitive. Thus, unless one can design a way to amplify the signal from the aptamer (or similar molecule) and do so in a way that takes advantage of the cellular environment, it might be more effective to stick with a non-cellular context for using these tools to detect extracellular analytes if the goal is to do so in a technology-limited environment.

6 Response and Readout

The biological part of the transduction process of a whole-cell biosensor ends at the response, which often consists of production or activation of a reporter, a biomolecule or process that directly or indirectly creates a signal that is detectable (through readout) by the outside world (Table 2).

6.1 Color-Based Reporters

One early reporter for DNA-binding intracellular receptors was beta-galactosidase, a cytosolic protein from bacteria that cleaves an exogenous chemical called X-gal to release a dark blue product. This reporter is functional in the context of whole-cell biosensors (see e.g., (Weaver et al. 2015)), but has several drawbacks, especially for applications in technology-limited environments. One major concern is the need for X-gal itself, a rather expensive and unstable compound. Another is that X-gal does not cross the cell membrane, which means that cell lysis (e.g., by liquid nitrogen) is required to enable the enzyme to access the X-gal substrate (Möckli and Auerbach 2004; Weaver et al. 2015). An additional drawback is the time required for signal development: any reporter based on transcription/translation is likely to need several hours to accumulate sufficient amounts of the reporter protein. This beta-galactosidase reporter system needs several additional hours to develop the color after the reporter protein is expressed (Möckli and Auerbach 2004; Weaver et al. 2015).

A variety of researchers have tried to enable improved color-producing reporters by using other combinations of enzymes and substrates or by secreting the reporter enzyme out of the cell so that the color-producing substrate can be provided in the exterior environment. The recent development of technologies for producing betalains (a family of colored compounds including beet-red betanin) in yeast is promising (Grewal et al. 2018). Regardless, many approaches for colorimetric biosensors still rely on the lacZ reporter system (Martin-Yken 2020). Due to the

Table 2 Examples of reporting methodologies for yeast biosensors

Type of response process	Examples	Advantages	Disadvantages	Selected examples from this review
Color-based	X-gal, betalains, lycopene	Readout visible to the naked eye	Expensive and unstable substrate for reaction; cell lysis must occur for X-gal; color development can take a long time (~hours-days)	Möckli and Auerbach (2004), Weaver et al, (2015),Ostrov (2017), Grewal et al. (2018), Chen et al. (2018), Martin-Yken (2020)
Light-based	GFP, YFP, RFP Luciferase	Relatively quick (~hours); noninvasive readout; sensitive; fluorescent proteins do not require substrate	Requires specific filters/light wavelengths (this may be helped with advances in portable fluorescence readers); luciferase requires an expensive substrate	Leskinen et al. (2003), Allard (2008), Adeniran et al. (2015), Borse et al. (2017), Miller et al. (2020a), Martin-Yken (2020)
Olfactory	Isoamyl-acetate	Readout only using human nose; minimal training required	Difficult to quantify	Miller et al. (2020b)
Probe-based measurement	Amperomentry (current), potentiometry (pH)	Quick (~hours or faster); noninvasive readout; sensitive	Requires immobilization and specialized equipment	Renneberg et al. (2004), Seo et al. (2009), Yudina et al. (2015), Yudina et al. (2015)
Metabolism	Cell growth/death	No bioengineering required	Less specific than other typical reporting systems; systems based on nutrient depletion or colony formation can take several days for final readout	Hung et al. (2018), Shepherd et al. (2021), Trentman et al. (2021)
Hybrid process	Cell growth following analyte-induced expression of required enzyme	Readout visible to the naked eye	Systems based on colony formation can take several days for final readout	Lehmann et al. (2000)

lower protein expression levels and the longer replication times of yeast compared to other potential whole-cell biosensors (e.g., *E. coli*), signals produced by these color-based systems are relatively weak and can take a long time (e.g.,>24 h) to develop (Chen et al. 2018).

6.2 Light-Based Reporters

Because of these challenges with color-based systems, a popular set of reporters for transcription/translation-based transduction processes is GFP (green fluorescent protein). GFP and similar polypeptides emit fluorescent light in response to excitation. Readout of the reporter can be assessed quickly, non-invasively, and without additional manipulation simply by exposing cells to the appropriate excitation wavelength and assessing the light level at the expected emission wavelength (Lei et al. 2006). A potential drawback of these systems is that filters and lights of specific wavelengths are generally needed to assess the level of fluorescence. However, recent work has shown that inexpensive, portable fluorescence readers have made the use of fluorescence as a reporter much cheaper and has allowed for data collection in the field (Borse et al. 2017; Miller et al. 2020a). Decreasing background noise/interference and enhancing the fluorescence signal can improve fluorescence detection.

Another popular reporter is luciferase, which produces light when provided with the substrate luciferin. A disadvantage of luciferase is that the luciferin substrate is both somewhat expensive and has a limited shelf-life (Leskinen et al. 2003). However, because the substrate is dark, there is a very low background, meaning that readout of luciferase levels is 10–1000 × more sensitive than a comparable assay with GFP (Allard 2008).

6.3 Olfactory Reporters

The reporters discussed thus far are light-based: they produce light or color during the readout process. A more recently developed reporter for transcription/translation-based transduction processes in yeast is an enzyme that produces a scent: "scentsor" assays involve the induction of a reporter enzyme that produces banana smell (Miller et al. 2020b). Aromatic compounds and their properties have been paired with yeast for years through bioengineering and fermentation (Van Wyk et al. 2018), making diverse olfactory reporters in whole-cell yeast-based biosensors a realistic possibility. The benefit of these assays is that the readout requires no equipment other than the human nose and minimal training or resources outside of the biosensor itself. However, it is difficult to quantify a scent, limiting scentsor-based assays to yes/no assessments. Despite these challenges, the ease of use suggests that olfactory-based reporter systems could be ideal for technology-limited environments (Miller et al. 2020b).

6.4 Other Potentially Useful Reporters

As noted above, cell growth and/or metabolism can be used as reporters for biosensors used to detect nutrients (e.g., (Shepherd et al. 2021; Trentman et al. 2021)) or toxins (Hung et al. 2018).

In addition, while the readouts of the reporters above follow logically from the activities of the reporter proteins themselves, it is possible to have situations where the responses, reporters, and readouts are less directly connected and/or have additional layers. For example, Lehmann et al. (2000) engineered a yeast strain in which a lactose-metabolizing enzyme (LacZ) was placed under control of a copper-responsive promoter. Because yeast cannot otherwise metabolize lactose, this created cells that require copper to grow on lactose. Thus, with this strain, one can use growth on media lacking a carbon source other than lactose to detect the presence of copper. In this case, LacZ is a reporter, and induction of LacZ is one level of response, but cell growth resulting from the copper-induced ability to metabolize lactose is another level of response. The overall readout for detection of copper would be whatever method is convenient to assess cell growth and/or metabolism; Lehmann and colleagues actually used amperometry to measure oxygen consumption. Using a similar strain in which LacZ was induced by copper, Tag et al. (2007) measured changes in lactose concentration in response to heavy metals. This lactose-based method works in combination with flow injection analysis to allow for semi-continuous measurements and to increase the sensor's capability as an in-line sensor for heavy metal waste produced by manufacturers.

7 Whole-Cell Biosensor Housing and Reading Devices

The text thus far focuses on the characteristics of the yeast cells used for whole-cell biosensor work. To make a useful whole-cell biosensor-based device, one generally needs to design a type of "housing" to hold the cells, expose them to a potential analyte, and develop the signal. Also needed is a reading device to interpret (recognize and/or quantify) the signal, though in some cases the senses of the biosensor-users themselves (e.g., eyes to detect color, nose to detect smell) can provide this functionality. An additional issue is that one also needs a way to distribute the biosensor cells. Because this review focuses on whole-cell biosensors for use in technology-limited settings, we concentrate here on reading and housing devices that require minimal skills or access to specialized equipment for use; such devices are generally based on large populations of cells. However, it is important to recognize that yeast-based whole-cell biosensors have a long history of use in laboratory settings (see e.g., (Routledge and Sumpter 1996)), and applications based on small populations (as in microtiter plates) or individual cells (as in microfluidic devices) have been proliferating. We discuss some microtiter plate-based approaches below, but for a deeper discussion of lab-based yeast whole-cell biosensor-devices, see (Adeniran et al. 2015; Jarque et al. 2016a; Martin-Yken 2020).

First, it is important to clarify what we mean by "low-tech, field-friendly" housing devices. Such devices should be able to be distributed and employed at the point of use without sterile technique and preferably without micro-pipettes. While the end use of such sensor devices may have few requirements, it is important to emphasize that the manufacturing of these devices generally will require a laboratory setting. Similarly, low-tech field-friendly reading devices should be small, inexpensive, require minimal training, and have minimal power requirements (i.e., preferably they would require no power or be battery-operable).

Given these restrictions, three basic categories of low-tech devices for housing whole-cell yeast-based biosensors become particularly apparent: bio-paper analytical devices (bioPADs), microtiter plates, and plastic culture tubes. In a yeast bioPAD, biosensor yeast cells are immobilized in a hydrogel on paper, and are exposed to the potential analyte by soaking the PAD in an aqueous solution containing the potential analyte. In present implementations of yeast PADs, this solution also contains some yeast nutrients as well as antibiotics to reduce the growth of bacteria (Weaver et al. 2015; Miller et al. 2020a). However, one could imagine lateral flow versions where nutrients, antibiotics, and any additional reagents necessary for function of the reporter (e.g., color-generating substrates) are placed on the PAD below the yeast spot.

Advantages to bioPADs include transportability (they are small and light), ease of use, and stability (bioPADs have been shown to be shelf-stable for > 1 year at −20 °C, > six months at 4 °C, and at least 56 days at 37 °C) (Weaver et al. 2015; Miller et al. 2020a). However, one challenge thus far has been reading them. An initially published color-based reporter (LacZ) is not field-friendly because it requires that the yeast be lysed with liquid nitrogen to release the LacZ enzyme and subsequently "developed" by exposing the bioPADs to a solution containing the color-developing substrate X-gal (Weaver et al. 2015). If a suitably intense and rapidly developing color-based reporter could be developed, then a cell phone might be an ideal reader (see e.g., (Ballard et al. 2020)). More recently, we have developed a bioPAD that utilizes a fluorescent reporter (Miller et al. 2020a). For reading this bioPAD, we utilized a small and inexpensive home-built reader device, which is collapsible and light for ease of transport (Miller et al. 2020a). Others have worked to develop cell-phone-based fluorescence readers, suggesting this will be an option in the future (Wei et al. 2013; Vashist et al. 2014; Berg et al. 2015).

Another type of housing device that should be adaptable to work in a field environment is the microtiter plate. Microtiter plate-based whole-cell biosensor assays have been long-established in laboratory environments (see e.g., (Cevenini et al. 2018; Miller et al. 2020a)). As presently described, most microtiter plate assays require sterile technique, incubating shakers, and use of micro-pipettes. However, we have developed an assay where yeast can be grown in microtiter plates without shaking, at room temperature (Shepherd et al. 2021; Trentman et al. 2021). Reading these microtiter plates is then the remaining challenge. Remarkably, a number of authors have worked to develop cell-phone based microtiter plate readers (Berg et al. 2015; Cevenini et al. 2018; Wang et al. 2018), meaning that microtiter-based assays may become suitable for technology-limited settings. In the interim, many relatively modest laboratories have access to standard plate-reading devices. One question is how yeast would be distributed across the microtiter plate. If access to micropipettes is lacking, dry yeast could perhaps be preloaded.

For cases where microtiter plates are problematic, another possibility is simply growing the yeast in capped plastic centrifuge tubes; for this approach, tubes might be distributed after preloading them with freeze-dried yeast and media (more about yeast long-term storage below). For any particular assay developed under more standard lab conditions, controls would have to be performed to ascertain if/how growing in a capped tube at room temperature would affect the biosensor performance. However, based on our experience, yeast-based biosensors can be remarkably resilient to changes in aeration and/or temperature: the main effect is an alteration to the time needed for signal development. While housing biosensors in pre-loaded plastic tubes is not currently common practice, one that fvcould imagine that pre-loading plastic tubes for whole-cell yeast-based biosensors could increase their utility in the field or use in LMICs where materials may be scarce. Two examples of assays that could benefit from such a process would be the "Scentsor" (Miller et al. 2020b) and the phosphorus centrifuge tube assay (Shepherd et al. 2021; Trentman et al. 2021). While neither system relies on complex signal reading devices (they use scent and optical density respectively) they do rely on culturing the yeast with their analyte of interest. Through the use of preloaded tubes containing the components necessary for the assays to function, these biosensors could be made available for use in community testing sites or other technology-limited settings by only requiring the addition of the sample of interest.

8 Long-Term Storage

Whole-cell yeast-based biosensors must be stored in a way that the biosensors can be transported and used at a later time. When biosensor yeast are used in paper-based assays, yeast cultures are typically stored using a sugar-based hydrogel to protect the yeast until the biosensor is ready for use (Fine et al. 2006; Weaver et al. 2015; Jarque et al. 2016b; Miller et al. 2020a). As noted above, bioPADs constructed with yeast in hydrogel have been shown to be shelf-stable for > 1 year at -20 °C, > six months at 4 °C, and at least 56 days at 37 °C (Weaver et al. 2015; Miller et al. 2020a). Storage of yeast spotted onto paper-based "dipstick" biosensors in argon has also been reported (Ostrov et al. 2017). For long-term storage, drying methods such as lyophilization (freeze-drying) or storage in hydrogel have been effective. Depending on when the biosensors will need to be used, freeze-dried yeast can be stored for 2 months at 4 °C or 10 months at -18 °C and still retain normal activity and sensitivity in assays (Jarque et al. 2016b). Immobilization in polyvinyl alcohol hydrogel particles increased yeast storage time to 1 year (Herkommerová et al. 2018). More high-tech treatments may also have utility in enhancing long-term storage. For example, the application of different coatings (e.g. polyelectrolytes, biomolecules, metal nanoparticles, or oxides) has been found to improve reproducibility and cell sensitivity, viability, and stability as compared to bare cells (Bittner et al. 2015; Dai et al. 2018). In particular, bio-silica sol-gels can act as protective shells against environmental factors. These gels have even been found to protect cells from exposure to heavy metals and UV radiation for up to 28 days (Ponamoreva et al. 2015).

9 Conclusions

Whole-cell yeast-based biosensors are affordable and transportable devices that can detect analytes of interest in an environment with limited resources. These biosensors excel at detecting biologically relevant analytes, including pharmaceuticals, toxins, and chemically undefined or complex mixtures ranging from environmental mutagens to bioavailable phosphorous. They can be tailored to detect a desired analyte using a range of possible readout mechanisms by harnessing the large existing (and continuously expanding) toolbox of modular parts assembled from across the biological world. Recent work on whole-cell yeast-based biosensors has been focused on expanding the applications of the biosensors and improving the sensors in terms of signal response, sensitivity, and reporter/readout mechanisms. The largest roadblock to utilization of whole-cell yeast-based biosensors is the time associated with the transduction and response processes; harnessing other natural biological processes (e.g., using signal transduction pathways instead of transcription/translation) or creating new processes through synthetic biology should provide significant improvements in analyte detection time. Ideally, improved transduction/response processes would be paired with a readout mechanism that requires minimal additional technology (e.g., color-changing reporters requiring only the user's eyes for analyte detection) so that the target analyte could be detected quickly in a field setting. Despite these challenges, yeast cell hardiness and biosensor modularity suggests these obstacles can be overcome, affirming the utility of whole-cell yeast-based biosensors as an analyte detection tool in the lab and field.

References

Adachi T, Nakamura Y (2019) Aptamers: a review of their chemical properties and modifications for therapeutic application. Molecules 23:4229. https://doi.org/10.3390/molecules24234229

Adeniran A, Sherer M, Tyo KEJ (2015) Yeast-based biosensors: design and applications. FEMS Yeast Res 15:1–15. https://doi.org/10.1111/1567-1364.12203

Adeniran A, Stainbrook S, Bostick JW, Tyo KEJ (2018) Correction to "Detection of a peptide biomarker by engineered yeast receptors." ACS Synth Biol 7:1973–1973. https://doi.org/10.1021/acssynbio.8b00291

Ahmad R, Dalziel JE (2020) G protein-coupled receptors in taste physiology and pharmacology. Front Pharmacol 11:587664. https://doi.org/10.3389/fphar.2020.587664

Allard STM (2008) Bioluminescent reporter genes. https://www.promega.com/resources/pubhub/enotes/bioluminescent-reporter-genes/

Aranda-Díaz A, Mace K, Zuleta I et al (2017) Robust synthetic circuits for two-dimensional control of gene expression in yeast. ACS Synth Biol 6:545–554. https://doi.org/10.1021/acssynbio.6b00251

Ballard ZS, Joung H-A, Goncharov A et al (2020) Deep learning-enabled point-of-care sensing using multiplexed paper-based sensors. NPJ Digital Medicine 3:1–8. https://doi.org/10.1038/s41746-020-0274-y

Baron U, Bujard H (2000) Tet repressor-based system for regulated gene expression in eukaryotic cells: principles and advances. Methods Enzymol 327:401–421. https://doi.org/10.1016/s0076-6879(00)27292-3

Berg B, Cortazar B, Tseng D et al (2015) Cell phone-based hand-held microplate reader for point-of-care testing of enzyme-linked immunosorbent assays. ACS Nano 9:7857–7866. https://doi.org/10.1021/acsnano.5b03203

Billinton N, Barker MG, Michel CE et al (1998) Development of a green fluorescent protein reporter for a yeast genotoxicity biosensor. Biosens Bioelectron 13:831–838. https://doi.org/10.1016/s0956-5663(98)00049-9

Bittner M, Jarque S, Hilscherová K (2015) Polymer-immobilized ready-to-use recombinant yeast assays for the detection of endocrine disruptive compounds. Chemosphere 132:56–62. https://doi.org/10.1016/j.chemosphere.2015.02.063

Borse V, Patil AS, Srivastava R (2017) Development and testing of portable fluorescence reader (PorFloR^{TM}). In: 2017 9th international conference on communication systems and networks (COMSNETS). pp 498–501

Bui VN, Nguyen TTH, Bettarel Y et al (2015) Genotoxicity of chemical compounds identification and assessment by yeast cells transformed with GFP reporter constructs regulated by the PLM2 or DIN7 promoter. Int J Toxicol 34:31–43. https://doi.org/10.1177/1091581814566870

Burrill DR, Silver PA (2011) Synthetic circuit identifies subpopulations with sustained memory of DNA damage. Genes Dev 25:434–439. https://doi.org/10.1101/gad.1994911

Cevenini L, Lopreside A, Calabretta MM et al (2018) A novel bioluminescent NanoLuc yeast-estrogen screen biosensor (nanoYES) with a compact wireless camera for effect-based detection of endocrine-disrupting chemicals. Anal Bioanal Chem 410:1237–1246. https://doi.org/10.1007/s00216-017-0661-7

Chen B, Lee HL, Heng YC et al (2018) Synthetic biology toolkits and applications in *Saccharomyces cerevisiae*. Biotechnol Adv 36:1870–1881. https://doi.org/10.1016/j.biotechadv.2018.07.005

Conroy PJ, Hearty S, Leonard P, O'Kennedy RJ (2009) Antibody production, design and use for biosensor-based applications. Semin Cell Dev Biol 20:10–26. https://doi.org/10.1016/j.semcdb.2009.01.010

Dai B, Wang L, Wang Y et al (2018) Single-cell nanometric coating towards whole-cell-based biodevices and biosensors. ChemistrySelect 3:7208–7221. https://doi.org/10.1002/slct.201800963

Dameron CT, Winge DR, George GN et al (1991) A copper-thiolate polynuclear cluster in the ACE1 transcription factor. PNAS 88:6127–6131. https://doi.org/10.1073/pnas.88.14.6127

Emr SD, Schauer I, Hansen W et al (1984) Invertase beta-galactosidase hybrid proteins fail to be transported from the endoplasmic reticulum in *Saccharomyces cerevisiae*. Mol Cell Biol 4:2347–2355. https://doi.org/10.1128/MCB.4.11.2347

Fine T, Leskinen P, Isobe T et al (2006) Luminescent yeast cells entrapped in hydrogels for estrogenic endocrine disrupting chemical biodetection. Biosens Bioelectron 21:2263–2269. https://doi.org/10.1016/j.bios.2005.11.004

Fuentes N, Silveyra P (2019) Estrogen receptor signaling mechanisms. Adv Protein Chem Struct Biol 116:135–170. https://doi.org/10.1016/bs.apcsb.2019.01.001

Gao CY, Pinkham JL (2000) Tightly regulated, β-estradiol dose-dependent expression system for yeast. Biotechniques 29:1226–1231. https://doi.org/10.2144/00296st02

Gardner JM, Jaspersen SL (2014) Manipulating the yeast genome: deletion, mutation, and tagging by PCR. Methods Mol Biol 1205:45–78. https://doi.org/10.1007/978-1-4939-1363-3_5

Gnügge R, Rudolf F (2017) *Saccharomyces cerevisiae* shuttle vectors. Yeast 34:205–221. https://doi.org/10.1002/yea.3228

Gong L, Yang G, Yang B, Gu J (2020) Development of the yeast *Saccharomyces cerevisiae* as a biosensor for the toxicity detection of toxic substances. bioRxiv 07 Jan 2020. 898106. https://doi.org/10.1101/2020.01.07.898106

Grewal PS, Modavi C, Russ ZN et al (2018) Bioproduction of a betalain color palette in *Saccharomyces cerevisiae.* Metab Eng 45:180–188. https://doi.org/10.1016/j.ymben.2017.12.008

Gruber CJ, Gruber DM, Gruber IML et al (2004) Anatomy of the estrogen response element. Trends Endocrinol Metab 15:73–78. https://doi.org/10.1016/j.tem.2004.01.008

Hartner FS, Glieder A (2006) Regulation of methanol utilisation pathway genes in yeasts. Microb Cell Fact 5:39. https://doi.org/10.1186/1475-2859-5-39

Herkommerová K, Zemančíková J, Sychrová H, Antošová Z (2018) Immobilization in polyvinyl alcohol hydrogel enhances yeast storage stability and reusability of recombinant laccase-producing *S. cerevisiae.* Biotechnol Lett 40:405–411. https://doi.org/10.1007/s10529-017-2485-0

Hilger D, Masureel M, Kobilka BK (2018) Structure and dynamics of GPCR signaling complexes. Nat Struct Mol Biol 25:4–12. https://doi.org/10.1038/s41594-017-0011-7

Hollis RP, Killham K, Glover LA (2000) Design and application of a biosensor for monitoring toxicity of compounds to eukaryotes. Appl Environ Microbiol 66:1676–1679

Hori K, Sen A, Artavanis-Tsakonas S (2013) Notch signaling at a glance. J Cell Sci 126:2135–2140. https://doi.org/10.1242/jcs.127308

Hung C-W, Martínez-Márquez JY, Javed FT, Duncan MC (2018) A simple and inexpensive quantitative technique for determining chemical sensitivity in *Saccharomyces cerevisiae.* Sci Rep 8:11919. https://doi.org/10.1038/s41598-018-30305-z

Ito-Harashima S, Matano M, Onishi K et al (2020) Construction of reporter gene assays using CWP and PDR mutant yeasts for enhanced detection of various sex steroids. Genes Environ 42:20. https://doi.org/10.1186/s41021-020-00159-x

Jarque S, Bittner M, Hilscherová K (2016) Freeze-drying as suitable method to achieve ready-to-use yeast biosensors for androgenic and estrogenic compounds. Chemosphere 148:204–210. https://doi.org/10.1016/j.chemosphere.2016.01.038

Jarque S, Bittner M, Blaha L, Hilscherova K (2016) Yeast biosensors for detection of environmental pollutants: current state and limitations. Trends Biotechnol 34:408–419. https://doi.org/10.1016/j.tibtech.2016.01.007

Jeyaprakash A, Welch JW, Fogel S (1991) Multicopy CUP1 plasmids enhance cadmium and copper resistance levels in yeast. Mol Gen Genet 225:363–368. https://doi.org/10.1007/BF00261675

Klauser B, Atanasov J, Siewert LK, Hartig JS (2015) Ribozyme-based aminoglycoside switches of gene expression engineered by genetic selection in *S. cerevisiae.* ACS Synth Biol 4:516–525. https://doi.org/10.1021/sb500062p

Lehmann M, Riedel K, Adler K, Kunze G (2000) Amperometric measurement of copper ions with a deputy substrate using a novel *Saccharomyces cerevisiae* sensor. Biosens Bioelectron 15:211–219. https://doi.org/10.1016/S0956-5663(00)00060-9

Lengger B, Jensen MK (2020) Engineering G protein-coupled receptor signalling in yeast for biotechnological and medical purposes. FEMS Yeast Res 20. https://doi.org/10.1093/femsyr/foz087

Leskinen P, Virta M, Karp M (2003) One-step measurement of firefly luciferase activity in yeast. Yeast 20:1109–1113. https://doi.org/10.1002/yea.1024

Lei Y, Chen W, Mulchandani A (2006) Microbial biosensors. Analytica Chimica Acta 568(1):200–210. https://doi.org/10.1016/j.aca.2005.11.065

Louvion JF, Havaux-Copf B, Picard D (1993) Fusion of GAL4-VP16 to a steroid-binding domain provides a tool for gratuitous induction of galactose-responsive genes in yeast. Gene 131:129–134. https://doi.org/10.1016/0378-1119(93)90681-r

Lu Y, Tian Y, Wang R et al (2015) Dual fluorescent protein-based bioassay system for the detection of genotoxic chemical substances in *Saccharomyces cerevisiae.* Toxicol Mech Methods 25:698–707. https://doi.org/10.3109/15376516.2015.1070305

Malcı K, Walls LE, Rios-Solis L (2020) Multiplex genome engineering methods for yeast cell factory development. Front Bioeng Biotechnol 8:1264. https://doi.org/10.3389/fbioe.2020.589468
Martin-Yken H (2020) Yeast-based biosensors: current applications and new developments. Biosensors 10:51. https://doi.org/10.3390/bios10050051
Miller RA, Brown G, Barron E et al (2020a) Development of a paper-immobilized yeast biosensor for the detection of physiological concentrations of doxycycline in technology-limited settings. Anal Methods 12:2123–2132. https://doi.org/10.1039/D0AY00001A
Miller RA, Lee S, Fridmanski EJ et al (2020b) "Scentsor": a whole-cell yeast biosensor with an olfactory reporter for low-cost and equipment-free detection of pharmaceuticals. ACS Sensors 5:3025–3030. https://doi.org/10.1021/acssensors.0c01344
Möckli N, Auerbach D (2004) Quantitative β-galactosidase assay suitable for high-throughput applications in the yeast two-hybrid system. Biotechniques 36:872–876. https://doi.org/10.2144/04365PT03
Mukherjee K, Bhattacharyya S, Peralta-Yahya P (2015) GPCR-based chemical biosensors for medium-chain fatty acids. ACS Synth Biol 4:1261–1269. https://doi.org/10.1021/sb500365m
Nakamura Y, Hashimoto T, Ishii J, Kondo A (2016) Dual-color reporter switching system to discern dimer formations of G-protein-coupled receptors using Cre/loxP site-specific recombination in yeast. Biotechnol Bioeng 113:2178–2190. https://doi.org/10.1002/bit.25974
Nelson DL, Cox MM (2017) Lehninger principles of biochemistry. In: Freeman WH, 7th edn. New York, NY
Ong JY, Pence JT, Molik DC et al (2021) Yeast grown in continuous culture systems can detect mutagens with improved sensitivity relative to the Ames test. PLoS ONE 16:e0235303. https://doi.org/10.1371/journal.pone.0235303
Ostrov N, Jimenez M, Billerbeck S et al (2017) A modular yeast biosensor for low-cost point-of-care pathogen detection. Sci Adv 3:e1603221. https://doi.org/10.1126/sciadv.1603221
Paetkau DW, Riese JA, MacMorran WS et al (1994) Interaction of the yeast RAD7 and SIR3 proteins: implications for DNA repair and chromatin structure. Genes Dev 8:2035–2045. https://doi.org/10.1101/gad.8.17.2035
Peltomaa R, Benito-Peña E, Barderas R, Moreno-Bondi MC (2019) Phage display in the quest for new selective recognition elements for biosensors. ACS Omega 4:11569–11580. https://doi.org/10.1021/acsomega.9b01206
Ponamoreva ON, Kamanina OA, Alferov VA et al (2015) Yeast-based self-organized hybrid bio-silica sol–gels for the design of biosensors. Biosens Bioelectron 67:321–326. https://doi.org/10.1016/j.bios.2014.08.045
Redden H, Morse N, Alper HS (2015) The synthetic biology toolbox for tuning gene expression in yeast. FEMS Yeast Res 15:1–10. https://doi.org/10.1111/1567-1364.12188
Renneberg T, Kwan RCH, Chan C et al (2004) A salt-tolerant yeast-based microbial sensor for 24 hour community wastewater monitoring in coastal regions. Microchim Acta 148:235–240. https://doi.org/10.1007/s00604-004-0266-7
Roney IJ, Rudner AD, Couture J-F, Kærn M (2016) Improvement of the reverse tetracycline transactivator by single amino acid substitutions that reduce leaky target gene expression to undetectable levels. Sci Rep 6:27697. https://doi.org/10.1038/srep27697
Routledge EJ, Sumpter JP (1996) Estrogenic activity of surfactants and some of their degradation products assessed using a recombinant yeast screen. Environ Toxicol Chem 15:241–248. https://doi.org/10.1002/etc.5620150303
Routledge SJ, Mikaliunaite L, Patel A et al (2016) The synthesis of recombinant membrane proteins in yeast for structural studies. Methods 95:26–37. https://doi.org/10.1016/j.ymeth.2015.09.027
Schindler D (2020) Genetic engineering and synthetic genomics in yeast to understand life and boost biotechnology. Bioengineering 7:137. https://doi.org/10.3390/bioengineering7040137
Seo KS, Choo KH, Chang HN, Park JK (2009) A flow injection analysis system with encapsulated high-density *Saccharomyces cerevisiae* cells for rapid determination of biochemical oxygen demand. Appl Microbiol Biotechnol 83:217–223. https://doi.org/10.1007/s00253-008-1852-0

Shaw WM, Yamauchi H, Mead J et al (2019) Engineering a model cell for rational tuning of GPCR signaling. Cell 177:782-796.e27. https://doi.org/10.1016/j.cell.2019.02.023

Shepherd HAM, Trentman MT, Tank JL, et al (2021) Development of a yeast-based assay for bioavailable phosphorous. bioRxiv 28 Feb 2021. 433264. https://doi.org/10.1101/2021.02.28.433264

Smith DF, Toft DO (1993) Steroid receptors and their associated proteins. Mol Endocrinol 7:4–11. https://doi.org/10.1210/mend.7.1.8446107

Smith AD, Logeman BL, Thiele DJ (2017) Copper acquisition and utilization in fungi. Annu Rev Microbiol 71:597–623. https://doi.org/10.1146/annurev-micro-030117-020444

Stainbrook SC, Yu JS, Reddick MP et al (2017) Modulating and evaluating receptor promiscuity through directed evolution and modeling. Protein Eng Des Sel 30:455–465. https://doi.org/10.1093/protein/gzx018

Tamai KT, Liu X, Silar P et al (1994) Heat shock transcription factor activates yeast metallothionein gene expression in response to heat and glucose starvation via distinct signalling pathways. Mol Cell Biol 14:8155–8165

Tag K, Riedel K, Bauer H-J, Hanke G, Baronian KHR, Kunze G (2007) Amperometric detection of Cu2+ by yeast biosensors using flow injection analysis (FIA). Sensors and Actuators B: Chemical 122(2):403–409. https://doi.org/10.1016/j.snb.2006.06.007

Trentman MT, Tank JL, Shepherd HAM, et al (2021) Characterizing bioavailable phosphorus concentrations in an agricultural stream during hydrologic and streambed disturbances. Biogeochemistry. https://doi.org/10.1007/s10533-021-00803-w

Urlinger S, Baron U, Thellmann M et al (2000) Exploring the sequence space for tetracycline-dependent transcriptional activators: novel mutations yield expanded range and sensitivity. PNAS 97:7963–7968. https://doi.org/10.1073/pnas.130192197

Van Wyk N, Kroukamp H, Pretorius IS (2018) The smell of synthetic biology: engineering strategies for aroma compound production in yeast. Fermentation 4:54. https://doi.org/10.3390/fermentation4030054

Vashist SK, Mudanyali O, Schneider EM et al (2014) Cellphone-based devices for bioanalytical sciences. Anal Bioanal Chem 406:3263–3277. https://doi.org/10.1007/s00216-013-7473-1

Venkatesh AG, Sun A, Brickner H et al (2015) Yeast dual-affinity biobricks: progress towards renewable whole-cell biosensors. Biosens Bioelectron 70:462–468. https://doi.org/10.1016/j.bios.2015.03.044

Versele M, Lemaire K, Thevelein JM (2001) Sex and sugar in yeast: two distinct GPCR systems. EMBO Rep 2:574–579. https://doi.org/10.1093/embo-reports/kve132

Vieira Gomes AM, Souza Carmo T, Silva Carvalho L, et al (2018) Comparison of yeasts as hosts for recombinant protein production. Microorganisms 6:38. https://doi.org/10.3390/microorganisms6020038

Vopálenská I, Váchová L, Palková Z (2015) New biosensor for detection of copper ions in water based on immobilized genetically modified yeast cells. Biosens Bioelectron 72:160–167. https://doi.org/10.1016/j.bios.2015.05.006

Wang L-J, Naudé N, Demissie M et al (2018) Analytical validation of an ultra low-cost mobile phone microplate reader for infectious disease testing. Clin Chim Acta 482:21–26. https://doi.org/10.1016/j.cca.2018.03.013

Wang J, Yang D, Guo X et al (2020) A novel RNA aptamer-modified riboswitch as chemical sensor. Anal Chim Acta 1100:240–249. https://doi.org/10.1016/j.aca.2019.11.071

Weaver AA, Halweg S, Joyce M et al (2015) Incorporating yeast biosensors into paper-based analytical tools for pharmaceutical analysis. Anal Bioanal Chem 407:615–619. https://doi.org/10.1007/s00216-014-8280-z

Wei Q, Qi H, Luo W et al (2013) Fluorescent imaging of single nanoparticles and viruses on a smart phone. ACS Nano 7:9147–9155. https://doi.org/10.1021/nn4037706

Yoon HK, Jung ST, Kim J-H, Yoo TH (2012) Recent development of highly sensitive protease assay methods: signal amplification through enzyme cascades. Biotechnol Bioproc E 17:1113–1119. https://doi.org/10.1007/s12257-012-0545-9

Yu D, Liao L, Zhang J et al (2018) A novel, easy and rapid method for constructing yeast two-hybrid vectors using in-fusion technology. Biotechniques 64:219–224. https://doi.org/10.2144/btn-2018-0007

Yudina NY, Arlyapov VA, Chepurnova MA et al (2015) A yeast co-culture-based biosensor for determination of waste water contamination levels. Enzyme Microb Technol 78:46–53. https://doi.org/10.1016/j.enzmictec.2015.06.008

Zhang CC, Glenn KA, Kuntz MA, Shapiro DJ (2000) High level expression of full-length estrogen receptor in *Escherichia coli* is facilitated by the uncoupler of oxidative phosphorylation, CCCP. J Steroid Biochem Mol Biol 74:169–178. https://doi.org/10.1016/s0960-0760(00)00120-5

Zhang J, Jensen MK, Keasling JD (2015) Development of biosensors and their application in metabolic engineering. Curr Opin Chem Biol 28:1–8. https://doi.org/10.1016/j.cbpa.2015.05.013

Zhao H, Zhang Y, Pan M et al (2019) Dynamic imaging of cellular pH and redox homeostasis with a genetically encoded dual-functional biosensor, pHaROS, in yeast. J Biol Chem 294:15768–15780. https://doi.org/10.1074/jbc.RA119.007557

Zhou X, Vink M, Klaver B et al (2006) Optimization of the Tet-On system for regulated gene expression through viral evolution. Gene Ther 13:1382–1390. https://doi.org/10.1038/sj.gt.3302780

Applications of Yeast Synthetic Biology

Yeast Synthetic Biology Approaches for the Production of Valuable Polyphenolic Compounds

Daniela Gomes, João Rainha, Ligia R. Rodrigues, and Joana L. Rodrigues

Abstract

Polyphenols are secondary metabolites isolated from plants that are known for their biological and therapeutical properties. However, the extraction from plants renders low yields, besides being expensive and not environmentally friendly. Therefore, the use of heterologous microorganisms, so-called microbial chassis, became an interesting alternative approach to produce polyphenols. With the advances in the metabolic engineering and synthetic biology fields, the development of such microbial chassis able to produce these compounds with higher yields and productivities became easier. Several yeast species, such as *Saccharomyces cerevisiae*, *Yarrowia lipolytica*, and *Pichia pastoris*, have already been engineered to produce some polyphenols. However, there is still a long way to go before these compounds can be produced by heterologous organisms at an industrial scale. In this chapter, we review the recent advances in the production of polyphenolic compounds using yeasts as heterologous hosts, as well as several synthetic biology approaches used to improve such production.

1 Introduction

Polyphenolic compounds or polyphenols are valuable compounds that are naturally present in plants. These secondary metabolites have several recognized health benefits due to their antioxidant, anti-inflammatory, anticancer, antiviral, and wound healing properties (Hussain et al. 2016; Panda et al. 2017; Park 2015; Rodrigues et al. 2015c; Tsao 2010). However, polyphenols are present in very low amounts in plants. For this reason, their isolation from plants besides being difficult and expensive is not enough to support their increasing demand (Krivoruchko and Nielsen

D. Gomes · J. Rainha · L. R. Rodrigues · J. L. Rodrigues (✉)
Centre of Biological Engineering, University of Minho, 4710-057 Braga, Portugal
e-mail: joanarodrigues@ceb.uminho.pt

F. Darvishi Harzevili (ed.), *Synthetic Biology of Yeasts*,
https://doi.org/10.1007/978-3-030-89680-5_5

2015; Rodrigues et al. 2015c). The production of polyphenolic compounds using microorganisms as heterologous chassis could be an interesting alternative method to produce them in a fast, cheap, and more environmentally friendly way (Rainha et al. 2020a). To achieve the production of polyphenols in microorganisms, it is necessary to design and construct the biosynthetic pathways responsible for their production. This step is challenging since it requires the introduction of several heterologous genes into the chassis. Moreover, these genes should be efficiently expressed and produce functional enzymes (Braga et al. 2018, 2019). With the exploitation of metabolic engineering and synthetic biology tools, the construction of complete biosynthetic pathways in microorganisms became easier. These tools have also been used to improve the host chassis toward an increase in the production of the desired compounds. The heterologous production of plant polyphenolic compounds was mostly explored using *Saccharomyces cerevisiae* as yeast chassis (Rainha et al. 2020a). Nevertheless, in the last years, other yeasts such as *Pichia pastoris* and *Yarrowia lipolytica* have been also explored as alternatives to produce relevant polyphenolic compounds. In the following subchapters, the biological relevance and the biosynthetic pathways responsible for the production of different polyphenolic compounds are described. Moreover, several examples of heterologous production of plant polyphenols in yeasts will be explored.

2 Polyphenolic Compounds

Polyphenolic compounds are naturally produced in plants and interfere in several physiological processes, namely in plant growth and development, seed germination, and response to biotic and abiotic stresses (Sharma et al. 2019; Tanase et al. 2019). Polyphenolic compounds have been widely used in the pharmaceutical industry due to their relevant biological activities. Polyphenols have also been widely used in the food industry as food additives and in active food packaging to improve food preservation and stability (de Araújo et al. 2020; Martillanes et al. 2017). Moreover, these compounds have also other commercial applications, such as in the production of cosmetics, fertilizers, paints, surfactants, among others (de Araújo et al. 2020). Due to these potential applications, the global market size for polyphenols is expected to reach USD 2.26 billion by 2027 with a CAGR of 7.2% (Proficient Market 2020).

Phenolic compounds are composed of one aromatic ring containing one or more hydroxyl groups. These compounds are divided into two subclasses: hydroxybenzoic acids and hydroxycinnamic acids. Hydroxycinnamic acids are the precursors of polyphenolic compounds. These compounds are constituted by multiple phenol building blocks. Due to their large diversity, polyphenols are also organized in different subclasses (Fig. 1). The largest subclass of polyphenols is flavonoids. However, polyphenols can also be divided into stilbenoids, curcuminoids, lignans, polyphenolic amides, and coumarins (Singla et al. 2019).

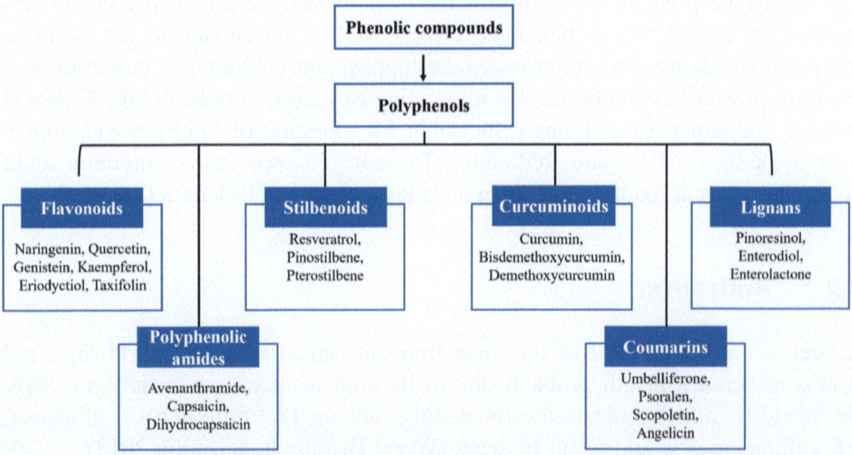

Fig. 1 Different classes and examples of polyphenolic compounds

3 Biological Activities

Polyphenolic compounds are widely distributed in nature, namely in plants, fruits, and vegetable species. Moreover, more than 8000 different structures are known (Tsao 2010). In the last years, polyphenols have been widely studied as biologically active compounds and have shown health benefits in the prevention and treatment of several diseases due to their recognized antioxidant, anticancer, antiviral, anti-inflammatory, and wound healing properties (Braga et al. 2018; Cassidy et al. 2020; Dayem et al. 2016; Rodrigues et al. 2015c; Sobhani et al. 2020; Van de Velde et al. 2019).

3.1 Antioxidant

Polyphenols are recognized as natural antioxidant molecules due to their ability to scavenge, inhibit, and neutralize reactive species. The antioxidant properties of these compounds are affected by the number of hydroxyl groups and their arrangement in the aromatic ring. For example, the flavonoids 3′-hydroxygenistein, 3′-hydroxyldaidzein, 6-hydroxyldaidzein, and 8-hydroxyldaidzein were found to be more potent antioxidants than quercetin and ascorbic acid due to the presence of one more hydroxyl group (Rüfer and Kulling 2006). Moreover, polyphenols are also able to activate antioxidant enzymes and to perform the inhibition of enzymes involved in the oxidation process (Hussain et al. 2016).

The application of several polyphenolic compounds in the treatment of diseases caused by oxidative stress has been widely studied (Huang et al. 2018; Panda et al. 2017; Rashmi et al. 2017; Sökmen and Akram Khan 2016). The protective effect of naringenin in Alzheimer's disease was studied. Ghofrani et al. (2015) showed that naringenin pretreatment improves the memory and learning capacities in Aβ-injected rat model. These beneficial effects were achieved due to the decrease in lipid peroxidation and apoptosis in the hippocampus. Moreover, curcumin was found to protect biomembranes by scavenging oxygen free radicals like hydroxyl radicals and superoxide anions responsible for initiation of lipid peroxidation in liver and kidneys of diabetic db/db mice. These results showed that curcumin could be used to prevent oxidative stress in diabetes patients (Soto-Urquieta et al. 2014).

3.2 Anticancer

Cancer is considered one of the most frequent causes of death worldwide, and it is considered a health problem due to its high incidence in society. In 2018, the World Health Organization estimated that among 18.1 million cases of cancer, 9.6 million cases would result in death (World Health Organization 2018).

To reduce cancer mortality, the discovery of new therapeutic compounds became a priority. Polyphenols have been reported as potential therapeutic compounds for several types of cancer. These compounds can interfere in cell proliferation, tumor initiation and proliferation, metastasis, angiogenesis, and apoptosis (Niedzwiecki et al. 2016). These effects are attributed to the ability of polyphenols to interact with transcription factors, signaling molecules, growth and apoptotic regulators, and adhesion molecules (Avtanski and Poretsky 2018; Khan et al. 2020; Park 2015). For example, it was shown that curcumin could decrease the Warburg effect in different cancer cell lines. The Warburg effect is characterized by the high productions of lactate and high glucose uptake by tumor cells even when oxygen is available. The reduction of the Warburg effect was achieved due to the ability of curcumin to downregulate pyruvate kinase M2 expression by regulating mammalian target of rapamycin (mTOR)/ hypoxia-inducible factor 1–α (HIF1α) pathway. Additionally, it was also demonstrated that the viability of cancer cells was reduced after curcumin treatment (Siddiqui et al. 2018). Eriodyctiol also showed anticancer effect against human lung cancer cell line A549. This effect was characterized by downregulating the expression of the apoptotic protein Bcl-2 and upregulating the expression of Bax protein. Consequently, apoptosis was induced. Additionally, phosphatidylinositol 3 kinase (PI3K)/protein kinase B (AKT)/mTOR signaling pathway was also inhibited. All these modifications in the protein expression have contributed to the inhibition of the cancer cell's growth (Zhang et al. 2020). Another polyphenolic compound that has been widely studied as an anticancer agent is quercetin. This flavonoid showed activity against several types of cancer cell lines. For example, Hashemzaei et al. (2017) have studied the effect of quercetin treatment in nine different tumor cell lines. In all the tested tumor cell lines, it was concluded that quercetin induces apoptosis. Additionally, the effect of quercetin in vivo was also assessed in mice bearing estrogen receptor-positive breast cancer MCF-7 and colon carcinoma CT-26 tumors. After the treatment with quercetin, a significant reduction in the tumor volume was found. The authors concluded that quercetin has the potential to be used as an anticancer agent due to its in vitro and in vivo effects. The anticancer properties of resveratrol have been also studied. For example, Zeng et al. (2017) studied the effect of resveratrol in human colon cancer cells. The authors have found that resveratrol inhibits the proliferation and induces the apoptosis of these cells. Moreover, it was also reported that resveratrol upregulates the activity of bone morphogenetic protein 7 (BMP7). BMP7 is considered an effective tumor suppressor that interacts with the PI3K/AKT signaling pathway. This signaling pathway was inactivated by resveratrol, and its inactivation was potentiated by BMP7, thus suggesting that resveratrol could be applied in the treatment of colon cancer. In addition to these studies, there are many more that demonstrate the ability of polyphenols to act on several cancer types (Abbaszadeh et al. 2019). For this reason, these compounds have a lot of potentials to be used as anticancer drugs.

3.3 Antiviral

Beyond the antioxidant and anticancer activities, polyphenols have also been reported as natural compounds with antiviral properties due to the presence of hydroxyl and ester groups (Kamboj et al. 2012). Several studies have reported the antiviral activity of polyphenols against a large variety of virus, such as herpes simplex virus (HSV), dengue virus (DENV), zika virus, human immunodeficiency virus (HIV), severe acute respiratory syndrome coronavirus-2 (SARS-CoV-2), among others (Ali and Banerjea 2016; Chen et al. 2012; Jasso-Miranda et al. 2019; Lee et al. 2017; Paraiso et al. 2020; Vázquez-Calvo et al. 2017). For example, the in vitro effect of a mixture of polyphenols extracted from almond skin in HSV type I was tested. The treatment with 0.4 mg/mL of the mixture containing catechin, naringenin-7-O-glucoside, kaempferol-3-O-glucoside, epicatechin, isorhamnetin-3-O-rutinoside, and isorhamnetin-3-O-glucoside led to a reduction in the expression of viral proteins ICP0, UL42, and Us11. Moreover, the viral DNA accumulation was lower compared to the control (infected cells without any treatment). The results showed that the mixture of polyphenols tested is efficient in the treatment of HSV type I infected cells (Musarra-Pizzo et al. 2019). Quercetin has also shown activity against HSV type I. In this study, Raw 264.7 cells were infected with HSV type I in the presence of quercetin at different concentrations. Quercetin reduced the expression of HSV-1 glycoprotein D that is essential for viral entry into the host cells and ICP0 that is expressed in the HSV-1 replication cycle. The effect on the viral proteins responsible for the virus replication was also studied. The levels of expression of ICP0, UL13, and UL52 were significantly reduced in cells treated with quercetin. With these results, it was possible to conclude that quercetin affects the virus entrance in the host cells and its further replication (Lee et al. 2017). Another compound with reported antiviral properties is curcumin. For example, this compound has shown activity against HIV by the degradation of the Tat protein that is responsible for viral replication. Moreover, the treatment of infected cells with this compound resulted in significant inhibition of virus production (Ali and Banerjea 2016). Polyphenols, like fisetin and quercetin, were also studied as efficient antiviral compounds in DENV infection. Fisetin and quercetin were found to inhibit DENV-2 and DENV-3 infections. Both compounds exhibited an effect in the inflammatory response by decreasing the production and secretion of tumor necrosis factor-alpha (TNF-α) that is related to severe DENV infections (Jasso-Miranda et al. 2019). Recently, polyphenols have also shown potential activity against SARS-CoV2 (Paraiso et al. 2020). It was found that angiotensin-converting enzyme-2 (ACE2) cellular receptor is downregulated during SARS-Cov2 infection, and this downregulation could be the reason for the pathogenesis and progression of SARS-CoV2. Horne and Vohl (2020) have suggested that resveratrol has the potential to control the pathogenesis of SARS-Cov2 by upregulating the ACE2 cellular receptor.

3.4 Anti-Inflammatory and Wound Healing Agents

Inflammation is a natural response of the body to the presence of pathogens or non-infectious factors like irradiation, toxic compounds, damaged cells, and among others. This inflammatory response is characterized by the production and secretion of pro-inflammatory cytokines and microbial products. During the inflammatory response, inflammatory cytokines, like TNF-α, interleukin (IL) 1β and 6, are synthesized and secreted and their interaction with different receptors leads to the activation of signaling pathways (Chen et al. 2018).

Polyphenols have been reported as natural compounds with the ability to modulate inflammatory responses. These compounds can interfere in the regulation of pro-inflammatory gene expression and the synthesis of pro-inflammatory cytokines (Yahfoufi et al. 2018). For example, genistein showed an ability to suppress the skin inflammatory response induced by psoriasis. This flavonoid was able to decrease the production of inflammatory cytokines, such as TNF-α, IL-1β, IL-6, IL-17, and IL-23, in skin mouse and human keratinocyte HaCaT cells with induced psoriasis-like inflammation. Moreover, genistein also inhibited STAT3 and nuclear factor kappa-β (NF-kB) phosphorylation. Consequently, it was concluded that genistein reduced the inflammatory response in psoriasis-like models (Wang et al. 2019). Naringenin has also shown anti-inflammatory activity in murine models of induced liver injury. After the treatment with naringenin, the authors have found a decrease in the production of TNF-α and IL-6 in the macrophages and T-cells. Moreover, the NF-kB activation was reduced. For this reason, it was concluded that naringenin could be useful to attenuate the inflammatory response in acute inflammatory diseases (Jin et al. 2017). Scopoletin has also shown a capacity to decrease the inflammatory response in cerulein-induced acute pancreatitis (AP) and associated lung injury in mice. This coumarin was found to downregulate NF-kB as well as TNF-α, IL-1β, IL33, and monocyte chemoattractant protein 1 (MCP-1). The downregulation of the NF-kB pathway and the production of inflammatory cytokines had a protective effect against inflammation (Leema and Tamizhselvi 2018).

Polyphenols have also wound healing properties. The wound healing process involves three different phases that are the inflammatory phase, proliferative phase, and remodeling phase (Ibrahim et al. 2018). For example, curcumin has been found to accelerate the wound closure in mice with excisional wound healing. This effect is related to a reduction in the expression of TNF-α and matrix metallopeptidase 9, which are up-regulated in the presence of wounds, by inhibiting the NF-kB signaling pathway. Moreover, this compound was also able to increase cell proliferation and collagen synthesis (Yen et al. 2018). Quercetin has shown positive effects on the wound healing process in pressure ulcer lesions. This compound was also found to accelerate wound closure. Moreover, the production of TNF-α and IL-1β was also significantly reduced through suppression of the MAPK signaling pathway (Yin et al. 2018).

4 Biosynthetic Pathway

The biosynthetic pathway responsible for polyphenols production in plants is complex given that several different enzymes are involved. The polyphenols biosynthesis includes the shikimate pathway and phenylpropanoid pathway.

Shikimate pathway is responsible for the production of chorismate through successive enzymatic reactions (Fig. 2). This pathway uses two molecules derived from the primary metabolism as starter substrates: Erythrose-4-phosphate (E4P) and phosphoenol pyruvate (PEP) (Zabalza et al. 2017). These molecules are condensed to produce 3-deoxy-D-arabino-heptulosonic acid 7-phosphate (DAHP) through the action of the DAHP synthase (DAHPS) (Munack et al. 2016). DAHP is converted into 3-dehydroquinate (DHQ) by DHQ synthase (DHQS). Then, DHQ is converted into shikimate through the successive action of the bifunctional enzyme DHQ dehydratase/shikimate dehydrogenase (DHQD/SDH) (Ding et al. 2007). In the next step, shikimate is phosphorylated by shikimate kinase (SK) originating shikimate 3-phosphate (S3P). Through the action of 5-enolpyruvylshikimate 3-phosphate (EPSP) synthase (EPSPS), S3P, and other molecules of PEP are condensed and one molecule of EPSP is produced. Afterward, EPSP is dephosphorylated by chorismate synthase (CS) to produce chorismate (Marchiosi et al. 2020). Chorismate is the main precursor of the aromatic amino acids *L*-phenylalanine, *L*-tyrosine, and *L*-tryptophan. Polyphenols are only derived from *L*-phenylalanine and *L*-tyrosine, and these two amino acids can be obtained from chorismate by two alternative routes. The first step is common in both routes and consists of the production of prephenate from chorismate by chorismate mutase (CM) (Parthasarathy et al. 2018). To produce *L*-phenylalanine, prephenate could be converted into phenylpyruvate or arogenate by prephenate dehydratase (PDT) or prephenate aminotransferase (PAT), respectively. Then, phenylpyruvate and arogenate are converted into *L*-phenylalanine by the action of aromatic amino acid aminotransferase (AAT) or arogenate dehydratase (ADT), respectively. Prephenate can also be converted into *p*-hydroxyphenylpyruvate by the action of prephenate dehydrogenase (PDH). Afterward, *L*-tyrosine could be produced from *p*-hydroxyphenylpyruvate

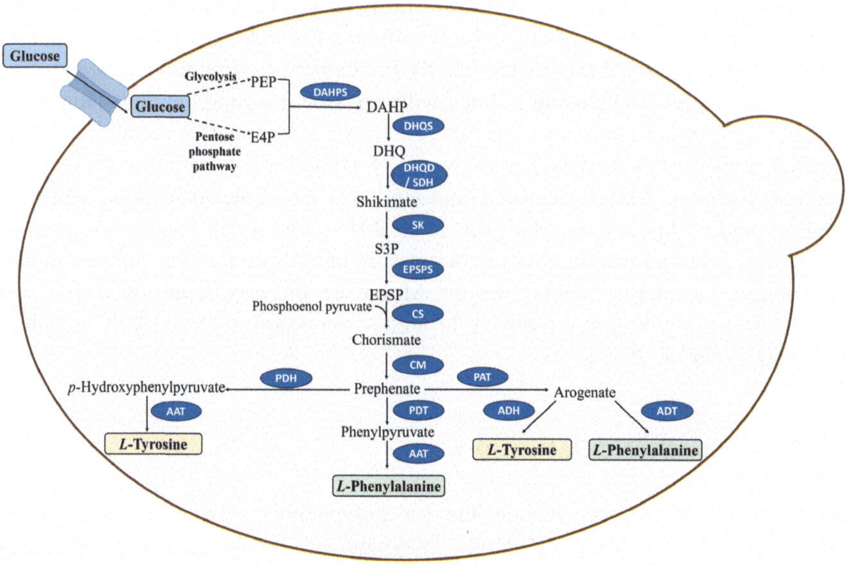

Fig. 2 Schematic representation of the shikimate pathway involved in the synthesis of the aromatic amino acids *L*-tyrosine and *L*-phenylalanine. AAT, aromatic amino acid aminotransferase; ADH, arogenate dehydrogenase; ADT, arogenate dehydratase; CM, chorismate mutase; CS, chorismate synthase; DAHP, 3-deoxy-D-arabino-heptulosonic acid 7-phosphate; DAHPS, 3-deoxy-D-arabino-heptulosonic acid 7-phosphate synthase; DHQ, 3-dehydroquinate; DHQD, 3-dehydroquinate dehydratase; DHQS, 3-dehydroquinate synthase; E4P, erythrose-4-phosphate; EPSP, 5-enolpyruvylshikimate 3-phosphate; EPSPS, 5-enolpyruvylshikimate 3-phosphate synthase; PAT, prephenate aminotransferase; PDH, prephenate dehydrogenase; PDT, prephenate dehydratase; PEP, phosphoenol pyruvate; S3P, shikimate 3-phosphate; SDH, shikimate dehydrogenase; SK, shikimate kinase

and arogenate by the action of aromatic amino acid aminotransferase (AAT) and arogenate dehydrogenase (ADH), respectively (Tzin and Galili 2010).

L-tyrosine and *L*-phenylalanine are converted into the hydroxycinnamic acids through the phenylpropanoid pathway (Fig. 3). When *L*-phenylalanine is used as a substrate, the first enzyme of this biosynthetic pathway is phenylalanine ammonia-lyase (PAL). This enzyme performs the deamination of *L*-phenylalanine to produce cinnamic acid. Afterward, cinnamic acid is hydroxylated by cinnamic acid 4-hydroxylase (C4H). In this step, *p*-coumaric acid is produced (Fraser and Chapple 2011). Alternatively, *p*-coumaric acid could be produced in a single enzymatic step from *L*-tyrosine. This alternative route uses tyrosine ammonia lyase (TAL) to perform the deamination of *L*-tyrosine (Watts et al. 2006).

Subsequently, *p*-coumaric acid could be used to produce other hydroxycinnamic acids. By the action of 4-coumarate 3-hydroxylase (C3H), *p*-coumaric acid is converted into caffeic acid. Moreover, caffeic acid could be converted into ferulic acid by caffeic acid 3-*O*-methyltransferase (COMT). These hydroxycinnamic acids could be activated to the correspondent coenzyme A (CoA) esters by

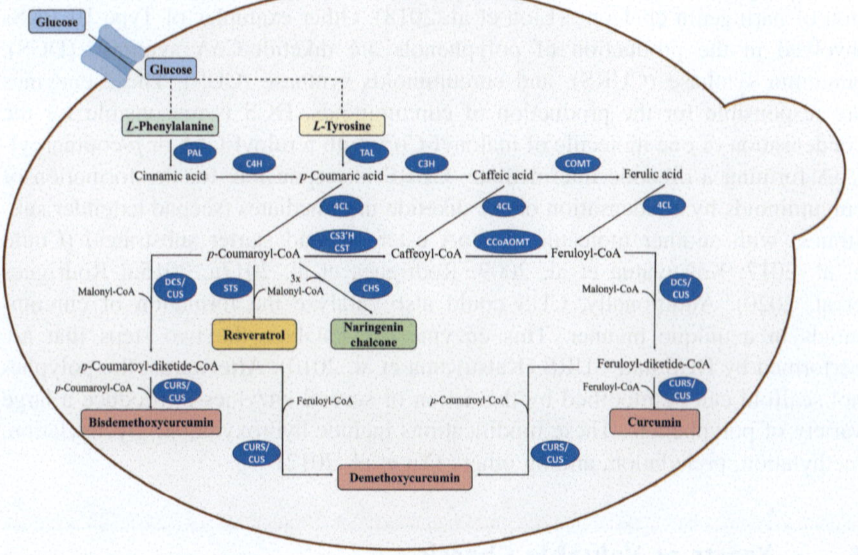

Fig. 3 Schematic representation of the biosynthetic pathway responsible for the production of polyphenols (curcuminoids, resveratrol, and naringenin chalcone) using aromatic amino acids as substrates. The enzymes involved in the phenylpropanoid pathway are phenylalanine ammonia-lyase (PAL), tyrosine ammonia-lyase (TAL), cinnamic acid 4-hydroxylase (C4H), 4-coumarate 3-hydroxylase (C3H), caffeic acid 3-*O*-methyltransferase (COMT), and 4-coumarate-CoA ligase (4CL). *p*-coumaroyl 5-O-shikimate 3'-hydroxylase (CS3'H), *p*-coumaroyl shikimate transferase (CST), and caffeoyl-CoA 3-*O* methyltransferase (CCoAOMT) catalyze the conversion of *p*-coumaric acid into other hydroxycinnamic acids. The Type III polyketide synthase (PKS) enzymes responsible for curcuminoids, stilbenoids, and flavonoids scaffold production, which are underlined, are diketide-CoA synthase (DCS), curcumin synthase (CURS), curcuminoids synthase (CUS), chalcone synthase (CHS), and stilbene synthase (STS)

4-coumarate-CoA ligase (4CL). Moreover, *p*-coumaroyl-CoA could also be converted into caffeoyl-CoA by *p*-coumaroyl 5-*O*-shikimate 3'-hydroxylase (CS3'H) or *p*-coumaroyl shikimate transferase (CST). Subsequently, caffeoyl-CoA could be converted into feruloyl-CoA by caffeoyl-CoA 3-*O* methyltransferase (CCoAOMT). After the CoA esters biosynthesis, the pathway deviates toward the formation of the different polyphenols. The enzymes involved in these steps are different, depending on the type of polyphenol that is produced. Generally, these steps include the action of Type III polyketide synthase (PKS) enzymes (Fig. 3). Type III PKSs are involved in the condensation of the CoA esters (starter units) with malonyl-CoA (extender unit) leading to the polyphenol scaffold (Yu et al. 2012). For example, stilbene synthase (STS) is responsible for the condensation of three molecules of malonyl-CoA with *p*-coumaroyl-CoA. By the action of this enzyme, it is obtained resveratrol which is considered the stilbene backbone (Parage et al. 2012). Chalcone synthase (CHS) is the Type III PKS involved in the formation of the flavonoid scaffold. This enzyme also catalyzes the condensation of three molecules of malonyl-CoA with *p*-coumaroyl-CoA leading to the formation of naringenin chalcone (Liou et al. 2018). Other examples of Type III PKSs involved in the production of polyphenols are diketide-CoA synthase (DCS), curcumin synthase (CURS), and curcuminoids synthase (CUS). These enzymes are responsible for the production of curcuminoids. DCS is responsible for the condensation of one molecule of malonyl-CoA with feruloyl-CoA or *p*-coumaroyl-CoA forming a diketide intermediate. CURS is responsible for the formation of curcuminoids by condensation of the diketide intermediates (second extender substrates) with another molecule of CoA ester (second starter substrates) (Couto et al. 2017; Katsuyama et al. 2009; Rodrigues et al. 2017a, 2015a; Rodrigues et al. 2020). Additionally, CUS could also catalyze the formation of curcuminoids in a unique manner. This enzyme can catalyze the two steps that are performed by DCS and CURS (Katsuyama et al. 2010). Afterwards, the polyphenol scaffold can be modified by the action of several enzymes to produce a large variety of polyphenols. These modifications include hydroxylation, glycosylation, methylation, prenylation, among others (Yu et al. 2012).

5 Yeasts as Valuable Chassis

Polyphenols, like other plant secondary metabolites, are produced and accumulated in very low amounts in plants. Due to the low amounts produced in plants, the extraction of polyphenols from their natural resources is very difficult and the yields are considered low (Krivoruchko and Nielsen 2015; Rainha et al. 2020a; Rodrigues et al. 2015c; Rodrigues and Rodrigues 2020). Moreover, the amounts of these compounds are affected by climatic and seasonal variations. The chemical synthesis could be an alternative to produce some polyphenols. However, some of these compounds have complex structures being hard to chemically synthesize. Additionally, chemical synthesis involves the use of toxic compounds and expensive substrates. For this reason, the process is considered non-environmentally friendly and expensive (Liu et al. 2017). Due to their potential applications and to try to satisfy their industrial demand, there has been an increased interest in the heterologous production of polyphenols.

The heterologous production in plants has the advantage of only requiring the introduction of one or two genes of the biosynthetic pathway (Rodrigues et al. 2015c). The other genes from the pathway, such as the ones responsible for the phenylpropanoid pathway, are already present in the plant. However, heterologous production in plants has also disadvantages. One of these disadvantages is the long period that the plant takes to grow. The growth is also dependent on the climacteric conditions and fertile land occupation. Additionally, the production yields are usually very low and variable due to cellular heterogeneity. Beyond this, the downstream purification process is more difficult since several similar compounds are also produced in plants (Ebrahimi and Mokhtari 2017; Wilson and Roberts 2012).

The use of microorganisms as chassis has emerged as an alternative solution to heterologously produce several compounds of interest. Heterologous production has several advantages comparing to plant and chemical synthesis. Microorganisms have rapid growth cycles and, consequently, it allows short production times compared to plants. Moreover, they can grow in inexpensive substrates. For this reason, heterologous production is considered a more efficient and inexpensive method (Krivoruchko and Nielsen 2015). As with plants, microorganisms have also the capacity to produce the required precursors involved in the synthesis of plant natural compounds, namely aromatic amino acids and malonyl-CoA (Milke et al. 2018). Nevertheless, the reconstruction of large biosynthetic pathways in microorganisms is challenging since it requires the introduction of several genes of the pathway in the chassis, their efficient expression, and the production of functional enzymes (Braga et al. 2019).

The heterologous production became easier with all the advances achieved in the metabolic engineering and synthetic biology fields. Within bacteria, *Escherichia coli* has been the most used to produce polyphenolic compounds. In the case of yeasts, *S. cerevisiae* has been the most widely explored. These microorganisms have been widely used since they are easy to grow and manipulate and well-characterized. Moreover, there are available several genetic tools to genetically modify them (Krivoruchko and Nielsen 2015; Rainha et al. 2020b; Rodrigues et al. 2017b; Rodrigues and Rodrigues 2017). Comparing to *E. coli*, *S. cerevisiae* has several advantages to produce plant compounds. *S. cerevisiae*, unlike *E. coli*, can perform post-translational modifications, and it contains intracellular compartments resembling the ones present in plant cells such as the endomembrane system. This characteristic allows the functional expression of genes derived from plants, namely the eukaryotic cytochrome P450 enzymes that are involved in the synthesis of polyphenolic compounds (C4H and C3H). Moreover, this microorganism is generally recognized as safe (GRAS) facilitating its use in the production of nutritional and pharmaceutical compounds (Xu et al. 2020). However, other yeasts like *P. pastoris* and *Y. lipolytica* have been also explored for the production of polyphenols. *P. pastoris* has also several characteristics that make it a good host for heterologous production of relevant compounds. Like *S. cerevisiae*, *P. pastoris* is also easy to grow in a simple and low-cost medium. Moreover, it is known that this microorganism can produce proteins with high yields, perform the appropriate folding, and secret them to the extracellular medium. *P. pastoris* is also able to perform post-translational modifications (Karbalaei et al. 2020). *Y. lipolytica* is also considered a good chassis for the heterologous production of polyphenols. This oleaginous yeast has a high metabolic flux of the tricarboxylic acid (TCA) cycle being able to produce high amounts of acetyl-CoA and malonyl-CoA that are essential precursors in the synthesis of polyphenols. *Y. lipolytica* is also able to grow on inexpensive substrates like renewable feedstocks and wastes making the industrial production process more economic and environmentally friendly. This microorganism also contains intracellular compartments facilitating the expression of plant-derived genes and it is considered food-grade safe (Gu et al. 2020; Xu et al. 2020). Due to these unique features, *S. cerevisiae*, *Y. lipolytica*, and *P. pastoris* were exploited as a microbial chassis for the heterologous production of several compounds, including polyphenols.

6 Case Studies: Production of Hydroxycinnamic Acids and Polyphenolic Compounds in Yeasts

To achieve the industrial production of polyphenols in yeasts, it is essential to optimize several steps such as improve the precursor and extender substrates availability, improve the expression of the enzymes involved in the biosynthetic pathway, decrease or inhibit the expression of enzymes involved in competing pathways, among others (Chen et al. 2020). The emergence of synthetic biology and metabolic engineering tools made this process easier. Within the subclasses of polyphenolic compounds, the production of flavonoids and stilbenoids using yeasts as heterologous hosts has been the most explored. Additionally, there are some reports of heterologous production of coumarins, curcuminoids, and polyphenolic amides. The production of lignans was the only one that was never reported in any yeast strain. The highest production titers of hydroxycinnamic acids and polyphenols using *S. cerevisiae*, *Y. lipolytica*, and *P. pastoris* as heterologous hosts are summarized in Table 1.

6.1 Heterologous Production of Hydroxycinnamic Acids

Hydroxycinnamic acids belong to the group of the phenolic compounds. These compounds could be found in several vegetables, fruits, and beverages such as coffee and tea (Abramovič 2015). *p*-Coumaric acid, caffeic acid, and ferulic acid are the hydroxycinnamic acids involved in the synthesis of polyphenols, and they have a C6-C3 structure (Fig. 4). These compounds have also associated strong antioxidant properties as well as anti-inflammatory, anticancer and antimicrobial properties, among others (Rodrigues et al. 2015a; Taofiq et al. 2017).

The development of yeast chassis with the ability to produce industrial significant amounts of polyphenolic compounds became a priority. The production of significant amounts of hydroxycinnamic acids, especially *p*-coumaric acid, is important since these compounds are the main precursors of polyphenols. In the last years, several attempts were performed to construct engineered strains with the ability to produce relevant amounts of hydroxycinnamic acids. *S. cerevisiae* and *Y. lipolytica* were explored to produce hydroxycinnamic acids, especially *p*-coumaric acid since this compound is the main precursor of all the subclasses of polyphenols. Until now, to the best of our knowledge, there are no reports of hydroxycinnamic acid production in *P. pastoris*.

In *S. cerevisiae*, the construction of one strain able to produce *p*-coumaric acid from glucose was reported. Several metabolic engineering approaches were explored to prevent the pathway deviation and to improve the flux toward amino

Table 1 Highest values reported of heterologous production of hydroxycinnamic acids and polyphenols in *Saccharomyces cerevisiae*, *Yarrowia lipolytica*, and *Pichia pastoris*. The strain, genes, modifications in the host chassis, substrate, and production titers obtained are presented

Class	Compound	Strain	Genes[a]	Modifications in the chassis	Substrate	Production (mg/L)	References
Hydroxycinnamic acids	*p*-Coumaric acid	*S. cerevisiae* IMX	*At*PAL and *At*C4H	Construction of aromatic amino acids overproducing strain	Glucose (20 g/L)	12,500	Liu et al. (2019a)
	Caffeic acid	*S. cerevisiae* BY4741	*Rt*TAL, *Pa*HpaB, and *Se*HpaC	–	Tyrosine (0.5 g/L)	289.4	Liu et al. (2019b)
	p-Coumaric acid	*Y. lipolytica* W29	*Rt*TAL	Construction of aromatic amino acids overproducing strain	Glucose (40 g/L)	595.3	Gu et al. (2020)
	Ferulic acid	*Y. lipolytica* P01g	*Tf*AXE	–	Corncob biomass (2%)[b]	77	Huang et al. (2011)

(continued)

Table 1 (continued)

Class	Compound	Strain	Genes[a]	Modifications in the chassis	Substrate	Production (mg/L)	References
Flavonoids	Naringenin	*S. cerevisiae* BY4741	*Fj*TAL, *At*4CL, *Ha*CHS, and *Ph*CHI	Construction of aromatic amino acids overproducing strain	Glucose (20 g/L)	220	Lyu et al. (2019)
	Kaempferol	*S. cerevisiae* BY4741	*Fj*TAL, *At*4CL, *At*CHS, *At*CHI, *Nt*F3H, and *At*FLS	Construction of aromatic amino acids and malonyl-CoA overproducing strain	Glucose (20 g/L)	168.1	Du et al. (2020)
	Quercetin		*Fj*TAL, *At*4CL, *At*CHS, *At*CHI, *Nt*F3H, *Ph*F3'H, and *At*FLS			154.2	
	Myrcetin		*Fj*TAL, *At*4CL, *At*CHS, *At*CHI, *Nt*F3H, *Ph*F3'H, *Sl*F3'5'H, and *At*FLS			145	
	Delphinidin		*Fj*TAL, *At*4CL, *At*CHS, *At*CHI, *Nt*F3H, *Ph*F3'H, *Sl*F3'5'H, *Aa*DFR, and *Ge*ANS			26.1	
	Pelargonidin		*Fj*TAL, *At*4CL, *At*CHS, *At*CHI, *Nt*F3H, *Aa*DFR, and *Ge*ANS			33.3	
	Cyanidin		*Fj*TAL, *At*4CL, *At*CHS, *At*CHI, *Nt*F3H, *Ph*F3'H, *Aa*DFR, and *Ge*ANS			31.7	

(continued)

Table 1 (continued)

Class	Compound	Strain	Genes[a]	Modifications in the chassis	Substrate	Production (mg/L)	References
Flavonoids	Naringenin	*Y. lipolytica* Po1f	*Se*TAL, *Nt*4CL, and *Hs*CHS	Construction of aromatic amino acids and malonyl-CoA overproducing strain	Glucose (80 g/L)	898	Palmer et al. (2020)
	Eriodictyol	*Y. lipolytica* Po1f	*Rt*TAL, *Pc*4CL, *Ph*CHS, *Ms*CHI, *Gh*F3'H, and *Cr*CPR	Construction of a strain with improved chorismate and malonyl-CoA biosynthesis	Glucose (40 g/L)	134.2	Lv et al. (2019b)
	Taxifolin		*Rt*TAL, *Pc*4CL, *Ph*CHS, *Ms*CHI, *Gh*F3'H, *Cr*CPR, and *Sl*F3H			110.5	
	Liquiritigenin	*Y. lipolytica* (ATCC201249)	*Zm*PAL, *Pc*C4H, *Pc*4CL, *Ph*CHS, *Ms*CHR, and *Ms*CHI	–	Glucose (20 g/L) + *p*-Coumaric acid (100 mg/L)	62.4	Akram et al. (2020)
	8-hydroxydaidzein	*P. pastoris* X-33	*Ao*CYP57B3 and *Bm*BM3R	–	Daidzein (25.4 mg/L)	0.58	Chang et al. (2013)
	3′-hydroxydaidzein					0.23	
	6-hydroxydaidzein					9.1	
	3'-hydroxygenistein	*P. pastoris* X-33	*Ao*CYP57B3 and *Sc*CPR	Strain modified by treatment with hydrogen peroxide to induce oxidative stress and, consequently, select the most resistant strains	Genistein (135 mg/L)	20.3	Wang et al. (2016)

(continued)

Table 1 (continued)

Class	Compound	Strain	Genes[a]	Modifications in the chassis	Substrate	Production (mg/L)	References
Stilbenoids	Resveratrol	*S. cerevisiae* CEN.PK102-5B	*At*PAL, *At*C4H, *At*4CL, and *Vv*VST	Construction of aromatic amino acids and malonyl-CoA overproducing strain	Glucose (88 g/L)	800	Li et al. (2016)
	Pterostilbene		*At*PAL, *At*C4H, *At*4CL, *Vv*VST, and *Vv*ROMT			34.9	
	Pinostilbene		*At*PAL, *At*C4H, *At*4CL, *Vv*VST, and *Sb*ROMT			5.52	
	Resveratrol	*Y. lipolytica* W29	*Fj*TAL, *At*4CL, and *Vv*STS	Construction of aromatic amino acids and malonyl-CoA overproducing strain	Glucose (40 g/L)	12,400	Sáez-Sáez et al. (2020)

(continued)

Table 1 (continued)

Class	Compound	Strain	Genes[a]	Modifications in the chassis	Substrate	Production (mg/L)	References
Coumarins	Scopoletin	*S. cerevisiae* BY4742	*Pc*4CL, *At*F6′H1, *Pa*HpaB, and *En*HpaC	–	Lignin hydrolysate containing *p*-coumaric acid (2.3 g/L) and ferulic acid (0.5 g/L)	4.79	Zhao et al. (2020)
Curcuminoids	Bisdemethoxycurcumin	*Y. lipolytica* Po1f	*Nt*4CL and *Os*CUS	Construction of aromatic amino acids and malonyl-CoA overproducing strain	*p*-Coumaric acid (0.33 g/L)	0.17	Palmer et al. (2020)
Polyphenolic amides	*p*-Coumaroyl-3-hydroxyanthranilic acid	*S. cerevisiae* CENPK113-5d	*Nt*4CL and *cc*HCT	–	*p*-Coumaric acid (462 mg/L) and 3-Hydroxyanthranilic acid (77 mg/L)	120	Moglia et al. (2015)
	Caffeoyl-3-hydroxyanthranilic acid				Caffeic acid (540 mg/L) and 3-Hydroxyanthranilic acid (77 mg/L)	22	

[a] *Aa*—*Anthurium andraeanum, At*—*Arabidopsis thaliana, Ao*—*Aspergillus oryzae, Bm*—*Bacillus megaterium, Cc*—*Cynara cardunculus, Cr*—*Catharanthus roseus, En*—*Enterobacteriaceae, Fj*—*Flavobacterium johnsoniaeu, Ge*—*Gerbera* specie, *Gh*—*Gerbera hybrid, Ha*—*Hypericum androsaemum, Hs*—*Huperiza serrata, Ms*—*Mendicago sativa, Nt*—*Nicotiana tabacum, Os*—*Oryza sativa, Pa*—*Pseudomonas aeruginosa, Pc*—*Petroselinum crispum, Ph*—*Populus hybrid, Rc*—*Rhodobacter capsulatus, Rt*—*Rhodosporidium toruloides, Sb*—*Sorghum bicolor, Sc*—*Saccharomyces cerevisiae, Se*—*Salmonella enterica, Sl*—*Solanum lycopersicum,* Tf—*Thermobifida fusca* NTU22, *Vv*—*Vitis vinifera, Zm*—*Zea mays*; 4CL—4-coumarate-CoA ligase, ANS—anthocyanidin synthase, AXE—thermostable esterase, BM3R—reductase domain of the CYP102A1 gene, C3H—4-coumarate 3-hydroxylase, C4H— cinnamic acid 4-hydroxylase, CHR—chalcone reductase, CHI—chalcone isomerase, CHS—chalcone synthase, CPR- cytochrome P450 reductase, CUS—curcuminoid synthase, CYP57B3—cytochrome P450 monooxygenase, DFR—dihydroflavonol 4-reductase, F3H—flavanone 3-hydroxylase, F3'H—flavonoid 3'-hydroxylase, F3′5′H—flavonoid 3',5'-hydroxylase, F6'H1—feruloyl-CoA 6'-hydroxylase, FLS—flavonol synthase, FSI—flavone synthase I, HpaB and HpaC—4-hydroxyphenylacetate 3-hydroxylase, PAL— phenylalanine ammonia lyase, ROMT—resveratrol *O*-methyltransferase, STS—stilbene synthase, TAL—tyrosine ammonia lyase, VST—resveratrol synthase; [b] Corncob biomass (2%) contains feruloyl-polysaccharide that is AXE substrate

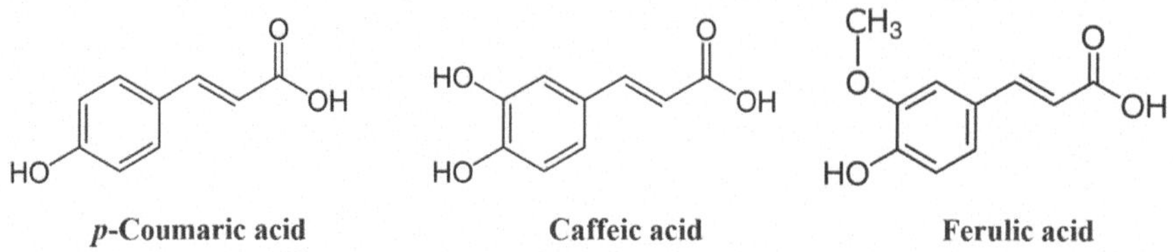

Fig. 4 Structure of the hydroxycinnamic compounds *p*-coumaric acid, caffeic acid, and ferulic acid

acids since they are present in limited amounts in the cells. The constructed strain holds a knockout in the pyruvate decarboxylase (PDC) genes, PDC5 and Aro10. These genes are responsible for the conversion of the precursor of phenylalanine, that is, phenylpyruvate, into phenylacetaldehyde contributing to the pathway deviation. Moreover, DAHP synthase and chorismate mutase genes were replaced by a feedback-insensitive version to prevent the inhibition of aromatic amino acids synthesis since these genes are strongly affected by the amounts of aromatic amino acids produced. Moreover, the authors have overexpressed a shikimate kinase from *E. coli* since this could be a limiting step in the production of higher amounts of *p*-coumaric acid. TAL from *Flavobacterium johnsoniae* was also expressed in this engineered strain that was able to produce almost 2000 mg/L of *p*-coumaric acid in fed-batch fermentation (Rodriguez et al. 2015). Later, the same group studied the influence of the production of *p*-coumaric acid in the metabolism of *S. cerevisiae*-producing strains to construct more efficient and robust strains (Rodriguez et al. 2017a). After transcriptomic and metabolomic analysis, the authors have found that genes involved in the sugars and amino acids transport were downregulated in the producer strain. When these downregulated genes were knocked out, it was found that the production of *p*-coumaric acid was improved. The highest production of *p*-coumaric acid (2400 mg/L) was obtained in the *S. cerevisiae* CEN.PK strain that holds the deletion of the tyrosine and tryptophan amino acid transporter 1 (TAT1), which is probably involved in the transport of tyrosine to the outside of the cells being less available for *p*-coumaric acid production (Rodriguez et al. 2017a). More recently, Liu et al. (2019a) have also constructed a *S. cerevisiae* strain able to produce higher amounts of *p*-coumaric acid. In this strain, PAL and C4H genes from *Arabidopsis thaliana* were expressed to produce *p*-coumaric acid. To increase the cytochrome P450 activity, cytochrome B5 (CYB5) from *S. cerevisiae* and cytochrome P450 reductase (CPR1) from *A. thaliana* were overexpressed. Moreover, the strain was also modified to overexpress a feedback-insensitive version of the DAHP synthase and chorismate mutase genes. The TAL gene from *F. johnsoniae* was also expressed, and the production of *p*-coumaric acid was verified using both tyrosine and phenylalanine branches. Prephenate dehydrogenase (PDH) genes from different microorganisms were also tested, and it was found that PDH from *Medicago truncatula* significantly improves the shikimate pathway flux and, consequently, the production of *p*-coumaric acid. The production of E4P was also improved by expressing a heterologous phosphoketolase pathway in conjugation with the expression of the gene that encodes glycerol-1-phosphatase. To prevent the pathway deviation and formation of by-products, PDC5 and Aro10 were also deleted. Additionally, the expression levels of several genes were also optimized by promoter replacement. After all these modifications, the yeast chassis was able to produce 12,500 mg/L of *p*-coumaric acid from glucose in a fed-batch fermentation experiment. Currently, this was the highest production of *p*-coumaric acid reported and this strain could be considered an excellent starting strain for the production of polyphenolic compounds with high titers and yields (Liu et al. 2019a).

The heterologous production of caffeic acid was also achieved in *S. cerevisiae.* To produce caffeic acid using tyrosine as substrate, a strain holding the TAL gene from *Rhodosporidium toruloides,* 4-hydroxyphenylacetate 3-monooxygenase (HpaB), and NADPH-flavin oxidoreductase (HpaC) genes derived from *Pseudomonas aeruginosa* and *Salmonella enterica,* respectively, was constructed. HpaB and HpaC can catalyze the transformation of *p*-coumaric acid into caffeic acid. This strain was able to produce 289.4 mg/L of caffeic acid (Liu et al. 2019b).

Y. lipolytica was also explored as a heterologous host to produce *p*-coumaric acid. Gu et al. (2020) reported the construction of a strain able to produce *p*-coumaric acid from glucose. The platform strain has modifications toward an increase in the aromatic amino acids biosynthesis, namely the substitution of the native DAHP synthase for a feedback-insensitive version of the gene. In order to increase the flux toward the shikimate pathway, transketolase gene was overexpressed in this strain to convert fructose-6-phosphate into E4P, and pyruvate kinase gene was deleted to prevent PEP consumption. Moreover, TAL gene was expressed in this strain and 593.5 mg/L of *p*-coumaric acid was produced (Gu et al. 2020). The production of ferulic acid was also obtained in *Y. lipolytica* by expressing an esterase gene from *Thermobifida fusca.* This esterase can release the ferulic acid that is linked to lignin and other polysaccharides in the cell wall of plants. In this study, the esterase was able to convert corncob biomass residue, that is a lignocellulose residue, into 77 mg/L of ferulic acid (Huang et al. 2011). However, the production of ferulic acid from tyrosine or glucose was not explored yet.

6.2 Heterologous Production of Flavonoids

Within polyphenolic compounds, flavonoids are the class with more variety of structures. These compounds are responsible for organoleptic properties and for the pigments that confer color to vegetables, fruits, and herbs. Due to their medical and industrial applications, the market value of these compounds is estimated to reach USD 1.06 billion by 2025 (Grand View Research 2016). The flavonoid skeleton is characterized by two benzene rings that are connected by one pyrene ring that contains an oxygen molecule (Fig. 5). Depending on the differences in the structure of these compounds, flavonoids could be grouped into different subclasses. These subclasses are flavonols, flavones, flavanones, flavonols, isoflavonoids, and anthocyanins (Panche et al. 2016).

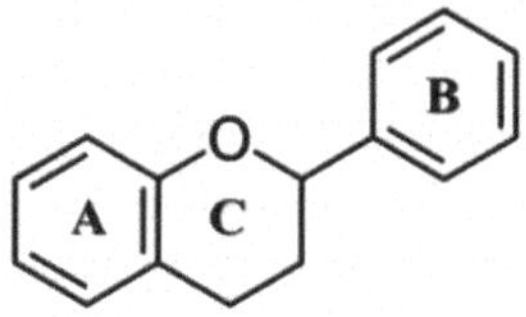

Fig. 5 Basic structure of flavonoids compounds. A and B correspond to two benzene rings and C corresponds to a pyrene ring containing an oxygen molecule

In the last years, several research efforts have been made to construct yeast cell factories with the ability to produce flavonoids with high titers and yields. *S. cerevisiae* has been the most explored yeast to produce these compounds. Several *S. cerevisiae* strains were constructed to produce flavonoids using *p*-coumaric acid or amino acids as precursors (Jiang et al. 2005; Lyu et al. 2017; Trantas et al. 2009; Yan et al. 2005). However, these substrates are more expensive, and the industrial production process was unfeasible (Santos et al. 2011). For this reason, *de* novo biosynthesis of these compounds from cheap carbon sources became a priority. Koopman et al. (2012) have constructed a *S. cerevisiae* strain able to produce naringenin from glucose. This strain harbors the genes responsible for the naringenin biosynthetic pathway (PAL1, C4H, CPR1, 4CL3, CHS3, and chalcone isomerase (CHI1)) from *A. thaliana.* The production of naringenin was evaluated in shake flasks fermentations, and it was found that low amounts of naringenin were produced (1.47 mg/L). The authors have modified the *S. cerevisiae* strain to construct a new yeast chassis able to overproduce amino acids and, consequently, optimize the production yields and titers. In this strain, the DAHP synthase gene was replaced for a feedback-insensitive version of the gene. PDC gene was also deleted to prevent the pathway deviation. Additionally, the number of copies of CHS gene was also increased to improve its catalytic efficiency. To improve the production of *p*-coumaric acid, TAL from *R. capsulatus* was also expressed. These modifications have resulted in an improvement of naringenin production reaching 54.5 mg/L in shake flask cultures and 108.9 mg/L in aerated, pH-controlled batch reactors from 20 g/L of glucose (Koopman et al. 2012). Rodriguez et al. (2017b) have also constructed *S. cerevisiae* strains able to produce several flavonoids using glucose as a substrate. In this work, the genes responsible for the biosynthesis of naringenin, liquiritigenin, kaempferol, resokaempferol, quercetin, and fisetin were integrated into the yeast genome. These strains contained modifications to improve the flux toward amino acids, namely the knockout of the PDC5 and Aro10 genes and the overexpression of DAHP synthase and chorismate mutase. The highest productions obtained in this study were 26.57 mg/L of kaempferol and 20.38 mg/L of quercetin (Rodriguez et al. 2017b). Duan et al. (2017) also constructed a *S. cerevisiae* strain to produce kaempferol de novo. Since malonyl-CoA low availability represents a limiting step in the high-level production of polyphenols, the authors overexpressed the endogenous genes responsible for the synthesis of malonyl-CoA from ethanol to increase its availability. Using a fed-batch fermentation strategy, the production of kaempferol has reached 66.29 mg/L. More recently, Lyu et al. (2019) also optimized a *S. cerevisiae* strain to improve the flux toward amino acids and prevent the pathway deviation. As performed by other research groups, the authors constructed a strain with a knockout in the PDC5 and Aro10 genes. Additionally, the genes responsible for the biosynthetic pathway were overexpressed and the promotors were replaced by the glucose-inducible strong promoter pGAL1. This strain was able to produce 86 mg/L of kaempferol. Moreover, this strain was also able to produce 220 mg/L of naringenin when Tween 80, that is a surfactant that could increase the enzymatic activity and cell biomass growth, was added to the fermentation medium. This is the highest de novo production of naringenin using *S. cerevisiae* as chassis reported so far (Lyu et al. 2019).

The biosynthetic pathway responsible for the synthesis of anthocyanins has been firstly engineered in *S. cerevisiae* by Eichenberger et al. (2018). In this study, the genes responsible for the biosynthetic pathway were integrated into a single copy in the yeast genome. The constructed strains were able to produce 0.85 mg/L of pelargonidin-3-O-glucosidase, 1.55 mg/L of cyanidin-3-O-glucosidase and 1.86 mg/L of delphinidin-3-O-glucosidase (Eichenberger et al. 2018). Afterward, a co-culture strategy was designed by Du et al. (2020) to produce naringenin, kaempferol, quercetin, myricetin, delphinidin, pelargonidin, and cyanidin. In this co-culture strategy, one strain harbors the biosynthetic pathway responsible for the production of naringenin, and the other strains harbor the other genes responsible for the production of the three flavonols and the three anthocyanidins from naringenin. The first strain contains several modifications to improve the production of aromatic amino acids and malonyl-CoA and to prevent the deviation of the pathway for competing pathways. The first strain was able to produce 144.1 mg/L of naringenin. The co-culture of both strains resulted in the production of 168.1 mg/L of kaempferol, 154.2 mg/L of quercetin, 145 mg/L of myrcetin, 26.1 mg/L delphinidin, 33.3 mg/L of pelargonidin, and 31.7 mg/L of cyanidin. This study demonstrated the importance of using co-culture strategies to reduce the metabolic burden caused by the presence of large biosynthetic pathways (Du et al. 2020; Rodrigues et al. 2020). Additionally, these studies strengthened the relevance of improving the synthesis of aromatic amino acids and the extender substrates availability to significantly increase the production of the desired compounds.

Although *Y. lipolytica* has been less used than *S. cerevisiae*, it has been recently explored as a yeast chassis for the production of flavonoids. Lv et al. (2019a) have constructed the biosynthetic pathways responsible for the production of naringenin, eriodyctiol, and taxifolin. The authors have developed a method for integration of the genes responsible for the biosynthetic pathway in multiple copies in random sites of the *Y. lipolytica* genome by combining 26 s ribosomal DNA (rDNA) multi-copy integration with Cre-loxP system. For the integration into the yeast genome, this methodology uses the sequence of the repetitive 26 s rDNA sites of *Y. lipolytica* genome as regions of homology for the cassette integration. Beyond the genes of the biosynthetic pathway, the cassette also contains the selection marker URA3 with loxP flanking sequences. After integration in the genome and selection of the transformants, Cre recombinase is expressed and recognizes the loxP sequences and, consequently, the selection marker is removed. The development of this method of integration was important because this microorganism has few auxotrophic markers available. Moreover, the use of plasmids can lead to genomic instability as well as copy number variations. After integrating the genes responsible for the biosynthetic pathways, the authors have evaluated the ability of the constructed strains to produce naringenin, eriodyctiol, and taxifolin in Yeast Extract–Peptone–Dextrose (YPD)-rich medium. The best performing strains were able to produce 71.2 mg/L, 54.2 mg/L, and 48.1 mg/L of naringenin, eriodyctiol, and taxifolin, respectively (Lv et al. 2019a). Later, the same research group has performed several steps to optimize the production of these compounds. The authors have optimized the copy number of the CHS and CPR genes that are responsible for limiting steps in the production of these flavonoids. Moreover, Lv et al. (2019b) also enhanced the availability of the chorismate and malonyl-CoA precursors by overexpressing the genes responsible for the synthesis of both compounds. Moreover, fermentation parameters such as pH and C/N ratio were also optimized. In the best conditions, the optimized strains were able to produce 252.4 mg/L, 134.2 mg/L, and 110.5 mg/L of naringenin, eriodictyol, and taxifolin, respectively. Comparing with the previous study, these optimizations led to 3.5-fold, 2.5-fold, and 2.3-fold improvement in the production of naringenin, eriodictyol, and taxifolin, respectively (Lv et al. 2019b). More recently, a mixture of glucose and xylose was used as a substrate to produce naringenin (Wei et al. 2020). Due to the need to produce flavonoids in an inexpensive way, the search for low-cost substrates, such as agricultural residues, has become a priority to reduce the production costs of these compounds at an industrial scale (Gudiña et al. 2020). Since xylose is one of the sugars that is present in a higher amount in agricultural residues, Wei et al. (2020) engineered the *Y. lipolytica* strain toward the use of xylose as a substrate to produce naringenin since the wild-type strain is not able to metabolize xylose. This strain was constructed by integration into the genome of a xylose-inducible module for activation of the expression of the genes responsible for xylose utilization. By activating the expression of these genes, xylose is metabolized and originates E4P which is a precursor of the shikimate pathway. Moreover, the genes responsible for the naringenin biosynthetic pathway were also integrated into the genome by Clustered Regularly Interspaced Short Palindromic Repeats—associated Cas9 endonuclease (CRISPR-Cas9). The production of naringenin was evaluated in the strain that only contains the naringenin biosynthetic pathway and in the strain that contains the naringenin biosynthetic pathway and the xylose-inducible utilization module. The first strain was able to produce 239.1 mg/L of naringenin in YPD fermentation medium. The strain that also contains the xylose-inducible utilization module was able to produce 715.3 mg/L of naringenin in YPD fermentation medium containing 60 g/L of xylose. This study showed that the xylose utilization pathway is a good alternative to be explored to increase the precursor availability and, consequently, improve the production of the desired polyphenolic compound. Moreover, this study opens doors to explore the production of polyphenols using different bio-wastes as low-cost substrates. In the same year, Palmer et al. (2020) also engineered a *Y. lipolytica* strain toward the production of naringenin. The genes responsible for the biosynthetic pathway were codon-optimized for *Y. lipolytica* and integrated into random sites of the yeast genome. The constructed strain was also modified to increase the production of naringenin. The modifications performed in this strain include the replacement of DAHP synthase gene for a feedback-insensitive version, expression of the peroxisome biogenesis factor gene (PEX10) involved in the peroxisome synthesis and improves the formation of acetyl-CoA, and integration of another copy of the CHS gene. This strain was able to produce 898 mg/L of naringenin from 80 g/L of glucose, being the highest production of naringenin reported so far in any yeast, thus demonstrating the high potential of using *Y. lipolytica* as heterologous host to produce polyphenols. Moreover, *Y. lipolytica* was also recently engineered toward de novo production of liquiritigenin. Different genes from diverse sources were tested to evaluate the best combination to produce higher amounts of liquiritigenin. The gene combination that results in the best production of liquiritigenin was PAL from *Zea mays*, C4H and 4CL from *Petroselinum crispum*, CHS from *Pterolophia hybrida*, and CHR and CHI from *Medicago sativa*. CHR and CHI were fused to increase the production of liquiritigenin. Additionally, the TEF promoter that initially regulated CHI expression was replaced by the lipase promoter. All these modifications allowed the improvement of the production of liquiritigenin reaching 8.2 mg/L from glucose. Moreover, feeding *p*-coumaric acid led to the production of 62.4 mg/L (Akram et al. 2020).

P. pastoris has been less explored than the other yeasts in the heterologous production of polyphenols. The yeast was used to produce hydroxylated isoflavones. However, this production was achieved by bioconversion of the non-hydroxylated compound (Chang et al. 2013; Ding et al. 2015; Wang et al. 2016). Chang et al. (2013) reported the production of 8-hydroxydaidzein, 3′-hydroxydaidzein, and 6-hydroxydaidzein from daidzein. In this study, the authors constructed a fusion gene by combining the gene encoding CYP57B3, which is responsible for the hydroxylation of genistein, with the reductase domain of the CYP102A1 gene (BM3R) from *Bacillus megaterium*. The fusion gene was inserted into the *P. pastoris* genome using an integrative vector. The conversion of daidzein by the constructed strain was evaluated and 0.58 mg/L, 0.23 mg/L, and 9.1 mg/L of 8-hydroxydaidzein, 3′-hydroxydaidzein, and 6-hydroxydaidzein were produced, respectively (Chang et al. 2013). Moreover, the production of 3’-hydroxygenistein from genistein was also achieved in *P. pastoris* through the expression of a cytochrome P450 hydroxylase (CYP57B3) from *Aspergillus oryzae* fused with a cytochrome P450 oxidoreductase gene (*Sc*CPR) from *S. cerevisiae*. The production of 3’-hydroxygenistein was evaluated in this strain, and it was verified that 3.5 mg/L of 3’-hydroxygenistein were produced (Ding et al. 2015). Later, Wang et al. (2016) have obtained mutants of the strain constructed by Ding et al. (2015) upon treatment with periodic hydrogen peroxide. This treatment was used to induce oxidative stress. Afterward, the most resistant strains were selected since it was hypothesized that these strains produced higher amounts of

3'-hydroxygenistein that has recognized antioxidant properties. The best mutant could produce 20.3 mg/L of 3'-hydroxygenistein (Wang et al. 2016). These are the only studies that report heterologous production of flavonoids in *P. pastoris*.

6.3 Heterologous Production of Stilbenoids

Stilbenoids are polyphenolic compounds with limited distribution in plant species. These compounds are mostly found in berries and grapes, and they are involved in the plant's response against several stress factors including pathogens (Rivière et al. 2012). Due to their interesting biological activities, resveratrol has been the most extensively studied and explored stilbenoid (Shen et al. 2013). The global market size of resveratrol was estimated to be USD 65 million in 2020, and it is expected to reach USD 106 million by 2026 (Market Watch 2020). Stilbenoids are composed of a skeleton with a C6-C2-C6 structure (Fig. 6). However, this skeleton could be decorated to produce several stilbenoids by hydroxylation, methylation, glycosylation, among others (Pecyna et al. 2020).

The first reports of stilbenoids heterologous production in *S. cerevisiae* used *p*-coumaric acid as a substrate to obtain resveratrol (Becker et al. 2003; Beekwilder et al. 2006; Shin et al. 2011; Sydor et al. 2010; Wang et al. 2011; Wang and Yu 2012). The highest production of resveratrol using *p*-coumaric acid as substrate was achieved by Sydor et al. (2010). In this study, an industrial Brazilian *S. cerevisiae* strain, that was isolated from a sugar cane plantation, carrying the plasmids containing 4CL1 from *A. thaliana* and STS from *Vitis vinifera* showed a higher ability to produce resveratrol compared to the well-described *S. cerevisiae* CEN.PK2-1C strain. This industrial strain was able to produce 391 mg/L of resveratrol from 2.46 g/L of *p*-coumaric acid in a rich fermentation medium. However, other studies showed the application of interesting metabolic engineering, as well as synthetic biology approaches to improve production titers and yields. For example, Wang et al. (2011) have codon-optimized a TAL gene since the native gene from *Rhodobacter sphaeroides* did not demonstrate in vivo activity in *S. cerevisiae*. By using a codon-optimized gene, the probability of occurring translation errors or the termination of the translation is lower. Moreover, codon-optimized genes are usually highly expressed. Also, the authors have also expressed arabinose-H

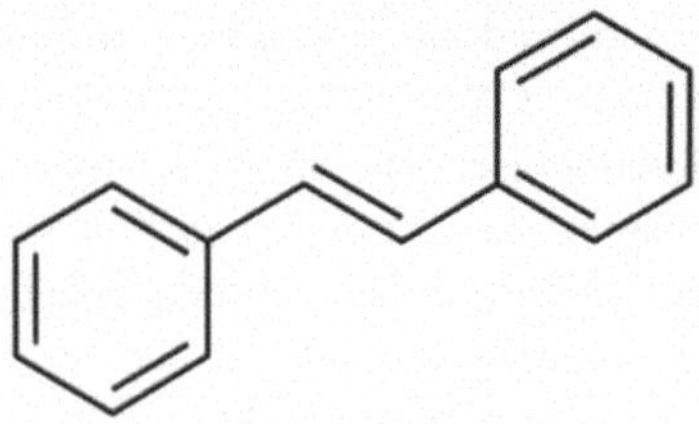

Fig. 6 Basic structure of stilbenoids

+ transport protein (AraE). This protein is involved in arabinose transport. However, it was verified that resveratrol could also be transported when its intracellular concentration reaches the threshold. For this reason, it was hypothesized that AraE probably increases the lipid membrane permeability for the diffusion of resveratrol. These modifications resulted in a 2.44-fold increase in the production of resveratrol (2.3 mg/L) (Wang et al. 2011). The same research group also constructed one synthetic scaffold between 4CL1 and STS to improve the resveratrol production. The scaffold interacts with the enzymes by small ligand peptides and improved the substrate flux between both heterologous enzymes. Compared to the non-scaffold strategy, the production of resveratrol increased 5-fold reaching 14.4 mg/L (Wang and Yu 2012). Moreover, resveratrol was also produced from tyrosine by Shin et al. (2012). In this study, PAL, C4H, 4CL1, and STS were expressed in *S. cerevisiae* W303-1A. In order to increase the extender substrate availability, the promoter of the native ACC gene was replaced by pGAL1. This modification resulted in a 1.3-fold improvement in the resveratrol production reaching 5.8 mg/L from 2.17 g/L of tyrosine (Shin et al. 2012). Later, de novo production of resveratrol in *S. cerevisiae* was achieved by Li et al. (2015) and Li et al. (2016). First, it was evaluated the influence of different isoforms of the 4CL gene from *A. thaliana* in the resveratrol production (Li et al. 2015). The authors have found that the expression of *At*4CL1 resulted in higher productions of resveratrol comparing with *At*4CL2, thus being possible to conclude that the choice of the enzymes is a step that can greatly influence the production titers and yields. The construction of aromatic amino acids overproducing strain was also evaluated. This strain was constructed by overexpressing feedback-insensitive versions of the DAHP synthase and chorismate mutase genes. This modification resulted in a 1.8-fold improvement in the production of resveratrol. The authors have also overexpressed a version of the ACC gene with a double mutation into the serine residues 659 and 1157 to increase malonyl-CoA availability. These mutations prevent the ACC phosphorylation and, consequently, its inactivation by the action of sucrose non-fermenting protein 1. The overexpression of the ACC mutant version has resulted in a 1.3-fold improvement in the production of resveratrol. In order to increase the expression levels and, consequently, the production of resveratrol, the genes responsible for the resveratrol biosynthetic pathway were integrated into multiple copies in the Ty repetitive sites that are present in the yeast genome using an integrative plasmid. The determination of the copy number was performed by qPCR. The maximum number of integrations found in the selected transformants was eight copies. The strain that holds the eight copies of the genes responsible for the biosynthetic pathway was able to produce 36-fold more resveratrol than the strains holding only one copy of the genes. In fed-batch fermentation, this strain was able to produce 531.4 mg/L of resveratrol from 40 g/L of glucose (Li et al. 2015). Later, the same research group optimized the resveratrol production to 800 mg/L from glucose in fed-batch fermentation. This improvement was obtained by changing the heterologous pathway for phenylalanine utilization by expressing PAL and C4H genes instead of TAL. CPR from *A. thaliana* and CYB5 from *S. cerevisiae* were overexpressed toward an improvement in the cytochrome P450 activity. Moreover, the flux toward the aromatic amino acids production was improved by replacing DAHP synthase and chorismate mutase by feedback-resistant versions of these genes. Malonyl-CoA synthesis was also improved by expressing the acetyl-CoA synthase from *Salmonella enterica* and the mutated version of the ACC gene. Beyond these modifications, the pathway deviation was also prevented by deleting the PDC gene. In this study, two stilbenoids that are derived from resveratrol were also produced. By expressing resveratrol O-methyltransferase (ROMT) from *Sorghum bicolor*, the constructed strain produced 5.52 mg/L of pinostilbene. By expressing ROMT from *V. vinifera*, the strain produced 34.92 mg/L of pterostilbene (Li et al. 2016). Pterostilbene was also previously produced using *p*-coumaric acid as a substrate in *S. cerevisiae* expressing 4CL, STS, and ROMT (Wang et al. 2015a). However, the production was even lower (2.2 mg/L) than the one reported by Li et al. (2016).

The heterologous production of resveratrol was also achieved using *Y. lipolytica* as microbial chassis. Huang et al. (2006) reported for the first time the production of resveratrol in *Y. lipolytica.* The strain expressing PAL, C4H, 4CL, and STS was able to produce 1.46 mg/L of resveratrol from 36.2 mg/L of tyrosine. More recently, there have been many advances to produce larger amounts of resveratrol in *Y. lipolytica.* Beyond the construction of the *p*-coumaric acid biosynthetic pathway, Gu et al. (2020) also constructed the pathway responsible for the resveratrol production. The authors expressed 4CL and STS in the strain previously constructed to produce *p*-coumaric acid. Although this strain has several modifications to improve the flux toward the synthesis of aromatic amino acids and the shikimate pathway, only 12.67 mg/L of resveratrol were produced from 40 g/L of glucose. Palmer et al. (2020) have also reported the construction of a *Y. lipolytica* strain able to de novo produce resveratrol. This strain also contains modifications to improve the synthesis of malonyl-CoA and amino acids, and it was able to produce 48.7 mg/L of resveratrol from 20 g/L of glucose and 32.8 mg/L of *p*-coumaric acid. Afterward, the heterologous production of resveratrol in *Y. lipolytica* was optimized by Sáez-Sáez et al. (2020). In this study, TAL from *Flavobacterium johnsoniae*, 4CL from *A. thaliana* and STS from *V. vinifera* were successfully integrated into the *Y. lipolytica* genome using integrative plasmids. DAHP synthase and chorismate mutase were replaced by feedback-insensitive versions. Moreover, PCD5 and Aro10 were also deleted to prevent the pathway deviation. Two mutations were introduced in the ACC gene to prevent the phosphorylation and inactivation of malonyl-CoA synthesis. This strain was able to produce 85 mg/L of resveratrol. In order to improve the resveratrol titer, five copies of the genes responsible for the biosynthetic pathway were introduced using integrative plasmids in the previously constructed strain reaching the production of 409 mg/L of resveratrol. In fed-batch fermentation, this strain was able to produce 12.400 mg/L of resveratrol from glucose. This is the highest production of resveratrol reported so far in any microorganism (Sáez-Sáez et al. 2020). More recently, He et al. (2020) also constructed a strain for de novo production of resveratrol. The authors studied the production using the tyrosine or the phenylalanine routes or the conjugation of both routes. The biosynthetic pathways were firstly integrated into a single copy in the *Y. lipolytica* genome. Afterward, one extra copy of the biosynthetic pathways was also integrated to increase the metabolic flux toward the production of resveratrol. When glucose was used as a substrate, the strain carrying two copies of the tyrosine route was the best resveratrol producer (85 mg/L). Additionally, the performance of the three strains was also tested using glycerol as an alternative substrate. In these experiments, the strain carrying two copies of both tyrosine and phenylalanine routes showed the best results of resveratrol production. In bioreactor fermentation, this strain was able to produce 430 mg/L of resveratrol from 100 g/L of glycerol without any supplementation of amino acids (He et al. 2020). All these studies show the great potential to use *Y. lipolytica* as a chassis to industrially produce resveratrol and other resveratrol-derived compounds.

6.4 Heterologous Production of Coumarins

Coumarins are polyphenolic compounds that are largely found in many plant species being fundamentally present in essential oils and fruits (Jain and Joshi 2012). These compounds act in the defense mechanism of plants against stress factors such as fungal infections (Stringlis et al. 2019). The core structure of coumarins is composed of one benzene ring and one α-pyrone ring (Fig. 7). These compounds are divided into two main classes that are simple coumarins and complex coumarins. In the group of complex coumarins, the compounds are classified as furanocoumarins, pyranocoumarins, phenylcoumarins, and bicoumarins (Matos et al. 2015).

Relatively to the flavonoids and stilbenoids, the heterologous production of coumarins has been less explored. Simple coumarins, such as umbelliferone, scopoletin, and esculetin, were already produced in *E. coli* (Lin et al. 2013; Yang et al. 2015; Zhao et al. 2019). However, the use of yeasts as microbial chassis to produce these compounds practically has not been explored. Until now, only scopoletin was already produced in *S. cerevisiae* (Zhao et al. 2020). The genes 4CL from *Petroselinum crispum* and feruloyl-CoA 6′-hydroxylase (F6′H1) from *A. thaliana* responsible for the synthesis of scopoletin from ferulic acid were expressed into *S. cerevisiae* BY4742. This strain was able to produce 0.364 mg/L of scopoletin. Since these results of production were very low, the authors fused 4CL and F6'H1 with different linkers. The best production result was obtained when the genes were fused with the linker $(GGGGS)_4$, resulting in a 3-fold improvement in the production of scopoletin. After integrating the fusion gene

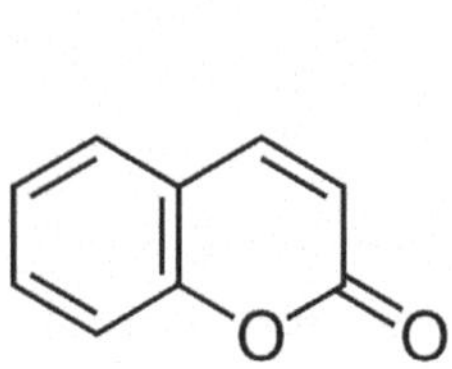

Fig. 7 Basic structure of coumarins

into the *S. cerevisiae* genome, 3.42 mg/L of scopoletin were produced using ferulic acid as substrate. The authors also tested the production of scopoletin using *p*-coumaric acid as substrate. HpaB from *P. aeruginosa*, HpaC from *Enterobacteriaceae* and COMT from *A. thaliana* were integrated into the genome of the strain carrying the fusion gene 4CL-F6'H1. This strain was able to produce 4.98 mg/L of scopoletin using *p*-coumaric acid as substrate. Due to the interest of using alternative substrates to produce compounds of interest, the production was also evaluated using lignin hydrolysate as substrate (Gudiña et al. 2020). Since ferulic acid and *p*-coumaric acid are monomers of lignin, this strain produced 4.79 mg/L of scopoletin (Zhao et al. 2020). As far as we know, this is the only report of heterologous production of coumarins in yeasts. Although the production titers are lower compared with the productions obtained in *E. coli*, we believe that this production could be successfully improved and reach higher levels. Moreover, the construction of the biosynthetic pathway to produce the other simple and complex coumarins should also be explored using *S. cerevisiae* and other yeasts as chassis.

6.5 Heterologous Production of Curcuminoids

Curcuminoids are natural polyphenolic compounds that can be found in the rhizome of the plant *Curcuma longa* which is widely known as turmeric. The rhizome of turmeric is mainly composed of curcumin, demethoxycurcumin, and bisdemethoxycurcumin. These compounds have been largely used as a food additive, as well as therapeutic agents due to their several recognized therapeutical activities (Amalraj et al. 2017). For this reason, the market value of curcumin, which is present in higher amounts in turmeric and exhibits the most potent biological activities, is expected to reach USD 151.9 million by 2027 (Hewlings and Kalman 2017; Grand View Research 2020). The core structure of curcuminoids is composed of a C6-C7-C6 structure being classified as diarylheptanoids (Fig. 8) (Rodrigues et al. 2015c).

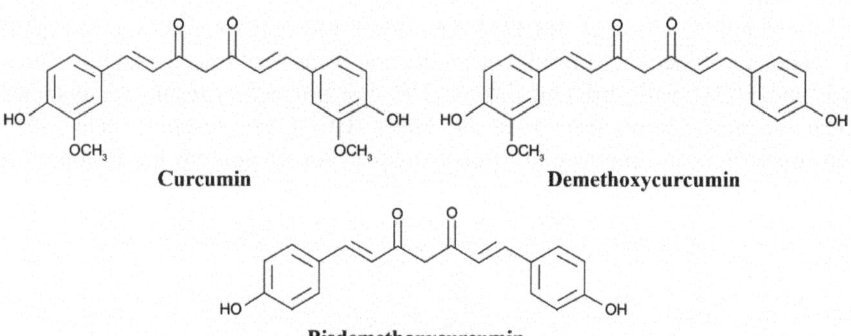

Fig. 8 Structure of the three curcuminoids compounds: curcumin, demethoxycurcumin, and bisdemethoxycurcumin

The heterologous production of curcuminoids has been widely explored in *E. coli.* These compounds were already produced using hydroxycinnamic acids and tyrosine as starter substrates (Couto et al. 2017; Fang et al. 2018; Katsuyama et al. 2008; 2010; Kim et al. 2017; Rodrigues et al. 2017a, 2015a, 2020; Wang et al. 2013, 2015b). However, to the best of our knowledge, the construction of the biosynthetic pathway responsible for the curcuminoids production was not achieved until now in *S. cerevisiae* and *P. pastoris.* The only report of the production of curcuminoids in yeast species was the production of bisdemethoxycurcumin using *Y. lipolytica* as yeast chassis (Palmer et al. 2020). In this study, the authors constructed a *Y. lipolytica* strain expressing 4CL gene from *Nicotiana tabacum* and a codon-optimized version of CUS gene from *Oryza sativa.* PEX10 and ACC1 genes were overexpressed toward an improvement in the flux toward malonyl-CoA biosynthesis. This strain was able to produce 0.17 mg/L of bisdemethoxycurcumin from 2 mM of *p*-coumaric acid (Palmer et al. 2020). Although this production value is lower than the productions reported in *E. coli*, we believe that this study can be a starting point for producing larger amounts of these compounds in *Y. lipolytica* and also in other yeasts. The construction of the whole biosynthetic pathway to produce curcuminoids from glucose as well as the modification of the constructed strains using synthetic biology and metabolic engineering strategies could be an interesting approach to produce these compounds with a high yield and purity.

6.6 Heterologous Production of Polyphenolic Amides

Polyphenolic amides are polyphenolic compounds that could be found in foods such as peppers and oats (Tsao 2010). Avenanthramide, which belongs to the polyphenolic amides group, is present in oats and is involved in the plants' response to the infection with pathogens (Meydani 2009). The structure of these compounds consists of an anthranilic acid and a hydroxycinnamic acid linked by an amide bond (Fig. 9). More than 40 different structures are known for avenanthramide compounds due to the presence of radical substituent groups in the structure (Perrelli et al. 2018).

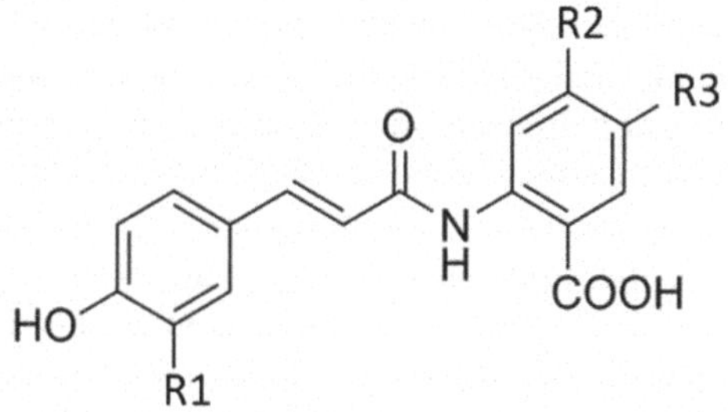

Fig. 9 Basic structure of avenanthramides

As occurred with coumarins and curcuminoids, the heterologous production of polyphenolic amides using yeasts as chassis has been also less explored. There is only one report of the heterologous production of two avenanthramide-derived compounds in *S. cerevisiae*. 4CL from *N. tabacum* and hydroxycinnamoyl-CoA:shikimate/quinate hydroxycinnamoyl transferase (HCT) from *Cynara cardunculus* were successfully expressed in *S. cerevisiae* for the conversion of the hydroxycinnamic acids into CoA esters and the CoA esters into the respective amide. In this study, 120 mg/L of *p*-coumaroyl-3-hydroxyanthranilic (avenanthramide I) were produced from 462 mg/L of *p*-coumaric acid and 77 mg/L of 3-hydroxyanthranilic acid. Moreover, 22 mg/L of caffeoyl-3-hydroxyanthranilic acid (avenanthramide II) was produced from 540 mg/L of caffeic acid and 77 mg/L of 3-hydroxyanthranilic acid (Moglia et al. 2015). This study only tested the production from the hydroxycinnamic acids. The next step could be the construction of the complete biosynthetic pathway for the de novo production of these compounds.

7 Conclusions

Polyphenolic compounds have been recognized as promising compounds to be used in the treatment of several diseases. Due to the increasing demand for these compounds by the pharmaceutical and nutraceutical industry, the development of methodologies to industrially produce them became a priority. In the last years, several research efforts have been made toward the production of polyphenolic compounds with high yields and titers using microorganisms. The use of microorganisms, namely yeast species, became an attractive solution since the process is faster and more economic and environmentally friendly. Regarding flavonoids, the heterologous production of naringenin has been the most explored. The maximal production of this compound was 898 mg/L in *Y. lipolytica* (Palmer et al. 2020). Within stilbenoids, the production of resveratrol using yeasts as chassis has been the most studied. This compound was already produced in a significant amount in *Y. lipolytica* (12.5 g/L) (Sáez-Sáez et al. 2020). Although these reports are a good starting point, there is still a long way to go to produce these compounds at an industrial scale. Nevertheless, it was possible to conclude that *Y. lipolytica* is potentially an excellent heterologous host for the industrial production of these compounds due to the higher yields reported. Moreover, the production of coumarins, curcuminoids, and polyphenolic amides remains much less explored. In this case, exploiting different enzyme combinations performs a step-by-step optimization, as well as constructing the complete biosynthetic pathway for the de novo production of these compounds are important steps to improve and achieve relevant productions of these polyphenols. The main bottlenecks in the production of polyphenols using microbial cell factories are the substrates and precursors availability and the deviation of these compounds for competing pathways. The use of metabolic engineering and synthetic biology approaches are useful to overcome these issues and, consequently, to construct more robust and adapted strains.

In summary, we believe that the production of polyphenols could reach industrial levels after performing several optimizations in the yeast chassis and also in the operational methods of the fermentation process.

Acknowledgements This study was supported by the Portuguese Foundation for Science and Technology (FCT) under the scope of the strategic funding of UIDB/BIO/04469/2020 unit and BioTecNorte operation (NORTE-01-0145- FEDER-000004) funded by the European Regional Development Fund (ERDF) under the scope of Norte2020—North Portugal Regional Program. DG and JR are recipients of a fellowship supported by a doctoral advanced training (SFRH/BD/04433/2020 and SFRH/BD/138325/2018, respectively) funded by FCT.

References

Abbaszadeh H, Keikhaei B, Mottaghi S (2019) A review of molecular mechanisms involved in anticancer and antiangiogenic effects of natural polyphenolic compounds. Phytother Res 33:2002–2014

Abramovič H (2015) Antioxidant properties of hydroxycinnamic acid derivatives: a focus on biochemistry, physicochemical parameters, reactive species, and biomolecular interactions. In: Preedy VR (ed) Coffee in health and disease prevention. Elsevier Inc, pp 843–852

Akram M, Rasool A, An T, Feng X, Li C (2020) Metabolic engineering of *Yarrowia lipolytica* for liquiritigenin production. Chem Eng Sci 230(2021):116177

Ali A, Banerjea AC (2016) Curcumin inhibits HIV-1 by promoting tat protein degradation. Sci Rep 6(27539):1–9

Amalraj A, Pius A, Gopi S, Gopi S (2017) Biological activities of curcuminoids, other biomolecules from turmeric and their derivatives—a review. J Tradit Complement Med 7(2):205–233

Avtanski D, Poretsky L (2018) Phyto-polyphenols as potential inhibitors of breast cancer metastasis. Mol Med 24(1):1–17

Becker JVW, Armstrong GO, Van Der Merwe MJ, Lambrechts MG, Vivier MA, Pretorius IS (2003) Metabolic engineering of *Saccharomyces cerevisiae* for the synthesis of the wine-related antioxidant resveratrol. FEMS Yeast Res 4(1):79–85

Beekwilder J, Wolswinkel R, Jonker H, Hall R, De Rie Vos CH, Bovy A (2006) Production of resveratrol in recombinant microorganisms. Appl Environ Microbiol 72(8):5670–5672

Braga A, Ferreira P, Oliveira J, Rocha I, Faria N (2018) Heterologous production of resveratrol in bacterial hosts: current status and perspectives. World J Microbiol Biotechnol 34(8):1–11

Braga A, Rocha I, Faria N (2019) Microbial hosts as a promising platform for polyphenol production. In: Akhtar M, Swamy M, Sinniah U (eds) Natural bio-active compounds: volume 1: production and applications vol 1. Springer, pp 71–103

Cassidy L, Fernandez F, Johnson JB, Naiker M, Owoola AG (2020) Broszczak DA (2020) Oxidative stress in Alzheimer's disease: a review on emergent natural polyphenolic therapeutics. Complement Ther Med 49:102294

Chang TS, Chao SY, Chen YC (2013) Production of ortho-hydroxydaidzein derivatives by a recombinant strain of *Pichia pastoris* harboring a cytochrome P450 fusion gene. Process Biochem 48(3):426–429

Chen X, Qiao H, Liu T, Yang Z, Xu L, Xu Y, Ge HM, Tan RX, Li E (2012) Inhibition of herpes simplex virus infection by oligomeric stilbenoids through ROS generation. Antiviral Res 95(1):30–36

Chen L, Deng H, Cui H, Fang J, Zuo Z, Deng J, Li Y, Wang X, Zhao L (2018) Inflammatory responses and inflammation-associated diseases in organs. Oncotarget 9(6):7204–7218

Chen R, Yang S, Zhang L, Zhou YJ (2020) Advanced Strategies for Production of Natural Products in Yeast. Cell Press Reviews 23(3):100879

Couto MR, Rodrigues JL, Rodrigues LR (2017) Optimization of fermentation conditions for the production of curcumin by engineered *Escherichia coli*. J R Soc Interface 14(133):20170470

Dayem AA, Choi HY, Yang GM, Kim K, Saha SK, Cho SG (2016) The anti-cancer effect of polyphenols against breast cancer and cancer stem cells: molecular mechanisms. Nutrients 8(9):1–37

de Araújo FF, de Paulo Farias D, Neri-Numa IA, Pastore GM (2020) Polyphenols and their applications: an approach in food chemistry and innovation potential. Food Chem 338(2021):1–15

Ding L, Hofius D, Hajirezaei MR, Fernie AR, Börnke F, Sonnewald U (2007) Functional analysis of the essential bifunctional tobacco enzyme 3-dehydroquinate dehydratase/shikimate dehydrogenase in transgenic tobacco plants. J Exp Bot 58(8):2053–2067

Ding HY, Chiang CM, Tzeng WM, Chang TS (2015) Identification of 3′-hydroxygenistein as a potent melanogenesis inhibitor from biotransformation of genistein by recombinant *Pichia pastoris*. Process Biochem 50(10):1614–1617

Du Y, Yang B, Yi Z, Hu L, Li M (2020) Engineering *Saccharomyces cerevisiae* coculture platform for the production of flavonoids. J Agric Food Chem 68(7):2146–2154

Duan L, Ding W, Liu X, Cheng X, Cai J, Hua E, Jiang H (2017) Biosynthesis and engineering of kaempferol in *Saccharomyces cerevisiae*. Microb Cell Fact 16(1):1–10

Ebrahimi M, Mokhtari A (2017) Engineering of secondary metabolites in tissue and cell culture of medicinal plants: an alternative to produce beneficial compounds using bioreactor technologies. In: Abdullah SNA, Chai-Ling H, Wagstaff C (eds) Crop improvement. Springer, Cham, pp 137–167

Eichenberger M, Hansson A, Fischer D, Dürr L, Naesby M (2018) De novo biosynthesis of anthocyanins in *Saccharomyces cerevisiae*. FEMS Yeast Res 18(4):1–13

Fang Z, Jones JA, Zhou J, Koffas MAG (2018) Engineering *Escherichia coli* co-cultures for production of curcuminoids from glucose. Biotechnol J 1700576(13):1–8

Fraser CM, Chapple C (2011) The phenylpropanoid pathway in arabidopsis. The Arabidopsis Book, vol 9. p e0152

Ghofrani S, Joghataei MT, Mohseni S, Baluchnejadmojarad T, Bagheri M, Khamse S, Roghani M (2015) Naringenin improves learning and memory in an Alzheimer's disease rat model: insights into the underlying mechanisms. Eur J Pharmacol 764(2015):195–201

Grand View Research (2016) Flavonoids market size worth $1.06 billion by 2025. https://www.grandviewresearch.com/press-release/global-flavonoids-market

Grand View Research (2020) Curcumin Market Size Worth $151.9 Million By 2027. https://www.grandviewresearch.com/press-release/curcumin-market

Gu Y, Ma J, Zhu Y, Ding X, Xu P (2020) Engineering *Yarrowia lipolytica* as a chassis for de novo synthesis of five aromatic-derived natural products and chemicals. ACS Synth Biol 9(8):2096–2106

Gudiña EJ, Amorim C, Braga A, Costa Â, Rodrigues JL, Rodrigues SS, Rodrigues LR (2020) Biotech green approaches to unravel the potential of residues into valuable products. In: Inamuddin (ed) Green chemistry for the sustainable development of chemical industry. Springer, pp 97–150

Hashemzaei M, Far AD, Yari A, Heravi RE, Tabrizian K, Taghdisi SM, Sadegh SE, Tsarouhas K, Kouretas D (2017) Anticancer and apoptosis—inducing effects of quercetin in vitro and in vivo. Oncol Rep 38:819–828

He Q, Szczepańska P, Yuzbashev T, Lazar Z, Ledesma-Amaro R (2020) De novo production of resveratrol from glycerol by engineering different metabolic pathways in *Yarrowia lipolytica*. Metab Eng Commun 11:0–5

Hewlings S, Kalman D (2017) Curcumin: a review of its effects on human health. Foods 6(10):92

Horne JR, Vohl M-C (2020) Biological plausibility for interactions between dietary fat, resveratrol, ACE2, and SARS-CoV illness severity. Am J Physiol Endocrinol Metab 318:830–833

Huang YC, Chen YF, Chen CY, Chen WL, Ciou YP, Liu WH, Yang CH (2011) Production of ferulic acid from lignocellulolytic agricultural biomass by *Thermobifida fusca* thermostable esterase produced in *Yarrowia lipolytica* transformant. Biores Technol 102(17):8117–8122

Huang ZQ, Chen P, Su WW, Wang YG, Wu H, Peng W, Li PB (2018) Antioxidant activity and hepatoprotective potential of quercetin 7-rhamnoside in vitro and in vivo. Molecules 23(5):1–13

Huang LL, Xue Z, Zhu QQ (2006) Method for the production of resveratrol in a recombinant oleaginous microorganism (Patent No. PCT/US2006/019085)
Hussain T, Tan B, Yin Y, Blachier F, Tossou MCB, Rahu N (2016) Oxidative stress and inflammation: what polyphenols can do for us? Oxidative Med Cell Longevity 2016
Ibrahim NI, Wong SK, Mohamed IN, Mohamed N, Chin KY, Ima-Nirwana S, Shuid AN (2018) Wound healing properties of selected natural products. Int J Environ Res Public Health 15(11):1–23
Jain PK, Joshi H (2012) Coumarin: chemical and pharmacological profile. J Appl Pharm Sci 2(6):236–240
Jasso-Miranda C, Herrera-Camacho I, Flores-Mendoza LK, Dominguez F, Vallejo-Ruiz V, Sanchez-Burgos GG, Pando-Robles V, Santos-Lopez G, Reyes-Leyva J (2019) Antiviral and immunomodulatory effects of polyphenols on macrophages infected with dengue virus serotypes 2 and 3 enhanced or not with antibodies. Infect Drug Resist 2019(12):1833–1852
Jiang H, Wood KV, Morgan JA (2005) Metabolic engineering of the phenylpropanoid pathway in *Saccharomyces cerevisiae.* Appl Environ Microbiol 71(6):2962–2969
Jin L, Zeng W, Zhang F, Zhang C, Liang W (2017) Naringenin ameliorates acute inflammation by regulating intracellular cytokine degradation. J Immunol 199(10):3466–3477
Kamboj A, Saluja AK, Kumar M, Atri P (2012) Antiviral activity of plant polyphenols. J Pharm Res 5(5):2402–2412
Karbalaei M, Rezaee SA, Farsiani H (2020) *Pichia pastoris*: a highly successful expression system for optimal synthesis of heterologous proteins. J Cell Physiol 235(9):5867–5881
Katsuyama Y, Matsuzawa M, Funa N, Horinouchi S (2008) Production of curcuminoids by *Escherichia coli* carrying an artificial biosynthesis pathway. Microbiology 154(9):2620–2628
Katsuyama Y, Kita T, Horinouchi S (2009) Identification and characterization of multiple curcumin synthases from the herb *Curcuma longa.* FEBS Lett 583(17):2799–2803
Katsuyama Y, Hirose Y, Funa N, Ohnishi Y, Horinouchi S (2010) precursor-directed biosynthesis of curcumin analogs in *Escherichia coli.* Biosci Biotechnol Biochem 74(3):641–645
Khan H, Reale M, Ullah H, Sureda A, Tejada S, Wang Y, Zhang ZJ, Xiao J (2020) Anti-cancer effects of polyphenols via targeting p53 signaling pathway: updates and future directions. Biotechnol Advan 38:107385
Kim EJ, Cha MN, Kim BG, Ahn JH (2017) Production of curcuminoids in engineered *Escherichia coli.* J Microbiol Biotechnol 27(5):975–982
Koopman F, Beekwilder J, Crimi B, van Houwelingen A, Hall RD, Bosch D, van Maris AJA, Pronk JT, Daran JM (2012) *De novo* production of the flavonoid naringenin in engineered *Saccharomyces cerevisiae.* Microb Cell Fact 11:1–15
Krivoruchko A, Nielsen J (2015) Production of natural products through metabolic engineering of *Saccharomyces cerevisiae.* Curr Opin Biotechnol 35:7–15
Lee S, Lee HH, Shin YS, Kang H, Cho H (2017) The anti-HSV-1 effect of quercetin is dependent on the suppression of TLR-3 in Raw 264.7 cells. Arch Pharm Res 40(5):623–630
Leema G, Tamizhselvi R (2018) Protective effect of scopoletin against cerulein-induced acute pancreatitis and associated lung injury in mice. Pancreas 47(5):577–585
Li M, Kildegaard KR, Chen Y, Rodriguez A, Borodina I, Nielsen J (2015) *De novo* production of resveratrol from glucose or ethanol by engineered *Saccharomyces cerevisiae.* Metab Eng 32(2015):1–11
Li M, Schneider K, Kristensen M, Borodina I, Nielsen J (2016) Engineering yeast for high-level production of stilbenoid antioxidants. Sci Rep 6(36827):1–8
Lin Y, Sun X, Yuan Q, Yan Y (2013) Combinatorial biosynthesis of plant-specific coumarins in bacteria. Metab Eng 18(2013):69–77
Liou G, Chiang YC, Wang Y, Weng JK (2018) Mechanistic basis for the evolution of chalcone synthase catalytic cysteine reactivity in land plants. J Biol Chem 293(48):18601–18612
Liu X, Ding W, Jiang H (2017) Engineering microbial cell factories for the production of plant natural products: from design principles to industrial-scale production. Microb Cell Fact 16(1):1–9

Liu L, Liu H, Zhang W, Yao M, Li B, Liu D, Yuan Y (2019a) Engineering the biosynthesis of caffeic acid in *saccharomyces cerevisiae* with heterologous enzyme combinations. Engineering 5(2):287–295

Liu Q, Yu T, Li X, Chen Y, Campbell K, Nielsen J, Chen Y (2019b) Rewiring carbon metabolism in yeast for high level production of aromatic chemicals. Nat Commun 10(1):1–13

Lv Y, Edwards H, Zhou J, Xu P (2019a) Combining 26s rDNA and the Cre-loxP system for iterative gene integration and efficient marker curation in *Yarrowia lipolytica*. ACS Synth Biol 8(3):568–576

Lv Y, Marsafari M, Zhou J, Xu P (2019b) Optimizing oleaginous yeast cell factories for flavonoids and hydroxylated flavonoids biosynthesis. ACS Synth Biol 8:2514–2523

Lyu X, Ng KR, Lee JL, Mark R, Chen WN (2017) Enhancement of naringenin biosynthesis from tyrosine by metabolic engineering of *Saccharomyces cerevisiae*. J Agric Food Chem 65(31):6638–6646

Lyu X, Zhao G, Ng KR, Mark R, Chen WN (2019) Metabolic engineering of *Saccharomyces cerevisiae* for *de Novo* production of kaempferol. J Agric Food Chem 67(19):5596–5606

Marchiosi R, dos Santos WD, Constantin RP, de Lima RB, Soares AR, Finger-Teixeira A, Mota TR, de Oliveira DM, Foletto-Felipe MP, Abrahão J, Ferrarese-Filho O (2020) Biosynthesis and metabolic actions of simple phenolic acids in plants. Phytochem Rev 19(4):865–906

Market Watch (2020) Global trans resveratrol market 2020—indepth analysis of current market trends including industry share, size, manufacturers and future prospects. https://www.marketwatch.com/press-release/global-trans-resveratrol-market-2020—sindepth-analysis-of-current-market-trends-including-industry-share-size-manufacturers-and-future-prospects-2020-09-22?tesla=y

Martillanes S, Rocha-Pimienta J, Cabrera-Bañegil M, Martín-Vertedor D, Delgado-Adámez J (2017) Application of phenolic compounds for food preservation: food additive and active packaging. In: Soto-Hernández M (ed) Phenolic compounds—biological activity. InTechOpen, pp 39–56

Matos MJ, Santana L, Uriarte E, Abreu OA, Molina E, Yordi EG (2015) Coumarins—an important class of phytochemicals. Rao V (eds) Phytochemicals—isolation, characterisation and role in human health. InTechOpen, pp 114–140

Meydani M (2009) Potential health benefits of avenanthramides of oats. Nutr Rev 67(12):731–735

Milke L, Aschenbrenner J, Marienhagen J, Kallscheuer N (2018) Production of plant-derived polyphenols in microorganisms: current state and perspectives. Appl Microbiol Biotechnol 102(4):1575–1585

Moglia A, Goitre L, Gianoglio S, Baldini E, Trapani E, Genre A, Scattina A, Dondo G, Trabalzini L, Beekwilder J, Retta SF (2015) Evaluation of the bioactive properties of avenanthramide analogs produced in recombinant yeast. BioFactors 41(1):15–27

Munack S, Roderer K, Ökvist M, Kamarauskaite J, Sasso S, Van Eerde A, Kast P, Krengel U (2016) Remote control by inter-enzyme allostery: a novel paradigm for regulation of the shikimate pathway. J Mol Biol 428(6):1237–1255

Musarra-Pizzo M, Ginestra G, Smeriglio A, Pennisi R, Sciortino MT, Mandalari G (2019) The antimicrobial and antiviral activity of polyphenols from almond (*Prunus dulcis* L.) skin. Nutrients 11(10):1–11

Niedzwiecki A, Roomi MW, Kalinovsky T, Rath M (2016) Anticancer efficacy of polyphenols and their combinations. Nutrients 8(9):1–17

Palmer CM, Miller KK, Nguyen A, Alper HS (2020) Engineering 4-coumaroyl-CoA derived polyketide production in *Yarrowia lipolytica* through a β-oxidation mediated strategy. Metab Eng 57(2020):174–181

Panche AN, Diwan AD, Chandra SR (2016) Flavonoids: an overview. J Nutr Sci 5(47):1–15

Panda AK, Chakraborty D, Sarkar I, Khan T, Sa G (2017) New insights into therapeutic activity and anticancer properties of curcumin. J Exp Pharmacol 9:31–45

Parage C, Tavares R, Réty S, Baltenweck-Guyot R, Poutaraud A, Renault L, Heintz D, Lugan R, Marais GAB, Aubourg S, Hugueney P (2012) Structural, functional, and evolutionary analysis

of the unusually large stilbene synthase gene family in grapevine. Plant Physiol 160(3):1407–1419
Paraiso IL, Revel JS, Stevens JF (2020) Potential use of polyphenols in the battle against COVID-19. Curr Opin Food Sci 32:149–155
Park S (2015) Polyphenol compound as a transcription factor inhibitor. Nutrients 7(11):8987–9004
Parthasarathy A, Cross PJ, Dobson RCJ, Adams LE, Savka MA, Hudson AO (2018) A three-ring circus: metabolism of the three proteogenic aromatic amino acids and their role in the health of plants and animals. Front Mol Biosci 5:1–30
Pecyna P, Wargula J, Murias M, Kucinska M (2020) More than resveratrol: new insights into stilbene-based compounds. Biomolecules 10(8):1–40
Perrelli A, Goitre L, Salzano AM, Moglia A, Scaloni A, Retta SF (2018) Biological activities, health benefits, and therapeutic properties of avenanthramides: from skin protection to prevention and treatment of cerebrovascular diseases. Oxid Med Cell Longev 2018(6015351):1–17
Proficient Market (2020). Global polyphenols market will register a CAGR of around 9.0% by 2027. https://proficientmarket.com/press-release/1225/global-polyphenols-market
Rainha J, Gomes D, Rodrigues LR, Rodrigues JL (2020a) Synthetic biology approaches to engineer *Saccharomyces cerevisiae* towards the industrial production of valuable polyphenolic compounds. Life 10(56):1–26
Rainha J, Rodrigues JL, Rodrigues LR (2020b) CRISPR-Cas9: a powerful tool to efficiently engineer *Saccharomyces cerevisiae*. Life 11(13):1–16
Rashmi R, Magesh SB, Ramkumar KM, Suryanarayanan S, SubbaRao MV (2017) Antioxidant potential of naringenin helps to protect liver tissue from streptozotocin-induced damage. R Biochem Mol Biol 7(1):76–84
Rivière C, Pawlus AD, Mérillon JM (2012) Natural stilbenoids: distribution in the plant kingdom and chemotaxonomic interest in *Vitaceae*. Nat Prod Rep 29(11):1317–1333
Rodrigues JL, Rodrigues LR (2017) Synthetic biology: perspectives in industrial biotechnology. In: Pandey A, Teixeira J (eds) Current developments in biotechnology and bioengineering: foundations of biotechnology and bioengineering. Elsevier, pp 239–269
Rodrigues JL, Prather KLJ, Kluskens LD, Rodrigues LR (2015a) Heterologous production of curcuminoids. Microbiol Mol Biol Rev 79(1):39–60
Rodrigues JL, Araújo RG, Prather KLJ, Kluskens LD (2015b) Production of curcuminoids from tyrosine by a metabolically engineered *Escherichia coli* using caffeic acid as an intermediate. Biotechnol J 10(4):1–27
Rodrigues JL, Araújo RG, Prather KLJ, Kluskens LD, Rodrigues LR (2015c) Heterologous production of caffeic acid from tyrosine in *Escherichia coli*. Enzyme Microb Technol 71:36–44
Rodrigues JL, Couto MR, Araújo RG, Prather KLJ, Kluskens L, Rodrigues LR (2017a) Hydroxycinnamic acids and curcumin production in engineered *Escherichia coli* using heat shock promoters. Biochem Eng J 125:41–49
Rodrigues JL, Ferreira D, Rodrigues LR (2017b) Synthetic biology strategies towards the development of new bioinspired technologies for medical applications. In: Rodrigues LR, Mota M (eds) Bioinspired materials for medical applications. Elsevier, pp 451–497
Rodrigues JL, Gomes D, Rodrigues LR (2020) A combinatorial approach to optimize the production of curcuminoids from tyrosine in *Escherichia coli*. Front Bioeng Biotechnol 8(59):1–15
Rodrigues JL, Rodrigues LR (2020) Biosynthesis and heterologous production of furanocoumarins: perspectives and current challenges. Natural Product Rep
Rodriguez A, Kildegaard KR, Li M, Borodina I, Nielsen J (2015) Establishment of a yeast platform strain for production of *p*-coumaric acid through metabolic engineering of aromatic amino acid biosynthesis. Metab Eng 31(2015):181–188
Rodriguez A, Chen Y, Khoomrung S, Özdemir E, Borodina I, Nielsen J (2017a) Comparison of the metabolic response to over-production of *p*-coumaric acid in two yeast strains. Metab Eng 44(2017):265–272
Rodriguez A, Strucko T, Stahlhut SG, Kristensen M, Svenssen DK, Forster J, Nielsen J, Borodina I (2017b) Metabolic engineering of yeast for fermentative production of flavonoids. Biores Technol 245:1645–1654

Rüfer CE, Kulling SE (2006) Antioxidant activity of isoflavones and their major metabolites using different in vitro assays. J Agric Food Chem 54(8):2926–2931

Sáez-Sáez J, Wang G, Marella ER, Sudarsan S, Cernuda Pastor M, Borodina I (2020) Engineering the oleaginous yeast *Yarrowia lipolytica* for high-level resveratrol production. Metab Eng 62(2020):51–61

Santos CNS, Koffas M, Stephanopoulos G (2011) Optimization of a heterologous pathway for the production of flavonoids from glucose. Metab Eng 13(4):392–400

Sharma A, Shahzad B, Rehman A, Bhardwaj R, Landi M, Zheng B (2019) Response of phenylpropanoid pathway and the role of polyphenols in plants under abiotic stress. Molecules 24(13):1–22

Shen T, Zie C-F, Wang X-N, Lou H-X (2013) Stilbenoids. In: Ramawat KG, Mérillon J-M (eds) Natural products: phytochemistry, botany and metabolism of alkaloids, phenolics and terpenes. Springer, pp 1902–1948

Shin SY, Han NS, Park YC, Kim MD, Seo JH (2011) Production of resveratrol from p-coumaric acid in recombinant *Saccharomyces cerevisiae* expressing 4-coumarate:coenzyme A ligase and stilbene synthase genes. Enzyme Microb Technol 48(1):48–53

Shin SY, Jung SM, Kim MD, Han NS, Seo JH (2012) Production of resveratrol from tyrosine in metabolically engineered *Saccharomyces cerevisiae.* Enzyme Microb Technol 51(4):211–216

Siddiqui FA, Prakasam G, Chattopadhyay S, Rehman AU, Padder RA, Ansari MA, Irshad R, Mangalhara K, Bamezai RNK, Husain M, Ali SM, Iqbal MA (2018) Curcumin decreases warburg effect in cancer cells by down-regulating pyruvate kinase M2 via mTOR-HIF1α inhibition. Sci Rep 8(1):2–10

Singla RK, Dubey AK, Garg A, Sharma RK, Fiorino M, Ameen SM, Haddad MA, Al-Hiary M (2019) Natural polyphenols: chemical classification, definition of classes, subcategories, and structures. J AOAC Int 102(5):1397–1400

Sobhani M, Farzaei MH, Kiani S, Khodarahmi R (2020) Immunomodulatory; anti-inflammatory/antioxidant effects of polyphenols: A comparative review on the parental compounds and their metabolites. Food Rev Intl 00(00):1–53

Sökmen M, Akram Khan M (2016) The antioxidant activity of some curcuminoids and chalcones. Inflammopharmacology 24(2–3):81–86

Soto-Urquieta MG, López-Briones S, Pérez-Vázquez V, Saavedra-Molina A, González-Hernández GA, Ramírez-Emiliano J (2014) Curcumin restores mitochondrial functions and decreases lipid peroxidation in liver and kidneys of diabetic db/db mice. Biol Res 47(74):1–8

Stringlis IA, De Jonge R, Pieterse CMJ (2019) The age of coumarins in plant-microbe interactions. Plant Cell Physiol 60(7):1405–1419

Sydor T, Schaffer S, Boles E (2010) Considerable increase in resveratrol production by recombinant industrial yeast strains with use of rich medium. Appl Environ Microbiol 76(10):3361–3363

Tanase C, Bujor O-C, Popa VI (2019) Phenolic natural compounds and their influence on physiological processes in plants. In: Watson RR (ed) Polyphenols in plants, 2nd edn. Elsevier Inc, pp 45–58

Taofiq O, González-Paramás AM, Barreiro MF, Ferreira ICFR, McPhee DJ (2017) Hydroxycinnamic acids and their derivatives: cosmeceutical significance, challenges and future perspectives, a review. Molecules 22(2):1–24

Trantas E, Panopoulos N, Ververidis F (2009) Metabolic engineering of the complete pathway leading to heterologous biosynthesis of various flavonoids and stilbenoids in *Saccharomyces cerevisiae.* Metab Eng 11(6):355–366

Tsao R (2010) Chemistry and biochemistry of dietary polyphenols. Nutrients 2(12):1231–1246

Tzin V, Galili G (2010) New Insights into the shikimate and aromatic amino acids biosynthesis pathways in plants. Mol Plant 3(6):956–972

Van de Velde F, Esposito D, Grace MH, Pirovani ME, Lila MA (2019) Anti-inflammatory and wound healing properties of polyphenolic extracts from strawberry and blackberry fruits. Food Res Int 121(2019):453–462

Vázquez-Calvo Á, de Oya NJ, Martín-Acebes MA, Garcia-Moruno E, Saiz JC (2017) Antiviral properties of the natural polyphenols delphinidin and epigallocatechin gallate against the flaviviruses West Nile virus, Zika virus, and Dengue virus. Front Microbiol 8(1314):1–8

Wang Y, Yu O (2012) Synthetic scaffolds increased resveratrol biosynthesis in engineered yeast cells. J Biotechnol 157(1):258–260

Wang Y, Halls C, Zhang J, Matsuno M, Zhang Y, Yu O (2011) Stepwise increase of resveratrol biosynthesis in yeast *Saccharomyces cerevisiae* by metabolic engineering. Metab Eng 13(5):455–463

Wang Y, Bhuiya MW, Zhou R, Yu O (2015a) Pterostilbene production by microorganisms expressing resveratrol O-methyltransferase. Ann Microbiol 65(2):817–826

Wang S, Zhang S, Xiao A, Rasmussen M, Skidmore C (2015b) Metabolic engineering of *Escherichia coli* for the biosynthesis of various phenylpropanoid derivatives. Metab Eng 29:153–159

Wang TY, Tsai YH, Yu IZ, Chang TS (2016) Improving 3′-hydroxygenistein production in recombinant *Pichia pastoris* using periodic hydrogen peroxide-shocking strategy. J Microbiol Biotechnol 26(3):498–502

Wang S, Zhang S, Zhou T, Zeng J, Zhan J (2013) Design and application of an in vivo reporter assay for phenylalanine ammonia-lyase. Appl Microbiol Biotechnol 97(17):7877–7885

Wang A, Wei J, Lu C, Chen H, Zhong X, Lu Y, Li L, Huang H, Dai Z, Han L (2019) Genistein suppresses psoriasis-related inflammation through a STAT3–NF-κB-dependent mechanism in keratinocytes. Int Immunopharmacol 69(October 2018):270–278

Watts KT, Mijts BN, Lee PC, Manning AJ, Schmidt-Dannert C (2006) Discovery of a substrate selectivity switch in tyrosine ammonia-lyase, a member of the aromatic amino acid lyase family. Chem Biol 13(12):1317–1326

Wei W, Zhang P, Shang Y, Zhou Y, Ye BC (2020) Metabolically engineering of *Yarrowia lipolytica* for the biosynthesis of naringenin from a mixture of glucose and xylose. Bioresour Technol 314(2020):123726

Wilson SA, Roberts SC (2012) Recent advances towards development and commercialization of plant cell culture processes for synthesis of biomolecules. Plant Biotechnol J 10(3):249–268

World Health Organization (2018) Cancer. https://www.who.int/news-room/fact-sheets/detail/cancer

Xu X, Liu Y, Du G, Ledesma-Amaro R, Liu L (2020) Microbial chassis development for natural product biosynthesis. Trends Biotechnol 38(7):779–796

Yahfoufi N, Alsadi N, Jambi M, Matar C (2018) The immunomodulatory and anti-inflammatory role of polyphenols. Nutrients 10(11):1–23

Yan Y, Kohli A, Koffas MAG (2005) Biosynthesis of natural flavanones in *Saccharomyces cerevisiae*. Appl Environ Microbiol 71(9):5610–5613

Yang SM, Shim GY, Kim BG, Ahn JH (2015) Biological synthesis of coumarins in *Escherichia coli*. Microb Cell Fact 14(1):1–12. https://doi.org/10.1186/s12934-015-0248-y

Yen YH, Pu CM, Liu CW, Chen YC, Chen YC, Liang CJ, Hsieh JH, Huang HF, Chen YL (2018) Curcumin accelerates cutaneous wound healing via multiple biological actions: the involvement of TNF-α, MMP-9, α-SMA, and collagen. Int Wound J 15(4):605–617

Yin G, Wang Z, Wang Z, Wang X (2018) Topical application of quercetin improves wound healing in pressure ulcer lesions. Exp Dermatol 27(7):779–786

Yu D, Xu F, Zeng J, Zhan J (2012) Type III polyketide synthases in natural product biosynthesis. IUBMB Life 64(4):285–295

Zabalza A, Orcaray L, Fernández-Escalada M, Zulet-González A, Royuela M (2017) The pattern of shikimate pathway and phenylpropanoids after inhibition by glyphosate or quinate feeding in pea roots. Pestic Biochem Physiol 141(2017):96–102

Zhang Y, Zhang R, Ni H (2020) Eriodictyol exerts potent anticancer activity against A549 human lung cancer cell line by inducing mitochondrial-mediated apoptosis, G2/M cell cycle arrest and inhibition of m-TOR/PI3K/Akt signalling pathway. Arch Med Sci 16(2):446–452

Zhao Y, Jian X, Wu J, Huang W, Huang C, Luo J, Kong L (2019) Elucidation of the biosynthesis pathway and heterologous construction of a sustainable route for producing umbelliferone. J Biol Eng 13(1):1–13

Zhao CH, Zhang RK, Qiao B, Li BZ, Yuan YJ (2020) Engineering budding yeast for the production of coumarins from lignin. Biochem Eng J 160(May):107634

Zeng YH, Zhou LY, Chen QZ, Li Y, Shao Y, Ren WY, Liao YP, Wang H, Zhu JH, Huang M, He F, Wang J, Wu K, He BC (2017) Resveratrol inactivates PI3K/Akt signaling through upregulating BMP7 in human colon cancer cells. Oncol Rep 38(1):456–464

Yeast Synthetic Biology for Production of Artemisinin as an Antimalarial Drug

Arman Beyraghdar Kashkooli, Karim Farmanpour-Kalalagh, and Alireza Babaei

Abstract

Artemisinin, a sesquiterpene endoperoxide lactone derived from *Artemisia annua*, is highly effective against malaria parasite *Plasmodium falciparum*. Artemisinin and its derivatives (collectively termed artemisinins) demonstrate additional anticancer, anti-inflammation, antiviral, and anti-SARS-CoV-2 activity. Because of the expensive medicine of artemisinins and the low content of artemisinin biosynthesis in native host, researchers have been endeavored to produce artemisinin via alternative approaches. Previous attempts give attention to increase artemisinin biosynthesis via common plant breeding techniques and on engineering cultivated plants for wide production. However, current trends are focusing on the bioengineering of artemisinin biosynthetic pathway in heterologous expression platforms. To date, *in planta* artemisinin production in two model plants, tobacco (*Nicotiana benthamiana*) and moss (*Physcomitrella patens*), has been reported successfully. Nevertheless, to meet the large-scale demands, de novo reconstituting of artemisinin biosynthetic pathway in heterologous microorganisms such as *Escherichia coli* and *Saccharomyces cerevisiae* has had documentary achievements. Since there are challenges for the expression of cytochrome P450s (CYPs) from the eukaryotic origin (such as CYP71AV1) in *E. coli*, *S. cerevisiae* has been proposed as a heterologous host for production of artemisinin precursors such as amorphadiene, artemisinic acid, and dihydro artemisinic acid. Moreover, the presence of the MVA biosynthetic pathway for production of universal sesquiterpenes precursor, farnesyl pyrophosphate (FPP), and relative compatibility of *S. cerevisiae* cell environment for the expression of plant-origin genes (enzymes) have made this microorganism as a favorable platform for artemisinins production.

A. Beyraghdar Kashkooli (✉) · K. Farmanpour-Kalalagh · A. Babaei
Department of Horticultural Science, Faculty of Agriculture, Tarbiat Modares University, 14115-365, Tehran, Iran
e-mail: A.beyraghdar@modares.ac.ir

F. Darvishi Harzevili (ed.), *Synthetic Biology of Yeasts*,
https://doi.org/10.1007/978-3-030-89680-5_6

1 Introduction

Plants have developed the capacity to make a considerable diversity of specialized (secondary) metabolites. These metabolites have different classifications and groups which are produced via specific biosynthetic pathways. Terpenoids, containing the 5 carbon backbone in their structure, are the largest class of secondary metabolites. One of the most widely studied groups in this class is sesquiterpene lactones (SLs). SLs have different types and each has various applications in different industries. Artemisinin is an SL that is exclusively found in the plant *Artemisia annua*. Artemisinin, artemisinin derivatives (collectively termed as artemisinins), and artemisinin-based combination therapies (ACTs) are the first-line antimalarial drugs recommended by World Health Organization (Lin and Pakrasi 2018). In addition, artemisinins are effective in the treatment of various cancers, inflammations, viruses (Efferth 2018), and very recently in the fight against SARS-CoV-2 (Krishna et al. 2021; Nair et al. 2021; Uckun et al. 2021). Due to the low production of artemisinin in *A. annua*, classical methods such as nutrient manipulations, biotic effectors, enhancing glandular trichomes through plant breeding methods, and application of plant growth regulators have been done to increase artemisinin production. However, obviously, these attempts and the use of classical methods did not meet global artemisinin demand. Recent studies indicated that artemisinin production can be enhanced via plant manipulation techniques as well as the application of metabolic engineering methods of microorganisms. Plant biotechnology techniques including tissue culture (native host regeneration, callusgensis, etc.), hairy root and gall induction by *Agrobacterium rhizogenes* and *A. tumefaciens*, respectively, ploidy induction, adventitious roots induction, suspension culture, use of bioreactors, metabolic engineering in native and *in planta* heterologous host(s) have had significant success in artemisinin production. On the other hand, de novo reconstitution of artemisinin biosynthetic pathway in heterologous microorganisms such as *E. coli* and *S. cerevisiae* (baker's or brewer's yeast) increased artemisinin precursor production on large scale. In this chapter, we reviewed artemisinin production advances via *S. cerevisiae* using metabolic engineering and synthetic biology techniques.

2 Importance of Terpenoids Production

Terpenoids (isoprenoids) as the oldest and most diverse group of natural molecules represent the first instance of natural bioactive compounds (Pateraki et al. 2015). Most terpenes are isolated from plants which play an important role in some plant's metabolism (Davis and Croteau 2000). In plants, terpenes are involved in the biosynthesis of the phytohormones gibberellic acid, abscisic acid, and cytokinins and serve as precursors of steroid hormones. They also act as biosynthetic precursors for carotenoid pigments and phytol side chains in chlorophyll and components of electron-carrying coenzymes such as quinone, ubiquinone, and plastoquinone. The terpenes–environment interactions are more important as a chemical defense system against herbivores and pathogens, but they also have other roles such as absorption of pollinators and allopathic properties. Apart from their potential roles in plants, bioactive terpenoids have raised considerable interest as pharmaceuticals, nutraceuticals, and flavors and fragrances (Pateraki et al. 2015).

Structurally, terpenoids are various polymers created from the fundamental C5 block, the isoprene unit. They are organized and grouped based on the number of isoprene units. Therefore, hemiterpenes, monoterpenes, sesquiterpenes, triterpenes, diterpenes, and carotenoids contain 5, 10, 15, 20, 30, and 40 isoprene units, respectively. The long-chain polymers possess thousands of units and comprise natural products such as rubber. Regardless of their diversity, all terpenoids are biosynthesized from the universal precursors dimethylallyl pyrophosphate (DMAPP) and isopentenyl pyrophosphate (IPP, isopentenyl diphosphate, or IDP). This occurs via two different pathways, the methyl erythritol 4-phosphate (MEP; mevalonate-independent, or non-mevalonate) pathway and the mevalonate pathway (MVA) (Capell and Christou 2004).

3 Terpenoids Biosynthetic Pathway in Plants

Biosynthesis of terpenes can be divided into four stages. The first step involves the production of isopentenyl diphosphate (IPP) and its isomer, dimethylallyl pyrophosphate (DMAPP), which is carried out by isopentenyl diphosphate isomerase (IPP isomerase). In the second stage, IPP and DMAPP are structurally elongated to form more complex isoprenoid backbones called geranyl pyrophosphate (GPP C10), farnesyl pyrophosphate (FPP, C15), geranylgeranyl pyrophosphate (GGPP, C20). The third stage involves the conversion of GPP, FPP, and GGPP to the corresponding terpene groups producing monoterpenes, sesquiterpenes, and diterpenes, respectively. The final step in the biosynthesis of terpenes occurs in the cytosol (e.g., cyclization, hydroxylation, and carboxylation) which produces derivatives of the basic terpene groups (Davis and Croteau 2000; Moses et al. 2013).

In general, the major pathway for the biosynthesis of monoterpenes is within plastids. Monoterpenes can also be metabolically engineered to be produced in the cytosol, although these pathways can also produce intermediates (Bouwmeester 2006). In plants, two distinct but interacting pathways have been identified for IPP biosynthesis: the mevalonate (MVA) or cytosolic pathway, the methylerythritol 4-phosphate (MEP) pathway or the plastid pathway.

3.1 MEP (Methylerythritol 4-Phosphate) Pathway Versus the MVA (Mevalonate) Pathway

In MEP, pyruvate is first combined with glyceraldehyde-3-phosphate (G3P). After a series of reactions, 1-deoxy-d-xylulose-5-phosphate (DXP) is formed. This compound is then converted to methylerythritol-4-phosphate, and after a series of other steps, it is also converted to isopentyl diphosphate (IPP) (Moses et al. 2013).

The MEP pathway also referred to as the 1- deoxy-D-xylulose 5-phosphate (DXP) contains seven enzymatic steps and starts with the condensation of D-glyceraldehyde 3-phosphate (GAP) and pyruvate to generate 1-deoxy-D-xylulose 5-phosphate (DXP), which then undergoes isomerization/reduction with the formation of MEP. Five sequential steps are essential to shift MEP to IPP and DMAPP (Fig. 1). The MEP pathway depends on early metabolism for the provision of pyruvate and GAP, with the subsequent resulting from both glycolysis and the pentose phosphate pathway (PPP). Both IPP and DMAPP are substrates for short-chain prenyltransferases (PTs), which create prenyl diphosphate precursors, geranyl diphosphate (GPP), farnesyl diphosphate (FPP), and geranylgeranyl diphosphate (GGPP), for a numerous group of terpene synthases/cyclases (Dudareva et al. 2013; Moses et al. 2013; Vranova et al. 2013).

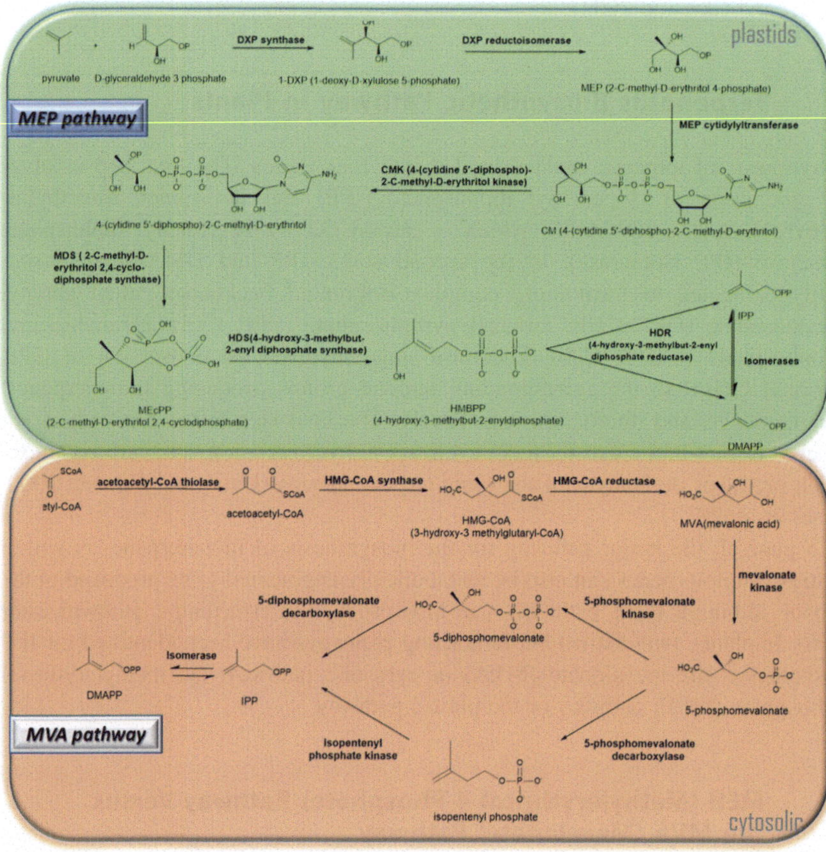

Fig. 1 MEP (methylerythritol 4-phosphate) and MVA (mevalonate) biosynthetic pathway in plants

In parallel to the MEP pathway, the mevalonate pathway (MVA) is one of the most important cellular metabolic pathways in all eukaryotes and many bacteria. This pathway aims to produce dimethylallyl diphosphate (DMAPP) as well as isopentenyl diphosphate (IPP), both of which act as a basis for molecular biosynthesis in a variety of pathways, including the synthesis of terpenoids, cell membrane preservation and regeneration, and many other cellular hormones. In the MVA pathway, two molecules of acetyl-CoA first combine to form acetoacetyl-CoA. Then, by adding another molecule of acetyl-CoA to acetoacetyl-CoA, and then by performing a series of other steps, mevalonate is obtained. Mevalonate is then converted to IPP under the influence of other steps and enzymes, and IPP is converted to DMAPP under the influence of isopentenyl diphosphate isomerase. IPP is elongated by the enzyme prenyltransferase, and geranyl diphosphate is produced by the enzyme geranyl diphosphate synthase, which is also called the precursor (monoterpene synthases) produces different types of monoterpenes (Vranova et al. 2012; Moses et al. 2013) (Fig. 1).

3.2 Sesquiterpene Lactones

Sesquiterpene lactones (SLs), a subgroup of the sesquiterpenoids, contain over 4000 various known structures. SLs are mostly colorless, bitter, and often aromatic constituents of plant essential oils found mostly in some plant species from *Asteraceae*. Their biological activities such as anticancer, antimalarial, antimigraine, antioxidant, antibacterial, antifungal, anti-inflammation, antiviral, and anti-SARS-CoV-2 have generated attention for medical uses. SLs are classified into six bicyclic or tricyclic subclasses including guaianolides, pseudoguaianolides, xanthanolides, eremophilanolides, eudesmanolides, and germacranolides. The subsequent modification of SLs is followed by double-bond reductases and glycosyltransferases (reviewed in Beyraghdar Kashkooli et al. 2018).

Sesquiterpenoids are biosynthesized from FPP, via the catalytic activity of terpene synthases (in this case, sesquiterpene synthases (STPS)). Molecular genetics and phytochemistry studies indicated that the biosynthesis of STLs in plants created a multiplex network of metabolic routes that compete for general precursors. Furthermore, it is also exclusively regulated by numerous temporal, spatial, and environmental factors. Some SLs biosynthetic pathways including costunolide, parthenolide, kauniolide (Liu et al. 2005), and artemisinin production have been fully characterized. On the other hand, a number of transcription factors that manage the transcription of biosynthetic genes have been characterized, which specified a great comprehension of how these STLs are biosynthesized. This information has led to a high contribution in characterization understanding of the other SL biosynthetic pathways in plants. More precisely, the regulation is specific for each SL biosynthetic pathway. Thus, it is imperative to identify the various factors that regulate and control the upstream (MVA and) and SLs-specific downstream biosynthetic pathway in every SL-producing plant (Perassolo et al. 2018).

3.3 Artemisinin Biosynthetic Pathway in *Artemisia annua*

It is important to deeply understand the key biosynthetic pathway routes and genes involved in artemisinin biosynthesis in order to be able to enhance artemisinin yield in the natural artemisinin producing plant, *A. annua,* as well as other heterologous host platforms. artemisinin is a sesquiterpene lactone, containing 15 carbon atoms with lipophilic properties (Lv et al. 2017). Both artemisinin and arteannuin B are produced via a biosynthetic pathway in the glandular trichomes (GTs) in leaves and oval of the *A. annua* (Wang et al. 2016). However, some reports have shown that the highest GTs densities are detected on flowers, buds, and young leaves of this plant (Olofsson et al. 2011). The biosynthetic pathway of GTs occurs via the terpenoids biosynthetic pathway. Isoprene units of terpenoids are obtained from two pathways: the MVA pathway, which occurs in the cytosol, MEP pathway in the plastids (Dewick 2009; Vranová et al. 2013). The artemisinin biosynthetic pathway starts with the conversion of 5-carbon units of IPP and MAPP units to FPP (Fuentes et al. 2016). A study on the artemisinin biosynthetic pathway shows that the first and key step in biosynthesis of this compound is the conversion of FPP to amorpha 4, 11-diene (amorphadiene), which is catalyzed by a well-known terpene cyclase, the amorphadiene synthase (ADS) (Bertea et al. 2005; Mercke et al. 2000). The cytochrome P450 hydroxylase (CYP71AV1) (Teoh et al. 2006) then converts amorphadiene to artemisinic alcohol. CYP71AV1 also oxidizes artemisinic alcohol to artemisinic aldehyde and artemisinic acid, respectively. Artemisinic aldehyde reductase (DBR2) and aldehyde dehydrogenase 1 (ALDH1) (Zhang et al. 2008), as the branching point, converts artemisinic aldehyde to dihydroartemisinic aldehyde (Bertea et al. 2005; Schramek et al. 2010). Bifurcation points of artemisinic aldehyde biosynthetic pathway can synthesize artemisinic acid by CYP71AV1 and ALDH1. Also, due to the non-enzymatic reaction, dihydroartemisinic acid and artemisinic acid are converted to artemisinin and arteannuin B, respectively (Brown 2010; Han et al. 2014; Ikram and Simonsen 2017). ALDH1 then produces dihydroartemisinic acid (DHAA) from dihydroartemisinic aldehyde. DHAA is considered as a precursor to the artemisinin (bio)synthesis (Tang et al. 2014), and finally its conversion to artemisinin is considered as a non-enzymatic reaction which occurs under the influence of UV light and oxygen (Bertea et al. 2005; Ikram et al. 2019; Schramek et al. 2010; Teoh et al. 2006) (Fig. 2).

Studies show that all genes responsible for biosynthesis of artemisinin are fully identified (Ikram et al. 2019). Identification of the genes involved in the biosynthetic pathway of artemisinin offers an opportunity to increase artemisinin production and subtle reduction in its production costs through plant breeding and/or the engineering of other putative hosts (Zhang et al. 2008).

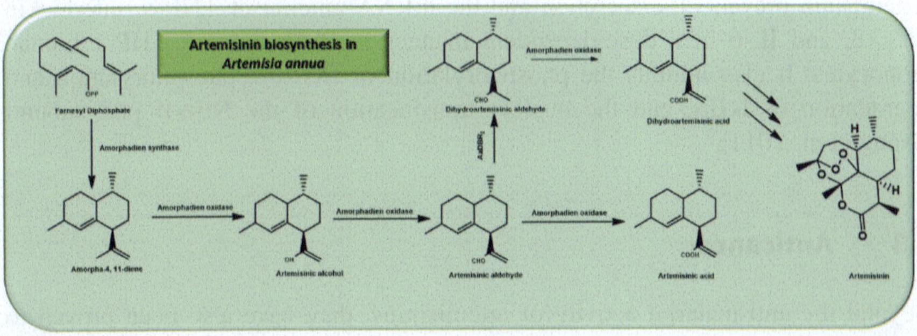

Fig. 2 Artemisinin biosynthetic pathway in *Artemisia annua* L

4 Application of Artemisinins in Medicine

4.1 Antimalarial

Malaria has been a worldwide outbreak health threat since old times. Scientific and hospital reports show that many patients die every year due to malaria infections. Artemisinins and artemisinin-based combination therapies (ACTs) are frontline antimalarial agents accepted for their potency and low toxicity (Yang et al. 2020). In the early 1970s, a group of Chinese scientists led by Professor Tu Youyou (winner of Nobel prize in physiology or medicine 2015) succeeded in obtaining the effective results of *A. annua* plant extracts on malaria (Liu and Liu 2016; Su and Miller 2015). By testing the natural extract from the *A. annua*, Professor Tu found that the compound could have 100% inhibitory effect on malaria parasites in vivo experiments (Tu 2011). In the late 1970s, artemisinin was effectively recognized as an anti-malarial drug (Klayman 1985; Tu 2011; Xie et al. 2016). So far, several mechanisms of action such as interfering with the heme detoxification pathway, creating protein damage and inhibiting parasite proteasome function, inducing the alkylation of PfTCTP and inhibiting PfATPase, and membrane depolarizing of mitochondria via ROS production have been reported for artemisinin to fight malaria parasites (*Plasmodium falciparum* and *Plasmodium vivax*).

4.2 Anti-inflammation

The anti-inflammatory activity of artemisinins have been attributed to the regulating of pro-inflammatory cytokines expression, inhibition of nuclear factor-kappa B (NF-κB), toll-like receptors, Syk tyrosine kinase, phospholipase Cγ, PI3K/Akt, MAPK, STAT-1/3/5, Sp1, Nrf2/ARE signaling pathways, matrix metalloproteinases (MMPs), vascular endothelial growth factor (VEGF), promoting cell cycle arrest, driving reactive oxygen species (ROS) production, and inducing Bak or Bax-dependent or independent apoptosis (Ho et al. 2014; Cheong et al. 2020).

Artemisinin prevents the secretion and the mRNA amount of TNF-α, interleukin (IL)-1ß, and IL-6 in a dose-dependent manner in PMA-induced THP-1 human monocytes. It also inhibits the phosphorylation of IKKα/ß, phosphorylation and degradation of IκBα, and the nuclear translocation of the NF-κB p65 subunit (Wang et al. 2011).

4.3 Anticancer

Beyond the anti-malarial activity of artemisinins, they have also been proven to contain anticancer effects, indicating cytotoxic effects in combating numerous cancer types. These activities look to be done by artemisinin-induced modifications in several signaling pathways, interfering with various hallmarks of cancer at the same time. Considerable successes have been obtained to distinguish these pathways and to display their anticancer mechanisms of action (Wong et al. 2017). Recent investigations have documented the strong anticancer activity of artemisinins (Slezakova and Ruda-Kucerova 2017). The main mechanisms of action are reported for artemisinin in renal (Yu et al. 2019), breast (Cao et al. 2019; Li et al. 2019), ovarian (Liang et al. 2019), prostate (Willoughby et al. 2009), colon (Riganti et al. 2009), cervical (Mondal and Chatterji 2015), ishikawa endometrial (Tran et al. 2014), neuroblastoma (Zhu et al. 2014), nasopharyngeal (Wu et al. 2011), gastric (Zhang et al. 2014), leukemia (Ohgami et al. 2010), lung (Li et al. 2018), fibrosarcoma tumors (Singh and Lai 2004), for artesunate in cancer stem cells (Subedi et al. 2016), prostate (Nunes et al. 2017), head and neck (Roh et al. 2017), cervical (Thanaketpaisarn et al. 2011), colorectal (Chen et al. 2017), bladder (Zhao et al. 2020; Zhou et al. 2020), leukemia (Zhou et al. 2007), skin (Jiang et al. 2012), liver (Yin et al. 2020), laryngeal (Singh and Verma 2002), ovarian (McDowell et al. 2021), glioblastoma (Berdelle et al. 2011), rhabdomyosarcoma (Beccafico et al. 2015), breast (Pirali et al. 2020), endometrial (Yin et al. 2021), lung (Ma et al. 2011), colon (Hao et al. 2020), non-small cell lung (Wang et al. 2020), pancreatic (Wang et al. 2019), gastric (Zhang et al. 2015), beta-cell lymphoma (Våtsveen et al. 2018), and esophageal (Fei et al. 2018), for dihydroartemisinin in colorectal (Yao et al. 2018), cervical (Disbrow et al. 2005), esophageal (Li et al. 2018a, b), breast (Mao et al. 2013), hepatocellular (Wu et al. 2019), ovarian (Liu et al. 2018), cholangiocarcinoma and hepatocarcinoma (Chaijaroenkul et al. 2011), pancreatic (Chen et al. 2009), non-small-cell lung (Jiang et al. 2016), for artemisone in melanoma, breast, colon, and pancreatic (Gravett et al. 2011; Dwivedi et al. 2015), and for artemether in gastric (Alcântara et al. 2013) cancer cells.

4.4 Antiviral and Anti-SARS-CoV-2

Antiviral activity of artemisinins against DNA viruses of the Hepadnaviridae and Herpesviridae families such as Epstein-Barr virus, cytomegaloviruses, herpes simplex viruses 1 and 2, hepatitis B virus, and human herpesvirus 6 have been reported based on in vitro and in vivo studies. The confirmation is weaker for papilloma and polyomaviruses. Also, low or no prevention in vitro activity has been documented for RNA viruses such as influenza virus, human immunodeficiency viruses 1 and 2, and hepatitis C virus. (Efferth 2018). As mentioned above, artemisinins with anti-inflammatory properties contain wide-spectrum antiviral activity. Therefore, the inhibition of interleukins such as interleukin-6 (IL-6) plays a key role in combating SARS-CoV-2 (Krishna et al. 2021). Also, it is reported that the anti-SARS-CoV-2 activity of artemisinins has been attributed to its potency to prevent spike-protein-mediated and TGF-β-dependent primary phases in the inflammation process as well as its capability to destroy the post-entry intracellular events of the SARS-CoV-2 inflammation cycle needed for viral replication (Uckun et al. 2021).

5 Yeast Synthetic Biology for Artemisinin Precursors Production

Extensive application of the beneficial secondary metabolites has been hindered by little yield in vivo and the high expense of chemical synthesis in vitro. New production techniques are essential to provide the ever-growing requirements for these valuable metabolites. Previous attempts give attention to raising production via common breeding techniques, with low positive results, and on engineering cultivated plants for wide production in bioreactors. However, recent studies are focusing on bioengineering for these special biochemical routes (Arsenault et al. 2008). Also, the use of recombinant DNA technology in reconstituting metabolic pathways can increase the biosynthesis of metabolites by modification in the network rates and distribution (Farid et al. 2019). On the other hand, because of the economic importance of plant metabolites and the increased demand, researchers are looking for methods to produce extraordinary amounts of these compounds. Synthetic biology (SynBio) is a fascinating field in modern science, which is growing rapidly. There is no definite and accepted SynBio definition yet, but in general, SynBio is a new interdisciplinary field that includes the application of engineering in biology. Furthermore, recent advances in SynBio in microorganisms and metabolic engineering of compounds have made it possible to produce molecules that were previously impossible to be made heterologously (Paddon and Keasling 2014). The possibility of creating heterogeneous pathways in different microorganisms has led to an increase in the production of shrinking molecules (Tsuruta et al. 2009).

Due to the low performance of artemisinin biosynthesis, synthetic artemisinin biology started to eliminate these problems using various approaches. As previously mentioned, artemisinin biosynthesis pathway requires 11 different enzymatic steps which the pathway starts with the acetyl-CoA precursor. Also, newly recognized genes, such as cytochrome P450 and its related reductase, have been indicated to catalyze several phases in the artemisinin biochemical network. This has the promising results to create a semi-synthetic strategy to production that is both feasible and profitable. Artemisinin precursors de novo biosynthesis in engineered yeast is about many orders of the content above field-grown *A. annua* plant (Arsenault et al. 2008). Because cytochromes P450 enzymes (CYPs) from eukaryotic origin (such as CYP71AV1) cannot be perfectly expressed in *E. coli*, yeast was selected as a heterologous host for artemisinin precursor production (Paddon and Keasling 2014). In addition, this platform is more methodically and technically strong and less sensitive to phage contamination (Kong et al. 2013; Krivoruchko and Nielsen 2015; Perassolo et al., 2018). Attempts to raise flux by engineered networks are growing in yeast via combinations of engineering precursors networks and downstream improving gene expression (Arsenault et al. 2008).

Yeast is a single-celled fungi and one of the best-characterized eukaryotic organisms. Many cost-effective and beneficial varieties of yeast are available. Yeasts contain considerable functions in the fermentation of many food products such as bread and cheese. (Tamang and Fleet 2009). *S. cerevisiae* is one of the most studied eukaryotic platform organisms. Availability of a high density of elucidated information regarding its physiology, genetics, and biochemistry offers a significant improvement in the industrial application of this microorganism. The yeast model gives numerous similar advantages as prokaryotic platforms performed for terpene hydrocarbon biosynthesis, which also supplies the biosynthetic tools necessary for the appropriate functional expression of the downstream modifying enzymes such as cytochrome P450 hydroxylases (Kong et al. 2013). *S. cerevisiae* has been extensively used in biotechnology due to its generally regarded as safe (GRAS) is proper for large-scale performance. As a eukaryotic platform, cell and molecular biology of *S. cerevisiae* have been investigated comprehensively with many genetic engineering existing techniques. Additionally, *S. cerevisiae* displays high tolerance against inappropriate industrial conditions (Hong and Nielsen 2012). Thus, *S. cerevisiae* has been introduced as a model microorganism for synthetic biology (Lian et al. 2018). Therefore, de novo production of the artemisinin via metabolic engineering and SynBio in *S. cerevisiae* has relied on the complete knowledge of this organism. In this regard, an effective effort to produce semi-synthetic artemisinin was made in collaboration with the University of California, Amyris, and OneWorld Health, with the primary goal of host engineering to produce high content of artemisinin precursors that could be converted to the artemisinin by chemical processes (Paddon and Keasling 2014). Remarkably, it is noteworthy to imply that Amyris company is an industrial claimant for the production of artemisinic acid in the *S. cerevisiae* platform (Tang et al. 2014). The efficient application of SynBio for the semi-synthetic industrial production of artemisinin was reported in *S. cerevisiae* (Paddon et al. 2013), as one of the prominent points of synthetic artemisinin biology (Immethun et al. 2013). Here, we review and compartmentalize the de novo reconstitution and production of components involved in the artemisinin biosynthetic pathway in *S. cerevisiae*.

5.1 Amorphadiene Production

In order to increase amorphadiene, a volatile precursor of artemisinin, success reports have been documented. An amorphadiene synthase (ADS), a sesquiterpene synthase, can catalyze the cyclization of farnesyl pyrophosphate (FPP) to amorphadiene biosynthesis. The genomic ADS contains a complex structure containing six introns and seven exons and is classified as class III terpene synthase (Li et al. 2006). It is believed that transcription of ADS possesses tissue and cell specificity (Alejos-Gonzalez et al. 2011). ADS is detected at a bifurcation point of terpene metabolism, and it is regarded as an essential enzyme in the artemisinin biosynthesis. The little amount of the volatile amorphadiene in the *A. annua* and the high ADS activity was suggested to be potent confirmation that amorphadiene is an intermediate in artemisinin biosynthesis (Kong et al. 2013). FPP is a usual intermediate metabolite of various compounds in the MVA pathway of yeast (Baadhe et al. 2013). Therefore, the pivotal plan is to improve FPP biosynthesis and to regulate it toward the amorphadiene synthesis network instead of other competing pathways like sterol synthesis. Several genes in the biosynthesis of FPP in *S. cerevisiae* including *HMGR* and *ERG20* have been cloned successfully (Wang et al. 2013). Enhancing the expression of *HMGR* and *ERG20* can boost the titer of FPP (Han et al. 2008; Ram et al. 2010; Peng et al. 2011). The expression of the ADS gene plus GAL1 promoter in the BY4742 strain produces a low titer of amorphadiene (Ro et al. 2006). It is described that carbon flux extends toward IPP when the N-terminal regulatory region of HMGR is removed (Donald et al. 1997). In the WHT strain of *S. cerevisiae*, the expression of tHMGR, ADS, and ERG20 via GAPDH1, GAPDH1, and ADH1 promoters, respectively, enhances the amorphadiene production in yeast (Kong et al. 2007).

Because of some competing networks commencing with the universal FPP precursor, the overexpression of ERG20 and tHMGR genes may not exclusively enhance the carbon flux toward the biosynthesis of amorphadiene (Paddon et al. 2013). Squalene synthase (SQS), encoded by EGR9, which is responsible for the biosynthesis of squalene in the full process is the main competing branch point (Paradise et al. 2008). Results of other studies indicated that the FPP conversion to amorphadiene by expression of heterologous amorphadiene synthase (*ADS*) has been successful in yeast. First, ERG9 promoter of *S. cerevisiae* was replaced with repressible methionine (MET3) promoter via bipartite gene fusion method. Then, to dominate the reduction of the intermediate FPP by competitive networks in *S. cerevisiae*, fusion protein technique was used, and FPP synthase of yeast has been combined with ADS of *A. annua* where amorphadiene production was increased in yeast strains YCF-002 (11.2 mg/l) and YCF-005 (25.02 mg/l) compared with the control YCF-AD (5.5 mg/l) strain, respectively (Baadhe et al. 2013).

Totally, as mentioned above, since the expression of *ADS* alone leads to low levels of amorphadiene, some modifications were created in the MVA pathway intending to enhance FPP formation such as coupling the overexpression of two copies of *tHMGR*, the overexpression of the gene encoding FPS (*ERG20*), the overexpression of *UPC2–1*, which is an activated allele of UPC2 (a common transcription factor which influences the expression of several enzymes from the MVA pathway), and the downregulation of *ERG9* by a methionine repressible promoter (PMET3) amorphadiene production. (Perassolo et al. 2018).

5.2 Artemisinic Acid Production

To introduce an industrially potent yeast strain, several biological engineering attempts have been used to notably increase artemisinic acid production in the engineered microorganism (Kong et al. 2013). Regarding the cost and possibility in industry, *S. cerevisiae* is distinguished as a good model for artemisinic acid synthesis (reviewed in Li et al. 2016). To increase the biosynthesis of artemisinic acid in *S. cerevisiae* cells, some plans have been established, such as the overexpression of basic genes, and the repression (knock-down) of competing for biosynthetic pathway networks (Paddon and Keasling 2014). All these common approaches are mostly focused on enhancing carbon flux toward heterologous biosynthetic pathways in microorganism cells. It has been shown that the downregulation or upregulation of crucial genes can lead to malfunction of metabolism and consequently, the cell growth could be prevented by the accumulation of toxic intermediates (Sun et al. 2014).

The first heterologous artemisinic acid production in *S. cerevisiae* was reported by Ro et al., in 2006. In the engineered *S. cerevisiae*, production of artemisinic acid increased to 100 mg L^{-1} when all *A. annua*-derived genes (*ADS*, *CYP71AV1*, and *CPR*) were expressed on a single platform. The transgenic *S. cerevisiae* synthesized high artemisinic acid compared to relative native host biomass and in a shorter time. The effective conversion of amorphadiene to artemisinic acid by *S. cerevisiae* expressing the ER-bound CYP71AV1 convinced enough scientists to suggest it as an optimal model for expression of genes encoding membrane-bound enzymes.

In the last part of the pathway, although the complete biosynthetic pathway of artemisinin has been discovered, the final step, the conversion of dihydroartemisinic acid to artemisinin, which occurs non-enzymatically must be accomplished via chemical processes (Li et al. 2016) (Fig. 3).

6 Production of Artemisinin *in planta* Versus Application of Yeast SynBio

6.1 Metabolic Engineering of Artemisinin in *Artemisia annua*

The unstable yield and production of artemisinin is a major problem in the global need for this compound (Judd et al. 2019). So that, the highest amount of artemisinin was reported 2 % based on plant dry weight (Zhang et al. 2008). Metabolic engineering is considered a promising biotechnological technique and as a new method to improve the production of ARTs in native and heterologous

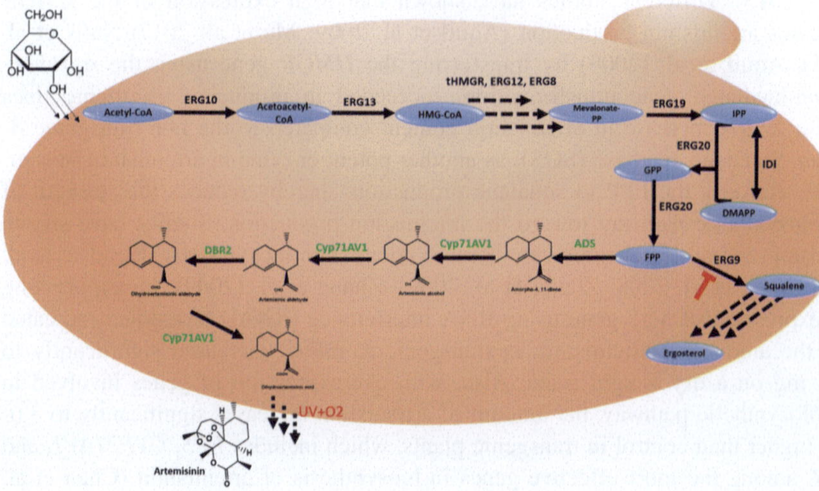

Fig. 3 De novo reconstitution of artemisinin biosynthetic pathway in *Saccharomyces cerevisiae*. **ERG10**, acetyl-CoA C-acetyltransferase; **ERG13**, hydroxymethylglutaryl-CoA synthase; **ERG12**, mevalonate kinase; **tHMGR**, truncated 3-hydroxy-3-methylglutaryl-CoA reductase; **ERG8**, phosphomevalonate kinase; **ERG19**, diphosphomevalonate decarboxylase; **IDI1**, isopentenyl-diphosphate delta-isomerase; **ERG20**, farnesyl diphosphate synthase/dimethylallyltranstransferase; **ERG9**, squalene synthase; **BTS1**, geranylgeranyl diphosphate synthase

hosts (Covello 2008; Ikram et al. 2019; Ma et al. 2015; Xie et al. 2016). Many efforts have been made to increase the production of artemisinin during various experiments. However, there are still lots of potential routes to increase the production of this valuable compound (Lv et al. 2017). To overcome the unsustainable production and low-yield of artemisinin in *A. annua*, attempts led to positive results in biological systems as native and heterologous expression systems (Paddon et al. 2013). One of the main approaches pursued in metabolic engineering is the sustainable production of artemisinin to benefit underdeveloped and more vulnerable communities (Ikram et al. 2017).

Insertion of *Agrobacterium*'s rol genes is considered an effective and efficient method for increasing the secondary metabolites in plants (Bulgakov 2008). Dilshad et al. (2015) indicated that transformation via *rol B* and *rol C* genes show high levels of gene expression and artemisinin production (9 and 4 times more than a non-transgenic plant, respectively). Also, the expression of the *ipt* gene transmitted by *A. tumefaciens* increases the content of artemisinin (30–70%) (Sa et al. 2001).

As mentioned earlier, ADS, Cytochrome P450 (CYP71AV1), double-bond reductase 2 (DBR2), and aldehyde dehydrogenase 1 (ALDH1) catalyze the different stages of artemisinin biosynthesis pathway. One of the strategies used in artemisinin metabolic engineering is to increase the expression of genes involved in the biosynthetic pathway. *3-hydroxy-3-methylglutaryl-CoA reductase* (*HMGR*) gene is known to be one of the most important factors in MVA pathway (Farhi et al. 2013). Different studies have shown that high expression of the *HMGR* increases artemisinin production (Aquil et al. 2009; Ma et al. 2017; Nafis et al. 2011). Aquil et al. (2009) by transferring the *HMGR* gene using the *Agrobacterium*-mediated gene transformation, succeeded in producing transgenic lines with a 22.5% increase in artemisinin content compared to the non-transgenic *A. annua*. Squalene synthase (SQS), as another potent enzyme in artemisinin biosynthesis, converts the FPP to squalene production, thereby reduces the strength of the biosynthetic pathway toward the artemisinin production. Studies have shown that suppressing this enzyme can increase artemisinin biosynthesis (Paradise et al. 2008; Yang et al. 2008; Zhang et al. 2009). Zhang et al. (2009), by suppressing the expression of *SQS* gene using RNA interference (RNAi) technique, indicated that the amount of artemisinin in transgenic *A. annua* increased significantly to 31.4 mg on a dry weight scale. Also, with overexpression of genes involved in the biosynthetic pathway, the amount of artemisinin increases significantly to 3.6-fold higher than control in transgenic plants, which includes *FPS*, *CYP71AV1*, and *CPR*, among the most effective genes in biosynthesis of artemisinin (Chen et al. 2013). Furthermore, in separate studies, the amount of artemisinin in transgenic plants was 38% higher than normal plants, and it was observed that the overexpression of the *farnesyl pyrophosphate synthase* (*FPS*) in *A. annua* significantly increased compared to the wild-type (Banyai et al. 2010; Han et al. 2006). It has also been reported that overexpression of the *chimeric farnesyl diphosphate synthase* in native plant strengthens the biosynthetic pathway of sesquiterpenes, leading to increased artemisinin biosynthesis in some lines (up to threefold) (Chen et al. 2000). In another study Kiani et al. (2016), with overexpression of *rol ABC* genes, also in diploid plants and *HMGR*, *FPS*, *ADS*, and *Aldh1* and upregulation of *ADS* in tetraploid plants of *A. annua* increased artemisinin level (Lin et al. 2011). Moreover, transcription factors (TFs) known as regulators of the location and time of secondary metabolites production (Yang et al. 2012), are one of the modules that can be used in the regulation of special metabolites production (Verpoorte and Memelink 2002). Identification of TFs involved in the regulation of artemisinin metabolism is important (Jiang et al. 2016; Shen et al. 2016). *AP2/ERF*, *bHLH*, *MYB*, and *WRKY* are members of various TFs in artemisinin (Lv et al. 2017) that influence the artemisinin biosynthesis in *A. annua*.

6.2 *In Planta* Artemisinin Production

To date, *in planta* artemisinin production in two model plants, tobacco (*Nicotiana benthamiana*) and moss (*Physcomitrella patens*), have been reported successfully. *N. benthamiana* is considered as one of the most widely used platforms for the production of various secondary metabolites. In addition to the possible disadvantages, the ability to change the pathway at tissue and intracellular levels, the possibility of rapid and transient expression, and no need for external primary sources are the main dominant features of *N. benthamiana* for the production of secondary metabolites (Carqueijeiro et al. 2020). Wang et al (2016), in the study of the role of *AaLTPs* (*Lipid Transfer Protein 3*) and *AaPDRs* (*Pleiotropic Drug Resistance 2*) from *A. annua* using transient expression in *N. benthamiana* leaves to increase the accumulation of dihydro artemisinic acid in the apoplastic space, indicated that the *AaLTP3* and *AaPDR2* prevent dihydroartemisinic acid reflux from the apoplast to the cell and strengthen the pathway. Recent reports demonstrated that transient expression is the most practical and successful method for the production of medicinal compounds, especially in this model plant (Reed et al. 2017).

Interestingly, in addition to *N. benthamiana*, *P. patens*, which is known as a vascular plant with an appropriate growth rate, is used as a model plant in applied biotechnology research (Büttner-Mainik et al. 2011; Ikram et al. 2015; Reski et al. 2015; Simonsen et al. 2009). Due to the complete sequencing of the moss genome and its haploid life cycle compared to other green hosts, it has become an attractive and efficient industrial production system for the production of secondary metabolites. On the other hand, studies have shown that *P. patens* have the potential to produce large-scale artemisinin (Ikram et al. 2017) (Fig. 4). So far, it has been possible to produce several recombinant drug proteins and other valuable molecules

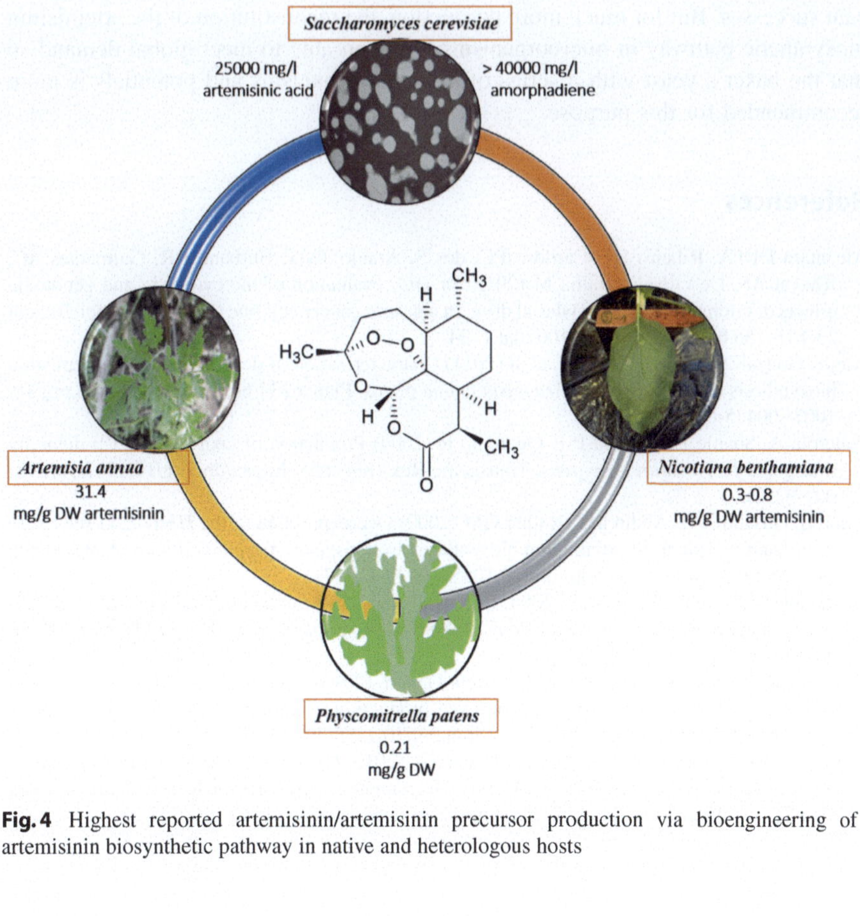

Fig. 4 Highest reported artemisinin/artemisinin precursor production via bioengineering of artemisinin biosynthetic pathway in native and heterologous hosts

in the moss system (Anterola et al. 2009; Pan et al. 2015; Reski et al. 2015; Zhan et al. 2014). Ikram et al. (2017) by engineering and inserting 5 genes involved in the biosynthetic pathway of artemisinin into *P. patens*, produced 0.21 mg/g of dry weight artemisinin after three days. Various reports indicate that the insertion of genes involved in the production of artemisinin has been successful using a constitutive expression in *P. patens*.

7 Conclusions

Given the fact that artemisinins, in addition to treating malaria, also play a key role in fighting other diseases, meeting global demands which has been a challenge for researchers and industry. Since the artemisinin levels in *Artemisia annua* plant is very low, its large-scale planting does not meet global needs. To this end, biotechnological tools have been able to eliminate these concerns to a large extent. In this regard, biosynthetic pathway engineering in the main host has been successfully performed. For further production, the engineering of the artemisinin biosynthetic pathway in two model plants (tobacco and moss) has led to significant successes. But for much more production, the reconstitution of the artemisinin biosynthetic pathway in microorganisms has been able to meet global demand so that the baker's yeast with a series of special mechanisms and potentials is more recommended for this purpose.

References

Alcântara DDFÁ, Ribeiro HF, Cardoso PC, dos S, Araújo TMT, Burbano RR, Guimarães AC, Khayat AS, De Oliveira Bahia M (2013) *In vitro* evaluation of the cytotoxic and genotoxic effects of artemether, an antimalarial drug, in a gastric cancer cell line (PG100). J Appl Toxicol 33:151–156.https://doi.org/10.1002/jat.1734

Alejos-Gonzalez F, Qu G, Zhou LL et al (2011) Characterization of development and artemisinin biosynthesis in self-pollinated *Artemisia annua* plants. Planta 234:685–697. https://doi.org/10.1007/s00425-011-1430-z

Anterola A, Shanle E, Perroud P-F, Quatrano R (2009) Production of taxa-4(5),11(12)-diene by transgenic *Physcomitrella patens*. Transgenic Res 18(4):655. https://doi.org/10.1007/s11248-009-9252-5

Aquil S, Husaini AM, Abdin MZ, Rather GM (2009) Overexpression of the *HMG-CoA reductase* gene leads to enhanced artemisinin biosynthesis in transgenic *Artemisia annua* plants. Planta Med 75:1453–1458. https://doi.org/10.1055/s-0029-1185775

Arsenault P, Wobbe K, Weathers P (2008) Recent advances in Artemisinin production through heterologous expression. Curr Med Chem 15:2886–2896. https://doi.org/10.2174/092986708786242813

Baadhe RR, Mekala NK, Parcha SR, Prameela Devi Y (2013) Combination of ERG9 repression and enzyme fusion technology for improved production of amorphadiene in *Saccharomyces cerevisiae*. J Anal Methods Chem.https://doi.org/10.1155/2013/140469

Banyai W, Kirdmanee C, Mii M, Supaibulwatana K (2010) Overexpression of *farnesyl pyrophosphate synthase* (*FPS*) gene affected Artemisinin content and growth of *Artemisia annua* L. Plant Cell Tissue Organ Cult 103:255–265. https://doi.org/10.1007/s11240-010-9775-8

Beccafico S, Morozzi G, Marchetti MC, Riccardi C, Sidoni A, Donato R, Sorci G (2015) Artesunate induces ROS- and p38 MAPK-mediated apoptosis and counteracts tumor growth *in vivo* in embryonal rhabdomyosarcoma cells. Carcinogenesis 36:1071–1083. https://doi.org/10.1093/carcin/bgv098

Berdelle N, Nikolova T, Quiros S, Efferth T, Kaina B (2011) Artesunate induces oxidative DNA damage, sustained DNA double-strand breaks, and the ATM/ATR damage response in cancer cells. Mol Cancer Ther 10:2224–2233. https://doi.org/10.1158/1535-7163.MCT-11-0534

Bertea CM, Freije JR, Van Der Woude H, Verstappen FWA, Perk L, Marquez V, De Kraker JW, Posthumus MA, Jansen BJM, De Groot A et al (2005) Identification of intermediates and enzymes involved in the early steps of artemisinin biosynthesis in *Artemisia annua*. Planta Med 71:40–47

Beyraghdar Kashkooli A, Van der Krol A, Bouwemeester HJ (2018) Terpenoid biosynthesis in plants. Flav Sci Proc XV WEURMAN FLAVOUR Res Symp 3–11. https://openlib.tugraz.at/download.php?id=5b6957d8e45b9%26location=browse

Brown GD (2010) The biosynthesis of artemisinin (Qinghaosu) and the phytochemistry of *Artemisia annua* L. (Qinghao). Molecules 15:7603–7698

Bulgakov VP (2008) Functions of rol genes in plant secondary metabolism. Biotechnol Adv 26:318–324. https://doi.org/10.1016/j.biotechadv.2008.03.001

Bouwmeester HJ (2006) Engineering the essence of plants. Nat Biotechnol 24:1359–1361. https://doi.org/10.1038/nbt1106-1359

Büttner-Mainik A, Parsons J, Jérôme H, Hartmann A, Lamer S, Schaaf A, Schlosser A, Zipfel PF, Reski R, Decker EL (2011) Production of biologically active recombinant human factor H in *Physcomitrella*. Plant Biotechnol J 9(3):373–383. https://doi.org/10.1111/j.1467-7652.2010.00552.x

Cao Y, Feng YH, Gao LW, Li XY, Jin QX, Wang YY, Xu YY, Jin F, Lu SL, Wei MJ (2019) Artemisinin enhances the anti-tumor immune response in 4T1 breast cancer cells *in vitro* and *in vivo*. Int Immunopharmacol 70:110–116. https://doi.org/10.1016/j.intimp.2019.01.041

Capell T, Christou P (2004) Progress in plant metabolic engineering. Curr Opin Biotechnol 15:148–154. https://doi.org/10.1016/j.copbio.2004.01.009

Carqueijeiro I, Langley C, Grzech D, Koudounas K, Papon N, O'Connor SE, Courdavault V (2020) Beyond the semi-synthetic artemisinin: metabolic engineering of plant-derived anti-cancer drugs. Curr Opin Biotechnol 65:17–24. https://doi.org/10.1016/j.copbio.2019.11.017

Chaijaroenkul W, Viyanant V, Mahavorasirikul W, Na-Bangchang K (2011) Cytotoxic activity of artemisinin derivatives against cholangiocarcinoma (CL-6) and hepatocarcinoma (Hep-G2) cell lines. Asian Pacific J Cancer Prev 12:55–59

Chen DH, Ye HC, Li GF (2000) Expression of a chimeric *farnesyl diphosphate synthase* gene in *Artemisia annua* L. transgenic plants via *Agrobacterium tumefaciens*-mediated transformation. Plant Sci 155:179–185. https://doi.org/10.1016/S0168-9452(00)00217-X

Chen H, Sun B, Pan S, Jiang H, Sun X (2009) Dihydroartemisinin inhibits growth of pancreatic cancer cells *in vitro* and *in vivo*. Anticancer Drugs 20:131–140. https://doi.org/10.1097/CAD.0b013e3283212ade

Chen X, Wong YK, Lim TK, Lim WH, Lin Q, Wang J, Hua Z (2017) Artesunate activates the intrinsic apoptosis of HCT116 cells through the suppression of fatty acid synthesis and the NF-κB pathway. Molecules 22.https://doi.org/10.3390/molecules22081272

Chen Y, Shen Q, Wang Y, Wang T, Wu S, Zhang L, Lu X, Zhang F, Jiang W, Qiu B, Gao E, Sun X, Tang K (2013) The stacked over-expression of *FPS*, *CYP71AV1* and *CPR* genes leads to the increase of artemisinin level in *Artemisia annua* L. Plant Biotechnol Rep 7:287–295. https://doi.org/10.1007/s11816-012-0262-z

Cheong DHJ, Tan DWS, Wong FWS, Tran T (2020) Anti-malarial drug, artemisinin and its derivatives for the treatment of respiratory diseases. Pharmacol Res 158:104901.https://doi.org/10.1016/j.phrs.2020.104901

Covello PS (2008) Making artemisinin. Phytochemistry. https://doi.org/10.1016/j.phytochem.2008.10.001

Davis EM, Croteau R (2000) Cyclization enzymes in the biosynthesis of monoterpenes, sesquiterpenes, and diterpenes 209:53–95. https://doi.org/10.1007/3-540-48146-x_2

Dewick PM (2009) Medicinal natural products: a biosynthetic approach, 3rd edn. Doi: 10 02/9780470742761.fmatter

Dilshad E, Cusido RM, Palazon J, Estrada KR, Bonfill M, Mirza B (2015) Enhanced artemisinin yield by expression of rol genes in *Artemisia annua*. Malar J 14:1–10. https://doi.org/10.1186/s12936-015-0951-5

Disbrow GL, Baege AC, Kierpiec KA, Yuan H, Centeno JA, Thibodeaux CA, Hartmann D, Schlegel R (2005) Dihydroartemisinin is cytotoxic to papillomavirus-expressing epithelial cells *in vitro* and *in vivo*. Cancer Res 65:10854–10861. https://doi.org/10.1158/0008-5472.CAN-05-1216

Donald KA, Hampton RY, Fritz IB (1997) Effects of overproduction of the catalytic domain of 3-hydroxy-3- methylglutaryl coenzyme A reductase on squalene synthesis in *Saccharomyces cerevisiae*. Appl Environ Microbiol 63:3341–3344

Dwivedi A, Mazumder A, du Plessis L, du Preez JL, Haynes RK, du Plessis J (2015) *In vitro* anti-cancer effects of artemisone nano-vesicular formulations on melanoma cells. Nanomedicine Nanotechnology. Biol Med 11:2041–2050. https://doi.org/10.1016/j.nano.2015.07.010

Dudareva N, Klempien A, Muhlemann JK, Kaplan I (2013) Biosynthesis, function and metabolic engineering of plant volatile organic compounds. New Phytol 198:16–32. https://doi.org/10.1111/nph.12145

Efferth T (2018) Beyond malaria: the inhibition of viruses by artemisinin-type compounds. Biotechnol Adv 36:1730–1737. https://doi.org/10.1016/j.biotechadv.2018.01.001

Farhi M, Kozin M, Duchin S, Vainstein A (2013) Metabolic engineering of plants for artemisinin synthesis. Biotechnol Genet Eng Rev 29(2):135–148. https://doi.org/10.1080/02648725.2013.821283

Farid F, Sideeq O, Khan F, Niaz K (2019) *Saccharomyces cerevisiae*, nonvitamin and nonmineral nutritional supplements. Elsevier Inc https://www.sciencedirect.com/science/article/pii/B9780128124918000667?via%3Dihub

Fei Z, Gu W, Xie R, Su H, Jiang Y (2018) Artesunate enhances radiosensitivity of esophageal cancer cells by inhibiting the repair of DNA damage. J Pharmacol Sci 138:131–137. https://doi.org/10.1016/j.jphs.2018.09.011

Fuentes P, Zhou F, Erban A, Karcher D, Kopka J, Bock R (2016) A new synthetic biology approach allows transfer of an entire metabolic pathway from a medicinal plant to a biomass crop. elife 5:1–26

Gravett AM, Liu WM, Krishna S, Chan WC, Haynes RK, Wilson NL, Dalgleish AG (2011) *In vitro* study of the anti-cancer effects of artemisone alone or in combination with other chemotherapeutic agents. Cancer Chemother Pharmacol 67:569–577. https://doi.org/10.1007/s00280-010-1355-4

Han JL, Liu BY, Ye HC, Wang H, Li ZQ, Li GF (2006) Effects of overexpression of the endogenous farnesyl diphosphate synthase on the artemisinin content in *Artemisia annua* L. J Integr Plant Biol 48(4):482–487. https://doi.org/10.1111/j.1744-7909.2006.00208.x

Han J, Wang H, Lundgren A, Brodelius PE (2014) Effects of overexpression of *AaWRKY1* on artemisinin biosynthesis in transgenic *Artemisia annua* plants. Phytochemistry 102:89–96

Han JL, Li ZQ, Ye HC (2008) Molecular cloning, prokaryotic expression, and enzyme activity assay of fps from *Artemisia annua*. J Agric Univ Hebei 31:71–75 (in Chinese with English abstract)

Hao DL, Xie R, De GJ, Yi H, Zang C, Yang MY, Liu L, Ma H, Cai WY, Zhao QH, Sui F, Chen YJ (2020) PH-responsive artesunate polymer prodrugs with enhanced ablation effect on rodent xenograft colon cancer. Int J Nanomedicine 15:1771–1786. https://doi.org/10.2147/IJN.S242032

Ho WE, Peh HY, Chan TK, Wong WSF (2014) Artemisinins: pharmacological actions beyond anti-malarial. Pharmacol Ther 142:126–139. https://doi.org/10.1016/j.pharmthera.2013.12.001

Hong KK, Nielsen J (2012) Metabolic engineering of *Saccharomyces cerevisiae*: a key cell factory platform for future biorefineries. Cell Mol Life Sci 69:2671–2690

Ikram NBK, Zhan X, Pan X-W, King BC, Simonsen HT (2015) Stable heterologous expression of biologically active terpenoids in green plant cells. Front Plant Sci 6:129. https://www.frontiersin.org/article/10.3389/fpls.2015.00129

Ikram NKBK, Simonsen HT (2017) A review of biotechnological artemisinin production in plants. Front Plant Sci 8:1–10

Ikram NKBK, Beyraghdar Kashkooli A, Peramuna A, van der Krol AR, Bouwmeester H, Simonsen HT (2019) Insights into heterologous biosynthesis of arteannuin B and artemisinin in *Physcomitrella patens*. Molecules 24:3822

Ikram NKBKh, Beyraghdar Kashkooli A, Peramuna AV, van der Krol AR, Bouwmeester H, Simonsen HT (2017) Stable production of the antimalarial drug artemisinin in the moss Physcomitrella patens. Frontiers in Bioengineering and Biotechnology 5:1–8. https://doi.org/10.3389/fbioe.2017.00047

Immethun CM, Hoynes-O'Connor AG, Balassy A, Moon TS (2013) Microbial production of isoprenoids enabled by synthetic biology. Front Microbiol. https://doi.org/10.3389/fmicb.2013.00075

Jiang J, Geng G, Yu X, Liu H, Gao J, An H, Cai C, Li N, Shen D, Wu X, Zheng L, Mi Y, Yang S (2016) Repurposing the anti-malarial drug dihydroartemisinin suppresses metastasis of non-small-cell lung cancer via inhibiting NF-κB/GLUT1 axis. Oncotarget 7:87271–87283. https://doi.org/10.18632/oncotarget.13536

Jiang Z, Chai J, Chuang HHF, Li S, Wang T, Cheng Y, Chen W, Zhou D (2012) Artesunate induces G0/G1 cell cycle arrest and iron-mediated mitochondrial apoptosis in A431 human epidermoid carcinoma cells. Anticancer Drugs 23:606–613. https://doi.org/10.1097/CAD.0b013e328350e8ac

Judd R, Bagley MC, Li M, Zhu Y, Lei C, Yuzuak S, Ekelöf M, Pu G, Zhao X, Muddiman DC, Xie DY (2019) Artemisinin biosynthesis in non-glandular trichome cells of *Artemisia annua*. Mol Plant 12:704–714. https://doi.org/10.1016/j.molp.2019.02.011

Kiani BH, Suberu J, Mirza B (2016) Cellular engineering of *Artemisia annua* and *Artemisia dubia* with the *rol ABC* genes for enhanced production of potent anti-malarial drug artemisinin. Malar J 15:1–17. https://doi.org/10.1186/s12936-016-1312-8

Klayman DL (1985) Qinghaosu (artemisinin): an antimalarial drug from China. Science 31:1049–1055

Kong J, Cheng K, Wang LN (2007) Increase of copy number of *HMG-CoA reductase* and *FPP synthase* genes improves the amorpha4,11-diene production in engineered yeast. Acta Pharm Sin 42:1314–1319

Kong J, Yang Y, Wang W, Cheng K, Zhu P (2013) Artemisinic acid: a promising molecule potentially suitable for the semi-synthesis of artemisinin. RSC Adv 3:7622–7641. https://doi.org/10.1039/c3ra40525g

Krishna S, Augustin Y, Wang J, Xu C, Staines HM, Platteeuw H, Kamarulzaman A, Sall A, Kremsner P (2021) Repurposing Antimalarials to tackle the COVID-19 pandemic. Trends Parasitol 37:8–11. https://doi.org/10.1016/j.pt.2020.10.003

Krivoruchko A, Nielsen J (2015) Production of natural products through metabolic engineering of *Saccharomyces cerevisiae*. Curr Opin Biotechnol 35:7–15. https://doi.org/10.1016/j.copbio.2014.12.004

Li C, Li J, Wang G, Li X (2016) Heterologous biosynthesis of artemisinic acid in Saccharomyces cerevisiae. J Appl Microbiol 120:1466–1478. https://doi.org/10.1111/jam.13044

Li ZQ, Liu Y, Liu BY, Wang H, Ye HC, Li GF (2006) Cloning, *E. coli* expression and molecular analysis of amorpha-4,11-diene synthase from a high-yield strain of *Artemisia annua* L. J Integr Plant Biol 48:1486–1492. https://doi.org/10.1111/j.1744-7909.2006.00381.x

Li J, Feng W, Lu H, Wei Y, Ma S, Wei L, Liu Q, Zhao J, Wei Q, Yao J (2019) Artemisinin inhibits breast cancer-induced osteolysis by inhibiting osteoclast formation and breast cancer cell proliferation. J Cell Physiol 234:12663–12675. https://doi.org/10.1002/jcp.27875

Li X, Gu S, Sun D, Dai H, Chen H, Zhang Z (2018) The selectivity of artemisinin-based drugs on human lung normal and cancer cells. Environ Toxicol Pharmacol 57:86–94. https://doi.org/10.1016/j.etap.2017.12.004

Li Y, Sui H, Jiang C, Li S, Han Y, Huang P, Du X, Du J, Bai Y (2018) Dihydroartemisinin increases the sensitivity of photodynamic therapy via NF-κB/HIF-1α/VEGF pathway in esophageal cancer cell *in vitro* and *in vivo*. Cell Physiol Biochem 48:2035–2045. https://doi.org/10.1159/000492541

Lian J, Mishra S, Zhao H (2018) Recent advances in metabolic engineering of *Saccharomyces cerevisiae*: New tools and their applications. Metab Eng 50:85–108. https://doi.org/10.1016/j.ymben.2018.04.011

Liang W, Liu J, Wu H, Qiao X, Lu X, Liu Y, Zhu H, Ma L (2019) Artemisinin induced reversal of EMT affects the molecular biological activity of ovarian cancer SKOV3 cell lines. Oncol Lett 18:3407–3414. https://doi.org/10.3892/ol.2019.10608

Lin X, Zhou Y, Zhang J, Lu X, Zhang F, Shen Q, Wu S, Chen Y, Wang T, Tang K (2011) Enhancement of artemisinin content in tetraploid *Artemisia annua* plants by modulating the expression of genes in artemisinin biosynthetic pathway. Biotechnol Appl Biochem 58:50–57. https://doi.org/10.1002/bab.13

Lin PC, Pakrasi HB (2018) Engineering cyanobacteria for production of terpenoids. Planta 249:145–154. https://cpb-us-w2.wpmucdn.com/sites.wustl.edu/dist/3/386/files/2018/11/Lin-Pakrasi2018-rg7fph.pdf

Liu W, Liu Y (2016) Youyou Tu: Significance of winning the 2015 nobel prize in physiology or medicine. Cardiovasc Diagn Ther 6:1–2

Liu Y, Gao S, Zhu J, Zheng Y, Zhang H, Sun H (2018) Dihydroartemisinin induces apoptosis and inhibits proliferation, migration, and invasion in epithelial ovarian cancer via inhibition of the hedgehog signaling pathway. Cancer Med 7:5704–5715. https://doi.org/10.1002/cam4.1827

Liu Y, Wang H, Ye HC, Li GF (2005) Advances in the plant isoprenoid biosynthesis pathway and its metabolic engineering. J Integr Plant Biol 47:769–782. https://doi.org/10.1111/j.1744-7909.2005.00111.x

Lv Z, Zhang L, Tang K (2017) New insights into artemisinin regulation. Plant Signal Behav 12:e1366398

Ma D, Li G, Alejos-Gonzalez F, Zhu Y, Xue Z, Wang A, Zhang H, Li X, Ye H, Wang H, Liu B, Xie DY (2017) Overexpression of a type-I isopentenyl pyrophosphate isomerase of *Artemisia annua* in the cytosol leads to high arteannuin B production and artemisinin increase. Plant J 91:466–479. https://doi.org/10.1111/tpj.13583

Ma DM, Wang Z, Wang L, Alejos-Gonzales F, Sun MA, Xie DY (2015) A genome-Wide scenario of terpene pathways in self-pollinated *Artemisia annua*. Mol Plant 8:1580–1598. https://doi.org/10.1016/j.molp.2015.07.004

Ma H, Yao Q, Zhang AM, Lin S, Wang XX, Wu L, Sun JG, Chen ZT (2011) The effects of artesunate on the expression of *EGFR* and *ABCG2* in A549 human lung cancer cells and a xenograft model. Molecules 16:10556–10569. https://doi.org/10.3390/molecules161210556

Mao H, Gu H, Qu X, Sun J, Song B, Gao W, Liu J, Shao Q (2013) Involvement of the mitochondrial pathway and Bim/Bcl-2 balance in dihydroartemisinin-induced apoptosis in human breast cancer *in vitro*. Int J Mol Med 31:213–218. https://doi.org/10.3892/ijmm.2012.1176

McDowell A, Hill KS, McCorkle JR, Gorski J, Zhang Y, Salahudeen AA, Ueland F, Kolesar JM (2021) Preclinical evaluation of artesunate as an antineoplastic agent in ovarian cancer treatment. Diagnostics 11:395. https://doi.org/10.3390/diagnostics11030395

Mercke P, Bengtsson M, Bouwmeester HJ, Posthumus MA, Brodelius PE (2000) Molecular cloning, expression, and characterization of *amorpha-4,11-diene synthase*, a key enzyme of artemisinin biosynthesis in *Artemisia annua* L. Arch Biochem Biophys 381:173–180

Mondal A, Chatterji U (2015) Artemisinin represses telomerase subunits and induces apoptosis in HPV-39 infected human cervical cancer cells. J Cell Biochem 116:1968–1981. https://doi.org/10.1002/jcb.25152

Moses T, Pollier J, Thevelein JM, Goossens A (2013) Bioengineering of plant (tri)terpenoids: From metabolic engineering of plants to synthetic biology *in vivo* and *in vitro*. New Phytol 200:27–43. https://doi.org/10.1111/nph.12325

Nair MS, Huang Y, Fidock DA, Polyak SJ, Wagoner J, Towler MJ, Weathers PJ (2021) Artemisia annua L. extracts prevent in vitro replication of SARS-CoV-2. https://doi.org/10.1101/2021.01.08.425825

Nafis T, Akmal M, Ram M, Alam P, Ahlawat S, Mohd A, Abdin MZ (2011) Enhancement of artemisinin content by constitutive expression of the *HMG-CoA reductase* gene in high-yielding strain of *Artemisia annua* L. Plant Biotechnol Rep 5:53–60. https://doi.org/10.1007/s11816-010-0156-x

Nunes JJ, Pandey SK, Yadav A, Goel S, Ateeq B (2017) Targeting NF-kappa B signaling by artesunate restores sensitivity of castrate-resistant prostate cancer cells to antiandrogens. Neoplasia 19:333–345. https://doi.org/10.1016/j.neo.2017.02.002

Ohgami Y, Elstad CA, Chung E, Shirachi DY, Quock RM, Lai HC (2010) Effect of hyperbaric oxygen on the anticancer effect of artemisinin on molt-4 human leukemia cells. Anticancer Res 30:4467–4470

Olofsson L, Engström A, Lundgren A, Brodelius PE (2011) Relative expression of genes of terpene metabolism in different tissues of *Artemisia annua* L. BMC Plant Biol 11:1–12

Pateraki I, Heskes AM, Hamberger B (2015) Cytochromes p450 for terpene functionalisation and metabolic engineering. Adv Biochem Eng Biotechnol 148:107–139. https://doi.org/10.1007/10_2014_301

Paddon CJ, Keasling JD (2014) Semi-synthetic artemisinin: A model for the use of synthetic biology in pharmaceutical development. Nat Rev Microbiol 12:355–367. https://doi.org/10.1038/nrmicro3240

Paddon CJ, Westfall PJ, Pitera DJ, Benjamin K, Fisher K, McPhee D, Leavell MD, Tai A, Main A, Eng D, Polichuk DR, Teoh KH, Reed DW, Treynor T, Lenihan J, Jiang H, Fleck M, Bajad S, Dang G, Dengrove D, Diola D, Dorin G, Ellens KW, Fickes S, Galazzo J, Gaucher SP, Geistlinger T, Henry R, Hepp M, Horning T, Iqbal T, Kizer L, Lieu B, Melis D, Moss N, Regentin R, Secrest S, Tsuruta H, Vazquez R, Westblade LF, Xu L, Yu M, Zhang Y, Zhao L, Lievense J, Covello PS, Keasling JD, Reiling KK, Renninger NS, Newman JD (2013) High-level semi-synthetic production of the potent antimalarial artemisinin. Nature 496:528–532. https://doi.org/10.1038/nature12051

Pan X-W, Han L, Zhang Y-H, Chen D-F, Simonsen HT (2015) Sclareol production in the moss *Physcomitrella patens* and observations on growth and terpenoid biosynthesis. Plant Biotechnol Rep 9:149–159. https://doi.org/10.1007/s11816-015-0353-8

Paradise EM, Kirby J, Chan R, Keasling JD (2008) Redirection of flux through the FPP branch-point in *Saccharomyces cerevisiae* by down-regulating squalene synthase. Biotechnol Bioeng 100:371–378

Peng MF, Chen M, Chen R (2011) The last gene involved in the MEP pathway of Artemisia annua: cloning and characterization and functional identification. J Med Plants Res 5:223–230

Perassolo M, Cardillo AB, Busto VD, Giulietti AM, Talou JR (2018) Biosynthesis of sesquiterpene lactones in plants and metabolic engineering for their biotechnological production. Chapter 4, pp 47–91. Sesquiterpene Lactones. https://doi.org/10.1007/978-3-319-78274-4_4

Pirali M, Taheri M, Zarei S, Majidi M, Ghafouri H (2020) Artesunate, as a HSP70 ATPase activity inhibitor, induces apoptosis in breast cancer cells. Int J Biol Macromol 164:3369–3375. https://doi.org/10.1016/j.ijbiomac.2020.08.198

Ram M, Khan MA, Jha P (2010) HMG-CoA reductase limits artemisinin biosynthesis and accumulation in *Artemisia annua* L. plants. Acta Physiol Plant 32:859–866

Reed J, Stephenson MJ, Miettinen K, Brouwer B, Leveau A, Brett P, Goss RJM, Goossens A, O'Connell MA, Osbourn A (2017) A translational synthetic biology platform for rapid access to gram-scale quantities of novel drug-like molecules. Metab Eng 42:185–193. https://doi.org/10.1016/j.ymben.2017.06.012

Reski R, Parsons J, Decker EL (2015) Moss-made pharmaceuticals: from bench to bedside. Plant Biotechnol J 13(8):1191–1198. Doi: 10.1111/pbi.12401

Riganti C, Doublier S, Viarisio D, Miraglia E, Pescarmona G, Ghigo D, Bosia A (2009) Artemisinin induces doxorubicin resistance in human colon cancer cells via calcium-dependent activation of HIF-1α and P-glycoprotein overexpression. Br J Pharmacol 156:1054–1066. https://doi.org/10.1111/j.1476-5381.2009.00117.x

Ro DK, Paradise EM, Ouellet M, Fisher KJ, Newman KL, Ndungu JM, Ho KA, Eachus RA (2006) Production of the antimalarial drug precursor artemisinic acid in engineered yeast. Nature 440:940–943

Roh JL, Kim EH, Jang H, Shin D (2017) Nrf2 inhibition reverses the resistance of cisplatin-resistant head and neck cancer cells to artesunate-induced ferroptosis. Redox Biol 11:254–262. https://doi.org/10.1016/j.redox.2016.12.010

Sa G, Mi M, He-chun Y, Ben-ye L, Guo-feng L, Kang C (2001) Effects of ipt gene expression on the physiological and chemical characteristics of *Artemisia annua* L. Plant Sci 160(4):691–698. https://doi.org/10.1016/S0168-9452(00)00453-2

Schramek N, Wang H, Römisch-Margl W, Keil B, Radykewicz T, Winzenhörlein B, Beerhues L, Bacher A, Rohdich F, Gershenzon J et al (2010) Artemisinin biosynthesis in growing plants of *Artemisia annua*. Phytochemistry 71:179–187

Shen Q, Yan T, Fu X, Tang K (2016) Transcriptional regulation of artemisinin biosynthesis in *Artemisia annua* L. Science Bulletin 61(1):18–25. https://doi.org/10.1007/s11434-015-0983-9

Simonsen HT, Drew DP, Lunde C (2009) Perspectives on using *Physcomitrella Patens* as an alternative production platform for thapsigargin and other terpenoid drug candidates. Perspect Med Chem 3, PMC.S2220. https://doi.org/10.4137/PMC.S2220

Singh NP, Lai HC (2004) Artemisinin induces apoptosis in human cancer cells. Anticancer Res 24:2277–2280

Singh NP, Verma KB (2002) Case report of a laryngeal squamous cell carcinoma treated with artesunate. Arch Oncol 10:279–280. https://doi.org/10.2298/AOO0204279S

Slezakova S, Ruda-Kucerova J (2017) Anticancer activity of artemisinin and its derivatives. Anticancer Res 37:5995–6003. https://doi.org/10.21873/anticanres.12046

Su XZ, Miller LH (2015) The discovery of artemisinin and the Nobel pize in physiology or medicine. Sci China Life Sci 58:1175–1179

Subedi A, Futamura Y, Nishi M, Ryo A, Watanabe N, Osada H (2016) High-throughput screening identifies artesunate as selective inhibitor of cancer stemness: involvement of mitochondrial metabolism. Biochem Biophys Res Commun 477:737–742. https://doi.org/10.1016/j.bbrc.2016.06.128

Sun Z, Meng H, Li J, Wang J, Li Q, Wang Y, Zhang Y (2014) Identification of novel knockout targets for improving terpenoids biosynthesis in *Saccharomyces cerevisiae*. PLoS One 9:e112615

Tamang JP, Fleet GH (2009) Yeasts diversity in fermented foods and beverages. In: Satyanarayana T, Kunze G (eds) Yeast biotechnology: diversity and applications. Springer, Dordrecht. https://doi.org/10.1007/978-1-4020-8292-4_9

Tang K, Shen Q, Yan T, Fu X (2014) Transgenic approach to increase artemisinin content in *Artemisia annua* L. Plant Cell Rep 33:605–615. https://doi.org/10.1007/s00299-014-1566-y

Teoh KH, Polichuk DR, Reed DW, Nowak G, Covello PS (2006) *Artemisia annua* L. (Asteraceae) trichome-specific cDNAs reveal CYP71AV1, a cytochrome P450 with a key role in the biosynthesis of the antimalarial sesquiterpene lactone artemisinin. FEBS Lett 580:1411–1416

Thanaketpaisarn O, Waiwut P, Sakurai H, Saiki I (2011) Artesunate enhances TRAIL-induced apoptosis in human cervical carcinoma cells through inhibition of the NF-κB and PI3K/Akt signaling pathways. Int J Oncol 39:279–285. https://doi.org/10.3892/ijo.2011.1017

Tran KQ, Tin AS, Firestone GL (2014) Artemisinin triggers a G1 cell cycle arrest of human Ishikawa endometrial cancer cells and inhibits cyclin-dependent kinase-4 promoter activity and expression by disrupting nuclear factor-κB transcriptional signaling. Anticancer Drugs 25:270–281. https://doi.org/10.1097/CAD.0000000000000054

Tsuruta H, Paddon CJ, Eng D, Lenihan JR, Horning T, Anthony LC, Regentin R, Keasling JD, Renninger NS, Newman JD (2009) High-level production of amorpha-4, 11-diene, a precursor of the antimalarial agent artemisinin, in *Escherichia coli*. PLoS One 4.https://doi.org/10.1371/journal.pone.0004489

Tu Y (2011) The discovery of artemisinin (qinghaosu) and gifts from Chinese medicine. Nat Med 17:1217–1220

Uckun FM, Saund S, Windlass H, Trieu V (2021) Repurposing anti-malaria phytomedicine artemisinin as a COVID-19 drug. Front Pharmacol 12:1–5. https://doi.org/10.3389/fphar.2021.649532

Vranová E, Coman D, Gruissem W (2013) Network analysis of the MVA and MEP pathways for isoprenoid synthesis. Annu Rev Plant Biol 64:665–700. https://doi.org/10.1146/annurev-arplant-050312-120116

Vranová E, Coman D, Gruissem W (2012) Structure and dynamics of the isoprenoid pathway network. Mol Plant 5:318–333. https://doi.org/10.1093/mp/sss015

Våtsveen TK, Myhre MR, Steen CB, Wälchli S, Lingjærde OC, Bai B, Dillard P, Theodossiou TA, Holien T, Sundan A, Inderberg EM, Smeland EB, Myklebust JH, Oksvold MP (2018) Artesunate shows potent anti-tumor activity in B-cell lymphoma. J Hematol Oncol 11:1–12. https://doi.org/10.1186/s13045-018-0561-0

Verpoorte R, Memelink J (2002) Engineering secondary metabolite production in plants. Curr Opin Biotechnol 13(2):181–187. https://doi.org/10.1016/S0958-1669(02)00308-7

Wang B, Beyraghdar Kashkooli A, Sallets A, Ting HM, de Ruijter NCA, Olofsson L, Brodelius P, Pottier M, Boutry M, Bouwmeester H et al (2016) Transient production of artemisinin in *Nicotiana benthamiana* is boosted by a specific lipid transfer protein from *A. annua*. Metab Eng 38:159–169

Wang Y, Huang Z, Wang L, Meng S, Fan Y, Chen T, Cao J, Jiang R, Wang C (2011) The antimalarial artemisinin inhibits pro-inflammatory cytokines via the NF-κB canonical signaling pathway in PMA-induced THP-1 monocytes. Int J Mol Med 27:233–241. https://doi.org/10.3892/ijmm.2010.580

Wang JS, Wang MJ, Lu X, Zhang J, Liu QX, Zhou D, Dai JG, Zheng H (2020) Artesunate inhibits epithelial-mesenchymal transition in non-small-cell lung cancer (NSCLC) cells by down-regulating the expression of BTBD7. Bioengineered 11:1197–1207. https://doi.org/10.1080/21655979.2020.1834727

Wang L, Li J, Shi X, Li S, Tang PM-K, Li Z, Li H, Wei C (2019) Antimalarial dihydroartemisinin triggers autophagy within HeLa cells of human cervical cancer through Bcl-2 phosphorylation at Ser70. Phytomedicine 52:147–156. https://doi.org/10.1016/j.phymed.2018.09.221

Wang W, Yang Y, Zheng XD, Huang SQ, Guo L, Kong JQ, Cheng KD (2013) The advance in synthetic biology: towards a microbe-derived paclitaxel intermediates. Acta Pharm Sin 48:187–192 (in Chinese with English abstract)

Willoughby JA, Sundar SN, Cheung M, Tin AS, Modiano J, Firestone GL (2009) Artemisinin blocks prostate cancer growth and cell cycle progression by disrupting Sp1 interactions with the cyclin-dependent kinase-4 (CDK4) promoter and inhibiting CDK4 gene expression. J Biol Chem 284:2203–2213. https://doi.org/10.1074/jbc.M804491200

Wong YK, Xu C, Kalesh KA, He Y, Lin Q, Wong WSF, Shen HM, Wang J (2017) Artemisinin as an anticancer drug: Recent advances in target profiling and mechanisms of action. Med Res Rev 37:1492–1517. https://doi.org/10.1002/med.21446

Wu J, Hu D, Yang G, Zhou J, Yang C, Gao Y, Zhu Z (2011) Down-regulation of BMI-1 cooperates with artemisinin on growth inhibition of nasopharyngeal carcinoma cells. J Cell Biochem 112:1938–1948. https://doi.org/10.1002/jcb.23114

Wu L, Cheng Y, Deng J, Tao W, Ye J (2019) Dihydroartemisinin inhibits proliferation and induces apoptosis of human hepatocellular carcinoma cell by upregulating tumor necrosis factor via JNK/NF-κB pathways. Evidence-Based Complement. Altern Med.https://doi.org/10.1155/2019/9581327

Xie DY, Ma DM, Judd R, Jones AL (2016) Artemisinin biosynthesis in *Artemisia annua* and metabolic engineering: questions, challenges, and perspectives. Phytochem Rev 15:1093–1114

Yang CQ, Fang X, Wu XM, Mao YB, Wang LJ, Chen XY (2012) Transcriptional regulation of plant secondary metabolism. J Integr Plant Biol 54:703–712. https://doi.org/10.1111/j.1744-7909.2012.01161.x

Yang R-Y, Feng L-L, Yang X-Q, Yin L-L, Xu X-L, Zeng Q-P (2008) Quantitative transcript profiling reveals down-regulation of A sterol pathway relevant gene and overexpression of artemisinin biogenetic genes in transgenic *Artemisia annua* plants. Planta Med 74:1510–1516. https://doi.org/10.1055/s-2008-1081333

Yang J, He Y, Li Y, Zhang X, Wong YK, Shen S, Zhong T, Zhang J, Liu Q, Wang J (2020) Advances in the research on the targets of anti-malaria actions of artemisinin. Pharmacol Ther 216:107697.https://doi.org/10.1016/j.pharmthera.2020.107697

Yao Z, Bhandari A, Wang Y, Pan Y, Yang F, Chen R, Xia E, Wang O (2018) Dihydroartemisinin potentiates antitumor activity of 5-fluorouracil against a resistant colorectal cancer cell line. Biochem Biophys Res Commun 501:636–642. https://doi.org/10.1016/j.bbrc.2018.05.026

Yin S, Yang H, Zhao X, Wei S, Tao Y, Liu M, Bo R, Li J (2020) Antimalarial agent artesunate induces G0/G1 cell cycle arrest and apoptosis via increasing intracellular ROS levels in normal liver cells. Hum Exp Toxicol 39:1681–1689. https://doi.org/10.1177/0960327120937331

Yin X, Liu Y, Qin J, Wu Y, Huang J, Zhao Q, Dang T, Tian Y, Yu P, Huang X (2021) Artesunate suppresses the proliferation and development of estrogen receptor-α-positive endometrial cancer in HAND2-Dependent pathway. Front Cell Dev Biol 8:1–14. https://doi.org/10.3389/fcell.2020.606969

Yu C, Sun P, Zhou Y, Shen B, Zhou M, Wu L, Kong M (2019) Inhibition of AKT enhances the anti-cancer effects of artemisinin in clear cell renal cell carcinoma. Biomed Pharmacother 118:109383.https://doi.org/10.1016/j.biopha.2019.109383

Zhan X, Zhang YH, Chen DF, Simonsen HT (2014) Metabolic engineering of the moss physcomitrella patens to produce the sesquiterpenoids patchoulol and α/β-santalene. Front Plant Sci 5(NOV), 1–10. https://doi.org/10.3389/fpls.2014.00636

Zhang Y, Teoh KH, Reed DW, Maes L, Goossens A, Olson DJH, Ross ARS, Covello PS (2008) The molecular cloning of artemisinic aldehyde Δ11(13) reductase and its role in glandular trichome-dependent biosynthesis of artemisinin in *Artemisia annua*. J Biol Chem 283:21501–21508

Zhang L, Jing F, Li F, Li M, Wang Y, Wang G, Sun X, Tang K (2009) Development of transgenic *Artemisia annua* (Chinese wormwood) plants with an enhanced content of artemisinin, an effective anti-malarial drug, by hairpin-RNA-mediated gene silencing. Biotechnol Appl Biochem 52:199. https://doi.org/10.1042/BA20080068

Zhang LX, Liu ZN, Ye J, Sha M, Qian H, Bu XH, Luan ZY, Xu XL, Huang AH, Yuan DL, Wu YQ, Wang XX, Wang J, Huang JX, Ye LH, Zhang HT, Wang YL, Zhang J, Zhang QX (2014) Artemisinin inhibits gastric cancer cell proliferation through upregulation of p53. Tumor Biol 38:639–646. https://doi.org/10.1002/cbin.10244

Zhang P, Luo HS, Li M, Tan SY (2015) Artesunate inhibits the growth and induces apoptosis of human gastric cancer cells by downregulating COX-2. Onco Targets Ther 8:845–854. https://doi.org/10.2147/OTT.S81041

Zhao F, Vakhrusheva O, Markowitsch SD, Slade KS, Tsaur I, Cinatl J, Michaelis M, Efferth T, Haferkamp A, Juengel E (2020) Artesunate impairs growth in cisplatin-resistant bladder cancer cells by cell cycle arrest, apoptosis and autophagy induction. Cells 9:1–19. https://doi.org/10.3390/cells9122643

Zhou HJ, Wang WQ, Wu GD, Lee J, Li A (2007) Artesunate inhibits angiogenesis and downregulates vascular endothelial growth factor expression in chronic myeloid leukemia K562 cells. Vascul Pharmacol 47:131–138. https://doi.org/10.1016/j.vph.2007.05.002

Zhou X, Chen Y, Wang F, Wu H, Zhang Y, Liu J, Cai Y, Huang S, He N, Hu Z, Jin X (2020) Artesunate induces autophagy dependent apoptosis through upregulating ROS and activating AMPK-mTOR-ULK1 axis in human bladder cancer cells. Chem Biol Interact 331:109273.https://doi.org/10.1016/j.cbi.2020.109273

Zhu S, Liu W, Ke X, Li J, Hu R, Cui H, Song G (2014) Artemisinin reduces cell proliferation and induces apoptosis in neuroblastoma. Oncol Rep 32:1094–1100. https://doi.org/10.3892/or.2014.3323

Yeast Synthetic Biology for the Production of Terpenoids Derived from Traditional Chinese Medicinal Plants

Yongjun Wei

Abstract

Traditional Chinese medicine has been widely used to cure diverse diseases in China. Modern medicinal chemistry and pharmaceutical research help to identify active natural products that functioned during disease treatment. Among them, terpenoids are one of the largest natural bioactive products in traditional Chinese medicine. Traditionally, these active terpenoids are extracted from traditional Chinese medicinal plants whose yield of the active terpenoids derived from the plants is low and the quality is not stable. The yield cannot satisfy the increasing terpenoid demand. Therefore, another sustainable supply of active terpenoids is of great interest. The development of synthetic biology enables yeasts to be ideal microbial cell factories for plant-derived natural product production. In the past few years, several typical active terpenoids derived from traditional Chinese medicinal plants have been produced in yeasts, including artemisinic acid, rare ginsenosides, rare licorice triterpenoids, and other value-added terpenoids. In this chapter, the terpenoid biosynthetic pathway and current synthetic biology strategies for the production of some typical terpenoids using yeasts are summarized. Moreover, future synthetic biology strategies for efficient terpenoid production are discussed.

Y. Wei (✉)
Key Laboratory of Advanced Drug Preparation Technologies, Ministry of Education, School of Pharmaceutical Sciences, Zhengzhou University, Zhengzhou 450001, Henan Province, China
e-mail: yongjunwei@zzu.edu.cn

Laboratory of Synthetic Biology, Zhengzhou University, Zhengzhou 450001, Henan Province, China

F. Darvishi Harzevili (ed.), *Synthetic Biology of Yeasts*,
https://doi.org/10.1007/978-3-030-89680-5_7

1 Introduction of Terpenoids and Their Production from Traditional Chinese Medicinal Plants

Terpenoids, also known as isoprenoids, represent a highly diverse group of natural products with wide use in cosmetics, pharmaceuticals, food, and other industries (Pichersky and Raguso 2018). The terpenoids are the most abundant plant natural products and have diverse structures (Zhou and Pichersky 2020). There are more than 55,000 different kinds of terpenoids derived from plants. The terpenoids are composed of the basic unit of isoprene and can be classified based on the isoprene numbers (Paramasivan and Mutturi 2017). Usually, the terpenoids can be divided into hemiterpenoids, such as isoprene; monoterpenoids, such as menthol, menthone, camphor, and eucalyptol; sesquiterpenoids, such as artemisinin (Qinghaosu青蒿素), humulene, caryophyllene, and zingiberene; diterpenoids, such as triptolide, oridonin, and taxol; triterpenoids, such as ginsenosides, glycyrrhetinic acid; and tetraterpenoids, such as lycopene, zeaxanthin, and astaxanthin (Jiang et al. 2016).

Traditional Chinese herbs harbor diverse terpenoids, and many of them show medicinal characteristics (Guan et al. 2020). The terpenoids have a wide array of clinical or other pharmacological properties, including antitumor, anti-inflammatory, antiviral, antimalarial effects, preventing and treating cardiovascular diseases, and other activities towards diverse metabolic diseases (Jaeger and Cuny 2016). The most famous terpenoids in the world might be artemisinin, which helps to cure malarial disease and saves many lives (Tu 2016). The medicinal plants harbor abundant terpenoids (Jaeger and Cuny 2016). With the development of modern pharmaceutical sciences, the demand for terpenoids increased (Paramasivan and Mutturi 2017). Traditionally, the terpenoids are extracted from plants, especially from herbs (Bergman et al. 2019). The herb plants need to grow for months or even years before they can be used for active terpenoid extraction, such as *Panax Ginseng* for five to seven years, *Glycyrrhiza uralensis* for several years, and *Artemisia annua* for several months (Fig. 1). Moreover, the terpenoids in plants are low; for example, the artemisinin composition in the biomass of *Artemisia annua* is 0.01–2% (Namuli et al. 2018); the rare ginsenoside compositions in *Panax* plants are less than 0.1% (Hwang et al. 2014). Besides, the plant growth and terpenoid content in the plant are affected by climate, soil types, and other environmental factors (Fig. 1) (Wei et al. 2017a). Chemical synthesis of these structurally complex terpenoids suffers from the high cost and low yield (Jansen and Shenvi 2014). Therefore, another sustainable supply of terpenoids is of great interest (Moser and Pichler 2019).

Synthetic biology approaches had been applied to produce value-added terpenoids, which provide a bright way to supply enough terpenoids for future commercial use (Bian et al. 2017; Liu et al. 2019b). The most commonly used model microorganism, *Saccharomyces cerevisiae*, grows rapidly on low-cost substrates and is robust to different adverse fermentation conditions (Wei et al. 2017b). Additionally, *S. cerevisiae* is easy for genetic manipulation, which has been widely used for the biosynthesis of novel non-native natural products (Liu et al. 2019b).

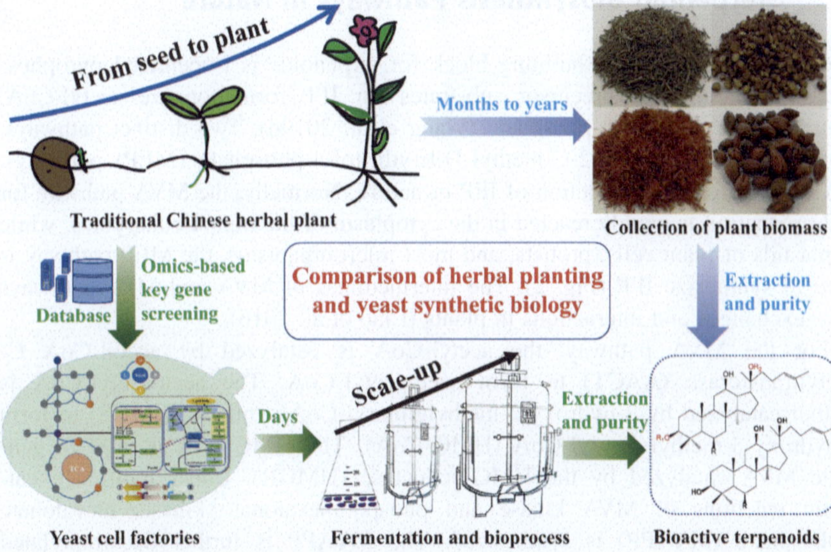

Fig. 1 Comparison of plant production of terpenoids and yeast synthetic biology for the production of terpenoids. The traditional Chinese herbal plants often grow months to years, and production of terpenoids requires plant biomass collection, extraction, and purity. The omics technologies and available database can help to identify key terpenoid biosynthetic genes, and these genes can be used for building yeast cell factories. It takes several days to ferment. Compared with plants, the native natural products in yeasts are few and simple, and it is easy to extract targeted bioactive terpenoids from yeast biomass

Other non-model yeasts, such as *Yarrowia lipolytica* and *Rhodosporidium toruloides*, had been applied in the biosynthesis of terpene and other natural products (Ma et al. 2020; Wang et al. 2020a; Yaegashi et al. 2017). The yeasts are considered to be ideal hosts for terpenoid biosynthesis, and they were easier to express some terpenoid biosynthetic genes than *Escherichia coli*, such as cytochrome P450s in the terpenoid biosynthetic pathway (Liu et al. 2020). Engineering yeasts for terpenoid production needs to build terpenoid precursor chassis cells, identify terpenoid biosynthetic pathways, and understand their metabolic networks in plants and yeasts (Zu et al. 2020). After the introduction of heterologous terpenoid biosynthetic genes in yeast terpenoid biosynthetic genes precursor chassis cell, strengthening the terpenoid biosynthetic pathway, downregulating competing pathways, alleviating the toxic of the targeted terpenoids, enhancing cofactor supply, and other strategies used to rewire the whole metabolism network for terpenoid production should be implemented to enable high titer, rate, and yield of targeted terpenoid production in yeasts (Nielsen and Keasling 2016; Liu et al. 2019b).

In this chapter, current knowledge of the terpenoid biosynthetic pathways in plants and yeasts, the omics technologies used for the discovering of key terpenoid biosynthetic enzymes, and engineering strategies for high-level terpenoid production in yeasts are summarized and discussed.

2 Terpenoid Biosynthesis Pathways in Nature

The carbon skeleton and building block for terpenoids is isopentenyl pyrophosphate (IPP), and the precursor substrates for IPP formation are acetyl-CoA, pyruvate, and glycerol-3-phosphate (Wang et al. 2019d). Two distinct pathways, mevalonate (MVA) and 2-C-methyl-D-Erythritol-4-phosphate (MEP) pathways, can be used for the production of IPP in nature. Normally, the MVA pathway for IPP production is mainly reacted in the cytoplasm of the higher eukaryotes; while in plastids of plant cells, protists, and most microorganisms, the MEP pathway is used to synthesize IPP (Fig. 2). The intermediates of MVA and MEP pathways have exchanges and interactions in plants (Liao et al. 2016).

For the MVA pathway, the acetyl-CoA is catalyzed by acetyl-CoA C-acetyltransferase (AACT) to form acetoacetyl-CoA. The acetoacetyl-CoA is further catalyzed by 3-hydroxy-3-methylglutaryl-CoA synthase (HMGS) to form 3-hydroxy-3-methylglutaryl-CoA (HMG-CoA). The HMG-CoA is used to generate MVA catalyzed by the HMG reductase (HMGR). Under multi-step catalytic reactions of MVA kinase and phosphomevalonate kinase, mevalonate diphosphate (MVAPP) is synthesized. The MVAPP is further decarboxylated to form IPP by mevalonate diphosphate decarboxylase. The IPP is used for the synthesis of dimethylallyl pyrophosphate (DMAPP) under the catalysis of isopentenyl-pyrophosphate delta isomerase (IDI) (Fig. 2) (Vranová et al. 2013). For the MEP pathway, the first step is the formation of D-xylulose-5-phosphate (DXP) catalyzed by DXP synthase, with pyruvate and glyceraldehyde-3-phosphate (G3P) as substrates. The second step is the reduction of DXP to MEP catalyzed by DXP reductoisomerase (Kim and Keasling 2001). With the help of multi-step enzymatic reactions, MEP is further catalyzed to a series of products of 4-diphosphocytidyl-2-*C*-methyl-D-erythritol (CDP-ME), 4-diphosphocytidyl-2-*C*-methyl-D-erythritol 2-phosphate (CDP-MEP), 2-*C*-methyl-D-erythritol 2,4-cyclopyrophosphate (MECPP), and 1-hydroxy-2-methyl-2-(*E*)-butenyl 4-pyrophosphate (HMBPP) (Liu et al. 2014). The HMBPP is transformed to IPP and DMAPP by 4-hydroxy-3-methylbut-2-en-1-yl diphosphate reductase (ispH) (Fig. 2).

The terpenoids are synthesized with the common five-carbon substrate of isoprenoid (Fig. 2) (Vranová et al. 2012). After the production of IPP and DMAPP, the geranyl pyrophosphate synthase (GPPS) uses IPP and DMAPP to synthesize geranyl diphosphate (GPP), and GPP is the precursor for monoterpenoid biosynthesis (Nagegowda and Gupta 2020). Farnesyl pyrophosphate synthase uses IPP and GPP to synthesize farnesyl diphosphate (FPP), and FPP is the precursor for sesquiterpenoid and triterpenoids biosynthesis. GPPS uses IPP and FPP to synthesize geranylgeranyl pyrophosphate (GGPP), and GGPP is the precursor for diterpenoid and tetraterpenoids biosynthesis (Nagegowda and Gupta 2020). In order to synthesize the final terpenoids, identification of the core structure modification enzymes, such as the terpene synthases, P450 and their redox partners of NADPH-cytochrome P450 reductases (CPRs), and glycosyltransferases, are essential (Wang et al. 2018).

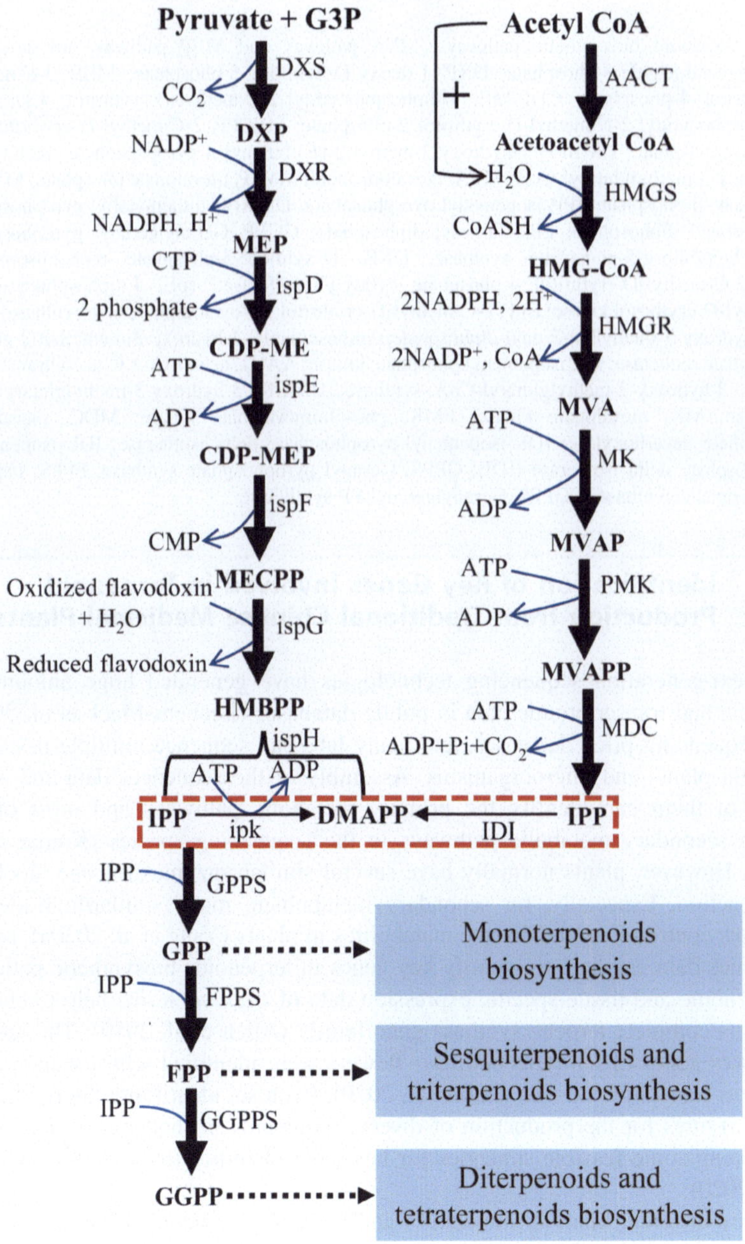

Pyruvate + G3P
DXS
CO_2
DXP
$NADP^+$
DXR
NADPH, H^+
MEP
CTP
ispD
2 phosphate
CDP-ME
ATP
ispE
ADP
CDP-MEP
ispF
CMP
MECPP
Oxidized flavodoxin
+ H_2O
ispG
Reduced flavodoxin
HMBPP
ispH
ATP
ADP
Acetyl CoA
+
AACT
Acetoacetyl CoA
H_2O
HMGS
CoASH
HMG-CoA
2NADPH, $2H^+$
HMGR
$2NADP^+$, CoA
MVA
ATP
MK
ADP
MVAP
ATP
PMK
ADP
MVAPP
ATP
MDC
ADP+Pi+CO_2
IPP
ipk
DMAPP
IDI
IPP
IPP
GPPS
GPP
Monoterpenoids biosynthesis
IPP
FPPS
FPP
Sesquiterpenoids and triterpenoids biosynthesis
IPP
GGPPS
GGPP
Diterpenoids and tetraterpenoids biosynthesis

◀**Fig. 2** Terpenoid biosynthetic pathways, MVA pathway and MEP pathway, are described. G3P, glyceraldehyde-3-phosphate; DXP, 1-deoxy-D-xylulose 5-phosphate; MEP, 2-*C*-methyl-D-erythritol 4-phosphate; CDP-ME, 4-diphosphocytidyl-2-*C*-methyl-D-erythritol; CDP-MEP, 4-diphosphocytidyl-2-*C*-methyl-D-erythritol 2-phosphate; MECPP, 2-*C*-methyl-D-erythritol 2,4-cyclopyrophosphate; HMBPP, 1-hydroxy-2-methyl-2-(*E*)-butenyl 4-pyrophosphate; HMG-CoA, 3-hydroxy-3-methylglutaryl-CoA; MVA, mevalonic acid; MVAP, mevalonate phosphate; MVAPP, mevalonate diphosphate; IPP, isopentenyl pyrophosphate; DMAPP, dimethylallyl pyrophosphate; GPP, geranyl diphosphate; FPP, farnesyl diphosphate; GGPP, Geranylgeranyl pyrophosphate. DXS, D-xylulose-5-phosphate synthase; DXR, D-xylulose-5-phosphate reductoisomerase; ispD, 2-C-methyl-D-erythritol 4-phosphate cytidylyltransferase; ispE, 4-diphosphocytidyl-2-C-methyl-D-erythritol kinase; ispF, 2-C-methyl-D-erythritol 2,4-cyclodiphosphate synthase; ispG, (E)-4-hydroxy-3-methylbut-2-enyl-diphosphate synthase; ispH, 4-hydroxy-3-methylbut-2-en-1-yl diphosphate reductase; ipk, isopentenyl phosphate kinase; AACT, acetyl-CoA C-acetyltransferase; HMGS, 3-hydroxy-3-methylglutaryl-CoA synthase; HMGR, 3-hydroxy-3-methylglutaryl-CoA reductase; MK, mevalonate kinase; PMK, phosphomevalonate kinase; MDC, mevalonate diphosphate decarboxylase; IDI, isopentenyl-pyrophosphate delta isomerase; IDI, isopentenyl-pyrophosphate delta isomerase (IDI); GPPS, Geranyl pyrophosphate synthase; FPPS, Farnesyl pyrophosphate synthase; GGPPS, Geranylgeranyl-PP synthetase

3 Identification of Key Genes Involved in Terpenoid Production from Traditional Chinese Medicinal Plants

The next-generation sequencing technologies have generated huge amounts of genomic and transcriptomic data in public databases (Leebens-Mack et al. 2019). The sequencing price decreases, and many labs can sequence multiple tissues of different plants and microorganisms. Assembly of the sequences data and annotation of them can identify the primary metabolic pathways and most of the known secondary metabolic pathways in the sequencing species (Kumar et al. 2016). However, plants normally have several similar enzymes for one biochemical reaction. Especially, for secondary metabolism, many similar/high-identity enzymes contribute to the diverse metabolites in plants (Yang et al. 2020a). Multiple omics data can help to identify key genes in terpenoids biosynthetic pathway. The genome and tissue-specific expression data of *Canabis sativa* help to characterize its complete terpene synthase gene family (Allen et al. 2019). The terpene synthases genes of *Lathyrus odoratus* flowers were identified with the help of *L. odoratus* transcriptome data (Bao et al. 2020). Even so, identifying the right modified enzymes for the production of diverse terpenoids is challenging. Therefore, developing some feasible strategies for key gene identification is necessary (Nett et al. 2020).

Many different kinds of annotation and screening strategies have been developed for genomic and transcriptomic data analyses (Nett et al. 2020; Srinivasan and Smolke 2020), and some of them have been used for key gene discovery (Scossa et al. 2018; Han et al. 2016). Triterpenoid saponins are the main bioactive components of licorice. The final step for the production of triterpenoid glycyrrhizin is to attach two glucuronic acids to the glycyrrhetinic acids. The glycosylation reactions normally occur in the cytosol, and the 3-*o*-glucuronosylation is believed to be a cytoplasmic glycosyltransferase. Recently, a cellulose synthase-like enzyme, an endoplasmic reticulum-membrane localization enzyme, can glucuronidate of specialized plant metabolites of triterpenoid saponins (Jozwiak et al. 2020). Nearly at the same time, a cellulose synthase-derived enzyme was identified to catalyze the 3-*o*-glucuronosylation in glycyrrhizin biosynthesis using gene co-expression analyses (Chung et al. 2020). All these two studies used the transcriptomic data of different plant tissues to identify the key genes for terpene biosynthesis, suggesting multi-omics data provide one efficient way for key gene identification.

Phylogenetic analyses, sequence similarity networks, homology-based model prediction, and other bioinformatics-aided strategies have been applied for key gene identification (Wang et al. 2020b). Key genes from the same plant genera normally have high identities with each other, and it provides the opportunities to use sequence similarity network and phylogenetic analyses to identify key genes in terpene biosynthesis. In order to identify the key glycosyltransferase genes for the production of triterpenoid saponin ginsenosides, all the potential glycosyltransferase genes were assembled from a *Panax* cDNA and EST database using the microbial OTU classification strategy (Wei et al. 2015b; Liang et al. 2019; Mai et al. 2020). The classified glycosyltransferase genes were further aligned, and the phylogenetic tree was built (Yan et al. 2014; Wang et al. 2015). Based on the phylogenetic analysis results, some glycosyltransferase genes were selected for microbial characterization, and the glycosyltransferases used for the ginsenosides of CK, Rh2, and Rg3 were identified (Yan et al. 2014; Wang et al. 2015). The sequence similarity network has been used to classify terpene synthase-like proteins. With the help of hidden Markov models, some putative terpene synthase genes from *Marchanita polymorpha* were identified (Kumar et al. 2016). Recently, a streamlined computational workflow was provided to explore natural product biosynthetic gene clusters, and it had potential application in terpene biosynthesis gene identification in the future (Navarro-Muñoz et al. 2020).

4 Engineering Yeasts for the Production of Terpenoid Precursors

Yeasts, including the model organisms of *S. cerevisiae* and *Y. lipolytica*, are attractive choices for microbial terpene production (Nielsen and Keasling 2016; Arnesen et al. 2020). In order to produce terpene at a high titer, rate, and yield (TRY), the terpenoid precursor biosynthesis in the yeasts were enhanced (Nielsen and Keasling 2016). The common metabolic engineering and newly developed synthetic biology strategies had been applied to overcome precursor limitation in yeast terpenoid precursor biosynthesis (Farhi et al. 2011; Zu et al. 2020). The yeasts are not the natural hosts for terpenoid production, and rewiring cellular metabolism and directing of metabolic flux for terpenoid precursor biosynthesis is necessary (Worland et al. 2020; Bian et al. 2017). The yeasts only have the MVA pathway for terpenoid biosynthesis. Based on the MVA pathway in yeast (Fig. 3), overexpression of some genes in the MVA pathway and optimization of the terpenoid precursor biosynthetic pathway by introducing exogenous efficient genes are the normal strategies (Paramasivan and Mutturi 2017). In fact, diverse genes have been selected for terpenoid precursor production (Fig. 3).

The rate-limiting enzymes in the MVA pathway are often selected for improving yeast terpenoid precursor synthesis. The HMGR is the first rate-limiting enzyme

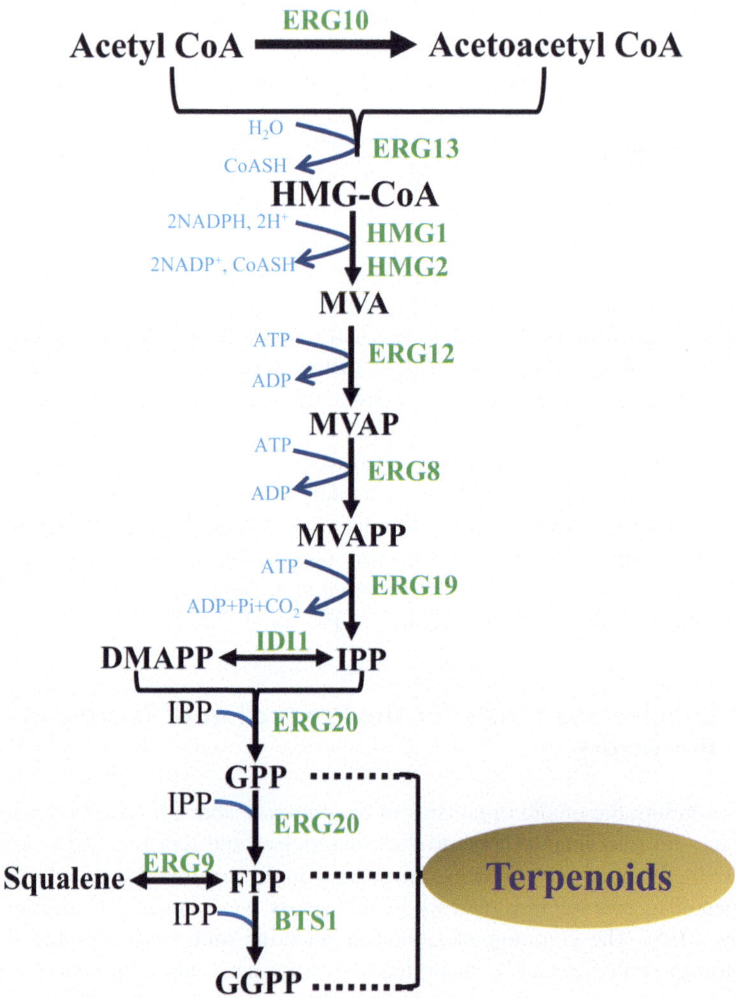

Fig. 3 Engineering targets for the production of terpenoids in *S. cerevisiae*. HMG-CoA, 3-hydroxy-3-methylglutaryl-CoA; MVA, mevalonic acid; MVAP, mevalonate phosphate; MVAPP, mevalonate diphosphate; IPP, isopentenyl pyrophosphate; DMAPP, dimethylallyl pyrophosphate; GPP, geranyl diphosphate; FPP, farnesyl diphosphate; GGPP, Geranylgeranyl pyrophosphate

in the MVA pathway. In order to enhance HMGR activity, the *tHMG1* (encoding the catalytic domains of yeast HMGR) is overexpressed in yeast, and the metabolic fluxes in the MVA pathway are increased and the final production of terpenoids is improved (Fig. 3) (Wu et al. 2018). The overexpression of *tHMG1* often leads to the improvement of terpenoid production, such as the limonene production using engineered *S. cerevisiae* (Amiri et al. 2016; Wu et al. 2018). HMG2 is the isoenzyme of HMG1 and is the dominant HMGR isoform under a low oxygen environment (Mantzouridou and Tsimidou 2010). Overexpression of *HMG2* variant (K6R) and *IDI1* can improve monoterpene production levels (Ignea et al. 2011, 2012). In fact, overexpression of *tHMG1*, *IDI1,* and *MAF1* increased geranyl acetate production in *S. cerevisiae* (Wu et al. 2018).

The ERG20 and ERG9 are two of the most essential enzymes for terpenoid precursor production (Fig. 3). Engineering ERG20 significantly improved the production of sclareol and other terpenoids (Ignea et al. 2015), and overexpression of fusion genes of *ERG20* and *BTS1* resulted in the increment of terpenoid production (Dai et al. 2012). Recently, the overexpression of *ERG20* and its mutant mERG20 (F96C mutation in ERG20, which might function as a geranylgeranyl diphosphate synthase) significantly improved geranylgeraniol content of *S. cerevisiae* (Dong et al. 2020). Dynamic control of *ERG20* expression level might improve terpenoid precursor production in yeast. The dynamic control of ERG20 combined with minimized endogenous downstream metabolism improved geraniol production (Zhao et al. 2017); moreover, dynamic control of the expression of *ERG20* and *ERG9* improved diterpenoid casbene production in *S. cerevisiae* (Callari et al. 2018). Downregulation of *ERG9* expression level leads to the high-level production of FPP (Asadollahi et al. 2008), and the common strategy is replacing native ERG9 promoter with *HXT1* or other promoters (Scalcinati et al. 2012). Besides, the introduction of exogenous rate-limiting genes of the MVA pathway in yeast could also improve terpenoid production, such as *PaGGPPS* and *SaGGPPS* (Cao et al. 2020). The lipid production would affect the terpenoid precursor of squalene by storage of squalene in the lipid body, and co-overexpression of *tHMG1* and *DGA1* coding for diacylglycerol acyltransferase led to over 250-fold higher squalene accumulation than a control strain (Wei et al. 2018). The transcriptional activators regulated terpenoid precursor synthesis in yeasts (Zhang et al. 2017; Davies et al. 2005). The overexpression of the transcription factor of *UPC2* led to the increase of terpenoid production (Shianna et al. 2001), and other transcription factors, such as *Rox1* and *Mot3*, repressed ergosterol biosynthesis in yeasts (Montañés et al. 2011).

5 Production of Artemisinin Precursor of Artemisinic Acid Derived from Artemisia Annua in Yeasts

The extraction of *Artemisia annua* had been used as an antimalarial drug in traditional Chinese medicine. In the 1970s, Youyou Tu discovered the active component artemisinin in the plant extraction of *Artemisia annua.* Since then, artemisinin was used as a potent antimalarial drug (Jung et al. 2004). Recent studies suggested that artemisinin and its derivatives had immune-modulation effects and antitumor activities (Yao et al. 2016; Kiani et al. 2020). Artemisinin is a sesquiterpene lactone endoperoxide, and it is mainly obtained by extracting from the leaves and other biomass of *A. annua* (Efferth 2017). The *A. annua* needs to grow several months before harvest, and the artemisinin component in wild *A. annua* is low (Cai et al. 2017). Moreover, planting *A. annua* occupies large amounts of fields, and the artemisinin component is affected by the climate. Therefore, artemisinin supply is uncertain and unstable, and the price fluctuates. The production of artemisinin using microorganisms is of great interest, for it might produce large amounts of artemisinin with limited space and short time.

Keasling's group developed bacterial chassis for terpenoid production, and the obtained *E. coli* strains are capable of synthesizing diverse terpenoid precursors (Martin et al. 2003, 2001). The cytochrome P450 is not easy to express in *E. coli*, and *S. cerevisiae* was selected for the production of high-level artemisinin. The FPP biosynthetic pathway was engineered to increase FPP production and decrease sterol production by overexpression of *tHMGR*, *UPC2-1*, and ERGs, and downregulation of *ERG9*. The amorphadiene synthase genes from *A. annua* was introduced to *S. cerevisiae*, in order to construct one strain with amorphadiene producing ability. Moreover, a novel cytochrome P450 (*CYP71AV1*) was identified to have the ability to perform three-step oxidation which can convert amorphadiene to artemisinic acid. The expression of the *CYP71AV1* and its redox partner from *A. annua* led to the production of artemisinic acid (Ro et al. 2006). As artemisinic acid can be easily oxidized to artemisinin using a chemical process, the production of artemisinic acid is a good proof of concept that using yeasts as microbial cell factories can synthesize plant natural products. The engineered yeast strains can produce artemisinic acid in a short time (4–5 days). Compared with the fact that planting *A. annua* needs months or years, these synthetic biology strategies save time and labor for valuable natural product synthesis (Fig. 1).

Further, overexpressing every gene in the MVA pathway doubled artemisinic acid production; however, the amporph-4,11-diene production is tenfold higher than artemisinic acid (Westfall et al. 2012). Therefore, they developed one optimized process to convert amporph-4,11-diene to dihydroartemisinic acid. The dihydroartemisinic acid can be further transformed to artemisinin (Westfall et al. 2012). In order to produce affordable artemisinic acid using yeasts, the artemisinic acid biosynthetic pathway was optimized by using proper promoters for genes and introducing efficient artemisinic acid biosynthetic genes. The fermentation titers of artemisinic acid reached as high as 25 g/L and a low-cost chemical process for the conversion of artemisinic acid to artemisinin was established, which provided a second semi-synthetic artemisinin source independent of botanical production (Paddon et al. 2013). The success of artemisinic acid production using yeasts paves the way for industrial-scale production of artemisinin derived from the traditional Chinese medicinal plant.

6 Production of Rare Ginsenosides Derived from *Panax* Plants in Yeasts

Ginsenosides are the main pharmacological components in *Panax ginseng* C. A. meyer which is one of the most famous traditional Chinese medicinal plants in East Asia. The ginsenosides are triterpenoids, and they are mainly distributed in *Panax* genus (Pace et al. 2015). The ginsenosides can be classified into two main types, dammarane-type and oleanate-type, and most ginsenosides are dammarane-type terpenoids (Shin et al. 2015). The ginsenosides have diverse biological activities, such as antitumor, anti-inflammatory, antioxidation, and antiaging. However, the ginsenoside compositions in *Panax* plant are lower, normally less than 4% (Schlag and McIntosh 2006). Moreover, the main functional ginsenosides are rare in natural *Panax* plants (Piao et al. 2020). Compound K (CK), a triterpenoid compound that has potential anti-inflammation, anti-diabetes and other biological activities, is only detected in animal or human blood (Yang et al. 2020b). Though CK is classified as ginsenoside and is proved to be transformed from natural ginsenosides, none CK was detected in *Panax* plants (Han et al. 2020). It takes five to seven years to obtain ginsenosides from *Panax* plants (Schlag and McIntosh 2006). Besides, the structure of ginsenosides is complex, and large-scale de novo chemical synthesis of ginsenosides is impossible until now. Microbial production of ginsenosides using synthetic biology strategies was applied (Chu et al. 2020).

The core skeletons of ginsenosides are protopanaxadiol (PPD) and protopanaxatriol (PPT). The precursors of PPD and PPT were synthesized via the MVA pathway in yeasts. The wild-type yeasts can synthesize 2,3-oxidosqualene. In order to produce ginsenosides in yeast, other heterologous genes need to be introduced to the yeasts. The compound of 2,3-oxidosqualene can be further converted to dammarenediol-II with Dammarenediol-II synthase (Fig. 4) (Han et al. 2006; Tansakul et al. 2006). The Cyt P450 enzyme of CYP716A47 was identified to be responsible for the conversion of dammarenediol-II to PPD (Han et al. 2011), and one P450 enzyme from *Arabidopsis thaliana* had the same function (Wang et al. 2015). Subsequently, the Cyt P450 enzyme of CYP716A53v2 can synthesize PPT with PPD as a substrate (Fig. 4) (Zhao et al. 2019).

Diverse UDP-glycosyltransferases (UGTs) are responsible for the glycosylation of PPD, PPT, and their glycosylation products (Fig. 4). The UGTPg1 was identified to glycosylate PPD and produced CK in yeasts. There are two possible pathways for CK production. One pathway is from the dammarenediol-II to PPD, and then to CK; the other pathway is from dammarenediol-II to 20S-O-β-(D-glucosyl)-dammarenediol II, and then to CK (Yan et al. 2014). Besides these catalysis abilities, UGTPg1 can also convert Rg3 to Rd, and Rh2 to F2, showing UGTPg1 might be responsible for the production of several different ginsenosides (Yan et al. 2014; Wei et al. 2015a). Moreover, several other *Panax* UGTs with high amino acid identities (>84%) indicated divergent functions, and they can convert PPT and PPT-type compounds to diverse ginsenosides (Wei et al. 2015a). The investigation of UGT94 family genes in *Panax* plants indicated that an unprecedented group of UGT94s with high amino acid identity had diverse

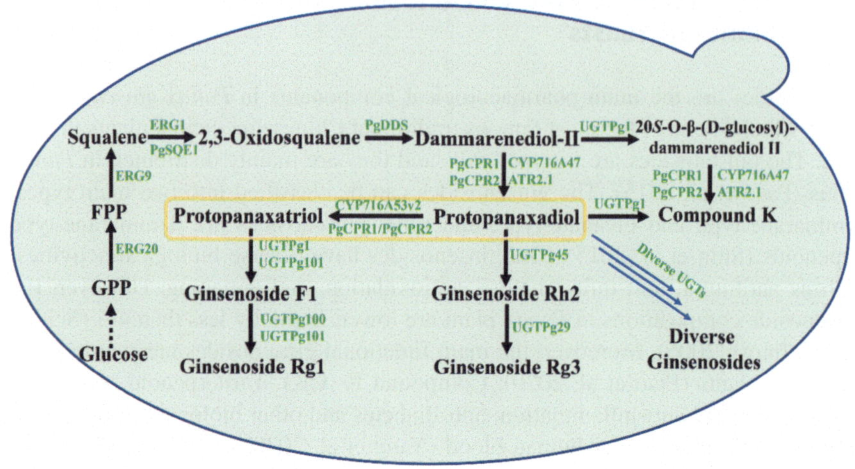

Fig. 4 Engineering for the production of ginsenosides in *S. cerevisiae*. GPP, geranyl diphosphate; FPP, farnesyl diphosphate

catalyzing activities in *Panax* plants (Yang et al. 2020a). These UGT94 enzymes were responsible for the synthesis of diverse ginsenosides (Yang et al. 2020a). Moreover, the UGT diversity in *Panax* plant was investigated, and two UGTs of UGTPg45 and UGTPg29 for ginsenosides Rh2 and Rg3 biosynthesis were identified (Wang et al. 2015). The production of Rh2 and Rg3 using synthetic biology strategies was applied in engineered yeasts (Wang et al. 2015). A semi-rationally designed UGT51 derived from yeast was also demonstrated to have the ability to synthesize Rh2, showing the synthetic biology strategies had great potential for terpenoid production in yeast (Zhuang et al. 2017). The Rh2 titer increased to 2.25 g/L via expressing multiple copies of UGTPg45 in a PPD-producing chassis, which provides the possible large-scale production of ginsenosides in yeasts (Wang et al. 2019c). Moreover, the non-conventional yeast, *Y. lipolytica*, had been engineered for CK production, and the final CK titer reached to 161.8 mg/L via fed-batch fermentation in a 5 L fermenter (Li et al. 2019a).

7 Production of Licorice Triterpenoid Derived from *Glycyrrhiza* Plants

Glycyrrhiza plants, including *Glycyrrhiza uralensis* Fisch., *Glycyrrhiza flata* Bat. and *Glycyrrhiza glabra* L., are widely used herbal plants in traditional Chinese medicine (Asl and Hosseinzadeh 2008). One of the main bioactive components of *Glycyrrhiza* plants is glycyrrhetinic acid (GA) (Li et al. 2014; Zhang and Ye 2009). GA has anti-inflammatory, antitumor, immunomodulatory, and other biological activities (Kowalska and Kalinowska-Lis 2019; Roohbakhsh et al. 2016). GA can be generated by the hydrolysis of glycyrrhizic acid. Glycyrrhizic acid is extracted from the roots of the licorice plants, and is one kind of triterpenoids. Glycyrrhizic acid is mainly used as a sweetening and flavoring agent for beverages, candies, chewing gum, and bitter drugs (Pastorino et al. 2018). The wild and cultivated *Glycyrrhiza* plants cannot satisfy the growing market demand for GA and glycyrrhizic acid. Moreover, the extraction of Glycyrrhizic acid from *Glycyrrhiza* plants pollutes the environment. For GA structure is complex, chemical synthesis yield is low and the cost is high (Wang et al. 2019a). Therefore, another sustainable supply of GA is necessary (Guan et al. 2020).

Compared with the ginsenosides biosynthetic pathway, the GA biosynthetic pathway is different. The precursors were synthesized with the MVA pathway, but the 2,3-oxidosqualene is converted to β-amyrin by β-amyrin synthase. The β-amyrin is oxidized to 11-oxo-β-amyrin with a licorice β-amyrin 11-oxidase which is a cytochrome P450 enzyme (Seki et al. 2008). Another licorice cytochrome P450 enzyme of CYP72A154 catalyzed three sequential oxidation steps at C-30 of 11-oxo-β-amyrin, and GA is obtained (Seki et al. 2011). In the catalysis reaction, CPRs provide electrons to the cytochrome P450s (Zhu et al. 2018). The UGTs (GuGT14 and UGT73P12) are responsible for the biosynthesis of glycyrrhizic acid (Nomura et al. 2019; Chen et al. 2019). In order to produce high-level GA in yeasts, the MVA pathway was strengthened and an acetyl-CoA competing pathway was disrupted. The final β-amyrin production reached 279 mg/L (Liu et al. 2019a). By introducing efficient β-amyrin synthase and P450 enzymes and optimizing metabolic flux to GA synthesis, the final 11-oxo-β-amyrin and GA titer in shake flask are 80 mg/L and 8.78 mg/L, respectively (Wang et al. 2019a).

8 Production of Other Terpenoids Derived from Traditional Chinese Medicinal Plants in Yeasts

Besides the sesquiterpenes and triterpenes, many monoterpenoids and diterpenoids have antitumor and other activities (Huang et al. 2012). The synthetic biology and metabolic engineering strategies had been used for the yeast biosynthesis of monoterpenoids and diterpenoids (Table 1) (Gao et al. 2020; Vickers et al. 2017). The *Nepeta cataria* is widely used in traditional Chinese medicine, and the main bioactive components are the monoterpenoids of menthone and pulegone (Gao et al. 2020). As the yeast Erg20p can function as the GPPS and FPPS, which limited monoterpenoid production by catalyzing GPP to FPP. In order to overcome this limitation, the GPP function of Erg20p was strengthened, and this led to a significant increase in monoterpenoid titers. By further engineering Erg20p to reduce its FPPS activity and fusion of the engineered ERG20 genes with the terpene synthase, the yield of the obtained engineered strain increased 340-fold over the starting strain (Ignea et al. 2014). The native terpene precursor GPP is tightly coupled with yeast sterol biosynthesis which is essential for yeast survival; therefore, engineering the native pathway is unable to direct the GPP metabolic flux to FPP. Ignea et al. established a synthetic orthogonal monoterpenoid biosynthetic pathway in *S. cerevisiae*. They identified the selectivity mechanism of monoterpene

Table 1 Some selected terpenoids derived from traditional Chinese medicinal plants which have been synthesized by yeasts

Terpenoid name	Traditional Chinese medicinal plant source	Yeast chassis	Titer	Terpenoid type	References
Geraniol	Diverse plants	*Saccharomyces cerevisiae* strain CEN.PK2-1C	1.68 g/L	Monoterpenoid	(Jiang et al. 2017)
α-Terpineol	Pine and cypress plants	*Saccharomyces cerevisiae* strain LCB08	21.88 mg/L	Monoterpenoid	(Zhang et al. 2019)
Limonene	Citrus fruits	*Saccharomyces cerevisiae* strain BY4741	166 mg/L	Monoterpenoid	(Ignea et al. 2019)
Miltiradiene	*Salvia miltiorrhiza* Bge	*Saccharomyces cerevisiae* strain S288C	3.5 g/L	Diterpenoid	(Hu et al. 2020)
Triptolide	*Tripterygium wilfordii*	*Saccharomyces cerevisiae* strain BY-HZ16	30.5 μg/g yeast biomass	Diterpenoid	(Tu et al. 2020)
(-)-β-elemene	*Curcuma wenyujin*	*Saccharomyces cerevisiae* strain SCIGS22a	190.7 mg/L	Sesquiterpenoid	(Hu et al. 2017)
Artemisinin	*Artemisia annua* L	*Saccharomyces cerevisiae* strain Y1284	25 g/L	Sesquiterpenoid	(Paddon et al. 2013)
Glycyrrhetinic acid	*Glycyrrhize glabra* L	*Saccharomyces cerevisiae* strain Y7	8.78 mg/L	Triterpenoid	(Wang et al. 2019a)
Ginsenoside compound K (CK)	*Panax* plants	*Saccharomyces cerevisiae* strain (WLN-3)	1.7 g/L	Triterpenoid	(Nan et al. 2020)
Ginsenoside F1	*Panax* plants	*Saccharomyces cerevisiae* strain (BY-F1)	450. 5 mg/L	Triterpenoid	(Wang et al. 2019b)
Ginsenoside Rh2	*Panax* plants	*Saccharomyces cerevisiae* strain ZWDRH2-10	2.25 g/L	Triterpenoid	(Wang et al. 2019c)

(continued)

Table 1 (continued)

Terpenoid name	Traditional Chinese medicinal plant source	Yeast chassis	Titer	Terpenoid type	References
Ginsenoside Rh1	*Panax* plants	*Saccharomyces cerevisiae* strain ZW-Rh1-20	92.8 mg/L	Triterpenoid	(Wei et al. 2015a)
Ginsenoside Rg3	*Panax* plants	*Saccharomyces cerevisiae* strain D20RG1	3.49 μmol/g dry cell weight	Triterpenoid	(Wang et al 2015)
Protopanaxadiol (PPD)	*Panax* plants	*Saccharomyces cerevisiae* strain GW10	152.37 mg/L	Triterpenoid	(Gao et al. 2018)

synthases and engineered enzymes to accept neryl diphosphate as the substrate. By combining the engineered monoterpene synthase and dynamic control of FPP and neryl diphosphate biosynthesis, the production of monoterpenes using the orthogonal monoterpenoid biosynthetic pathway increased sevenfold over using yeast canonical pathway (Ignea et al. 2019).

Some traditional Chinese medicinal plants can produce bioactive diterpenoid components, such as triptolide from *Tripterygium wilfordii* and oridonin from *Isodon rubescens* (Song et al. 2019; Cheng et al. 2019). The triptolide has anti-inflammatory, antitumor, and other biological activities, but its composition in the biomass of *T. wilfordii* is very low (from 0.0001 to 0.002%) (Zeng et al. 2013; Zhou et al. 2012). Integration analysis with the genomic, transcriptomic, and metabolomic data of *T. wilfordii* recovered the cytochrome P450 (CYP728B70) for triptolide intermediate of dehydroabietic acid. Introduction of CYP728B70 to engineered yeast achieved the production of diterpene alcohols and acids, which paves the way for triptolide biosynthesis in yeasts (Tu et al. 2020). The key genes of oridonin biosynthetic pathway are unclear (Pelot et al. 2017). Though paclitaxel (taxol) is not extracted from traditional Chinese medicinal plants, it is one of the most famous plant diterpenoids used as an effective antitumor drug (Gallego-Jara et al. 2020). The paclitaxel is mainly extracted from the bark of several *Taxus* species, and most *Taxus* species are used for taxol production which threatens the survival of these old trees (Expósito et al. 2009). The biosynthesis of paclitaxel has been realized in *Taxus baccata* suspension cultures and *Nicotiana benthamiana*, but its yield is low (Li et al. 2019b; Malik et al. 2011). Production of paclitaxel in yeast is still challenging, many essential enzymes need to be identified (Gallego-Jara et al. 2020). Diterpene biosynthesis involves the oxidation steps which requires identifying cytochrome P450 enzymes, and the total biosynthesis of diterpenes in yeast is difficult (Hu et al. 2020).

The multi-omics data can be collected from the available database or obtained by sequencing different fresh tissues of the plant at different growth phases or areas. By comparing the multi-omics data and constructing a network between different omics data, the genes in the terpenoid biosynthetic pathway can be identified. Using structural biology and protein engineering strategies, the activities of key enzymes would be improved and adapted to the yeast hosts. These key genes can be further used to redesign the heterologous pathway in engineered yeast, and the redesigned pathway can be applied to build yeast strains with enhanced terpenoid production. The synthetic biology and metabolic engineering strategies could direct yeast metabolic flux to the biosynthesis of terpenoids. Finally, yeast strains with a high titer, rate, and yield production of terpenoids can be obtained. Based on the proper scale-up fermentation strategies and bioprocess optimization, large-scale production of terpenoids in yeasts can be achieved (Fig. 1).

9 Conclusion and Perspective

The omics-based technologies can help to identify key genes easily from omics data of the traditional Chinese medicinal plants, and these key genes can be used for building yeast cell factories. Engineering for efficient yeast terpenoid precursor-producing chassis, introducing heterogenous terpenoid synthetic pathways, increasing key gene expression level, directing metabolic flux for terpenoid synthesis by downregulating competing pathway, enhancing cofactor supply for MVA synthesis, engineering sub-organelle for terpenoid production, and other frequently used strategies would significantly improve titer, rate, and yield of yeast terpenoid production.

With the development of genomic-scale models and high-throughput automated screening, engineered yeasts for efficient terpenoid production would be accelerated. The optimized fermentation and bioprocess could decrease the cost of yeast terpenoids. Besides, to produce affordable or commercial terpenoids, developing smart biomanufacturing with a low-cost, intelligent, and continuous process will lay the foundation for industrial-scale yeast production of valuable terpenoids. The production of many valuable terpenoids derived from traditional Chinese medicinal plants will be achieved in the near future.

Acknowledgements This work was supported by the National Natural Science Foundation of China (No. 31800079).

References

Allen KD, McKernan K, Pauli C, Roe J, Torres A, Gaudino R (2019) Genomic charsacterization of the complete terpene synthase gene family from *Cannabis sativa*. Plos One 14(9):e0222363.https://doi.org/10.1371/journal.pone.0222363

Amiri P, Shahpiri A, Asadollahi MA, Momenbeik F, Partow S (2016) Metabolic engineering of *Saccharomyces cerevisiae* for linalool production. Biotechnol Lett 38(3):503–508. https://doi.org/10.1007/s10529-015-2000-4

Arnesen JA, Kildegaard KR, Cernuda Pastor M, Jayachandran S, Kristensen M, Borodina I (2020) *Yarrowia lipolytica* Strains engineered for the production of terpenoids. Front Bioeng Biotechnol 8(945). https://doi.org/10.3389/fbioe.2020.00945

Asadollahi MA, Maury J, Møller K, Nielsen KF, Schalk M, Clark A, Nielsen J (2008) Production of plant sesquiterpenes in *Saccharomyces cerevisiae*: effect of ERG9 repression on sesquiterpene biosynthesis. Biotechnol Bioeng 99(3):666–677. https://doi.org/10.1002/bit.21581

Asl MN, Hosseinzadeh H (2008) Review of pharmacological effects of *Glycyrrhiza* sp. and its bioactive compounds. Phytotherapy Res 22(6):709–724

Bao T, Shadrack K, Yang S, Xue X, Li S, Wang N, Wang Q, Wang L, Gao X, Cronk Q (2020) Functional characterization of terpene synthases accounting for the volatilized-terpene heterogeneity in *Lathyrus odoratus* cultivar flowers. Plant Cell Physiol 61(10):1733–1749. https://doi.org/10.1093/pcp/pcaa100

Bergman ME, Davis B, Phillips MA (2019) Medically useful plant terpenoids: biosynthesis, occurrence, and mechanism of action. Molecules (Basel, Switzerland) 24(21). https://doi.org/10.3390/molecules24213961

Bian G, Deng Z, Liu T (2017) Strategies for terpenoid overproduction and new terpenoid discovery. Curr Opin Biotechnol 48:234–241. https://doi.org/10.1016/j.copbio.2017.07.002

Cai T-Y, Zhang Y-R, Ji J-B, Xing J (2017) Investigation of the component in *Artemisia annua* L. leading to enhanced antiplasmodial potency of artemisinin via regulation of its metabolism. J Ethnopharmacol 207:86–91. https://doi.org/10.1016/j.jep.2017.06.025

Callari R, Meier Y, Ravasio D, Heider H (2018) Dynamic control of ERG20 and ERG9 expression for improved casbene production in *Saccharomyces cerevisiae*. Front Bioeng Biotechnol 6(160). https://doi.org/10.3389/fbioe.2018.00160

Cao X, Yang S, Cao C, Zhou YJ (2020) Harnessing sub-organelle metabolism for biosynthesis of isoprenoids in yeast. Synth Syst Biotechnol 5(3):179–186. https://doi.org/10.1016/j.synbio.2020.06.005

Chen K, Hu Z-m, Song W, Wang Z-l, He J-b, Shi X-m, Cui Q-h, Qiao X, Ye M (2019) Diversity of o-glycosyltransferases contributes to the biosynthesis of flavonoid and triterpenoid glycosides in *Glycyrrhiza uralensis*. Acs Synth Biol 8(8):1858–1866. https://doi.org/10.1021/acssynbio.9b00171

Cheng W, Huang C, Ma W, Tian X, Zhang X (2019) Recent development of oridonin derivatives with diverse pharmacological activities. Mini Rev Med Chem 19(2):114–124. https://doi.org/10.2174/1389557517666170417170609

Chu LL, Montecillo JAV, Bae H (2020) Recent advances in the metabolic engineering of yeasts for ginsenoside biosynthesis. Front Bioeng Biotechnol 8(139). https://doi.org/10.3389/fbioe.2020.00139

Chung SY, Seki H, Fujisawa Y, Shimoda Y, Hiraga S, Nomura Y, Saito K, Ishimoto M, Muranaka T (2020) A cellulose synthase-derived enzyme catalyses 3-O-glucuronosylation in saponin biosynthesis. Nat Commun 11(1):5664. https://doi.org/10.1038/s41467-020-19399-0

Dai Z, Liu Y, Huang L, Zhang X (2012) Production of miltiradiene by metabolically engineered *Saccharomyces cerevisiae*. Biotechnol Bioeng 109(11):2845–2853. https://doi.org/10.1002/bit.24547

Davies BS, Wang HS, Rine J (2005) Dual activators of the sterol biosynthetic pathway of *Saccharomyces cerevisiae*: similar activation/regulatory domains but different response mechanisms. Mol Cell Biol 25(16):7375–7385. https://doi.org/10.1128/mcb.25.16.7375-7385.2005

Dong H, Chen S, Zhu J, Gao K, Zha W, Lin P, Zi J (2020) Enhance production of diterpenoids in Yeast by overexpression of the fused enzyme of ERG20 and its mutant mERG20. J Biotechnol 307:29–34. https://doi.org/10.1016/j.jbiotec.2019.10.019

Efferth T (2017) From ancient herb to modern drug: *Artemisia annua* and artemisinin for cancer therapy. Semin Cancer Biol 46:65–83. https://doi.org/10.1016/j.semcancer.2017.02.009

Expósito O, Bonfill M, Moyano E, Onrubia M, Mirjalili MH, Cusidó RM, Palazón J (2009) Biotechnological production of taxol and related taxoids: current state and prospects. Anticancer Agents Med Chem 9(1):109–121. https://doi.org/10.2174/187152009787047761

Farhi M, Marhevka E, Masci T, Marcos E, Eyal Y, Ovadis M, Abeliovich H, Vainstein A (2011) Harnessing yeast subcellular compartments for the production of plant terpenoids. Metab Eng 13(5):474–481. https://doi.org/10.1016/j.ymben.2011.05.001

Gallego-Jara J, Lozano-Terol G, Sola-Martínez RA, Cánovas-Díaz M, de Diego Puente T (2020) A compressive review about Taxol(®): History and future challenges. Molecules (Basel, Switzerland) 25(24). https://doi.org/10.3390/molecules25245986

Gao X, Caiyin Q, Zhao F, Wu Y, Lu W (2018) Engineering *Saccharomyces cerevisiae* for enhanced production of protopanaxadiol with cofermentation of glucose and xylose. J Agr Food Chem 66(45):12009–12016. https://doi.org/10.1021/acs.jafc.8b04916

Gao Q, Wang L, Zhang M, Wei Y, Lin W (2020) Recent advances on feasible strategies for monoterpenoid production in *Saccharomyces cerevisiae*. Front Bioeng Biotechnol 8:1372

Guan R, Wang M, Guan Z, Jin C-Y, Lin W, Ji X, Wei Y (2020) Metabolic engineering for glycyrrhetinic acid production in *Saccharomyces cerevisiae*. Front Bioeng Biotechnol 8:1318

Han JY, Kwon YS, Yang DC, Jung YR, Choi YE (2006) Expression and RNA interference-induced silencing of the dammarenediol synthase gene in *Panax ginseng*. Plant Cell Physiol 47(12):1653–1662. https://doi.org/10.1093/pcp/pcl032

Han J-Y, Kim H-J, Kwon Y-S, Choi Y-E (2011) The Cyt P450 Enzyme CYP716A47 catalyzes the formation of protopanaxadiol from dammarenediol-II during ginsenoside biosynthesis in *Panax ginseng*. Plant Cell Physiol 52(12):2062–2073. https://doi.org/10.1093/pcp/pcr150

Han R, Rai A, Nakamura M, Suzuki H, Takahashi H, Yamazaki M, Saito K (2016) De novo deep transcriptome analysis of medicinal plants for gene discovery in biosynthesis of plant natural products. Methods Enzymol 576:19–45. https://doi.org/10.1016/bs.mie.2016.03.001

Han X, Li W, Duan Z, Ma X, Fan D (2020) Biocatalytic production of compound K in a deep eutectic solvent based on choline chloride using a substrate fed-batch strategy. Bioresour Technol 305:123039.https://doi.org/10.1016/j.biortech.2020.123039

Hu Y, Zhou YJ, Bao J, Huang L, Nielsen J, Krivoruchko A (2017) Metabolic engineering of *Saccharomyces cerevisiae* for production of germacrene A, a precursor of beta-elemene. J Ind Microbiol Biotechnol 44(7):1065–1072. https://doi.org/10.1007/s10295-017-1934-z

Hu T, Zhou J, Tong Y, Su P, Li X, Liu Y, Liu N, Wu X, Zhang Y, Wang J, Gao L, Tu L, Lu Y, Jiang Z, Zhou YJ, Gao W, Huang L (2020) Engineering chimeric diterpene synthases and isoprenoid biosynthetic pathways enables high-level production of miltiradiene in yeast. Metab Eng 60:87–96. https://doi.org/10.1016/j.ymben.2020.03.011

Huang M, Lu JJ, Huang MQ, Bao JL, Chen XP, Wang YT (2012) Terpenoids: natural products for cancer therapy. Expert Opin Investig Drugs 21(12):1801–1818. https://doi.org/10.1517/13543784.2012.727395

Hwang CR, Lee SH, Jang GY, Hwang IG, Kim HY, Woo KS, Lee J, Jeong HS (2014) Changes in ginsenoside compositions and antioxidant activities of hydroponic-cultured ginseng roots and leaves with heating temperature. J Ginseng Res 38(3):180–186. https://doi.org/10.1016/j.jgr.2014.02.002

Ignea C, Cvetkovic I, Loupassaki S, Kefalas P, Johnson CB, Kampranis SC, Makris AM (2011) Improving yeast strains using recyclable integration cassettes, for the production of plant terpenoids. Microb Cell Fact 10:4. https://doi.org/10.1186/1475-2859-10-4

Ignea C, Trikka FA, Kourtzelis I, Argiriou A, Kanellis AK, Kampranis SC, Makris AM (2012) Positive genetic interactors of HMG2 identify a new set of genetic perturbations for improving sesquiterpene production in *Saccharomyces cerevisiae*. Microb Cell Fact 11:162. https://doi.org/10.1186/1475-2859-11-162

Ignea C, Pontini M, Maffei ME, Makris AM, Kampranis SC (2014) Engineering monoterpene production in yeast using a synthetic dominant negative geranyl diphosphate synthase. Acs Synth Biol 3(5):298–306. https://doi.org/10.1021/sb400115e

Ignea C, Trikka FA, Nikolaidis AK, Georgantea P, Ioannou E, Loupassaki S, Kefalas P, Kanellis AK, Roussis V, Makris AM, Kampranis SC (2015) Efficient diterpene production in yeast by engineering Erg20p into a geranylgeranyl diphosphate synthase. Metab Eng 27:65–75. https://doi.org/10.1016/j.ymben.2014.10.008

Ignea C, Raadam MH, Motawia MS, Makris AM, Vickers CE, Kampranis SC (2019) Orthogonal monoterpenoid biosynthesis in yeast constructed on an isomeric substrate. Nat Commun 10(1):3799. https://doi.org/10.1038/s41467-019-11290-x

Jaeger R, Cuny E (2016) Terpenoids with special pharmacological significance: a review. Nat Prod Commun 11(9):1373–1390

Jansen DJ, Shenvi RA (2014) Synthesis of medicinally relevant terpenes: reducing the cost and time of drug discovery. Future Med Chem 6(10):1127–1148. https://doi.org/10.4155/fmc.14.71

Jiang Z, Kempinski C, Chappell J (2016) Extraction and analysis of terpenes/terpenoids. Curr Protoc Plant Biol 1:345–358. https://doi.org/10.1002/cppb.20024

Jiang G-Z, Yao M-D, Wang Y, Zhou L, Song T-Q, Liu H, Xiao W-H, Yuan Y-J (2017) Manipulation of GES and ERG20 for geraniol overproduction in *Saccharomyces cerevisiae*. Metab Eng 41:57–66. https://doi.org/10.1016/j.ymben.2017.03.005

Jozwiak A, Sonawane PD, Panda S, Garagounis C, Papadopoulou KK, Abebie B, Massalha H, Almekias-Siegl E, Scherf T, Aharoni A (2020) Plant terpenoid metabolism co-opts a component of the cell wall biosynthesis machinery. Nat Chem Biol 16(7):740–748. https://doi.org/10.1038/s41589-020-0541-x

Jung M, Lee K, Kim H, Park M (2004) Recent advances in artemisinin and its derivatives as antimalarial and antitumor agents. Curr Med Chem 11(10):1265–1284. https://doi.org/10.2174/0929867043365233

Kiani BH, Kayani WK, Khayam AU, Dilshad E, Ismail H, Mirza B (2020) Artemisinin and its derivatives: a promising cancer therapy. Mol Biol Rep 47(8):6321–6336. https://doi.org/10.1007/s11033-020-05669-z

Kim SW, Keasling JD (2001) Metabolic engineering of the nonmevalonate isopentenyl diphosphate synthesis pathway in *Escherichia coli* enhances lycopene production. Biotechnol Bioeng 72(4):408–415. https://doi.org/10.1002/1097-0290(20000220)72:4%3c408::aid-bit1003%3e3.0.co;2-h

Kowalska A, Kalinowska-Lis U (2019) 18β-Glycyrrhetinic acid: its core biological properties and dermatological applications. Int J Cosmet Sci 41(4):325–331. https://doi.org/10.1111/ics.12548

Kumar S, Kempinski C, Zhuang X, Norris A, Mafu S, Zi J, Bell SA, Nybo SE, Kinison SE, Jiang Z, Goklany S, Linscott KB, Chen X, Jia Q, Brown SD, Bowman JL, Babbitt PC, Peters RJ, Chen F, Chappell J (2016) Molecular diversity of terpene synthases in the liverwort *Marchantia polymorpha*. Plant Cell 28(10):2632–2650. https://doi.org/10.1105/tpc.16.00062

Leebens-Mack JH, Barker MS, Carpenter EJ, Deyholos MK, Gitzendanner MA, Graham SW, Grosse I, Li Z, Melkonian M, Mirarab S, Porsch M, Quint M, Rensing SA, Soltis DE, Soltis PS, Stevenson DW, Ullrich KK, Wickett NJ, DeGironimo L, Edger PP, Jordon-Thaden IE, Joya S, Liu T, Melkonian B, Miles NW, Pokorny L, Quigley C, Thomas P, Villarreal JC, Augustin MM, Barrett MD, Baucom RS, Beerling DJ, Benstein RM, Biffin E, Brockington SF, Burge DO, Burris JN, Burris KP, Burtet-Sarramegna V, Caicedo AL, Cannon SB, Çebi Z, Chang Y, Chater C, Cheeseman JM, Chen T, Clarke ND, Clayton H, Covshoff S, Crandall-Stotler BJ, Cross H, dePamphilis CW, Der JP, Determann R, Dickson RC, Di Stilio VS, Ellis S, Fast E, Feja N, Field KJ, Filatov DA, Finnegan PM, Floyd SK, Fogliani B, García N, Gâteblé G, Godden GT, Goh F, Greiner S, Harkess A, Heaney JM, Helliwell KE, Heyduk K, Hibberd JM, Hodel RGJ, Hollingsworth PM, Johnson MTJ, Jost R, Joyce B, Kapralov MV, Kazamia E, Kellogg EA, Koch MA, Von Konrat M, Könyves K, Kutchan TM, Lam V, Larsson A, Leitch AR, Lentz R, Li F-W, Lowe AJ, Ludwig M, Manos PS, Mavrodiev E, McCormick MK, McKain M, McLellan T, McNeal JR, Miller RE, Nelson MN, Peng Y, Ralph P, Real D, Riggins CW, Ruhsam M, Sage RF, Sakai AK, Scascitella M, Schilling EE, Schlösser E-M, Sederoff H, Servick S, Sessa EB, Shaw AJ, Shaw SW, Sigel EM, Skema C, Smith AG, Smithson A, Stewart CN, Stinchcombe JR, Szövényi P, Tate JA, Tiebel H, Trapnell D, Villegente M, Wang C-N, Weller SG, Wenzel M, Weststrand S, Westwood JH, Whigham DF, Wu S, Wulff AS, Yang Y, Zhu D, Zhuang C, Zuidof J, Chase MW, Pires JC, Rothfels CJ, Yu J, Chen C, Chen L, Cheng S, Li J, Li R, Li X, Lu H, Ou Y, Sun X, Tan X, Tang J, Tian Z, Wang F, Wang J, Wei X, Xu X, Yan Z, Yang F, Zhong X, Zhou F, Zhu Y, Zhang Y, Ayyampalayam S, Barkman TJ, Nguyen N-p, Matasci N, Nelson DR, Sayyari E, Wafula EK, Walls RL, Warnow T, An H, Arrigo N, Baniaga AE, Galuska S, Jorgensen SA, Kidder TI, Kong H, Lu-Irving P, Marx HE, Qi X, Reardon CR, Sutherland BL, Tiley GP, Welles SR, Yu R, Zhan S, Gramzow L, Theißen G, Wong GK-S (2019) One thousand plant transcriptomes and the phylogenomics of green plants. Nature 574(7780):679–685. https://doi.org/10.1038/s41586-019-1693-2

Li D, Wu Y, Zhang C, Sun J, Zhou Z, Lu W (2019a) Production of triterpene ginsenoside compound K in the non-conventional yeast *Yarrowia lipolytica*. J Agric Food Chem 67(9):2581–2588. https://doi.org/10.1021/acs.jafc.9b00009

Li J, Mutanda I, Wang K, Yang L, Wang J, Wang Y (2019b) Chloroplastic metabolic engineering coupled with isoprenoid pool enhancement for committed taxanes biosynthesis in *Nicotiana benthamiana*. Nat Commun 10(1):4850. https://doi.org/10.1038/s41467-019-12879-y

Li JY, Cao HY, Liu P, Cheng GH, Sun MY (2014) Glycyrrhizic acid in the treatment of liver diseases: literature review. Biomed Res Int 2014:872139. https://doi.org/10.1155/2014/872139

Liang J, Mai W, Tang J, Wei Y (2019) Highly effective treatment of petrochemical wastewater by a super-sized industrial scale plant with expanded granular sludge bed bioreactor and aerobic activated sludge. Chem Eng J 360:15–23. https://doi.org/10.1016/j.cej.2018.11.167

Liao P, Hemmerlin A, Bach TJ, Chye M-L (2016) The potential of the mevalonate pathway for enhanced isoprenoid production. Biotechnol Adv 34(5):697–713. https://doi.org/10.1016/j.biotechadv.2016.03.005

Liu H, Wang Y, Tang Q, Kong W, Chung W-J, Lu T (2014) MEP pathway-mediated isopentenol production in metabolically engineered *Escherichia coli*. Microb Cell Fact 13(1):135. https://doi.org/10.1186/s12934-014-0135-y

Liu H, Fan J, Wang C, Li C, Zhou X (2019a) Enhanced β-amyrin synthesis in *Saccharomyces cerevisiae* by coupling an optimal acetyl-CoA supply pathway. J Agr Food Chem 67(13):3723–3732

Liu Z, Zhang Y, Nielsen J (2019b) Synthetic biology of yeast. Biochemistry 58(11):1511–1520. https://doi.org/10.1021/acs.biochem.8b01236

Liu X, Zhu X, Wang H, Liu T, Cheng J, Jiang H (2020) Discovery and modification of cytochrome P450 for plant natural products biosynthesis. Synthetic and Systems Biotechnology 5(3):187–199. https://doi.org/10.1016/j.synbio.2020.06.008

Ma Y, Li W, Mai J, Wang J, Wei Y, Ledesma-Amaro R, Ji X-J (2020) Engineering *Yarrowia lipolytica* for sustainable production of the chamomile sesquiterpene (−)-α-bisabolol. Green Chem. https://doi.org/10.1039/D0GC03180A

Mai W, Hu T, Li C, Wu R, Chen J, Shao Y, Liang J, Wei Y (2020) Effective nitrogen removal of wastewater from vitamin B2 production by a potential anammox process. J Water Process Eng 37:101515.https://doi.org/10.1016/j.jwpe.2020.101515

Malik S, Cusidó RM, Mirjalili MH, Moyano E, Palazón J, Bonfill M (2011) Production of the anticancer drug taxol in *Taxus baccata* suspension cultures: a review. Process Biochem 46(1):23–34. https://doi.org/10.1016/j.procbio.2010.09.004

Mantzouridou F, Tsimidou MZ (2010) Observations on squalene accumulation in *Saccharomyces cerevisiae* due to the manipulation of HMG2 and ERG6. FEMS Yeast Res 10(6):699–707. https://doi.org/10.1111/j.1567-1364.2010.00645.x

Martin VJ, Yoshikuni Y, Keasling JD (2001) The in vivo synthesis of plant sesquiterpenes by *Escherichia coli*. Biotechnol Bioeng 75(5):497–503. https://doi.org/10.1002/bit.10037

Martin VJ, Pitera DJ, Withers ST, Newman JD, Keasling JD (2003) Engineering a mevalonate pathway in *Escherichia coli* for production of terpenoids. Nat Biotechnol 21(7):796–802. https://doi.org/10.1038/nbt833

Montañés FM, Pascual-Ahuir A, Proft M (2011) Repression of ergosterol biosynthesis is essential for stress resistance and is mediated by the Hog1 MAP kinase and the Mot3 and Rox1 transcription factors. Mol Microbiol 79(4):1008–1023. https://doi.org/10.1111/j.1365-2958.2010.07502.x

Moser S, Pichler H (2019) Identifying and engineering the ideal microbial terpenoid production host. Appl Microbiol Biotechnol 103(14):5501–5516. https://doi.org/10.1007/s00253-019-09892-y

Nagegowda DA, Gupta P (2020) Advances in biosynthesis, regulation, and metabolic engineering of plant specialized terpenoids. Plant Sci 294:110457.https://doi.org/10.1016/j.plantsci.2020.110457

Namuli A, Bazira J, Casim TU, Engeu PO (2018) A review of various efforts to increase artemisinin and other antimalarial compounds in *Artemisia Annua* L plant. Cogent Biol 4(1). https://doi.org/10.1080/23312025.2018.1513312

Nan W, Zhao F, Zhang C, Ju H, Lu W (2020) Promotion of compound K production in *Saccharomyces cerevisiae* by glycerol. Microb Cell Fact 19(1):41. https://doi.org/10.1186/s12934-020-01306-3

Navarro-Muñoz JC, Selem-Mojica N, Mullowney MW, Kautsar SA, Tryon JH, Parkinson EI, De Los Santos ELC, Yeong M, Cruz-Morales P, Abubucker S, Roeters A, Lokhorst W, Fernandez-Guerra A, Cappelini LTD, Goering AW, Thomson RJ, Metcalf WW, Kelleher NL, Barona-Gomez F, Medema MH (2020) A computational framework to explore large-scale biosynthetic diversity. Nat Chem Biol 16(1):60–68. https://doi.org/10.1038/s41589-019-0400-9

Nett RS, Lau W, Sattely ES (2020) Discovery and engineering of colchicine alkaloid biosynthesis. Nature. https://doi.org/10.1038/s41586-020-2546-8

Nielsen J, Keasling JD (2016) Engineering cellular metabolism. Cell 164(6):1185–1197. https://doi.org/10.1016/j.cell.2016.02.004
Nomura Y, Seki H, Suzuki T, Ohyama K, Mizutani M, Kaku T, Tamura K, Ono E, Horikawa M, Sudo H, Hayashi H, Saito K, Muranaka T (2019) Functional specialization of UDP-glycosyltransferase 73P12 in licorice to produce a sweet triterpenoid saponin, glycyrrhizin. Plant J 99(6):1127–1143. https://doi.org/10.1111/tpj.14409
Pace R, Martinelli EM, Sardone N, Combarieu ED (2015) Metabolomic evaluation of ginsenosides distribution in *Panax* genus (*Panax ginseng* and *Panax quinquefolius*) using multivariate statistical analysis. Fitoterapia 101:80–91. https://doi.org/10.1016/j.fitote.2014.12.013
Paddon CJ, Westfall PJ, Pitera DJ, Benjamin KR, Fisher K, Mcphee DJ, Leavell M, Tai A, Main A, Eng D (2013) High-level semi-synthetic production of the potent antimalarial artemisinin. Nature 496(7446):528–532
Paramasivan K, Mutturi S (2017) Progress in terpene synthesis strategies through engineering of *Saccharomyces cerevisiae*. Crit Rev Biotechnol 37(8):974–989. https://doi.org/10.1080/07388551.2017.1299679
Pastorino G, Cornara L, Soares S, Rodrigues F, Oliveira M (2018) Liquorice (*Glycyrrhiza glabra*): a phytochemical and pharmacological review. Phytotherapy Res PTR 32(12):2323–2339. https://doi.org/10.1002/ptr.6178
Pelot KA, Hagelthorn LM, Addison JB, Zerbe P (2017) Biosynthesis of the oxygenated diterpene nezukol in the medicinal plant *Isodon rubescens* is catalyzed by a pair of diterpene synthases. Plos One 12(4):e0176507.https://doi.org/10.1371/journal.pone.0176507
Piao XM, Huo Y, Kang JP, Mathiyalagan R, Zhang H, Yang DU, Kim M, Yang DC, Kang SC, Wang YP (2020) Diversity of ginsenoside profiles produced by various processing technologies. Molecules (Basel, Switzerland) 25(19). https://doi.org/10.3390/molecules25194390
Pichersky E, Raguso RA (2018) Why do plants produce so many terpenoid compounds? New Phytol 220(3):692–702. https://doi.org/10.1111/nph.14178
Ro DK, Paradise EM, Ouellet M, Fisher KJ, Newman KL, Ndungu JM, Ho KA, Eachus RA, Ham TS, Kirby J, Chang MC, Withers ST, Shiba Y, Sarpong R, Keasling JD (2006) Production of the antimalarial drug precursor artemisinic acid in engineered yeast. Nature 440(7086):940–943. https://doi.org/10.1038/nature04640
Roohbakhsh A, Iranshahy M, Iranshahi M (2016) Glycyrrhetinic acid and its derivatives: anti-cancer and cancer chemopreventive properties, mechanisms of action and structure—cytotoxic activity relationship. Curr Med Chem 23(5):498–517. https://doi.org/10.2174/0929867323666160112122256
Scalcinati G, Partow S, Siewers V, Schalk M, Daviet L, Nielsen J (2012) Combined metabolic engineering of precursor and co-factor supply to increase α-santalene production by *Saccharomyces cerevisiae*. Microb Cell Fact 11(1):117. https://doi.org/10.1186/1475-2859-11-117
Schlag EM, McIntosh MS (2006) Ginsenoside content and variation among and within American ginseng (*Panax quinquefolius* L.) populations. Phytochemistry 67(14):1510–1519. https://doi.org/10.1016/j.phytochem.2006.05.028
Scossa F, Benina M, Alseekh S, Zhang Y, Fernie AR (2018) The integration of metabolomics and next-generation sequencing data to elucidate the pathways of natural product metabolism in medicinal plants. Planta Med 84(12–13):855–873. https://doi.org/10.1055/a-0630-1899
Seki H, Ohyama K, Sawai S, Mizutani M, Ohnishi T, Sudo H, Akashi T, Aoki T, Saito K, Muranaka T (2008) Licorice β-amyrin 11-oxidase, a cytochrome P450 with a key role in the biosynthesis of the triterpene sweetener glycyrrhizin. P Natl Acad Sci USA 105(37):14204–14209
Seki H, Sawai S, Ohyama K, Mizutani M, Ohnishi T, Sudo H, Fukushima EO, Akashi T, Aoki T, Saito K (2011) Triterpene functional genomics in licorice for identification of CYP72A154 involved in the biosynthesis of glycyrrhizin. Plant Cell 23(11):4112–4123
Shianna KV, Dotson WD, Tove S, Parks LW (2001) Identification of a UPC2 homolog in *Saccharomyces cerevisiae* and its involvement in aerobic sterol uptake. J Bacteriol 183(3):830–834. https://doi.org/10.1128/JB.183.3.830-834.2001
Shin B-K, Kwon SW, Park JH (2015) Chemical diversity of ginseng saponins from *Panax ginseng*. J Ginseng Res 39(4):287–298. https://doi.org/10.1016/j.jgr.2014.12.005

Song W, Liu M, Wu J, Zhai H, Chen Y, Peng Z (2019) Preclinical pharmacokinetics of triptolide: a potential antitumor drug. Curr Drug Metab 20(2):147–154. https://doi.org/10.2174/1389200219666180816141506

Srinivasan P, Smolke CD (2020) Biosynthesis of medicinal tropane alkaloids in yeast. Nature. https://doi.org/10.1038/s41586-020-2650-9

Tansakul P, Shibuya M, Kushiro T, Ebizuka Y (2006) Dammarenediol-II synthase, the first dedicated enzyme for ginsenoside biosynthesis *Panax Ginseng*. Febs Lett 580(22):5143–5149. https://doi.org/10.1016/j.febslet.2006.08.044

Tu Y (2016) Artemisinin-A gift from traditional Chinese medicine to the world (Nobel Lecture). Angew Chem Int Ed Engl 55(35):10210–10226. https://doi.org/10.1002/anie.201601967

Tu L, Su P, Zhang Z, Gao L, Wang J, Hu T, Zhou J, Zhang Y, Zhao Y, Liu Y, Song Y, Tong Y, Lu Y, Yang J, Xu C, Jia M, Peters RJ, Huang L, Gao W (2020) Genome of *Tripterygium wilfordii* and identification of cytochrome P450 involved in triptolide biosynthesis. Nat Commun 11(1):971. https://doi.org/10.1038/s41467-020-14776-1

Vickers CE, Williams TC, Peng B, Cherry J (2017) Recent advances in synthetic biology for engineering isoprenoid production in yeast. Curr Opin Chem Biol 40:47–56. https://doi.org/10.1016/j.cbpa.2017.05.017

Vranová E, Coman D, Gruissem W (2012) Structure and dynamics of the isoprenoid pathway network. Mol Plant 5(2):318–333. https://doi.org/10.1093/mp/sss015

Vranová E, Coman D, Gruissem W (2013) Network analysis of the MVA and MEP pathways for isoprenoid synthesis. Annu Rev Plant Biol 64(1):665–700. https://doi.org/10.1146/annurev-arplant-050312-120116

Wang P, Wei Y, Fan Y, Liu Q, Wei W, Yang C, Zhang L, Zhao G, Yue J, Yan X, Zhou Z (2015) Production of bioactive ginsenosides Rh2 and Rg3 by metabolically engineered yeasts. Metab Eng 29:97–105. https://doi.org/10.1016/j.ymben.2015.03.003

Wang C, Liwei M, Park JB, Jeong SH, Wei G, Wang Y, Kim SW (2018) Microbial platform for terpenoid production: *Escherichia coli* and yeast. Front Microbiol 9:2460. https://doi.org/10.3389/fmicb.2018.02460

Wang C, Su X, Sun M, Zhang M, Wu J, Xing J, Wang Y, Xue J, Liu X, Sun W, Chen S (2019a) Efficient production of glycyrrhetinic acid in metabolically engineered *Saccharomyces cerevisiae* via an integrated strategy. Microb Cell Fact 18(1):95. https://doi.org/10.1186/s12934-019-1138-5

Wang P, Wei W, Ye W, Li X, Zhao W, Yang C, Li C, Yan X, Zhou Z (2019c) Synthesizing ginsenoside Rh2 in *Saccharomyces cerevisiae* cell factory at high-efficiency. Cell Discov 5:5. https://doi.org/10.1038/s41421-018-0075-5

Wang Q, Quan S, Xiao H (2019d) Towards efficient terpenoid biosynthesis: manipulating IPP and DMAPP supply. Bioresour Bioprocess 6(1):6. https://doi.org/10.1186/s40643-019-0242-z

Wang JH, Wang D, Li WX, Huang Y, Dai ZB, Zhang XL (2019b) Optimization of UDP-glucose supply module and production of ginsenoside F(1) in *Saccharomyces cerevisiae*. Zhongguo Zhong yao za zhi = Zhongguo zhongyao zazhi = China J Chin Materia Medica 44(21):4596–4604. https://doi.org/10.19540/j.cnki.cjcmm.20190829.101

Wang J, Ledesma-Amaro R, Wei Y, Ji B, Ji X-J (2020a) Metabolic engineering for increased lipid accumulation in *Yarrowia lipolytica*—a review. Bioresour Technol 313:123707.https://doi.org/10.1016/j.biortech.2020.123707

Wang M, Wei Y, Ji B, Nielsen J (2020b) Advances in metabolic engineering of *Saccharomyces cerevisiae* for cocoa butter equivalent production. Front Bioeng Biotechnol 8(1194). https://doi.org/10.3389/fbioe.2020.594081

Wei W, Wang P, Wei Y, Liu Q, Yang C, Zhao G, Yue J, Yan X, Zhou Z (2015a) Characterization of *Panax ginseng* UDP-glycosyltransferases catalyzing protopanaxatriol and biosyntheses of bioactive ginsenosides F1 and Rh1 in metabolically engineered Yeasts. Mol Plant 8(9):1412–1424. https://doi.org/10.1016/j.molp.2015.05.010

Wei Y, Gossing M, Bergenholm D, Siewers V, Nielsen J (2017a) Increasing cocoa butter-like lipid production of *Saccharomyces cerevisiae* by expression of selected cocoa genes. AMB Express 7(1):34–34

Wei Y, Siewers V, Nielsen J (2017b) Cocoa butter-like lipid production ability of non-oleaginous and oleaginous yeasts under nitrogen-limited culture conditions. Appl Microbiol Biotechnol 101(9):3577–3585. https://doi.org/10.1007/s00253-017-8126-7

Wei L-J, Kwak S, Liu J-J, Lane S, Hua Q, Kweon D-H, Jin Y-S (2018) Improved squalene production through increasing lipid contents in *Saccharomyces cerevisiae*. Biotechnol Bioeng 115(7):1793–1800. https://doi.org/10.1002/bit.26595

Wei Y, Zhou H, Zhang J, Zhang L, Geng A, Liu F, Zhao G, Wang S, Zhou Z, Yan X (2015b) Insight into dominant cellulolytic bacteria from two biogas digesters and their glycoside hydrolase genes. Plos One 10(6):e0129921.https://doi.org/10.1371/journal.pone.0129921

Westfall PJ, Pitera DJ, Lenihan JR, Eng D, Woolard FX, Regentin R, Horning T, Tsuruta H, Melis DJ, Owens A, Fickes S, Diola D, Benjamin KR, Keasling JD, Leavell MD, McPhee DJ, Renninger NS, Newman JD, Paddon CJ (2012) Production of amorphadiene in yeast, and its conversion to dihydroartemisinic acid, precursor to the antimalarial agent artemisinin. Proc Natl Acad Sci 109(3):E111–E118. https://doi.org/10.1073/pnas.1110740109

Worland AM, Czajka JJ, Li Y, Wang Y, Tang YJ, Su WW (2020) Biosynthesis of terpene compounds using the non-model yeast *Yarrowia lipolytica*: grand challenges and a few perspectives. Curr Opin Biotechnol 64:134–140. https://doi.org/10.1016/j.copbio.2020.02.020

Wu T, Li S, Zhang B, Bi C, Zhang X (2018) Engineering *Saccharomyces cerevisiae* for the production of the valuable monoterpene ester geranyl acetate. Microb Cell Fact 17(1):85. https://doi.org/10.1186/s12934-018-0930-y

Yaegashi J, Kirby J, Ito M, Sun J, Dutta T, Mirsiaghi M, Sundstrom ER, Rodriguez A, Baidoo E, Tanjore D, Pray T, Sale K, Singh S, Keasling JD, Simmons BA, Singer SW, Magnuson JK, Arkin AP, Skerker JM, Gladden JM (2017) *Rhodosporidium toruloides*: a new platform organism for conversion of lignocellulose into terpene biofuels and bioproducts. Biotechnol Biofuels 10(1):241. https://doi.org/10.1186/s13068-017-0927-5

Yan X, Fan Y, Wei W, Wang P, Liu Q, Wei Y, Zhang L, Zhao G, Yue J, Zhou Z (2014) Production of bioactive ginsenoside compound K in metabolically engineered yeast. Cell Res 24(6):770–773. https://doi.org/10.1038/cr.2014.28

Yang L, Zou H, Gao Y, Luo J, Xie X, Meng W, Zhou H, Tan Z (2020) Insights into gastrointestinal microbiota-generated ginsenoside metabolites and their bioactivities. Drug Metab Rev 52(1):125–138. https://doi.org/10.1080/03602532.2020.1714645

Yang C, Li C, Wei W, Wei Y, Liu Q, Zhao G, Yue J, Yan X, Wang P, Zhou Z (2020) The unprecedented diversity of UGT94-family UDP-glycosyltransferases in *Panax* plants and their contribution to ginsenoside biosynthesis. Sci Rep-Uk 10(1):15394. https://doi.org/10.1038/s41598-020-72278-y

Yao W, Wang F, Wang H (2016) Immunomodulation of artemisinin and its derivatives. Science Bulletin 61(18):1399–1406. https://doi.org/10.1007/s11434-016-1105-z

Zeng F, Wang W, Guan S, Cheng C, Yang M, Avula B, Khan IA, Guo DA (2013) Simultaneous quantification of 18 bioactive constituents in *Tripterygium wilfordii* using liquid chromatography-electrospray ionization-mass spectrometry. Planta Med 79(9):797–805. https://doi.org/10.1055/s-0032-1328596

Zhang Q, Ye M (2009) Chemical analysis of the Chinese herbal medicine Gan-Cao (licorice). J Chromatogr A 1216(11):1954–1969

Zhang C, Li M, Zhao G-R, Lu W (2019) Alpha-Terpineol production from an engineered *Saccharomyces cerevisiae* cell factory. Microb Cell Fact 18(1):160. https://doi.org/10.1186/s12934-019-1211-0

Zhang Y, Nielsen J, Liu Z (2017) Engineering yeast metabolism for production of terpenoids for use as perfume ingredients, pharmaceuticals and biofuels. FEMS Yeast Res 17(8). https://doi.org/10.1093/femsyr/fox080

Zhao J, Li C, Zhang Y, Shen Y, Hou J, Bao X (2017) Dynamic control of ERG20 expression combined with minimized endogenous downstream metabolism contributes to the improvement of geraniol production in *Saccharomyces cerevisiae*. Microb Cell Fact 16(1):17. https://doi.org/10.1186/s12934-017-0641-9

Zhao M, Lin Y, Wang Y, Li X, Han Y, Wang K, Sun C, Wang Y, Zhang M (2019) Transcriptome analysis identifies strong candidate genes for ginsenoside biosynthesis and reveals its underlying molecular mechanism in *Panax ginseng* C.A. Meyer. Sci Rep-Uk 9(1):615–615. https://doi.org/10.1038/s41598-018-36349-5

Zhou F, Pichersky E (2020) More is better: the diversity of terpene metabolism in plants. Curr Opin Plant Biol 55:1–10. https://doi.org/10.1016/j.pbi.2020.01.005

Zhou Z-L, Yang Y-X, Ding J, Li Y-C, Miao Z-H (2012) Triptolide: structural modifications, structure–activity relationships, bioactivities, clinical development and mechanisms. Nat Prod Rep 29(4):457–475. https://doi.org/10.1039/C2NP00088A

Zhu M, Wang C, Sun W, Zhou A, Wang Y, Zhang G, Zhou X, Huo Y, Li C (2018) Boosting 11-oxo-β-amyrin and glycyrrhetinic acid synthesis in *Saccharomyces cerevisiae* via pairing novel oxidation and reduction system from legume plants. Metab Eng 45:43–50. https://doi.org/10.1016/j.ymben.2017.11.009

Zhuang Y, Yang G-Y, Chen X, Liu Q, Zhang X, Deng Z, Feng Y (2017) Biosynthesis of plant-derived ginsenoside Rh2 in yeast via repurposing a key promiscuous microbial enzyme. Metab Eng 42:25–32. https://doi.org/10.1016/j.ymben.2017.04.009

Zu Y, Prather KLJ, Stephanopoulos G (2020) Metabolic engineering strategies to overcome precursor limitations in isoprenoid biosynthesis. Curr Opin Biotechnol 66:171–178. https://doi.org/10.1016/j.copbio.2020.07.005

Application of Yeast Synthetic Biology for the Production of Citrus Flavors

Karim Farmanpour-Kalalagh, Arman Beyraghdar Kashkooli, and Alireza Babaei

Abstract

Plants have the potential to produce an extensive variety of sesquiterpenoid compounds being used as flavor, fragrance, and medicine. Among the sesquiterpenoids of the citrus family, (+)-valencene and (+)-nootkatone are considered the most valuable compounds used in various industries. Due to the low yield of these compounds in their native hosts, commercial (+)-nootkatone is produced from (+)-valencene by oxidation, either chemically or enzymatically. Furthermore, chemical synthesis is complex, toxic, and harmful to the environment. Therefore, more attention should be paid to the environment-friendly and safe methods for (+)-nootkatone synthesis. Recently, however, significant progress has been achieved on enzymatic oxidation of (+)-valencene to (+)-nootkatone (or to its precursor nootkatol) in the biotransformation/bioconversion process. Nevertheless, ideally *de novo* (+)-nootkatone could also be produced from (+)-valencene via biotechnological approaches in heterologous hosts using the enzymes involved in its biosynthetic pathway. Among some of the organisms used for the production of (+)-valencene and (+)-nootkatone, the yeast platform has the highest potential for the production of these two valuable products. De novo reconstitution of related-biosynthetic pathways in *Saccharomyces cerevisiae, Yarrowia lipolytica,* and *Pichia pastoris* has led to high production of these compounds, expressing hope to meet the needs of various industries on large scales which will be described in the following chapter.

Keywords

(+)-valencene • (+)-nootkatone • Synthetic biology • *Saccharomyces cerevisiae* • *Yarrowia lipolytica* • *Pichia pastoris*

K. Farmanpour-Kalalagh · A. Beyraghdar Kashkooli (✉) · A. Babaei
Department of Horticultural Science, Faculty of Agriculture, Tarbiat Modares University, P.O. Box 14115-365, Tehran, Iran
e-mail: A.beyraghdar@modares.ac.ir

F. Darvishi Harzevili (ed.), *Synthetic Biology of Yeasts*,
https://doi.org/10.1007/978-3-030-89680-5_8

1 Introduction

Plants produce a vast diversity of secondary metabolites including terpenes (also termed as isoprenoids) that possess physiological and ecological properties (Chappell 2004). Low-molecular-weight terpenoids mostly exist as volatiles in different parts of plants (Lücker et al. 2004). Terpenoids include the largest group of secondary metabolites; lots of them show biological functions and are applied as pharma and nutraceutical ingredients and compounds used in food, flavor, fragrance, and cosmetics. The grouping of terpenoids is originated from the number of C5-subunits (isoprene) in their backbone (e.g., hemiterpenoids (C5), monoterpenoids (C10), sesquiterpenoids (C15), diterpenoids (C20), sesterterpenoids (C25), triterpenoids (C30), and tetraterpenoids (carotenoids) (C40)). The main hydrocarbon skeletons constructed from isoprene units are mainly altered by enzymes such as cytochromes P450s, glycosyltransferases, hydrogenases, and methyltransferase. The amount of terpenoids in natural origins is commonly low and extraction may result in by-products (Asmund Arnesen et al. 2020). Among the plant-derived terpenoids, sesquiterpenes which are classified as a large group of natural products show diverse properties for agriculture, food, pharmacology, and cosmetics industry (Beyraghdar Kashkooli 2018; Troost et al. 2019). (+)-Valencene and (+)-nootkatone are high-value ingredients for flavor, fragrance, and pharmaceutical industry (Cankar et al. 2015; Milhim et al. 2019). Because of valencene's fruity flavor, it is industrially applied as an additive in food and drinks. In addition to using as an additive, (+)-valencene can be oxidized to (+)-nootkatone. (+)-Nootkatone (oxidation product (+)-valencene which is found in grapefruit and heartwood of the Nootka cypress) is also a considerable added-value sesquiterpene with slightly bitter taste and low odor threshold (Girhard et al. 2009; Frohwitter et al. 2014). Since raw plant-derived material of nootkatone is insufficient on the global market, semisynthetic (+)-nootkatone is derived from (+)-valencene in commercial approaches. Chemical oxidation of (+)-valencene to (+)-nootkatone requires extensive use of tert-butyl chromate, which is a carcinogenic material. On the other hand, non-carcinogenic tert-butyl peracetate or tert-butyl hydroperoxide can be replaced which are extremely corrosive and are flammable (Cankar et al. 2011).

As mentioned above, both (+)-valencene and (+)-nootkatone are valuable compounds in the flavor and fragrance industry, but the low yield in nature and high cost of extraction restricts their application (Chen et al. 2019). Additionally, the quality, value, and availability of these two compounds are very much related to seasonal variations and the harvest time (Beekwilder et al. 2014). On the other hand, since commercial nootkatone is presently produced from valencene by oxidation, either by a chemical process or enzymatically, meeting the global demands seems difficult. Hence, increasing the production of (+)-valencene and (+)-nootkatone from (+)-valencene via bioengineering of their biosynthetic pathway in heterologous hosts is critical. In this chapter, we review all possible techniques for improving (+)-valencene and (+)-nootkatone production using chemical and enzymatical processes. In particular, we highlight heterologous advances in the production of these compounds using microorganism platforms, specifically yeast-based ones and their application flavor and fragrance industry.

2 Natural Origin(s) of Valencene and Nootkatone

The bicyclic sesquiterpene (+)-valencene, a natural sesquiterpene, is a constituent of the essential oils of different members of the citrus genus, including Valencia or sweet orange (*Citrus × sinensis*) as well as in grapevine (*Vitis vinifera* L.) and in small amounts from rhizomes of *Cyperus rotundus* (Tsoyi et al. 2011; Nam et al. 2016), Chinese bayberry (*Myrica rubra*) (Fujita et al. 2014; Langhasova et al. 2014; Ambrož et al. 2015), Welensali (*Croton flavens*) (Sylvestre et al. 2006), mandarins cultivars (Merle et al. 2004), dehydrated lime [*Citrus aurantifolia* (Christm.) Swingle] (Ramesh Yadav et al. 2004), Egyptian *Eucalyptus camaldulensis* var. brevirostris (El-Ghorab et al. 2002), heartwood of the Nootka cypress (also known as the Alaska yellow cedar, native to the Pacific Northwest coast of North America) (*Callitropsis nootkatensis*), and Leyland cypress (*Cupressoparis leylandii*) which is an inter-genetic hybrid of Nootka cypress (*Chamaecyparis nootkatensis*) and Monterey cypress (*Cupressus macrocarpa*) (Liu 2009).

(+)-Nootkatone is an important oxidized sesquiterpene for the flavor and fragrance industry. It has a characteristic grapefruit-like flavor and a low odor threshold. Natural (+)-nootkatone can be extracted in small amounts from flavedo of grapefruit (*Citrus × paradisi*), pomelo (*Citrus grandis*) (Tocmo et al. 2020), vetiver (or khus) grass (*Chrysopogon zizanioides*) (Mao et al. 2006; Filippi et al. 2013), rhizome of *Cyperus rotundus* (Tsoyi et al. 2011; Jaiswal et al. 2014), *Alpiniae Oxyphyllae* Fructus (Wang et al. 2018), and fruits of *Alpinia oxyphylla* Miquel (Xie et al. 2009) and in high amounts from heartwood of the Nootka cypress (*Callitropsis nootkatensis*) (Beekwilder et al. 2014) (Fig. 1.).

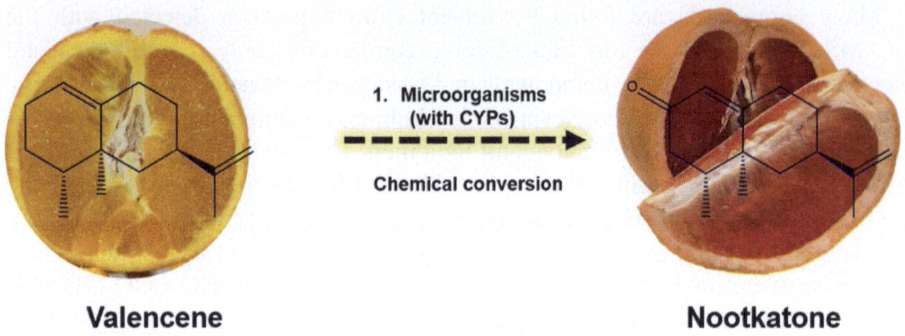

Fig. 1 Oxidation of valencene to nootkatone under different treatments/conditions via Cytochrome P450s (CYPs) or through chemical conversion

3 Yields of Valencene and Nootkatone in Native Hosts

Valencene is the main sesquiterpene in orange peel oil, and its titer has been commonly applied to specify the oil's commercial degree. To characterize the contribution of valencene to the aroma quality of a marketable orange oil by using multidimensional GC–O/GC–MS, thirty-seven aroma-active compounds have been identified in orange oil with valencene concentration between 54 and 68 μg/g (Elston et al. 2005). In addition to orange oil, valencene is reported in Chinese bayberry (*Myrica rubra*) with a relative abundance of 0.27–4.25% (Fujita et al. 2014), 5.97% (Langhasova et al. 2014), and 6% (Ambrož et al. 2015), 1.24% in Welensali (*Croton flavens*) (Sylvestre et al. 2006), 0.01–0.4% in mandarins cultivars (Merle et al. 2004), 0.1 (μL)/100 g in dehydrated lime [*Citrus aurantifolia* (Christm.) Swingle] (Ramesh Yadav et al. 2004), 0.44% in Egyptian *Eucalyptus camaldulensis* var. brevirostris (El-Ghorab et al. 2002).

Analyzing and comparison of essential oil components extracted from the heartwoods of Leyland cypress, Alaska yellow cedar, and Monterey cypress via steam distillation and solvent extraction by Liu (2009) indicated that the eighteen compounds are found both in Alaska cedar and Leyland cypress although there are only eight compounds found in Monterey cypress from steam distillation (6 and 12 h) of their heartwoods. The GC–MS analysis of volatile constituents from two fractions of steam distillation and solvent extraction, respectively, indicates that the main components in Alaska cedar, Monterey cypress, and Leyland cypress, are carvacrol, nootkatene, nootkatol, nootkatone, and nootkatin, which display about 60–90% of the whole capacity of the essential oil. These ingredients are from three diverse families which are p-Cymene, eremophilane, and tropolone (Fig. 2a and b). The evaluation of volatile constituents from Alaska cedar lumber and fresh wood exhibits that in parallel with the maturing and aging of the wood, the quantity of nootkatol and epi-nootkatol decreased, whereas the content of nootkatene increases significantly, which indirectly shows the conversion of nootkatol to nootkatene in the lumber boosted via the increasing of acidity of the heartwood (Fig. 2c).

Many compounds are found by solvent extraction when detected with the GC–MS. Because of the low titer of some components, only some of them are identified. There are twelve compounds in Alaska cedar, eleven in Leyland cypress, and only seven from Monterey cypress. Excluding two compounds, valencene-11, 12- diol and an unknown tropolone, all the others are detected in the steam distillation fractions. The lacking of valencene-11, 12- diol and the unknown tropolone are probably modified during steam distillation due to higher temperature (Fig. 2d). Nootkatone detection in other plants such as vetiver grass oil by applying comprehensive two-dimensional gas chromatography techniques (GC × GC-FID/MS) showed that the amount of this sesquiterpene is between 0.30 and 0.60 g/100 g oil among different varieties.

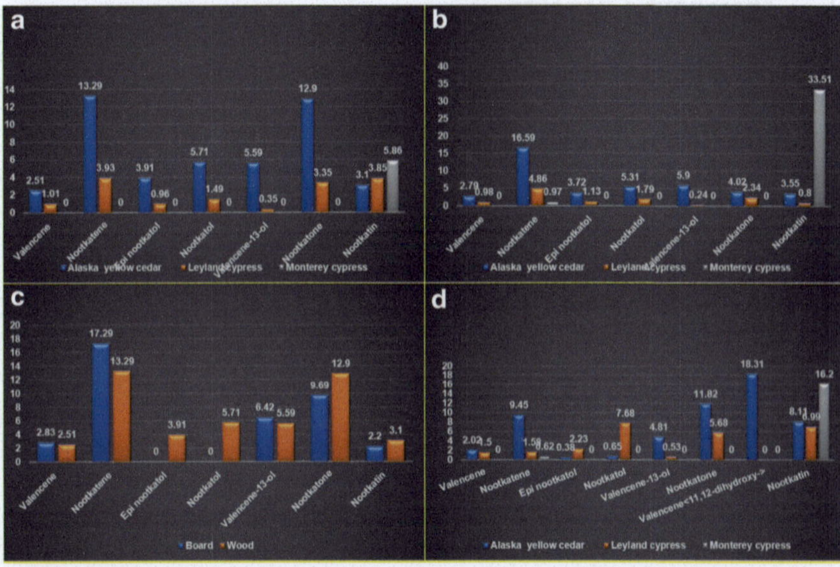

Fig. 2 **a** The components of the first 6 h distillation fraction from Alaska yellow cedar, Leyland cypress, and Monterey cypress; **b** The components of the 6-12 h distillation fraction from Alaska yellow cedar, Leyland cypress, and Monterey cypress; **c** The components of the 6 h distillation fraction from Alaska yellow cedar fresh wood and lumber; **d** The components of the 24 h solvent extraction from Alaska yellow cedar, Leyland cypress and, Monterey cypress analyzed by GC–MS. (Adapted from Liu 2009)

4 Importance and Application of Valencene and Nootkatone

4.1 Flavor and Fragrance Industry

Valencene is considered an important ingredient in consumer priority among a wide range of products, such as food, flavor, fragrance, beverage, personal, and home care products. Because of nootkatone's strong citrusy, sweet juicy notes, peely, woody, and grapefruit-like aroma characteristics as an oxidized sesquiterpene from valencene, it is involved in the formulation of citrus and grapefruit-flavored beverages. In fragrances, it is applied in the production of citrusy and dry components for men's perfumes. (+)-Nootkatone contains a φ value of 2.7×10^{11}, while (-)-nootkatone value estimated a φ 3.6×10^{8} with a slightly fresh, sour, green, organoleptic profile. Nootkatone is obtainable from various markets with diverse concentrations. This compound can be added in grapefruit drinks at a level of 2–6 ppm and can be used along with other citrus oils in fragrance formulations (such as bergamot oil, bitter orange oil, etc.,) to increase an interesting citrus note (Zviely 2009).

4.2 Therapeutic Benefits

Anti-cancer: The results of various studies show the anti-cancer activity of valencene and nootkatone. Essential oil from *Myrica rubra* leaves containing valencene plus seven dominant compounds prevents cancer cell increasing and induces apoptosis in some human intestinal lines. The anti-proliferative activity of extracted *M. rubra* essential oil (MEO) has been tested in human colon and ileocecal adenocarcinoma cell lines and results show that the MEO remarkably prevents cell proliferation in a concentration-dependent manner in all cell lines. In cancer cells, MEO creates apoptosis and leads to a considerable development of activities of the initiator as well as effector caspases (Langhasova et al. 2014).

The anti-proliferative effect of nootkatone on lung and bladder cancer cell lines at different concentrations has been investigated. Nootkatone showed a strong anti-proliferative effect on the A549 lung cancer cell line (Šunjerga 2019). In another successful study, nootkatone prevents v-Ki-ras2 Kirsten rat sarcoma (RAS) viral oncogene homolog (KRAS) downstream pathway and weakens non-small-cell lung cancer A549 cells to Adriamycin. Nootkatone activated AMP-activated protein kinase (AMPK) by liver kinase B1 (LKB1)-independent and calcium/calmodulin-dependent protein kinase kinase 2 (CAMKK2)-dependent pathways causes prevention of cell growth and development and induction of G1 cell limiting (Hung et al. 2019).

Efficacy of chemotherapy drug: Evaluation of the valencene, β-caryophyllene, β-caryophyllene oxide, α-humulene, and *trans*-nerolidol effects as essential compounds of the volatile constituents from *Myrica rubra* leaves against cancer cells and on the potency and toxicity of the anticancer drug doxorubicin in CaCo-2 cancer cells (an immortalized cell line of human colorectal adenocarcinoma cells) and in primary culture of rat hepatocytes display substantial antiproliferative activity in various intestinal cancer cell lines, with CaCo-2 cells as the most susceptible line. Results indicate that the valencene prevents the proliferation of CaCo-2 cancer cells and does not have an impact on the viability of hepatocytes. Furthermore, valencene can also strengthen the pro-oxidative activity of doxorubicin in CaCo-2 cells (Ambrož et al. 2015).

Anti-Alzheimer: Analyzing the therapeutic neuroprotective activity of nootkatone on an Alzheimer's disease mouse model caused by intracerebroventricular injection of lipopolysaccharide displays that nootkatone (10 mg/kg) has an effective efficiency in behavior trials including Y-maze and Morris water maze tests. The outcomes of the histopathological and immunohistochemical test exhibit that lipopolysaccharide leads to degeneration of neurons and initiation of microglia especially in the hippocampus and nootkatone (10 mg/kg) reverse these modifications. Enzyme-linked immunosorbent evaluation and western blot results also indicate that the model group enhances the expression of IL-1β, IL-6, TNF-α, NLRP3, and NF-κB p65, particularly in the hippocampus compared to the control group, and nootkatone (10 mg/kg) drops the high-level expression of inflammatory cytokines (Wang et al. 2018).

Anti-bacterial: Nootkatone displays anti-bacterial properties toward gram-positive bacteria such as *Corynebacterium diphtheriae* (most effective against this bacteria), *Enterococcus faecalis*, *Listeria monocytogenes*, *Bacillus cereus*, and *Staphylococcus aureus*. Nevertheless, no bactericidal activity has been reported against gram-negative bacteria. Additionally, high-concentration nootkatone shows anti-bacterial properties against gram-positive bacilli. These findings propound that the nootkatone may create a bactericidal property via attacking to cell wall or a specific metabolite. In addition, nootkatone can prevent the formation of biofilms by *Staphylococcus aureus* even at low concentrations (0.25 mM) (Yamaguchi 2019).

Skin protection: Electrophysiological assays in the evaluation of the antagonistic properties of *C. rotundus* extract and their compositions on TRPV1 and ORAI1 channels indicate that valencene extracted from the hexane fraction significantly prevents capsaicin-induced TRPV1 and ORAI1 flows at 90 μM ($69 \pm 15\%$ and $97 \pm 2\%$ at -60 and -120 mV, respectively). Also, valencene contains a preventive effect on cytoplasmic Ca^{2+} amounts and concentrations toward ORAI1 activation ($85 \pm 2\%$ at 50 μM). In addition, valencene concentration-dependently reduces the melanin quantity after UVB irradiation in murine B16F10 melanoma cells via $82.66 \pm 2.14\%$ at 15 μg/mL (Nam et al. 2016).

Insect repellent: Nootkatone can be applied as an insecticide or insect repellent, according to a decision from the US Environmental Protection Agency (EPA) in August 2020. Registration and introduction of a compound with industrial application make the process easier for producers to improve nootkatone-based products (Waltz 2020).

5 Nootkatone and Valencene Biosynthetic Pathway in Native Hosts

Terpenoids, also termed isoprenoids, are known as a wide and highly diverse group of natural products. They are regarded as one of the largest groups of secondary metabolites with various biological functions and valuable properties for diverse industrial applications (Kallscheuer et al. 2019). Classification of terpenoids is based on the number of carbon atoms forming the main skeleton. The general divisions of them are hemiterpenoids (C5), monoterpenoids (C10), sesquiterpenoids (C15), diterpenoids (C20), sesterterpenoids (C25), triterpenoids (C30), and tetraterpenoids (also known as carotenoids; C40) (Ashour et al. 2010; Beyraghdar Kashkooli et al. 2018). The central hydrocarbon backbone created from isoprene units is regularly changed via enzymes like cytochromes P450, hydrogenases, methyltransferases, and glycosyltransferases. Totally, terpenoids content in natural sources is typically low and extraction may result in by-products (Asmund Arnesen et al. 2020). Biosynthetically, all terpenoids are biosynthesized from the isoprene C5 backbones isopentenyl pyrophosphate (IPP) and dimethylallyl pyrophosphate (DMAPP), which are produced via the 2-C-methyl-D-erythritol 4-phosphate (MEP), also called 1-deoxy-D-xylulose 5-phosphate (DXP), and mevalonate (MVA) pathway. The MEP pathway, which commences from pyruvate and glyceraldehyde-3-phosphate (GAP), is commonly found in plastids of bacteria, cyanobacteria, green algae, and plants (Frank and Groll 2017). The cytosolic MVA pathway mainly exists in eukaryotes (mammals, plants, and fungi), but also archaea and a few bacteria. In the cytosol, sesquiterpenes are biosynthesized from isoprenoid precursors by the MVA pathway. The MVA pathway starts from acetyl-CoA and leads to the formation of isopentenyl diphosphate (IPP) and dimethyl allyl diphosphate (DMAPP). Then, farnesyl diphosphate synthase (FPS) accumulates DMAPP and IPP into farnesyl diphosphate (FPP). Terpene synthases (TPSs) use FPP to create terpene hydrocarbons, which can be further altered via oxidation into a diverse range of terpenoids (Chappell 2002). TPSs are classified based on their substrate characteristics. Notable advances have been achieved during the past decade in the biochemical reactions catalyzed by TPSs detail (Chappell 2004). Sesquiterpene synthases use the 15-carbon substrate farnesyl diphosphate (FPP) (Chappell 2004). Sharon-Asa et al. (2003) have isolated and characterized a sesquiterpene synthase (Cstps1) from *Citrus* that is presenting a new insight into the biochemical process of this enzyme, and bioengineering of (sesqui)terpenoids in native and heterologous hosts. Only a few functional (+)-valencene synthases were reported in previous studies including VvVal from *Vitis vinifera* (Lücker et al. 2004), Cstps1 from *Citrus sinensis* (Sharon-Asa et al. 2003), GFTpsD from *Citrus × paradisi*, and CnVS from Nootka cypress (*Callitropsis nootkatensis*) (Beekwilder et al. 2014).

The biosynthesis of (+)-valencene, as a precursor of (+)-nootkatone, is done through the MVA pathway. IPP is frequently condensed via prenyl transferases to create Farnesyl pyrophosphate (FPP). FPP is converted to (+)-valencene by the germacrenyl carbocation. So that the formation of internal bond and deprotonating is catalyzed by valencene synthase (Figs. 3 and 4). Not much information has been published about the enzymatic steps of converting (+)-valencene to (+)-nootkatone. It has been propounded that the biosynthetic pathway continues by a regioselective allylic hydroxylation of (+)-valencene to form nootkatol and then oxidized to (+)-nootkatone. Both phases can be accelerated by the catalytic function of a single multifunctional hydroxylase/oxidase or via a consecutive enzyme-mediated reaction. But the reason has not been determined yet why (+)-valencene accumulates in the peel of Valencia orange throughout fruit maturation while is further oxidized to (+)-nootkatone in grapefruit (Fraatz et al. 2009a).

6 Methods for Production of Valencene and Nootkatone

6.1 Extraction from Native Hosts

Valencene: In the extraction of valencene from citrus fruits, such as Valencia orange (*Citrus × sinensis*) peel, the following is estimated to have the required volume of the raw material: Approximately, 5 kg of active ingredient and oil is extracted from each ton of Valencia oranges. It is noteworthy to mention that the

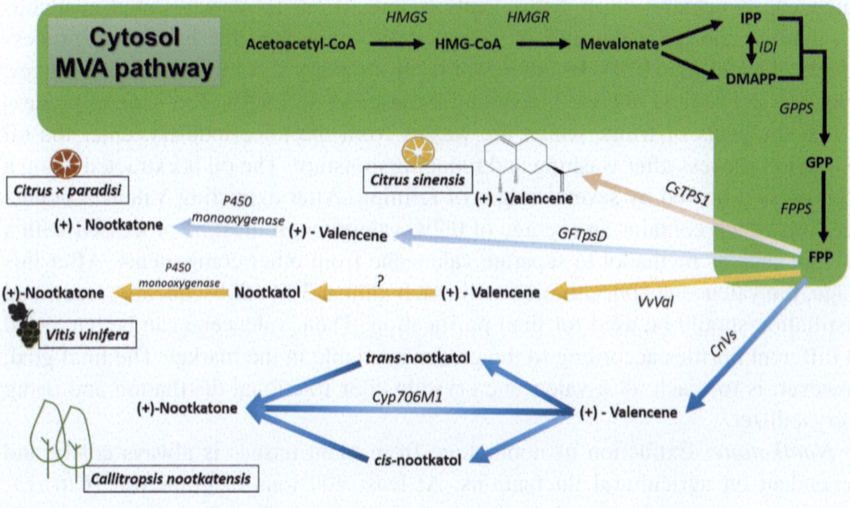

Fig. 3 Valencene and Nootkatone biosynthetic pathway in four major producing plants

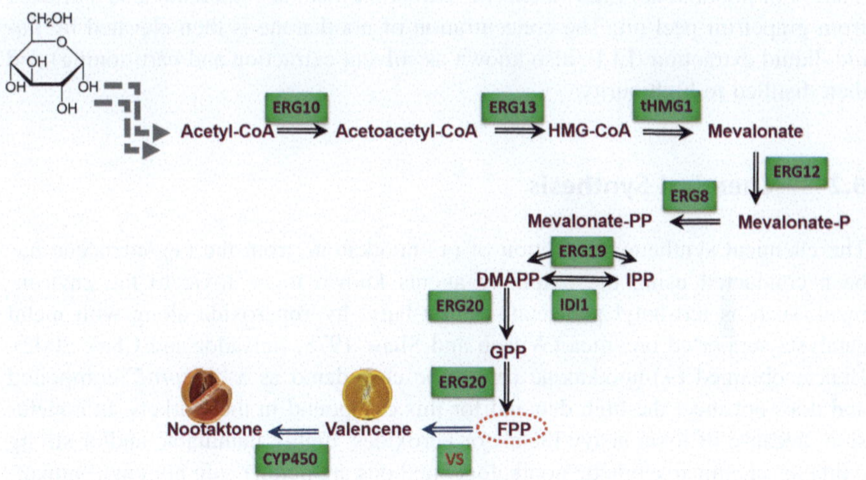

Fig. 4 Overview of yeast synbio for production of valencene and nootkatone. **ERG10**, acetyl-CoA C-acetyltransferase; **ERG13**, hydroxymethylglutaryl-CoA synthase; **ERG12**, mevalonate kinase; **tHMGR**, truncated 3-hydroxy-3-methylglutaryl-CoA reductase; **ERG8**, phosphomevalonate kinase; **ERG19**, diphosphomevalonate decarboxylase; **IDI1**, isopentenyl diphosphate delta-isomerase; **ERG20**, farnesyl diphosphate synthase/dimethylallyltranstransferase; **VS:** valencene synthase, **Cyp450**: cytochrome P450, valencene oxidase

valencene concentration in citrus fruits is low (0.2%–0.6% w:w), and availability, quality, and price depends on seasonal variations and the harvesting process (Beekwilder et al. 2014). In other words, from every 2.5 million kg of oranges, 1 kg of valencene is obtained (Asmund Arnesen et al. 2020). The general process is that the peels of fruits, which are wastes from the juice industry, enter the oil extraction process after washing and reducing moisture. The oil is extracted using a cold-press followed by several stages of refining. After extracting Valencia orange peel oil, which contains an average of 0.2% valencene, it must be processed with a solvent such as methanol to separate valencene from other components. After this stage, the valencene concentration will reach almost 30%. Subsequently, fractional distillation should be used for final purification. Then, valencene can be harvested in different purities according to the grades available in the market. The final goal, however, is to reach 98% valencene crystals after fractional distillation and using a crystallizer.

Nootkatone: Extraction of nootkatone from plant tissues is always costly and dependent on agricultural fluctuations. At least 400 tons of grapefruit (*Citrus × paradisi*) are needed to produce 1 kg of nootkatone. Therefore, the sustainable supply of raw materials for the industry is one of the biggest challenges in the production of nootkatone. Like valencene extraction methods, nootkatone is extracted from grapefruit peel oil. The concentration of nootkatone is then elevated by liquid–liquid extraction (LLE; also known as solvent extraction and partitioning) and then distilled to high purity.

6.2 Chemical Synthesis

The chemical synthetic production of (+)-nootkatone from the (+)-valencene has been conducted using the oxidizing agents known to be toxic to the environment, such as tert-butyl peracetate or tert-butyl hydroperoxide along with metal catalysts supported on silica (Wilson and Shaw 1978; Salvador and Clark 2002). Hence, obtained (+)-nootkatone cannot be considered as a "*natural*" compound and does not meet the high demand for this compound in the markets. In conclusion, because of toxic heavy metals or peroxides, highly flammable and/or strong oxidants, chemical synthetic nootkatone methods are neither safe nor environment-friendly. Therefore, more consideration is being paid to environment-friendly and unharmed techniques for (+)-nootkatone production (Wriessnegger et al. 2014).

6.3 Biotransformation (or Bioconversion)

Valencene biotransformation into nootkatol and nootkatone is believed to be catalyzed via the cytochrome P450 enzymes from both eukaryotic and prokaryotic (micro) organisms. Some (micro)organisms contain enzymes from the cytochrome P450 monooxygenase family, of which many were significantly proven for adding an oxygen atom into allyl groups and are the main candidates for biotransformation of (+)-valencene into (+)-nootkatone. However, finding regioselective P450 enzymes relevant to industrial demands is still a challenge in this regard (Gavira et al. 2013). A two-step enzymatic transformation of (+)-valencene has been suggested by a regioselective allylic hydroxylation of the 2-position of (+)-valencene to nootkatol and then by the oxidation to (+)-nootkatone. There are two hypotheses in this case. The first is that both reactions can be enhanced by a multifunctional cytochrome P450 enzyme. Second, the oxidative activity of cytochrome P450 leads to the production of nootkatol, followed by an alcohol dehydrogenase activity to produce (+)-nootkatone. Some enzymes originated from plants and microorganisms catalyzing the production of either nootkatol and/or (+)-nootkatone from (+)-valencene have been successfully identified (Cankar et al. 2014).

Cytochrome $P450_{cam}$ and $P450_{BM\text{-}3}$: (+)-valencene oxidation of the wild-type and mutants of $P450_{cam}$ from *Pseudomonas putida*, and of $P450_{BM\text{-}3}$ from *Bacillus megaterium*, have been studied as a promising way towards (+)-nootkatone production. Wild-type P450cam does not oxidize (+)-valencene but the mutants produce (+)-*trans*-nootkatol and (+)-nootkatone as the major products. Ignoring the non-selective properties, wild-type $P450_{BM\text{-}3}$ and its mutant forms contain dominant activities compared with the $P450_{cam}$. Among the many compounds, *cis*-(+)-valencene-1, 10-epoxide, *cis*- and *trans*-(+)-nootkatol, (+)-nootkatone, (+)-nootkatone-13S, 14-epoxide, and *trans*-(+)-nootkaton-9-ol were characterized from whole-cell reactions. The selectivity patterns express that (+)-valencene contains a single binding direction in $P450_{cam}$ but several directions in $P450_{BM\text{-}3}$ (Sowden et al. 2005).

CYP from ascomycete *Chaetomium globosum*: Oxidizing the exogenous (+)-valencene to nootkatone via the stereoselective production of alpha-nootkatol can be applied in submerged cultures of the ascomycete *Chaetomium globosum*. Inhibition tests propose that the insertion of the first oxygen atoms catalysis by cytochrome P450 monooxygenase. Hence in addition to nootkatone, non-volatile oxidation products along with flavor-active and inactive compounds can be identified. (+)-valencene, valencene-11, 12-epoxide, alpha-nootkatol, and nootkatone accumulated mostly inside the *C. globosum* cells. On the other hand, nootkatone-11, 12-epoxide is only in the culture medium. Therefore, active transporting of compounds into the extracellular sections is done during (+)-valencene detoxification (Kaspera et al. 2005).

CYP from algae and fungi: Studies indicate that the cytochrome P450s have an important role in (+)-valencene biotransformation by the green algae *Chlorella* species and fungi such as *Mucor* species, *Botryosphaeria dothidea*, and *Botryodiplodia theobromae* to yield nootkatone in high quantity (Furusawa et al. 2005).

CYP from *Gynostemma pentaphyllum*: It has been known that the suspension cultures of *Gynostemma pentaphyllum* can transform valencene into nootkatone as the major product. Also, biotransformation of valencene by *Caragana chamlagu* and *Hibiscus cannabinus* cultured plant cells show largely the same results (Sakamaki et al. 2005).

Premnaspirodiene oxygenase CYP71D55: Probing (+)-valencene oxidizing activity of cytochrome P450 enzymes from the CYP71 family shows that the premnaspirodiene oxygenase CYP71D55 from *Hyoscyamus muticus* contains catalytic oxidizing activity in conversion of (+)-valencene firstly to nootkatol in vitro (Takahashi et al. 2007).

Fungal dioxygenase from *Pleurotus sapidus*: A selective and strong allylic oxidation of the (+)-valencene to (+)-nootkatone can be obtained with lyophilization of the basidiomycete *Pleurotus sapidus*. Therefore, a novel oxygenase from *P. sapidus* can transform valencene to nootkatone. (Fraatz et al. 2009b).

CYP109B1 from *Bacillus subtilis*: It has been known that the cytochrome CYP109B1 from *Bacillus subtilis* can catalyze the oxidation of (+)-valencene to nootkatol and (+)- nootkatone. On the other hand, when (+)-valencene is bio-oxidized in vivo, a number of unfavorable multi-oxygenated by-products are produced. But as mentioned in Girhard et al. (2009) study, there are some techniques in the production of nootkatol and (+)-nootkatone up to 97% of the total product.

Dioxygenase of *Pleurotus sapidus*: A dioxygenase from *Pleurotus sapidus* converts (+)-valencene to (+)-nootkatone regio-specifically by a stereo-specific allylic hydroperoxidation. The isolation and characterizing of two allylic (+)-valencene-derived hydroperoxides along with homology data from amino acid sequencing expresses a lipoxygenase-like type of enzyme (Krügener et al. 2010).

CYP71AV8 from *Cichorium intybus* roots: A new P450 cDNA has been found in a chicory root EST library. Co-expression of the enzyme with a valencene synthase in yeast led to the formation of *trans*-nootkatol, *cis*-nootkatol, and (+)-nootkatone (Cankar et al. 2011).

Valencene dioxygenase from *Pleurotus sapidus*: Valencene dioxygenase from the edible basidiomycete *Pleurotus sapidus* transforms the (+)-valencene to the (+)-nootkatone and to nootkatol through intermediate hydroperoxides (Zelena et al. 2012).

CYP71D51v2 from tobacco and a P450 reductase from Arabidopsis: Recombinant yeast with P450 reductase from Arabidopsis and CYP71D51v2 from tobacco has been studied for optimizing the bioconversion mechanism. The bioconversion process produces β-nootkatol and nootkatone, with low efficiency which is reduced upon increasing the substrate concentration. Toxicity and harmfulness of the products for yeast platforms at concentrations higher than 100 mg/L, gathering of β-nootkatol in yeast endomembranes, and prevention of the CYP71D51v2 hydroxylation reaction by the obtained products are considered as the main reasons for low bioconversion yield. In addition, it has been distinguished that the generation of nootkatone from β-nootkatol is not a P450-dependent reaction. According to these achievements, researchers suggest new strategies for the performance of a P450-based bioconversion mechanism (Gavira et al. 2013).

CYP706M1 from Alaska yellow cedar: After co-expression of intended cytochrome P450s from Alaska yellow cedar in yeast with a valencene synthase, a *C. nootkatensis* valencene oxidase (CnVO) has been found to generate *trans*-nootkatol and (+)-nootkatone. Generation of (+)-nootkatone was observed at 144 ± 10 μg/L yeast culture. CnVO is one of the CYP706 family members of cytochrome P450 oxidases (Cankar et al. 2014).

CYP from *Botryodiplodia theobromae* 1368, *Yarrowia lipolytica* 2.2ab, and *Phanerochaete chrysosporium*: It has been observed that *Botryodiplodia theobromae* 1368, *Yarrowia lipolytica* 2.2ab, and *Phanerochaete chrysosporium* oxidize (+)-valencene to (+)-nootkatone, with concentrations of 231.7 ± 2.1, 216.9 ± 5.8, and 100.8 ± 2.6 mg/l (+)-nootkatone, respectively. Various biotransformation experiments (organic, aqueous, and biphasic) have been also tested which led to the same production levels of nootkatone (Palmerín-Carreño et al. 2015).

Peroxidase combined with a laccase from *Funalia trogii*: an incorporated platform of a dye-decolorizing peroxidase (Ftr-DyP) and a laccase isolated from the basidiomycete *Funalia trogii* transformed the (+)-valencene totally to the (+)-nootkatone, with a high concentration of 1100 mg/L (Kolwek et al. 2018).

6.4 Application of Synthetic Biology Using Yeast Platforms

Industrially, many (sesqui) terpenes have applications as a pharmaceutical flavor, fragrance, etc. (Cankar et al. 2015). Inserting related-gene(s), overexpression/knockdown of the enzyme(s) in the metabolic biosynthetic pathway, and use of diverse subcellular sections for production have led to high levels of (sesqui) terpenes (reviewed in Scholtmeijer et al. 2014). Microorganisms are used because of the variety of naturally existing strains, high metabolic diversity, and high production of related metabolite(s). On the other hand, production in microorganisms is a reliable technique to supply (sesqui) terpenoids for industrial applications. These processes, however, require the targeted transfer of the related biocatalytic pathways in an appropriate heterologous microorganism (Troost et al. 2019). Some yeast species are the most common microorganisms applied for the heterologous production of secondary metabolites. Here, we review the heterologous production of citrus flavors in three different yeast platforms as an attractive alternative for high production of (+)-valencene and (+)-nootkatone.

6.4.1 *Saccharomyces cerevisiae*

Saccharomyces cerevisiae, also known as baker's yeast, is one of the most common platforms used in the heterologous production of secondary metabolites. Studying, designing, and constructing *S. cerevisiae* cell factories for high-level production of citrus flavors is more appealing due to the short growth cycle and low costs (Chen et al. 2019). So far, successful studies have been performed on the production of citrus flavors in *S. cerevisiae* via bioengineering of related-pathway. After co-expression of candidate cytochrome P450 enzymes from Alaska yellow cedar with a valencene synthase in *S. cerevisiae*, a *C. nootkatensis* valencene oxidase (CnVO) has been identified to produce *trans*-nootkatol and (+)-nootkatone (144 ± 10 μg/L) (Cankar et al. 2014). Also, overexpression of CnVS in *S. cerevisiae* strain WAT11 only has produced 1.36 mg/L of (+)-valencene (Beekwilder et al. 2014). Whereas with overexpressing of (+)-valencene synthase GFTpsD and downregulating the squalene synthase, (+)-valencene is produced up to 3 mg/L (Asadollahi et al. 2008).

For the production of *trans*-nootkatol from (+)-valencene, two protocols (1): converting externally added (+)-valencene with resting cells or (2): cultivating engineered self-sufficient production strains) have been established and optimized with recombinant *S. cerevisiae* co-expressing CYP *Hyoscyamus muticus* premnaspirodiene oxygenase (*HPO*) and the cytochrome P450 reductase from *Arabidopsis thaliana* (*CPR*) to hydroxylate the C2 atom of (+)-valencene. The phase transfer process displays a substantial problem of whole-cell biotransformation of hydrophobic substrates. To solve the problem, Emmerstorfer et al. (2015) have designed an efficient *S. cerevisiae* strain containing the ability to produce valencene intracellularly. Furthermore, the self-sufficient production of *trans*-nootkatol has been carried out in biphasic systems via n-dodecane for catching the synthesized compounds to prevention of toxicity or inhibitory effects of high concentrations of compounds on yeasts (Emmerstorfer-Augustin and Pichler 2016).

In another successful study in order to overproduce valencene, *S. cerevisiae* was used as the suitable host for cell factory construction. In this regard, some bioengineering strategies were accomplished in the MVA biosynthetic pathway.

First of all, a renewable CRISPR/Cas9 system was designed to obtain multiple genome editing to enhance the FPP pool toward the valencene synthesis via the introduction of Cre/loxP. Furthermore, the main genes of the FPP biosynthetic pathway in the MVA pathway were overexpressed in the yeast genome to increase the metabolic flux to precursor FPP. To continue, the expression cassette of valencene synthase was developed by designing a promoter–terminator library, with PHXT7- VS-TTPI1 containing high performance in valencene production. (It is important to note that the identification of suitable expression cassettes containing promoter and terminator is critical for the improvement of valencene production). Finally, after fed-batch fermentation in a 3 L bioreactor, the valencene yield titer increased to 539.3 mg/L (Chen et al. 2019).

More recently, (+)-valencene was produced in high quantities (217.95 mg/L) in *S. cerevisiae* via overexpressing *Callitropsis nootkatensis* (+)-valencene synthase (*CnVS*) along with the expression of farnesyl diphosphate synthase (*ERG20*) as a fused protein, overexpression of *tHMG1*, and downregulating the squalene synthase enzyme (*ERG9*). Afterward, (+)-valencene oxidation by overexpressing the *Hyoscyamus muticus* premnaspirodiene oxygenase (*HPO*) variant (V482I/A484I) and cytochrome P450 reductase (*ATR1*) from *Arabidopsis thaliana* led to (+)-valencene oxidation to β-nootkatol and only low amounts of (+)-nootkatone (9.66 mg/L). Oxidation of β-nootkatol to (+)-nootkatone demonstrated that the short-chain dehydrogenase/reductase (SDR) superfamily dehydrogenases ZSD1 of *Zingiber zerumbet* and ABA2 of *Citrus sinensis* were effective in such an oxidation process. Lastly, overexpression of *ZSD1* and *ABA2* increased (+)-nootkatone yield to 59.78 mg/L and 53.48 mg/L, respectively (Meng et al. 2020).

6.4.2 *Yarrowia lipolytica*

Y. lipolytica is an ascomycetous yeast that is known as a safe microorganism. *Y. lipolytica* is part of the oleaginous yeast group, with cells containing more than 20% fat content (Nicaud 2012). Due to the endogenous MVA pathway, it is the potential platform for (sesqui) terpenoid production. Also as an oleaginous yeast, *Y. lipolytica* can utilize both hydrophilic and hydrophobic substrates as carbon sources to produce valuable compounds (Guo et al. 2018).

In the first heterologous production of (+)-nootkatone in *Y. lipolytica*, Guo et al., (2018) constructed (+)-nootkatone biosynthetic pathway in *Y. lipolytica* ATCC 201,249. Co-expressing (+)-valencene synthase (*CnVS*), codon-optimized (+)-nootkatone synthase opCYP706M1, and codon-optimized NADPH-cytochrome P450 reductase opAtCPR1 produced (+)-nootkatone with initial quantities of 45.6 g/L. Fusion of opCYP706M1 and opt46AtCPR1 (opAtCPR1 with 46 amino acids truncated at N-terminal) enhanced (+)-valencene transformation performance to (+)-nootkatone and (+)-nootkatone production raised to 312.2 g/L. In the next step, overexpression of *tHMG1* and *FPP synthase* (*ERG20*) enhanced the (+)-nootkatone production. Then, the last engineered strain led to the (+)-nootkatone production with a titer of 978.2 g/L. Also in another study, overexpression of *HMG1*, *ERG12*, *ACL1*, *SeACS*, *IDI*, *ERG20*, and *CnVS* along with *SQS* knock-down in *Y. lipolytica* strain ST9204 has led to 113.9 ± 6.6 mg/L (+)-valencene (Asmund Arnesen et al. 2020).

6.4.3 *Pichia pastoris*

P. pastoris is classified as a species of methylotrophic yeast. A comparative study including different yeast species and *Escherichia coli* has shown that the methylotrophic yeast *P. pastoris* is an appropriate platform for the functional expression of membrane-associated cytochrome P450 enzymes. The advantages of *P. pastoris* over *S. cerevisiae* include the capability to grow in high densities in simple media along with containing the potently regulated alcohol oxidase (AOX1) promoter (reviewed in Wriessnegger et al. 2014). Bearing these in mind, the first heterologous production of (+)-nootkatone in *P. pastoris* was done by Wriessnegger et al. (2014). In this study, *P. pastoris* was first used as a whole-cell biocatalyst for the production of (+)-nootkatone from (+)-valencene. Therefore, a strain co-expressing the premnaspirodiene oxygenase of *Hyoscyamus muticus* (*HPO*) and the *Arabidopsis thaliana* cytochrome P450 reductase (*CPR*) that hydroxylated extracellularly added (+)-valencene was generated. Then, phase transfer issues of (+)-valencene resolved via intracellular production of (+)-valencene by co-expression of valencene synthase from *Callitropsis nootkatensis*. Production of *trans*-nootkatol was performed via biphasic cultivations of *P. pastoris*; then oxidation process to (+)-nootkatone production was catalyzed by an intrinsic *P. pastoris* activity. Further, overexpression of a *P. pastoris alcohol dehydrogenase* and *tHmg1p* increased the (+)-nootkatone yield to 208 mg/L.

7 Conclusion

The flavor and fragrance industry has always been an attractive and rich industry in the world. Today, meeting the needs of the people and the market is one of the main challenges of these industries. (+)-Valencene and (+)-nootkatone are known as natural flavors and aromas and are produced in low amounts in their native hosts which requires a lot of raw material. On the other hand, although chemical synthesis of these compounds has been successful, due to the side effects and the fact of not being categorized as natural, this method is not very much accepted and popular. Significant success has also been achieved in the biotransformation/bioconversion method. In order to meet the market demand on a large scale, high amounts of these compounds have been obtained via the heterologous production of (+)-valencene and (+)-nootkatone in microorganisms, including yeasts *S. cerevisiae, Y. lipolytica,* and *P. pastoris.*

References

Ambrož M, Boušová I, Skarka A, Hanušová V, Králová V, Matoušková P, Szotáková B, Skálová L (2015) The influence of sesquiterpenes from *Myrica rubra* on the antiproliferative and pro-oxidative effects of doxorubicin and its accumulation in cancer cells. Molecules 20:15343–15358. https://doi.org/10.3390/molecules200815343

Asadollahi MA, Maury J, Møller K, Nielsen KF, Schalk M, Clark A, Nielsen J (2008) Production of plant sesquiterpenes in *Saccharomyces cerevisiae*: Effect of ERG9 repression on sesquiterpene biosynthesis. Biotechnol Bioeng 3:666–677. https://doi.org/10.1002/bit

Ashour M, Wink M, Gershenzon J (2010) Biochemistry of terpenoids: Monoterpenes, sesquiterpenes and diterpenes, biochemistry of plant secondary metabolism: 2nd edition. https://doi.org/10.1002/9781444320503.ch5

Asmund Arnesen J, Kildegaard KR, Cernuda Pastor M, Jayachandran S, Kristensen M, Borodina I (2020) *Yarrowia lipolytica* strains engineered for the production of terpenoids. Front Bioeng Biotechnol 8:1–14. https://doi.org/10.3389/fbioe.2020.00945

Beekwilder J, van Houwelingen A, Cankar K, van Dijk ADJ, de Jong RM, Stoopen G, Bouwmeester H, Achkar J, Sonke T, Bosch D (2014) Valencene synthase from the heartwood of Nootka cypress (*Callitropsis nootkatensis*) for biotechnological production of valencene. Plant Biotechnol J 12:174–182. https://doi.org/10.1111/pbi.12124

Beyraghdar Kashkooli A (2018) Biosynthesis, transport and combinatorial metabolic engineering of *Tanacetum parthenium* (feverfew) and *Artemisia annua* (sweet wormwood) sesquiterpene lactones. Ph.D. Thesis, pp 1–55. https://doi.org/10.18174/455405

Beyraghdar Kashkooli A, van der Krol A, Bouwmeester H (2018) Terpenoid biosynthesis in plants. Flavour Science, Verlag der Technischen Universität Graz, pp 3–12

Cankar K, Houwelingen AV, Bosch D, Sonke T, Bouwmeester H, Beekwilder J (2011) A chicory cytochrome P450 mono-oxygenase CYP71AV8 for the oxidation of (+)-valencene. FEBS Lett 585:178–182. https://doi.org/10.1016/j.febslet.2010.11.040

Cankar K, Van Houwelingen A, Goedbloed M, Renirie R, De Jong RM, Bouwmeester H, Bosch D, Sonke T, Beekwilder J (2014) Valencene oxidase CYP706M1 from Alaska cedar (*Callitropsis nootkatensis*). FEBS Lett 588:1001–1007. https://doi.org/10.1016/j.febslet.2014.01.061

Cankar K, Jongedijk E, Klompmaker M, Majdic T, Mumm R, Bouwmeester H, Bosch D, Beekwilder J (2015) (+)-Valencene production in *Nicotiana benthamiana* is increased by down-regulation of competing pathways. Biotechnol J 10:180–189. https://doi.org/10.1002/biot.201400288

Chappell J (2002) The genetics and molecular genetics of terpene and sterol origami. Curr Opin Plant Biol 5:151–157. https://doi.org/10.1016/S1369-5266(02)00241-8

Chappell J (2004) Valencene synthase-a biochemical magician and harbinger of transgenic aromas. Trends Plant Sci 9:265–269. https://doi.org/10.1016/j.tplants.2004.03.003

Chen H, Zhu C, Zhu M, Xiong J, Ma H, Zhuo M, Li S (2019) High production of valencene in *Saccharomyces cerevisiae* through metabolic engineering. Microb Cell Fact 18:1–14. https://doi.org/10.1186/s12934-019-1246-2

El-Ghorab AH, Fadel HM, El-Massry KF (2002) The Egyptian *Eucalyptus camaldulensis* var. brevirostris: Chemical compositions of the fruit volatile oil and antioxidant activity. Flavour Fragr J 17:306–312. https://doi.org/10.1002/ffj.1085

Elston A, Lin J, Rouseff R (2005) Determination of the role of valencene in orange oil as a direct contributor to aroma quality. Flavour Fragr J 20:381–386. https://doi.org/10.1002/ffj.1578

Emmerstorfer A, Wimmer-Teubenbacher M, Wriessnegger T, Leitner E, Müller M, Kaluzna I, Schürmann M, Mink D, Zellnig G, Schwab H, Pichler H (2015) Over-expression of ICE2 stabilizes cytochrome P450 reductase in *Saccharomyces cerevisiae* and *Pichia pastoris*. Biotechnol J 10:623–635. https://doi.org/10.1002/biot.201400780

Emmerstorfer-Augustin A, Pichler H (2016) Production of aromatic plant terpenoids in recombinant baker's yeast, in: Biotechnology of plant secondary metabolism: methods and protocols, methods in molecular biology. pp 79–89. https://doi.org/10.1007/978-1-4939-3393-8_8

Filippi JJ, Belhassen E, Baldovini N, Brevard H, Meierhenrich UJ (2013) Qualitative and quantitative analysis of vetiver essential oils by comprehensive two-dimensional gas chromatography and comprehensive two-dimensional gas chromatography/mass spectrometry. J Chromatogr 1288:127–148. https://doi.org/10.1016/j.chroma.2013.03.002

Fraatz MA, Berger RG, Zorn H (2009a) Nootkatone-a biotechnological challenge. Appl Microbiol Biotechnol 83:35–41. https://doi.org/10.1007/s00253-009-1968-x

Fraatz MA, Riemer SJL, Stöber R, Kaspera R, Nimtz M, Berger RG, Zorn H (2009b) A novel oxygenase from *Pleurotus sapidus* transforms valencene to nootkatone. J Mol Catal B Enzym 61:202–207. https://doi.org/10.1016/j.molcatb.2009.07.001

Frank A, Groll M (2017) The Methylerythritol phosphate pathway to isoprenoids. Chem Rev 117:5675–5703. https://doi.org/10.1021/acs.chemrev.6b00537

Frohwitter J, Heider SAE, Peters-Wendisch P, Beekwilder J, Wendisch VF (2014) Production of the sesquiterpene (+)-valencene by metabolically engineered *Corynebacterium glutamicum*. J Biotechnol 191:205–213. https://doi.org/10.1016/j.jbiotec.2014.05.032

Fujita S, Kajiyama K, Takabayashi M (2014) Volatile constituents of the leaf and fruit of *Myrica rubra*. J Nagoya Gakuin Univ Humanit Nat Sci 50, 25–33. http://doi.org/10.15012/00000352

Furusawa M, Hashimoto T, Noma Y, Asakawa Y (2005) Highly efficient production of nootkatone, the grapefruit aroma from valencene, by biotransformation. Chem Pharm Bull 53:1513–1514. https://doi.org/10.1248/cpb.53.1513

Gavira C, Höfer R, Lesot A, Lambert F, Zucca J, Werck-Reichhart D (2013) Challenges and pitfalls of P450-dependent (+)-valencene bioconversion by *Saccharomyces cerevisiae*. Metab Eng 18:25–35. https://doi.org/10.1016/j.ymben.2013.02.003

Girhard M, Machida K, Itoh M, Schmid RD, Arisawa A, Urlacher VB (2009) Regioselective biooxidation of (+)-valencene by recombinant *E. coli* expressing CYP109B1 from *Bacillus subtilis* in a two-liquid-phase system. Microb Cell Fact 8:1–12. https://doi.org/10.1186/1475-2859-8-36

Guo X, Sun J, Li D, Lu W (2018) Heterologous biosynthesis of (+)-nootkatone in unconventional yeast Yarrowia lipolytica. Biochem Eng J 137:125–131. https://doi.org/10.1016/j.bej.2018.05.023

Hung LVM, Moon JY, Ryu JY, Cho SK (2019) Nootkatone, an AMPK activator derived from grapefruit, inhibits KRAS downstream pathway and sensitizes non-small-cell lung cancer A549 cells to adriamycin. Phytomedicine 63:153000. https://doi.org/10.1016/j.phymed.2019.153000

Jaiswal Y, Liang Z, Guo P, Ho HM, Chen H, Zhao Z (2014) Tissue-specific metabolite profiling of *Cyperus rotundus* L. rhizomes and (+)-nootkatone quantitation by laser microdissection, ultra-high-performance liquid chromatography-quadrupole time-of-flight mass spectrometry, and gas chromatography-mass spectrometry. J Agric Food Chem 62:7302–7316. https://doi.org/10.1021/jf502494z

Kallscheuer N, Classen T, Drepper T, Marienhagen J (2019) Production of plant metabolites with applications in the food industry using engineered microorganisms. Curr Opin Biotechnol 56:7–17. https://doi.org/10.1016/j.copbio.2018.07.008

Kaspera R, Krings U, Nanzad T, Berger RG (2005) Bioconversion of (+)-valencene in submerged cultures of the ascomycete *Chaetomium globosum*. Appl Microbiol Biotechnol 67:477–483. https://doi.org/10.1007/s00253-004-1794-0

Kolwek J, Behrens C, Linke D, Krings U, Berger RG (2018) Cell-free one-pot conversion of (+)-valencene to (+)-nootkatone by a unique dye-decolorizing peroxidase combined with a laccase from *Funalia trogii*. J Ind Microbiol Biotechnol 45:89–101. https://doi.org/10.1007/s10295-017-1998-9

Krügener S, Krings U, Zorn H, Berger RG (2010) A dioxygenase of *Pleurotus sapidus* transforms (+)-valencene regio-specifically to (+)-nootkatone via a stereo-specific allylic hydroperoxidation. Bioresour Technol 101:457–462. https://doi.org/10.1016/j.biortech.2009.08.087

Langhasova L, Hanusova V, Rezek J, Stohanslova B, Ambroz M, Kralova V, Vanek T, Lou JD, Yun ZL, Yang J, Skalova L (2014) Essential oil from *Myrica rubra* leaves inhibits cancer cell proliferation and induces apoptosis in several human intestinal lines. Ind Crops Prod 59:20–26. https://doi.org/10.1016/j.indcrop.2014.04.018

Liu X (2009) Analysis and comparison of essential oil components extracted from the heartwoods of Leyland cypress, Alaska yellow cedar, and Monterey cypress. M.Sc. Thesis 1–55. http://hdl.handle.net/1957/11186

Lücker J, Bowen P, Bohlmann J (2004) *Vitis vinifera* terpenoid cyclases: Functional identification of two sesquiterpene synthase cDNAs encoding (+)-valencene synthase and (-)-germacrene D synthase and expression of mono- and sesquiterpene synthases in grapevine flowers and berries. Phytochemistry 65:2649–2659. https://doi.org/10.1016/j.phytochem.2004.08.017

Mao L, Henderson G, Bourgeois WJ, Vaughn JA, Laine RA (2006) Vetiver oil and nootkatone effects on the growth of pea and citrus. Ind Crops Prod 23:327–332. https://doi.org/10.1016/j.indcrop.2005.09.004

Meng X, Liu H, Xu W, Zhang W, Wang Z, Liu W (2020) Metabolic engineering *Saccharomyces cerevisiae* for *de novo* production of the sesquiterpenoid (+)-nootkatone. Microb Cell Fact 19:1–14. https://doi.org/10.1186/s12934-020-1295-6

Merle H, Morón M, Blázquez MA, Boira H (2004) Taxonomical contribution of essential oils in mandarins cultivars. Biochem Syst Ecol 32:491–497. https://doi.org/10.1016/j.bse.2003.09.010

Milhim M, Hartz P, Gerber A, Bernhardt R (2019) A novel short chain dehydrogenase from *Bacillus megaterium* for the conversion of the sesquiterpene nootkatol to (+)-nootkatone. J Biotechnol 301:52–55. https://doi.org/10.1016/j.jbiotec.2019.05.017

Nam JH, Nam DY, Lee DU (2016) Valencene from the rhizomes of *Cyperus rotundus* inhibits skin photoaging-related ion channels and UV-induced melanogenesis in B16F10 melanoma cells. J Nat Prod 79:1091–1096. https://doi.org/10.1021/acs.jnatprod.5b01127

Nicaud J-M (2012) Yarrowia lipolytica. Yeast 29:409–418. https://doi.org/10.1002/yea.2921

Palmerín-Carreño DM, Rutiaga-Quiñones OM, Verde Calvo JR, Prado-Barragán A, Huerta-Ochoa S (2015) Screening of microorganisms for bioconversion of (+)-valencene to (+)-nootkatone. LWT Food Sci Technol 64:788–793. https://doi.org/10.1016/j.lwt.2015.06.065

Ramesh Yadav A, Chauhan AS, Rekha MN, Rao LJM, Ramteke RS (2004) Flavour quality of dehydrated lime [*Citrus aurantifolia* (Christm.) Swingle]. Food Chem 85:59–62. https://doi.org/10.1016/j.foodchem.2003.06.002

Sakamaki H, Itoh KI, Taniai T, Kitanaka S, Takagi Y, Chai W, Horiuchi CA (2005) Biotransformation of valencene by cultured cells of *Gynostemma pentaphyllum*. J Mol Catal B Enzym 32:103–106. https://doi.org/10.1016/j.molcatb.2004.10.004

Salvador JAR, Clark JH (2002) The allylic oxidation of unsaturated steroids by tert-butyl hydroperoxide using surface functionalised silica supported metal catalysts. Green Chem 4:352–356. https://doi.org/10.1039/b201500p

Scholtmeijer K, Cankar K, Beekwilder J, Wösten HAB, Lugones LG, Bosch D (2014) Production of (+)-valencene in the mushroom-forming fungus *S. commune*. Appl Microbiol Biotechnol 98:5059–5068. https://doi.org/10.1007/s00253-014-5581-2

Sharon-Asa L, Shalit M, Frydman A, Bar E, Holland D, Or E, Lavi U, Lewinsohn E, Eyal Y (2003) Citrus fruit flavor and aroma biosynthesis: Isolation, functional characterization, and developmental regulation of *Cstps1*, a key gene in the production of the sesquiterpene aroma compound valencene. Plant J 36:664–674. https://doi.org/10.1046/j.1365-313X.2003.01910.x

Sowden RJ, Yasmin S, Rees NH, Bell SG, Wong LL (2005) Biotransformation of the sesquiterpene (+)-valencene by cytochrome $P450_{cam}$ and $P450_{BM-3}$. Org Biomol Chem 3:57–64. https://doi.org/10.1039/b413068e

Šunjerga A (2019) Antiproliferativno djelovanje nootkatona na dvije stanične linije raka. Undergraded Thesis, pp 1–43. https://urn.nsk.hr/urn:nbn:hr:167:215698

Sylvestre M, Pichette A, Longtin A, Nagau F, Legault J (2006) Essential oil analysis and anticancer activity of leaf essential oil of *Croton flavens* L. from Guadeloupe. J Ethnopharmacol 103:99–102. https://doi.org/10.1016/j.jep.2005.07.011

Takahashi S, Yeo YS, Zhao Y, O'Maille PE, Greenhagen BT, Noel JP, Coates RM, Chappell J (2007) Functional characterization of premnaspirodiene oxygenase, a cytochrome P450 catalyzing regio- and stereo-specific hydroxylations of diverse sesquiterpene substrates. J Biol Chem 282:31744–31754. https://doi.org/10.1074/jbc.M703378200

Tocmo R, Pena-Fronteras J, Calumba KF, Mendoza M, Johnson JJ (2020) Valorization of pomelo (*Citrus grandis* Osbeck) peel: A review of current utilization, phytochemistry, bioactivities, and mechanisms of action. Compr Rev Food Sci Food Saf 19:1969–2012. https://doi.org/10.1111/1541-4337.12561

Troost K, Loeschcke A, Hilgers F, Özgür AY, Weber TM, Santiago-Schübel B, Svensson V, Hage-Hülsmann J, Habash SS, Grundler FMW, Schleker ASS, Jaeger KE, Drepper T (2019) Engineered *Rhodobacter capsulatus* as a phototrophic platform organism for the synthesis of plant sesquiterpenoids. Front Microbiol 10:1–14. https://doi.org/10.3389/fmicb.2019.01998

Tsoyi K, Jang HJ, Lee YS, Kim YM, Kim HJ, Seo HG, Lee JH, Kwak JH, Lee DU, Chang KC (2011) (+)-Nootkatone and (+)-valencene from rhizomes of *Cyperus rotundus* increase survival rates in septic mice due to heme oxygenase-1 induction. J Ethnopharmacol 137:1311–1317. https://doi.org/10.1016/j.jep.2011.07.062

Waltz E (2020) Specter of eye toxicity looms over BCMA-targeted therapy. Nat Biotechnol 38:1363–1365. https://doi.org/10.1038/s41587-020-00757-8

Wang Y, Wang M, Xu M, Li T, Fan K, Yan T, Xiao F, Bi K, Jia Y (2018) Nootkatone, a neuroprotective agent from *Alpiniae oxyphyllae* Fructus, improves cognitive impairment in lipopolysaccharide-induced mouse model of Alzheimer's disease. Int Immunopharmacol 62:77–85. https://doi.org/10.1016/j.intimp.2018.06.042

Wilson CW, Shaw PE (1978) Synthesis of nootkatone from valencene. J Agric Food Chem 26:1430–1432. https://doi.org/10.1021/jf60220a054

Wriessnegger T, Augustin P, Engleder M, Leitner E, Müller M, Kaluzna I, Schürmann M, Mink D, Zellnig G, Schwab H, Pichler H (2014) Production of the sesquiterpenoid (+)-nootkatone by metabolic engineering of *Pichia pastoris*. Metab Eng 24:18–29. https://doi.org/10.1016/j.ymben.2014.04.001

Xie J, Sun B, Wang S, Ito Y (2009) Isolation and purification of nootkatone from the essential oil of fruits of *Alpinia oxyphylla* Miquel by high-speed counter-current chromatography. Food Chem 117:375–380. https://doi.org/10.1016/j.foodchem.2009.04.011

Yamaguchi T (2019) Antibacterial properties of nootkatone against gram-positive bacteria. Nat Prod Commun 14:1–5. https://doi.org/10.1177/1934578X19859999

Zelena K, Krings U, Berger RG (2012) Functional expression of a valencene dioxygenase from *Pleurotus sapidus* in *E. coli*. Bioresour Technol 108:231–239. https://doi.org/10.1016/j.biortech.2011.12.097

Zviely M (2009) Molecule of the month: Nootkatone. Perfum Flavorist 34:20–22. https://doi.org/10.2174/156802661107011301

Synthesis of Polyols and Organic Acids by Wild-Type and Metabolically Engineered *Yarrowia lipolytica* Strains

Chong Li, Weichao Lin, Khai Lun Ong, Jinhua Mou,
Carol Sze Ki Lin, and Patrick Fickers

Abstract

In the yeast *Yarrowia lipolytica*, sugar polyols and organic acids are derived from central metabolism, namely the citrate cycle and the pentose phosphate pathway. Although these metabolites have numerous applications in agro-food, chemical, and pharmaceutical industries, the main challenge is to gain productivity to obtain processes that are economically viable. *Y. lipolytica* is known for its ability to use industrial wastes or raw materials as feedstock and to grow at high cell density in large-scale bioreactor. Recent advances in metabolic engineering and synthetic biology allowed the development of efficient *Y. lipolytica*-based cell factories to bioconvert these feedstocks into added-value metabolites. This book chapter will focus on current knowledge on the synthesis of the most important polyols and organic acids in *Y. lipolytica*.

C. Li · W. Lin
Shenzhen Branch, Guangdong Laboratory for Lingnan Modern Agriculture, Shenzhen Key Laboratory of Agricultural Synthetic Biology, Genome Analysis Laboratory of the Ministry of Agriculture and Rural Affairs, Agricultural Genomics Institute at Shenzhen, Chinese Academy of Agricultural Sciences, Shenzhen, China

W. Lin
School of Life Sciences, Henan University, Kaifeng 47500, China

K. L. Ong · J. Mou · C. S. K. Lin
School of Energy and Environment, City University of Hong Kong, Kowloon Tong, Hong Kong

P. Fickers (✉)
Microbial Process and Interaction, TERRA Teaching and Research Centre, University of Liege – Gembloux Agro-Bio Tech, Gembloux, Belgium
e-mail: pfickers@uliege.be

F. Darvishi Harzevili (ed.), *Synthetic Biology of Yeasts*,
https://doi.org/10.1007/978-3-030-89680-5_9

1 Introduction

Several polyols and organic acids have been included in the top 12 platform chemical list of the US Department of Energy (Werpy and Petersen 2004). This highlights their importance in nowadays industries. Historically, most of these compounds have been produced industrially by chemical synthesis in a fossil fuel-dependent fashion. For instance, succinic acid (SA) could be produced by the catalytic hydrogenation of maleic anhydride (Ong et al. 2020) while erythritol, four-carbon sugar alcohol, could be obtained from dialdehyde starch in the presence of nickel as a catalyst at high temperatures (Carly and Fickers 2018).

In microorganisms, these compounds are also intermediates of biochemical pathways, such as SA that is involved in the Krebs cycle, or cell metabolites such as erythritol that is produced by certain osmotolerant yeasts in response to osmotic stress (Carly and Fickers 2018). Actually, there is an increasing demand from global drivers for greener and more sustainable processes to be developed in the context of circular economy. With the advent of molecular technologies such as metabolic engineering and synthetic biology, industrial microorganisms could now be engineered to bioconvert organic and industrial wastes into value-added chemicals using the so-called biorefinery concept. In this book chapter, we will focus on the non-conventional yeast *Yarrowia lipolytica* as a cell factory and detail recent advances made in strain and process engineering to produce polyols and organic acid of biotechnological interest.

2 Yarrowia lipolytica

Y. lipolytica is an Hemiascomycetes yeast isolated from dairy products, dry sausages, lipid-rich or alkane-containing matrix (polluted soil and sewage) or hypersaline environments such as seawater (Nicaud 2012). This yeast has specific metabolisms for alkanes and triglycerides catabolism (Fickers et al. 2005). Also, it presents other metabolic features such as the ability to sustain high osmotic pressure owing to the synthesis of a specific metabolite, namely erythritol, or the ability to synthesize and secrete enzymes (lipase, protease) with high efficiency (up to several g/L). Interest in *Y. lipolytica* started in the mid-60s for the production of single-cell proteins (SCP) from alkanes and later the yeast was a research focus for heterologous protein production. It is also considered as a model organism to study dimorphism as *Y. lipolytica* could grow either as yeast-like cell and pseudo-filamentous forms (Barth and Gaillardin 1996). *Y. lipolytica* has gained a GRAS status (Generally Recognized As Safe) that is of interest for the development of food-related applications (Groenewald et al. 2014). The complete annotated genome from strain E150 was released in 2004 (Dujon et al. 2004) and the genome of several *Yarrowia* strains are now publicly available (http://gryc.inra.fr/). Numerous molecular tools have been also developed such as efficient promoters and vectors for fine-tuned gene expression (Shabbir Hussain et al. 2016;

Sassi et al. 2016; Trassaert et al. 2017; Park et al. 2019), DNA assembly methods (Wong et al. 2017; Vandermies et al. 2017; Celińska et al. 2017; Larroude et al. 2019), gene disruption systems (Fickers et al. 2003), and genome editing tools including CRISPR/Cas9 (Larroude et al. 2020). All these technologies have been recently reviewed (Abdel-Mawgoud et al. 2018; Larroude et al. 2018; Bilal et al. 2020). Efficient strategies have been also developed to cultivate *Y. lipolytica* in bioreactors with the aim to maximize recombinant protein or metabolite production titers (Do et al. 2019; Vandermies and Fickers 2019). Also, review papers related to *Y. lipolytica* physiology have been published (Barth and Gaillardin 1996; Fickers et al. 2005; Nicaud 2012).

3 Organic Acid

Y. lipolytica is considered as a model organism to study the production of organic acids from different carbon sources including glycerol, glucose but also waste carbon sources (waste cooking oil, pineapple waste, olive-mill wastewater, or rapeseed oil, Table 1). Organic acids such as citric acid, iso-citric acid, succinic acid, itaconic acid, and α-ketoglutaric acid which are all intermediates of Krebs cycle will be discussed in this section. The main production data for these organic acids are summarized in Table 1.

3.1 Citric Acid

Citric acid (CA) is a tricarboxylic organic acid that plays a central role in the metabolism of all aerobic organisms. It is an intermediate of the Krebs cycle (Fig. 1). Based on its non-toxic, pH buffering, and chelating properties, CA has become an important industrial product. Applications for CA include the food, cosmetic, and pharmaceutical industries, where it is used as a flavoring agent, antioxidant, preservative, or pH buffering system. About 70% of the global total CA production is used by the food industry, 20% by detergent, and the remaining 10% by the chemical and pharmaceutical industries (Carsanba et al. 2019). The global CA production exceeded 2 million tons in 2018. The market is projected to reach a volume of nearly 3 million tons by 2024, with an expected Compound Annual Growth Rate (CAGR) of 4% between 2019 and 2024 (Fickers et al. 2020; Ciriminna et al. 2017). Historically, CA was isolated from citrus fruits, but this method was displaced by fermentation of sugars using the filamentous fungus *Aspergillus niger*. In bench-top bioreactors, a production titer of 130 kg/m^3 could be obtained after five to eight days of batch cultivation (Ciriminna et al. 2017). Because of its high productivity, *A. niger* dominated the market for CA production. However, the processes developed for CA production from molasses are multi-stage, are not environmentally friendly, and are being limited by the raw material sources used (Morgunov et al. 2013).

Table 1 Some examples of organic acid produced in *Y. lipolytica*

Y. lipolytica strain	Genetic modification	Medium	Organic acid, production (g/L)	References
AJD-pADUTGut1/2	$GUT1^{OE}$, $GUT2^{OE}$	Glycerol	93, CA	Mironczuk et al. (2016)
SWJ-1b	$\Delta acl1$, ICL^{OE}	Inulin	84, CA	Forster et al. (2007b)
		Cooking oil	80, CA	Papanikolaou et al. (2008)
W29	$YHM2^{OE}$, $AMPD^{OE}$	Glucose	97, CA	Liu et al. (2013)
NG40/UV7	MT	Glycerol-containing wastes	122.2, CA	Carsanba et al. (2019)
NCIM 3589	WT	Pineapple waste	202.4, CA	Wojtatowicz et al. (1991)
A101-1.22	MT	Glycerol	124.2, CA	Yuzbasheva et al. (2019)
A101-1.44	MT	Glucose	100, CA	Morgunov et al. (2013)
K57	WT	Glucose	72.1, CA	Rymowicz et al. (2010)
VKM Y-2723	WT	Rapeseed oil	70.6, ICA	Forster et al. (2007a)
		Ethanol	109.6, ICA	Kamzolova et al. (2016)
704-UV4-A/NG50	MT	Rapeseed oil	86, ICA	Forster et al. (2007a)
Strain 20	$ACO1^{OE}$	Rapeseed oil	72.6, ICA	Kamzolova et al. (2018)
PSA3.0	EO	Glucose	76.8, SA	Tan et al. (2013)
PSA02004	WT	Sugarcane bagasse hydrolysates	33.2, SA	Yang et al. (2017)
		Fruit and vegetable waste	140.6, SA	Li et al. (2018a)
H222-AZ2	$Ylshd5^{OE}$	Glycerol	25, SA	Bondarenko et al. (2016)
PGC01003	*Ylshd5*	YPD medium	65.7, SA	Jost et al. (2014)
		Food waste hydrolysate	87.9, SA	Jost et al. (2014)
		Glycerol	160, SA	Yuzbashev et al. (2010)

(continued)

Table 1 (continued)

Y. lipolytica strain	Genetic modification	Medium	Organic acid, production (g/L)	References
PGC202	*Ylshd5, ach; PCKEO, SCS2*^{EO}	Glycerol	110.7, SA	Cui et al. (2017)
		Mix food waste	31.7, SA	Li et al. (2016)
PO1f-CAD	*CAD*^{EO}	Glucose	0.033, IA	Blazeck et al. (2015)
PCM-ACO-MFS	*MTT*^{EO}, *CAD*^{EO}	Glucose	22, IA	Zhao et al. (2019)
H355	WT	N-alkanes	195, KG	Otten et al. (2015)
H355A	*IDH1*^{EO}, *PYC1*^{EO}	Raw glycerol	186, KG	Chin et al. (2015)
VKM Y -2412	WT	Ethanol	172, KG	Chernyavskayaá et al. (2000)
		Rapeseed oil	102.5, KG	Yin et al. (2012)
RoPYC2	*RoPYC2*^{EO}	Glycerol	62.5, KG	Kamzolova et al. (2012)

Abbreviations: OE: gene overexpression, WT: wild-type, MT: mutant strain (obtained by chemical and/or UV mutagenesis), ME: evolved strain, CA: citric acid, ICA: iso-citric acid, SA: succinic acid, IA: itaconic acid, KG: ketoglutarate. *GUT1*: glycerokinase, *GUT2*: glyceraldehyde-3-phosphate dehydrogenase, *ACL1*: ATP citrate lyase, *ICL*: isocitrate lyase, *YHM2:* mitochondrial citrate carrier, *YlAMPD*: adenosine monophosphate deaminase, *ACO1*: aconitase, *Ylshd5:* succinate dehydrogenase subunit 5, *ACH:* acetyl-coenzyme A hydrolase, *PCK:* phosphoenolpyruvate carboxykinase, *SCS2:* succinyl CoA synthetase, *CAD:*cis-aconitate decarboxylase, *MTT:* mitochondrial cis-aconitate transporter, *IDH1:* isocitrate dehydrogenase, *PYC1:* pyruvate carboxylase, *RoPYC2:* pyruvate carboxylase

In a study by Kamzolova and Morgunov, 43 different wild yeast isolates were evaluated for CA production (Kamzolova and Morgunov 2017). Among these, *Y. lipolytica* was found as the best CA producer on minimal medium with titer and yield of 85 g/L and 0.70 g/g, respectively (Kamzolova and Morgunov 2017). Based on this special ability and other biochemical characteristics, *Y. lipolytica* has been recognized as a potential host to produce Krebs cycle intermediates, including CA. In addition, cultivation conditions such as pH 4.5–5.0, temperature of 28 °C, and glycerol concentration in culture medium from 20 to 80 g/L were found the optimal values for CA production (Morgunov et al. 2013). The main drawback of using *Y. lipolytica* to produce CA is the co-production of iso-CA (ICA) in high amounts (Cavallo et al. 2017). However, several authors have reported that nitrogen limitation during yeast growth could alleviate this disadvantage. Indeed, nitrogen starvation activates adenosine monophosphate (AMP) deaminase (Beopoulos et al. 2009a; b). This leads to a decrease in the mitochondrial AMP concentration that inhibits the iso-citrate dehydrogenase. This finally blocks the Krebs cycle at the

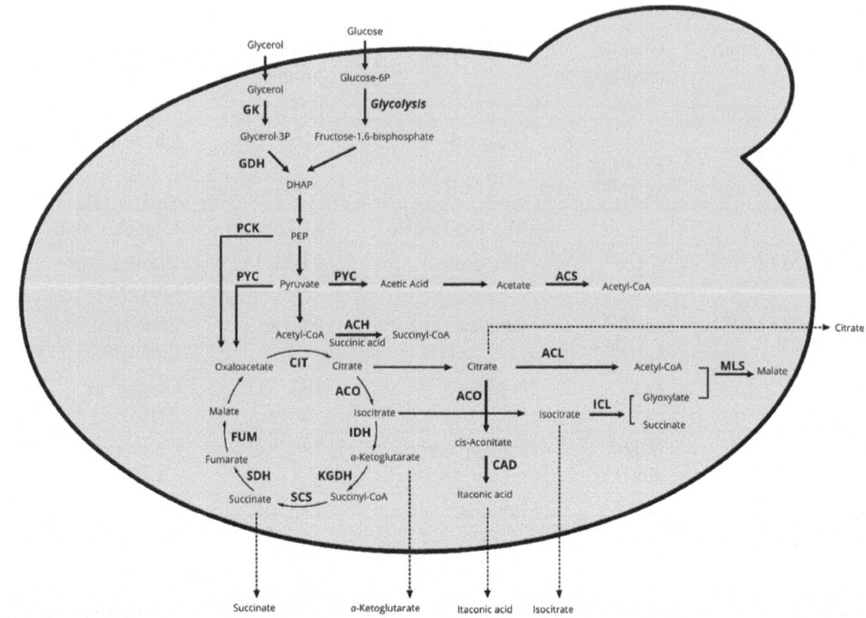

Fig. 1 **Overview of the principal metabolic pathways for organic acid synthesis in *Y. lipolytica*.** GK: glycerokinase; GDH: Glyceraldehyde-3-phosphate dehydrogenase; PCK: Phosphoenolpyruvate carboxykinase; PYC: pyruvate carboxylase; PDC: pyruvate decarboxylase; ACS: acetyl-coenzyme A synthetase; ACH: acetyl-coenzyme A hydrolase; CIT: citrate synthetase; ACO: aconitase; IDH: isocitrate dehydrogenase; KGDH: αketoglutarate dehydrogenase; SCS: succinyl CoA synthetase; SDH: succinate dehydrogenase; FUM: fumarase; ACL: ATP citrate lyase; ICL: isocitrate lyase; MLS: malate synthase; PEP: phosphoenol pyruvate; CAD: cis-aconitate decarboxylase

ICA stage allowing accumulation of early metabolites in the mitochondria, such as CA (Fig. 1) (Beopoulos et al. 2009a; b). Compared to other limiting factors, nitrogen starvation yielded to high accumulation of CA. Indeed, upon limitation of nitrogen, phosphorus, or sulfur, CA could accumulate in the culture medium up to 18.0 g/L, 16.0 g/L, and 16.8 g/L, respectively (Morgunov et al. 2013).

Low pH value is considered as a limiting factor for CA production by *Y. lipolytica* wild-type strains (Tomaszewska et al. 2014). However, the engineered strain AJD pADUTGut1/2 was obtained by overexpression of *GUT1* and *GUT2* genes encoding, respectively, glycerol kinase (*YALI0F00384g*) and glycerol-3-phosphate dehydrogenase (*YALI0B02948g*) were able to produce CA efficiently from glycerol at pH 3 (Mironczuk et al. 2016). Initially, the aim of these modifications was to improve erythritol production (see below), which was indeed achieved when strain AJD pADUTGut1/2 was grown under high osmotic pressure (Mironczuk et al. 2016). However, in the absence of osmotic stress, CA becomes the main metabolite produced during culture. Moreover, it was shown that overexpression *GUT1* alone significantly increases glycerol uptake rate, whereas in the strain overexpressing *GUT2* gene, higher quantities of CA were obtained. This suggested that

GUT2 could be crucial for CA production (Mironczuk et al. 2016). With strain *Y. lipolytica* AJD pADUTGut1/2, CA titer of 63.9 g/L was obtained at pH 3.0 which corresponds to a 14.5-fold increase as compared to the wild-type strain A101 under the same experimental conditions. On glycerol-based medium, a CA titer of 93 g/L was obtained at pH 6.0 (Mironczuk et al. 2016). The purification of glycerol is an expensive process, leading to CA production process unprofitable. As an economical alternative, crude glycerol, which is a direct byproduct of biodiesel production or saponification could be used. However, the utilization of such substrates could be limited by their high content of impurities such as salts, sodium hydroxide, methanol, or other organic compounds, which could be toxic for yeasts or hinder the production process. Despite these facts, *Y. lipolytica* can easily grow on such substrates and was found to produce CA with titer of 76 g/L at pH 3 (Rzechonek et al. 2018a). Overexpression of gene *ICL1* encoding iso-citrate lyase led to a decrease in ICA production from 10–12 to 3–6% (Forster et al. 2007b). However, the overexpression of *ICL1* did not influence the total production of CA and ICA. When *ICL1* overexpression was coupled with the disruption of *ACL1* gene (*YALI0E34793g*) encoding ATP citrate lyase and overexpression of the *K. marxianus INU1* gene encoding inulinase, CA and ICA titers of 84.0 g/L and 1.8 g/L were obtained, respectively, from inulin (100 g/L) after 214 h of culture in a 2L scale bioreactor (Liu et al. 2013). Besides this, overexpression of genes encoding mitochondrial citrate carrier (*YHM2, YALI0B10736g*) and adenosine monophosphate deaminase (*AMDP, YALI0E11495g*) in *Y. lipolytica* wild-type strain W29, a CA titer with productivity and yield from glucose of 97.1 g/L, 0.93 g/(L h) and 0.5 g/g were obtained, respectively, during culture in bioreactor (Yuzbasheva et al. 2019).

The CA production by different *Y. lipolytica* wild-type strains has been also reported. With an acetate-negative mutant strain *Y. lipolytica* A101-1.22 grown on glycerol-containing waste from the biodiesel industry, a CA titer of 124.2 g/L was obtained in repeated batch culture (Rymowicz et al. 2010). From a screening study, strain K57 was found able to produce CA with titer of 72.1 g/L and yield of 0.77 g/g on glucose-based medium during batch culture (Carsanba et al. 2019).

Industrial wastes have been also considered as possible carbon substrates for CA production (Wojtatowicz et al. 1991; Imandi et al. 2008; Papanikolaou et al. 2008; Morgunov et al. 2013; Liu et al. 2014; Morgunov and Kamzolova 2015). Using mutant strain *Y. lipolytica* NG40/UV7 and glycerol-containing wastes, CA production was found to increase by 40.6% (122.2 g/L) as compared to that obtained with pure glycerol (Morgunov et al. 2013; Morgunov and Kamzolova 2015). Pineapple wastes were also used successfully with a production titer of 202 g/kg of waste in optimized conditions (namely, 0.34% yeast extract, 70.71% moisture content of the substrate, 0.64% KH_2PO_4, and 0.69% Na_2HPO_4) (Imandi et al. 2008). Olive-mill wastewater (OMW) blended with glucose was used also for CA production in nitrogen-limited conditions. The titer obtained was 28.9 g/L (Papanikolaou et al. 2008). With *Y. lipolytica* SWJ-1b grown in the presence of waste cooking oil (80 g/L), a CA titer of 31.7 g/L was obtained within 336 h of culture in a 10 L bioreactor (Liu et al. 2014). The highest CA concentration,

100 g/L, was obtained by the mutant strain A-101–1.14 following the shortest growth (24 h) and production (80 h) phases from potato starch (Wojtatowicz et al. 1991).

3.2 Iso-citric Acid

Iso-citric acid (ICA) exists in the form of four stereoisomers, of which only the threo-Ds-form is an intermediate of Krebs cycle (Kamzolova et al. 2016, 2018). A promising application of ICA is sports medicine. ICA has a marked energetic and anti-hypoxic effect and can be used as a physiological stimulant of sportsmen undergoing intensive long-term physical training (Kamzolova et al. 2016). Recently, ICA has been tested as a natural prophylactic and therapeutic agent. It has been shown efficient for the treatment of iron-deficiency anemia and in the resorption of blood clots (Morgunov et al. 2019).

At present, ICA is produced via chemical synthesis. However, this results in a racemic mixture of stereoisomers that cannot be separated by chemical methods. ICA can be found also in the leaves and stems of some plants from the Crassulaceae family or in fruits such as blackberry and blackcurrant. However, the purification of ICA from plant extracts or fruit juices that contain a wide range of organic acids and other components is a complicated and very expensive technological process. Therefore, the development of biotechnological methods for ICA production is extremely important. At present, the most promising method for ICA production is considered to be microbiological synthesis (Kamzolova et al. 2016; Laptev et al. 2017).

It is known that *Y. lipolytica* accumulates ICA and CA in the culture medium when cell growth is nutrients limited. Moreover, the ICA/CA ratio greatly depends on the carbon source used for cell growth. For example, wild-type strains of *Y. lipolytica* secrete mainly CA and about 8–16% ICA on carbohydrates or glycerol while approximately 50–65% CA and 35–50% ICA on substrates like alkanes, triglycerides, ethanol, or acetate. When using ethanol as a carbon source, an ICA proportion of 35–67% was found depending on the cultivation conditions (Kamzolova et al. 2016; Holz et al. 2009). Some authors have conducted several studies to increase the ICA production titer. The recombinant *Y. lipolytica* H222-S4 strain deleted for *ICL1* gene, when grown on glycerol or glucose, showed only a smaller enhancement (by 2–5%) of ICA /CA ratio. With *Y. lipolytica* H222-S4 T5 that overexpress *ICL1* gene, the relative ICA content in the medium was as low as 5–7% of the total acids (CA and ICA) (Forster et al. 2007a, b).

Y. lipolytica VKM Y-2723 and its mutant derivative 704-UV4-A/NG50 were selected from 60 yeast strains for their ability for ICA production. Under optimal culture conditions (i.e., iron concentration of 1.2 mg/L, temperature of 29 °C, pH 6.0, pO_2 50–55% of saturation, and 30 mM itaconic acid), *Y. lipolytica* VKM Y-2373 produced 70.6 g/L ICA and 22.4 g/L CA with an ICA/CA ratio of 1:0.32 when grown on rapeseed oil. In similar conditions, a titer of 86 g/L ICA and 20 g/L CA with an ICA/CA ratio of 1:0.23 was obtained with *Y. lipolytica* 704-UV4-A/NG50 (Kamzolova et al. 2013). Under the above optimal culture conditions, *Y. lipolytica* VKM Y-2723 produced 90.5 g/L ICA with a yield of 0.77 g/g (Kamzolova et al. 2018). Also, catalytic activities of enzymes such as alcohol dehydrogenase, citrate synthase, aconitate hydratase, iso-citrate dehydrogenase, and iso-citrate lyase but not aldehyde dehydrogenase were found higher in a repeated-batch cultivation of 748 h than in batch cultivation. Therefore, under optimal repeated-batch culture using ethanol as substrate, *Y. lipolytica* strain VKM Y-2723 produced 109.6 g/L ICA with a production rate of 1.35 g/L/h (Morgunov et al. 2019). Besides this, overexpression of *ACO1* (*YALI0D09361g*) encoding aconitase allowed to increase ICA titer (Laptev et al. 2017; Holz et al. 2009). For the wild-type strain H222 and H222-S4 grown on sunflower oil, the ICA proportion ranged between 35 and 49%, whereas it increased up to 71% with the *ACO1* multi-copy transformant T1 without any modification in the total organic acid titer (both CA and ICA). However, ICA production was only moderately increased from 8–12% up to 13–17% with carbon sources such as glucose, sucrose, and glycerol (Holz et al. 2009). When *ACO1* was expressed in multicopy, ICA and CA titers were, respectively, of 72.6 g/L and 29.0 g/L with an ICA/CA ratio of 2.5:1 during culture in 10L bioreactor in a rapeseed oil-based medium (Laptev et al. 2017).

3.3 Succinic Acid

As an almost ubiquitous metabolite in many organisms, succinic acid (SA) is an intermediate of the Krebs cycle. Due to its numerous potential applications, SA has been recognized as one of the most important and high value-added bio-based building block chemicals by the US Department of Energy (DOE) (Beauprez et al. 2010; Ahn et al. 2016). In 2004 and 2010, the US DOE reported SA as one of the five most promising bio-based platform chemicals. SA market worldwide is forecasted to grow at a compound annual growth rate (CAGR) of 15.7% between 2020 and 2026 (Ahn et al. 2016; Tan et al. 2014, 2020a). Among all the platform chemicals, SA is currently used as a surfactant, ion chelator, additive in agricultural and food, and pharmaceutical industries. It can be used as a precursor to synthesize γ-butyrolactone, 1,4-butanedioic acid, tetrahydrofuran, and other value-added chemicals (Ahn et al. 2016; Tan et al. 2014; McKinlay et al. 2007). Currently, succinate is produced mainly from petroleum-derived maleic anhydride. However, due to near future shortages of petroleum resources, severe environmental concerns related to chemical synthesis processes and to reach an economical bio-based production of SA, extensive research works have been focusing on the development of efficient microbial strains by metabolic engineering as well as optimized fermentation and downstream processes (Ahn et al. 2016; Tan et al. 2013). Besides, bio-based SA production technologies can reduce greenhouse gas emissions by 50% and energy demand by 30–40% as compared to the chemical production process (Tan et al. 2014).

By contrast to prokaryotes, some yeasts, such as *Y. lipolytica*, are highly tolerant to low pH, rendering them attractive as an industrial host for SA synthesis (Cui et al. 2017). A *Y. lipolytica* evolved strain, named PSA3.0, which can produce SA at low pH using glucose as substrate was selected from a long-lasting culture in in situ fibrous bed bioreactor (isFBB) (Li et al. 2018a). Strain PSA3.0 produced SA with a titer of 18.4 g/L and yield of 0.23 g/g at pH 3.0. These values are, respectively, 4.8 and 4.6-fold higher than those obtained with the parental strain PSA02004 at pH 3.0. Using fed-batch culture in bioreactor, a SA titer of 76.8 g/L was obtained, which is the highest value ever achieved from a glucose-based medium at low pH (Li et al. 2018a). Furthermore, *Y. lipolytica* PSA02004 produces 33.2 g/L SA with a yield of 0.58 g/g and productivity of 0.33 g/L/h from sugarcane bagasse hydrolysates. Using glucose-rich fruit and vegetable waste (FVW) hydrolysates, SA titer of 140.6 g/L with a productivity of 0.69 g/L/h has been obtained (Li et al. 2018b; Ong et al. 2019). Under optimal conditions of repeated-batch fermentation, SA titer of 55.3 g/L and the maximal productivity 2.6 g/L/h was reached with *Y. lipolytica* strain VKPM Y3753 (Bondarenko et al. 2016).

An increase in SA titer was also achieved by replacing the native promoter of the succinate dehydrogenase by strong promoters (Cui et al. 2017; Yuzbashev et al. 2010; Jost et al. 2014; Gao et al. 2016; Li et al. 2016, 2017; Yang et al. 2017). Under oxygen limitation, *Y. lipolytica* strain H222-AZ2 obtained by exchange of the native promoter of the succinate dehydrogenase subunit 2 encoding gene by the inducible promoter *POT1*, a SA productivity of 0.152 g/L/h and titer of 25 g/L after 165 h of culture in glycerol medium were obtained (Jost et al. 2014). With *Y. lipolytica* strain PGC01003, deleted for *Ylsdh5* genes encoding succinate dehydrogenase subunit 5, SA titer of 43 g/L was obtained during batch culture in 2.5 L bioreactor in medium containing crude glycerol. With a fed-batch strategy, the strain produced 160 g/L SA (Gao et al. 2016). After 21 days of evolution to enable the strain PGC01003 to catabolize glucose, the evolved strain produced 65.7 and 87.9 g/L SA using YPD medium and food waste hydrolysate, respectively (Yang et al. 2017).

Enhanced SA produced was also achieved by genetic modifications of the strains in succinate dehydrogenase or fermentation technology innovation. For example, when the native promoter of *SDH2* was replaced by an inducible promoter of *POT1*, SA titer at 25 g/L with productivity at 0.15 g/L/h was obtained by *Y. lipolytica* H222-AZ2 from glycerol under oxygen limitation (Jost et al. 2014) After inactivation of SDH5 that encodes succinate dehydrogenase subunit 5, SA titer at 43 g/L was obtained by batch fermentation from crude glycerol By PGC01003, and a much higher titer was achieved at 160 g/L SA via fed-batch fermentation strategy (Gao et al. 2016). Based on this, an in-situ fibrous bed bioreactor (isFBB) which could improve the initial cell density was developed, and SA titer at 198.2 g/L with average productivity of 1.46 g/L/h was achieved by PGC01003 via fed-batch strategy (Li et al. 2016). This value was further improved to 209.7 g/L when the immobilization matrix in isFBB was replaced by the more porous material (e.g., sugarcane bagasse), which is the highest value ever reported

(Li et al. 2017). However, it was reported that the accumulation of acetate was a major factor that impeded the fermentation at low pH, resulting in an obvious increase in downstream process cost (Cui et al. 2017). In order to solve this problem, the gene encoding the acetyl-coenzyme A hydrolase (*ACH1*) was deleted, and *PCK* from *S. cerevisiae* encodes phosphoenolpyruvate carboxykinase together with *SCS2* encoding endogenous succinyl CoA synthetase were overexpressed in PGC01003 to achieve the strain PGC202. As result, SA titer at 110.7 g/L with a yield of 0.53 g/g and productivity of 0.8 g/L/h was obtained by this strain via fed-batch fermentation with pH without control (final pH at 3.4) (Cui et al. 2017).

Nevertheless, all the metabolic evolution of *Y. lipolytica* for SA production has led to a partial or total loss of its ability to grow in glucose-based medium, which limits its industrial application (Yang et al. 2017). In this case, a strategy termed metabolic engineering was applied by Yang et al. to obtain a glucose-consuming *Y. lipolytica* (PSA02004) after a 21-day repeated fermentation of PGC01003 (Yang et al. 2017). As a result, the evolved strain PSA02004 could produce SA at 65.7 g/L and 87.9 g/L from the YPD medium and food waste hydrolysate at pH 6.0, respectively. Interestingly, with the help of isFBB, the pH for the cultivation of PSA02004 could be decreased to a level lower than 3.0 gradually by metabolic evolution, and the evolved strain named *Y. lipolytica* PSA3.0 that could produce SA with a titer of 19.3 g/L, productivity of 0.52 g/L/h, and yield of 0.29 g/g at pH 3.0 from YPD was achieved (Li et al. 2018a). The enzyme activity analysis demonstrated that the pathway from pyruvate to acetate was partially blocked in *Y. lipolytica* PSA3.0 after the evolution, which is beneficial to cell growth and SA production at low pH.

However, as for PGC01003, the accumulation of acetate is a limiting factor for further improvement of SA production as SA recovery request a pH adjustment leading to an increase of downstream process cost (Cui et al. 2017). To solve this issue, a new recombinant strain, named PGC202, was deleted for gene *ACH1 (YALI0E30965g)* encoding the acetyl-coenzyme A hydrolase and overexpressing *PCK* gene from *Saccharomyces cerevisiae* encoding phosphoenolpyruvate carboxykinase together with gene *SCS2* encoding endogenous succinyl CoA synthetase was constructed. It allowed improving the SA production process through the elimination of acetic acid overflow and by-products formation. Indeed, SA titer of 110.7 g/L with a maximum yield from glycerol of 0.53 g/g and a productivity of 0.8 g/L/h were obtained during fed-batch fermentation of strain PGC202 (Cui et al. 2017). Furthermore, the strain produces 31.7 g/L SA with a yield of 0.52 g/g and productivity of 0.60 g/L/h in isFBB fermentation when using glucose-containing MFWs hydrolysate as the carbon source supplemented with 3% of tryptone (Li et al. 2019).

3.4 Itaconic Acid

Itaconic acid (IA) is naturally produced by several *Aspergillus* species (Blazeck et al. 2015). The industrially versatile usability of IA and its derivatives are reflected in the wide range of applications such as in plastics, styrene-butadiene rubber, synthetic latex, super-absorbent polymers, unsaturated polyester resins, and detergents. The field of application of these products is widespread and ranges from paint, lacquer, paper industries, hygiene, and medical products as well as in the construction sector (Kuenz and Krull 2018). Thus, IA was recognized as one of 12 value-added chemicals from biomass by the US DOE in 2004 (Blazeck et al. 2015). IA market is estimated at US$28.4 million in the year 2020 in the USA while the world market is expected to reach $129.3 million by 2027 due to increasing demand for bio-based chemicals (2020b). To date, two strategies have been developed to produce IA, a chemical method based on the pyrolysis of citric acid and a biosynthesis process that relies on the decarboxylation of cis-aconitic acid by microorganisms (Kuenz and Krull 2018; Zhao et al. 2019).

Although current industrial production of IA is based on *A. terreus* fermentation allowing titer higher than 80 g/L, *A. terreus* suffers from poor growth in optimal media used for IA production and is negatively affected by shear stress in the bioreactor, thus precluding fermentations under conventional conditions (Blazeck et al. 2015). Moreover, a complex and expensive downstream process has been developed at an industrial scale (Kuenz and Krull 2018; Okamoto et al. 2014; Bellasio et al. 2015; Chin et al. 2015; Otten et al. 2015). In order to address these concerns, *Y. lipolytica* was considered as an alternative IA production host as this organism can natively grow at low pH and sustains a high flux toward the citric acid cycle (Blazeck et al. 2015). There are currently two reports on engineered *Y. lipolytica* strains used to produce IA. In the study of Blazeck et al., the native cis-aconitate decarboxylase (*CAD*) gene was overexpressed, and the resulting strain produced IA with a titer of 33 mg/L (Blazeck et al. 2015). Further strain optimizations of the metabolic pathway, enzyme localization, and media optimization strategies enabled IA titer of 4.6 g/L during culture in bioreactors, representing a 140-fold improvement. A recombinant strain which overexpresses genes encoding the mitochondrial cis-aconitate transporter MTT and *CAD* allowed an IA titer of 22.0 g/L in optimized conditions (Zhao et al. 2019). This represents a 60-fold improvement over the initial titer (0.36 g/L).

3.5 α-ketoglutaric Acid

α-ketoglutaric acid (α-KG), an important dicarboxylic acid, is an intermediate of the Krebs cycle. It is also involved in amino acid metabolism and has an important role in the regulation of the balance between carbon and nitrogen metabolism in many microorganisms. α-KG is widely applied in the industrial scope, e.g., as a building block for the chemical synthesis of heterocycles, dietary supplement, component of infusion solutions, and wound healing compounds (Otto et al. 2011; Yovkova et al. 2013). Currently, α-KG is synthesized by chemical processes or biosynthesis at an industrial scale. The main chemical routes utilized succinic acid and oxalic acid diethyl esters or relied on the oxidation of glyoxylic acid with sodium glutamate using copper as a catalyst. The drawbacks of these chemical routes are a lack of selectivity, a low yield, high risk from manipulation of harsh chemicals, and the generation of environmental hazards. These drawbacks sharply increased the downstream process cost and restricted the utilization of α-KG in food, medicine, and cosmetics applications.

Y. lipolytica is unable to synthesize the pyrimidine structure of the thiamine molecule. Thiamine is the cofactor of α-ketoglutarate dehydrogenase (*KGDH*) which is a key enzyme in the metabolism of α-KG; thus, the limitation of thiamine availability can reduce the activity of α-ketoglutarate dehydrogenase of the Krebs cycle. Under conditions of thiamine deficiency, the conversion of α-KG in the Krebs cycle is inhibited, which leads to its accumulation in the fermentation broth (Chernyavskayaá et al. 2000; Kamzolova et al. 2012; Yin et al. 2012).

Currently, different strains producing α-KG from different carbon sources have been reported. *Y. lipolytica* H355 allows α-KG titer of 195 g/L with a productivity of 1.3 g/L/h in the medium containing a mixture of n-alkanes (Guo et al. 2016b). Under optimal conditions, *Y. lipolytica* VKM Y-2412 produced 172 g/L of α-KG with a yield of 0.70 g/g and 102.5 g/L α-KG with the yield of 0.95 g/g from ethanol and rapeseed oil, respectively (Kamzolova et al. 2012; Kamzolova and Morgunov 2013). However, in order to overcome the disadvantages of low yield and accumulation of byproducts, several approaches including metabolic engineering strategies and different fermentation configurations have been investigated (Guo et al. 2016a). *Y. lipolytica* H355A that overexpress the iso-citrate dehydrogenase (*IDH1*) and pyruvate carboxylase (*PYC1*) genes can produce 186 g/L α-KG from raw glycerol. This represents a 19% increased production as compared to the control strain H355. *Y. lipolytica*-RoPYC2 which overexpress *RoPYC2* gene encoding pyruvate carboxylase produced α-KG with a titer of 62.5 g/L with only 13.5 g/L pyruvate (PA). As compared to the parental strain, α-KG production in strain RoPYC2 increased by 35.3% while PA production is reduced by 69.8% (Yovkova et al. 2013; Yin et al. 2012).

4 Polyols

In some biological systems, polyols are obtained by the reduction of their keto or aldo groups into hydroxyl groups (Rice et al. 2019). As polyols, *Y. lipolytica* synthesize mainly mannitol (MAN), which is suggested to provide cofactor NADPH for fatty acid synthesis and erythritol (EOL). The latter being produced in response to osmotic stress. Derivatives of erythritol, namely erythrulose (EOSE) and threitol (TOL), are also of industrial interest and will be also discussed below (Table 2).

4.1 Mannitol

MAN is a six-carbon alcohol involved in stress tolerance in microorganisms notably in the scavenging of reactive oxygen species (ROS) (Zhang et al. 2018; Sekova et al. 2019). A possible role of MAN in fatty acid metabolism has been

Table 2 Some examples of polyols produced in *Y. lipolytica*

Y. lipolytica strain	Genetic modification	Medium	Production, polyol	References
AIG	*GUT1*^{OE} *(YALI0F00384g)*	Crude glycerol Molasse	11 g/L, MAN	Rakicka et al. (2017a, b)
AMM	*ER*^{OE} *(YALI0F18590g)*	Glycerol	44 g/L, EOL	Janek et al. (2017)
HCY118	*ER10*^{OE} *(YALI0D07634g)*, *ER25*^{OE} *(YALI0C13508g)*, *ZWF1*^{EO} *(YALI0E22694g)*, *GND1*^{EO} *(YALI0B15598)*	Glucose	196 g/L, EOL	Cheng et al. (2018)
Po1d	*TKL1*^{OE} *(YALI0E06479g)*	Glycerol	43 g/L, EOL	Carly et al. (2017a, b)
MK1	*TKL1*^{OE} *(YALI0E06479g)*	Glycerol	51 g/l, EOL	Mironczuk et al. (2017)
MK1	*TKL1*^{OE} *(YALI0E06479g)*, *TAL*^{OE} *(YALI0F15587g)*	Glycerol	46 g/l, EOL	Mironczuk et al. (2017)
Po1d	*GUT1*^{OE} *(YALI0F00384g)*, *TKL1*^{OE} *(YALI0E06479g)*	Glycerol	78 g/l, EOL	Carly et al. (2017a, b)
M53	Δ*snf1 (YALI0D02101g)*	Peanuts cake	185 g/kg, EOL	Li et al. (2019)
Wratislavia K1	WT	Glycerol	180 g/L, EOL	Rakicka-Pustułka et al. (2020)
RIY210	*EYD1*^{OE} *(YALI0F01650g)*, Δ*eyk1 (YALI0F01606g)*	EOL	0.12 g/gDCW, EOSE	Carly et al. (2017a, b)
CGMCC7326	*Ss-XDH*^{OEcn}	Glucose	112 g/L, TOL	Chi et al. (2019)

Abbreviations: MAN: mannitol, EOL: erythritol, EOSE: erythrulose, TOL: threitol, *GUT1*: glycerokinase, *ER*: erythrose reductase, *ZWF1*: glucose-6P dehydrogenase, *GND1*: 6-phosphogluconate dehydrogenase, *TKL1*: transketolase, *TAL*: transaldolase, SNF1: regulator of lipid accumulation, *EYD1*: erythritol dehydrogenase, *EYK1*: erythritol kinase, *XDH*: xylitol dehydrogenase

also reported (Dulermo et al. 2015). MAN production has been investigated for several *Y. lipolytica* strains and different experimental conditions. For cells grown in shake-flasks in the presence of crude glycerol, mannitol was co-produced with EOL with titers ranging from 2.6 to 14.9 g/L. During glycerol fed-batch bioreactor culture, strains A-15 and A UV'1 synthetized MAN with a titer of, respectively, 41.4 and 38.1 g/L, corresponding to the productivity of 0.29 and 0.28 g/(L h) (Tomaszewska et al. 2012). With a derivative of strain AIB overexpressing *GUT1* gene encoding glycerol kinase grown in a batch bioreactor on a mixture of molasse and crude glycerol, maximal MAN titer of 11 g/L was obtained (Rakicka et al. 2017a). Similarly, using a mixture of crude glycerol and olive oil mill wastewater, the MAN titer reached 13 g/L within 140 h with a yield from glycerol of 0.21 g/g (Sarris et al. 2019).

4.2 Erythritol

Erythritol (EOL) is a four-carbon polyol. Its industrial interest relies on its sweetening property as it has the texture and taste of table sugar. EOL has also other interesting properties regarding its application. It is not metabolized by the human body, and thus, it is calories-free and does not modify glycemia (Carly and Fickers 2018). It could be synthesized by several osmotolerant microorganisms, including *Y. liploytica* and *Candida magnoliae,* as an osmoprotectant. In *Y. lipolytica*, EOL derives from erythrose-4P, an intermediate of the pentose phosphate pathway (PPP) (Fig. 2). The latter is dephosphorylated by an erythrose-4P phosphatase (E4PP) before being reduced by an erythrose reductase (ER) into erythritol. In this yeast, genes coding for erythrose reductase have been characterized (namely, *YALI0F18590g*, *YALI0D07634g,* and *YALI0C13508g*) (Janek et al. 2017; Cheng et al. 2018). The complete EOL catabolic pathway has been also

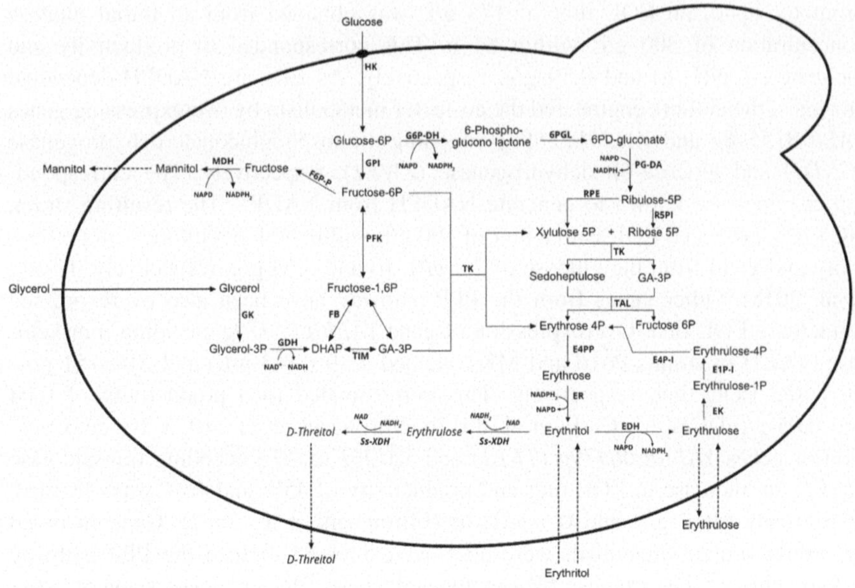

Fig. 2 Overview of the principal metabolic pathways for polyols and derivatives synthesis in *Y. lipolytica.* DHAP: dihydroxyacetone-P; E4PP: Erythrose 4P phosphatase; ER: erythrose reductase; EDH: erythritol dehydrogenase; EK: erythritol kinase; E1PI: erythrulose-1P isomerase; E4PI: erythrulose-4P isomerase; FBA: fructose-bisphosphate aldolase; F6P-P: Fructose-6P phosphatase; GK: glycerol kinase; G3PDH: glycerol-3P dehydrogenase; GPI: glucose-6P isomerase; G6PDH: glucose-6P dehydrogenase; HK: hexokinase; MDH: mannitol dehydrogenase; PFK: phosphofructokinase; PGDH: phopshogluconate dehydrogenase; RPE: Ribulose-P3 epimerase; R5PI: ribose-5P isomerase; TK: transketolase; TIM: triose isomerase; TAL: transaldolase; 6PGL: 6-phosphonogluconolactonase, Ss-XDH: xylitol dehydrogenase from *Scheffersomyces stipites.* Pathway in italics is heterologous

reported recently (Niang et al. 2020). It is first converted into erythrulose by an erythritol dehydrogenase (*EYD1, YALI0F01650g*) and subsequently phosphorylated into L-erythrulose-1P by an erythrulose kinase (*EYK1, YALI0F01606g*). Then, L-erythrulose-1P is isomerized into D-erythrulose-4P by an erythrulose-1P isomerase (*EYL1, YALI0F01584g*). Finally, erythrose-4P is generated by the activity of an erythrulose-4P isomerase (*EYL2, YALI0F01628g*).

Several review papers focusing on EOL have been published, and only the most striking information will be reported below (Carly and Fickers 2018; Regnat et al. 2018; Rzechonek et al. 2018b). An efficient strategy for EOL synthesis consisted of constitutively expressing genes encoding erythrose reductase (ER). Overexpression of ER gene *YALI0F18590g* in *Y. lipolytica* strain AMM led to a 20% increase in EOL titer as compared to the parental strain (44.4 g/L) (Janek et al. 2017). This corresponded to the productivity of 0.77 g/(L h) and a yield from glycerol of 0.44 g/g. More recently, two additional ERs have been characterized (*YALI0D07634g* and *YALI0C13508g*) (Cheng et al. 2018). Upon overexpression of these three ERs under the control of the strong constitutive promoter hp4d, an EOL titer of 178 g/L was obtained from an initial glucose concentration of 300 g/L within 84 h. This corresponded to productivity and yield of 2.1 g/(L h) and 0.59 g/g, respectively. As ERs are NADPH-dependent enzymes, the authors engineered the co-factor metabolism by overexpressing genes *YALI0B15598g* and *YALI0E22694g* encoding 6-phosphogluconate dehydrogenase (*GND1*) and glucose-6P dehydrogenase (*ZWF1*), respectively. The corresponding enzymes are known to generate NADPH from $NADP^+$. The resulting strain, HCY108, produced EOL with a titer of 190 g/L within 80 h of culture with productivity and yield from the glucose of 2.4 g/(L h) and 0.63 g/g, respectively (Cheng et al. 2018). Other genes from the PPP pathway have been also overexpressed to increase EOL titer. Overexpression of gene *YALI0E06479g* encoding transketolase (*TKL1*) in strains Po1d and MK1 yielded a 19% (43 g/L) and 51% (51 g/L) increased EOL titer, respectively. This corresponded to a productivity of 0.04 and 0.05 g/(gDCW h) (Carly et al. 2017b; Mirończuk et al. 2017). By coexpression of genes *YALI0E06479g* (*TKL1*) and *YALI0F15587g* encoding transaldolase (*TAL1*), an increase in EOL titer and productivity of 45% and 46% was obtained, respectively (46.7 g/L and 0.5 g/(L h) (Mirończuk et al. 2017). Gene involved in carbon source catabolism were also overexpressed to feed the PPP pathway with precursors (i.e., fructose-6P and glyceraldehyde-3P). Overexpression of genes *YALI0F00384g* (*GUT1*) and *YALI0B02948g* (*GUT2*) encoding, respectively, glycerol kinase (GK) and glycerol-3-P dehydrogenase (G3P-DH) allowed an EOL titer of 78 g/L from an initial glycerol concentration of 100 g/L within 72 h in a 5-L bioreactor (Mirończuk et al. 2016). Besides this, overexpression of genes *GUT1* and *TKL1* in a strain disrupted for gene *EYK1* allowed an erythritol titer, productivity, and yield from glycerol of 80 g/L, 1.03 g/(L h), and 0.53 g/g, respectively (Carly et al. 2017b). By overexpressing invertase encoding gene *SUC2* from *Saccharomyces cerevisiae* and native *GUT1* gene, an EOL titer of 100 g/L with a productivity and yield of 1.1 g/(L h) and 0.67 g/g, respectively, were obtained from a mixture of raw industrial molasses (60 g/L) and crude glycerol (100 g/L)

(Rakicka et al. 2017b). Disruption of gene *SNF1* (*YALI0D02101g*) coding for a regulator of lipid accumulation allowed an EOL production of 185.4 mg/g in solid-state fermentation (SSF) using peanut press cake mixed with 40% sesame meal and 10% waste cooking oil as substrate (Li et al. 2019). EOL production in large-scale bioreactor was recently reported. Using raw glycerol feeding, strain *Y. lipolytica* Wratislavia K1 produced EOL with titer and yield from glycerol of 180 g/L and 0.53 g/g, respectively, after 144 h of cultivation in 500 L bioreactor (Rakicka-Pustułka et al. 2020). With metabolically engineered *Y. lipolytica* strain HCY118, a production titer of 196 g/L with productivity and yield from glucose of 2.51 g/L h and 0.65 g/g, respectively, were obtained in a 30 m^3 bioreactor within 78 h (Wang et al. 2020).

4.3 Erythrulose

As mentioned already, erythrulose (EOSE) is an intermediate of the erythritol catabolic pathway. It has many applications in cosmetics as a sunless tanning agent and in chemistry as a precursor of several drugs such as anticancer (Bengamide E), antifungal (Tanikolide), substitute β-lactam, cytokine modulator (Cytoxazone), or cholesterol-lowering drugs (Crestor, Zetia) (Carly et al. 2018). In a *Y. lipolytica Δeyk1* derivative (RIY210), the conversion of EOSE into L-erythrulose-1P is impaired. Therefore, such a mutant strain can convert EOL into EOSE. The conversion rate could be increased by the overexpression of *EYD1* encoding erythritol dehydrogenase in a Δ*eyk1* genetic background. With such a strategy, an efficient fed-batch bioreactor process was developed to convert EOL into EOSE with a conversion rate and yield of 0.116 g/g DCW·h and 0.64 g/g, respectively (Carly et al. 2017a).

4.4 Threitol

Threitol (TOL) is a diastereoisomer of erythritol that is produced naturally as an antifreeze molecule by the fungus *Armillaria mellea* as well as the Alaskan beetle *Upis ceramboides*. TOL has application as a precursor for the synthesis of anticancer drugs (treosulfan and threitol ceramide). It is also a constitutive element of oxygen-sensitive pigments incorporated in a smart plastic film used for food packaging (Chi et al. 2019). Recently, xylitol dehydrogenase (Ss-*XDH*) from *Scheffersomyces stipites* CBS6054 was found able to oxidize EOL into EOSE irreversibly and then reduce EOSE into threitol (Chi et al. 2019). By overexpression of the corresponding gene in the *Y. lipolytica* strain CGMCC7326, a good producer of EOL from glucose, a TOL production titer of 112 g/L with a yield from the glucose of 0.37 g/g has been reported (Chi et al. 2019).

5 Conclusion and Prospects

Due to its potential for organic acids and polyols production and good tolerance to low pH, *Y. lipolytica* has gradually become a cell factory for their production. However, several key steps allowing to increase strain production titer and productivity remain to be identified. Process operations are still not optimal, especially on a large scale. Further investigations must be made to obtain cost-effective processes for most of the metabolites described herein.

Acknowledgements The authors would like to thank Andrew Zicler for his contribution in editing Figs. 1 and 2.

References

Abdel-Mawgoud AM, Markham KA, Palmer CM, Liu N, Stephanopoulos G, Alper HS (2018) Metabolic engineering in the host *Yarrowia lipolytica*. Metab Eng 50:192–208. https://doi.org/10.1016/j.ymben.2018.07.016

Ahn JH, Jang YS, Lee SY (2016) Production of succinic acid by metabolically engineered microorganisms. Curr Opinion Biotechnol 42:54–66. https://doi.org/10.1016/j.copbio.2016.02.034

Barth G, Gaillardin C (1996) Yarrowia lipolytica. In: Wolf K (ed) Nonconventional yeasts in biotechnology: a handbook. Springer, Heidelberg, pp 313–388

Beauprez JJ, De Mey M, Soetaert WK (2010) Microbial succinic acid production: natural versus metabolic engineered producers. Proc Biochem 45:1103–1114. https://doi.org/10.1016/j.procbio.2010.03.035

Bellasio M, Mattanovich D, Sauer M, Marx H (2015) Organic acids from lignocellulose: *Candida lignohabitans* as a new microbial cell factory. J Ind Microbiol Biotechnol 42:681–691. https://doi.org/10.1007/s10295-015-1590-0

Beopoulos A, Cescut J, Haddouche R, Uribelarrea JL, Molina-Jouve C, Nicaud JM (2009) *Yarrowia lipolytica* as a model for bio-oil production. Prog Lipid Res 48:375–387. https://doi.org/10.1016/j.plipres.2009.08.005

Beopoulos A, Chardot T, Nicaud JM (2009) *Yarrowia lipolytica*: a model and a tool to understand the mechanisms implicated in lipid accumulation. Biochimie 91:692–696. https://doi.org/10.1016/j.biochi.2009.02.004

Bilal M, Xu S, Iqbal HMN, Cheng H (2020) *Yarrowia lipolytica* as an emerging biotechnological chassis for functional sugars biosynthesis. Crit Rev Food Sci Nutr 1–18. https://doi.org/10.1080/10408398.2020.1739000

Bio-Succinic Acid Market Size is Expected to Grow at a CAGR of 15.7% - Valuates Reports (2020a). https://www.prnewswire.com/news-releases/bio-succinic-acid-market-size-is-expected-to-grow-at-a-cagr-of-15-7---valuates-reports-301116323.html

Blazeck J, Hill A, Jamoussi M, Pan A, Miller J, Alper HS (2015) Metabolic engineering of *Yarrowia lipolytica* for itaconic acid production. Metab Eng 32:66–73. https://doi.org/10.1016/j.ymben.2015.09.005

Bondarenko PY, Fedorov AS, Sineoky SP (2016) Optimization of repeated-batch fermentation of a recombinant strain of the yeast *Yarrowia lipolytica* for succinic acid production at low pH. Appl Biochem Microbiol 53:882–887. https://doi.org/10.1134/s0003683817090022

Carly F, Fickers P (2018) Erythritol production by yeasts: a snapshot of current knowledge. Yeast 35:455–463. https://doi.org/10.1002/yea.3306

Carly F, Gamboa-Melendez H, Vandermies M, Damblon C, Nicaud JM, Fickers P (2017) Identification and characterization of *EYK1*, a key gene for erythritol catabolism in *Yarrowia lipolytica*. Appl Microbiol Biotechnol 101:6587–6596. https://doi.org/10.1007/s00253-017-8361-y

Carly F, Steels S, Telek S, Vandermies M, Nicaud J-M, Fickers P (2018) Identification and characterization of *EYD1*, encoding an erythritol dehydrogenase in *Yarrowia lipolytica* and its application to bioconvert erythritol into erythrulose. Bioressour Technol 247:963–969. https://doi.org/10.1016/j.biortech.2017.09.168

Carly F, Vandermies M, Telek S, Steels S, Thomas S, Nicaud J-M, Fickers P (2017) Enhancing erythritol productivity in *Yarrowia lipolytica* using metabolic engineering. Metab Eng 42:19–24. https://doi.org/10.1016/j.ymben.2017.05.002

Carsanba E, Papanikolaou S, Fickers P, Erten H (2019) Screening various *Yarrowia lipolytica* strains for citric acid production. Yeast 36:319–327. https://doi.org/10.1002/yea.3389

Cavallo E, Charreau H, Cerrutti P, Foresti ML (2017) *Yarrowia lipolytica*: a model yeast for citric acid production. FEMS Yeast Res 17:1–16. https://doi.org/10.1093/femsyr/fox084

Celińska E, Ledesma-Amaro R, Larroude M, Rossignol T, Pauthenier C, Nicaud J-M (2017) Golden gate assembly system dedicated to complex pathway manipulation in *Yarrowia lipolytica*. Microb Biotechnol 10:450–455. https://doi.org/10.1111/1751-7915.12605

Cheng H, Wang S, Bilal M, Ge X, Zhang C, Fickers P, Cheng H (2018) Identification, characterization of two NADPH-dependent erythrose reductases in the yeast *Yarrowia lipolytica* and improvement of erythritol productivity using metabolic engineering. Microb Cell Fact 17:133. https://doi.org/10.1186/s12934-018-0982-z

Chernyavskayaá OGC, Shishkanova NV, Il'chenko AP, Finogenova TV (2000) Synthesis of alpha-ketoglutaric acid by *Yarrowia lipolytica* yeast grown on ethanol. Appl Microbiol Biotechnol 53:152–158

Chi P, Wang S, Ge X, Bilal M, Fickers P, Cheng H (2019) Efficient D-threitol production by an engineered strain of Yarrowia lipolytica overexpressing xylitol dehydrogenase gene from *Scheffersomyces stipitis*. Bioch Eng J 149. https://doi.org/10.1016/j.bej.2019.107259

Chin T, Sano M, Takahashi T, Ohara H, Aso Y (2015) Photosynthetic production of itaconic acid in *Synechocystis* sp. PCC6803. J Biotechnol 195:43–45, 107259. https://doi.org/10.1016/j.jbiotec.2014.12.016

Ciriminna R, Meneguzzo F, Delisi R, Pagliaro M (2017) Citric acid: emerging applications of key biotechnology industrial product. Chem Cent J 11:22–31. https://doi.org/10.1186/s13065-017-0251-y

Cui Z, Gao C, Li J, Hou J, Lin CSK, Qi Q (2017) Engineering of unconventional yeast *Yarrowia lipolytica* for efficient succinic acid production from glycerol at low pH. Metab Eng 42:126–133. https://doi.org/10.1016/j.ymben.2017.06.007

Do HD, Vandermies M, Fickers P, Theron CW (2019) Non-conventional yeast species for recombinant protein and metabolite production. Ref Mod Life Sci. Elsevier

Dujon B, Sherman D, Fischer G, Durrens P, Casaregola S, Lafontaine I, De Montigny J, Marck C, Neuvéglise C, Talla E, Goffard N, Frangeul L, Aigle M, Anthouard V, Babour A, Barbe V, Barnay S, Blanchin S, Beckerich JM, Beyne E, Bleykasten C, Boisramé A, Boyer J, Cattolico L, Confanioleri F, De Daruvar A, Despons L, Fabre E, Fairhead C, Ferry-Dumazet H, Groppi A, Hantraye F, Hennequin C, Jauniaux N, Ph, Joyet, Kachouri R, Kerrest A, Koszul R, Lemaire M, Lesur I, Ma L, Muller H, Nicaud JM, Nikolski M, Oztas S, Ozier-Kalogeropoulos O, Pellenz S, Potier S, Richard GF, Straub ML, Suleau A, Swennen D, Tekaia F, Wésolowski-Louvel M, Westhof E, Wirth B, Zeniou-Meyer M, Zivanovic I, Bolotin-Fukuhara M, Thierry A, Ch, Bouchier, Caudron B, Scarpelli C, Gaillardin C, Weissenbach J, Wincker P, Souciet JL (2004) Genome evolution in yeasts. Nature 430:35–44. https://doi.org/10.1038/nature02579

Dulermo T, Lazar Z, Dulermo R, Rakicka M, Haddouche R, Nicaud J-M (2015) Analysis of ATP-citrate lyase and malic enzyme mutants of Y*arrowia lipolytica* points out the importance of mannitol metabolism in fatty acid synthesis. Biochim Biophys Acta 1851:1107–1117. https://doi.org/10.1016/j.bbalip.2015.04.007

Fickers P, Benetti PH, Waché Y, Marty A, Mauersberger S, Smit MS, Nicaud JM (2005) Hydrophobic substrate utilisation by the yeast *Yarrowia lipolytica*, and its potential applications. FEMS Yeast Res 5:527–543. https://doi.org/10.1016/j.femsyr.2004.09.004

Fickers P, Cheng H, Sze Ki Lin C (2020) Sugar alcohols and organic acids synthesis in *Yarrowia lipolytica:* Where are we? Microorganisms 8:574. https://doi.org/10.3390/microorganisms8040574

Fickers P, Le Dall MT, Gaillardin C, Ph, Thonart, Nicaud JM (2003) New disruption cassettes for rapid gene disruption and marker rescue in the yeast *Yarrowia lipolytica*. J Microbiol Methods 55:727–737

Forster A, Aurich A, Mauersberger S, Barth G (2007) Citric acid production from sucrose using a recombinant strain of the yeast *Yarrowia lipolytica*. Appl Microbiol Biotechnol 75:1409–1417. https://doi.org/10.1007/s00253-007-0958-0

Forster A, Jacobs K, Juretzek T, Mauersberger S, Barth G (2007) Overexpression of the *ICL1* gene changes the product ratio of citric acid production by *Yarrowia lipolytica*. Appl Microbiol Biotechnol 77:861–869. https://doi.org/10.1007/s00253-007-1205-4

Gao C, Yang X, Wang H, Rivero CP, Li C, Cui Z, Qi Q, Lin CSK (2016) Robust succinic acid production from crude glycerol using engineered *Yarrowia lipolytica*. Biotechnol Biofuels 9:179–190. https://doi.org/10.1186/s13068-016-0597-8

Global Citric Acid Markets Report, 2011–2018 & 2019–2024 (2019) https://www.prnewswire.com/news-releases/global-citric-acid-markets-report-2011-2018--2019-2024-300814817.html

Global Itaconic Acid (IA) Industry (2020) https://www.reportlinker.com/p03646033/Global-Itaconic-Acid-IA-Industry.html?utm_source=PRN

Groenewald M, Boekhout T, Neuvéglise C, Gaillardin C, van Dijck P, Wyss M (2014) *Yarrowia lipolytica*: Safety assessment of an oleaginous yeast with a great industrial potential. Crit Rev Microbiol 40:187–206. https://doi.org/10.3109/1040841X.2013.770386

Guo H, Madzak C, Du G, Zhou J (2016) Mutagenesis of conserved active site residues of dihydrolipoamide succinyltransferase enhances the accumulation of alpha-ketoglutarate in *Yarrowia lipolytica*. Appl Microbiol Biotechnol 100:649–659. https://doi.org/10.1007/s00253-015-6995-1

Guo H, Su S, Madzak C, Zhou J, Chen H, Chen G (2016) Applying pathway engineering to enhance production of alpha-ketoglutarate in *Yarrowia lipolytica*. Appl Microbiol Biotechnol 100:9875–9884. https://doi.org/10.1007/s00253-016-7913-x

Holz M, Forster A, Mauersberger S, Barth G (2009) Aconitase overexpression changes the product ratio of citric acid production by *Yarrowia lipolytica*. Appl Microbiol Biotechnol 81:1087–1096. https://doi.org/10.1007/s00253-008-1725-6

Imandi SB, Bandaru VV, Somalanka SR, Bandaru SR, Garapati HR (2008) Application of statistical experimental designs for the optimization of medium constituents for the production of citric acid from pineapple waste. Bioresour Technol 99:4445–4450. https://doi.org/10.1016/j.biortech.2007.08.071

Janek T, Dobrowolski A, Biegalska A, Mirończuk AM (2017) Characterization of erythrose reductase from *Yarrowia lipolytica* and its influence on erythritol synthesis. Microb Cell Fact 16:118. https://doi.org/10.1186/s12934-017-0733-6

Jost B, Holz M, Aurich A, Barth G, Bley T, Muller RA (2014) The influence of oxygen limitation for the production of succinic acid with recombinant strains of *Yarrowia lipolytica*. Appl Microbiol Biotechnol 99:1675–1686. https://doi.org/10.1007/s00253-014-6252-z

Kamzolova SV, Allayarov RK, Lunina JN, Morgunov IG (2016) The effect of oxalic and itaconic acids on threo-Ds-isocitric acid production from rapeseed oil by *Yarrowia lipolytica*. Bioresour Technol 206:128–133. https://doi.org/10.1016/j.biortech.2016.01.092

Kamzolova SV, Chiglintseva MN, Lunina JN, Morgunov IG (2012) alpha-Ketoglutaric acid production by *Yarrowia lipolytica* and its regulation. Appl Microbiol Biotechnol 96:783–791. https://doi.org/10.1007/s00253-012-4222-x

Kamzolova SV, Dedyukhina EG, Samoilenko VA, Lunina JN, Puntus IF, Allayarov RL, Chiglintseva MN, Mironov AA, Morgunov IG (2013) Isocitric acid production from rapeseed oil by *Yarrowia lipolytica* yeast. Appl Microbiol Biotechnol 97:9133–9144. https://doi.org/10.1007/s00253-013-5182-5

Kamzolova SV, Morgunov IG (2017) Metabolic peculiarities of the citric acid overproduction from glucose in yeasts *Yarrowia lipolytica*. Bioresour Technol 243:433–440. https://doi.org/10.1016/j.biortech.2017.06.146

Kamzolova SV, Morgunov IG (2013) alpha-Ketoglutaric acid production from rapeseed oil by *Yarrowia lipolytica* yeast. Appl Microbiol Biotechnol 97:5517–5525. https://doi.org/10.1007/s00253-013-4772-6

Kamzolova SV, Shamin RV, Stepanova NN, Morgunov GI, Lunina JN, Allayarov RK, Samoilenko VA, Morgunov IG (2018) Fermentation conditions and media optimization for isocitric acid production from ethanol by *Yarrowia lipolytica*. BioMed Res Int 2018:2543210. https://doi.org/10.1155/2018/2543210

Kuenz A, Krull S (2018) Biotechnological production of itaconic acid-things you have to know. Appl Microbiol Biotechnol 102:3901–3914. https://doi.org/10.1007/s00253-018-8895-7

Laptev IA, Filimonova NA, Allayarov RK, Kamzolova SV, Samoilenko VA, Sineoky SP, Morgunov IG (2017) New recombinant strains of the yeast *Yarrowia lipolytica* with overexpression of the aconitate hydratase gene for the obtainment of isocitric acid from rapeseed oil. Appl Biochem Microbiol 52:699–704. https://doi.org/10.1134/s000368381607005x

Larroude M, Park Y-K, Soudier P, Kubiak M, Nicaud J-M, Rossignol T (2019) A modular golden gate toolkit for *Yarrowia lipolytica* synthetic biology. Microb Biotechnol 12:1249–1259. https://doi.org/10.1111/1751-7915.13427

Larroude M, Rossignol T, Nicaud J-M, Ledesma-Amaro R (2018) Synthetic biology tools for engineering *Yarrowia lipolytica*. Biotechnol Adv 36:2150–2164. https://doi.org/10.1016/j.biotechadv.2018.10.004

Larroude M, Trabelsi H, Nicaud J-M, Rossignol T (2020) A set of *Yarrowia lipolytica* CRISPR/Cas9 vectors for exploiting wild-type strain diversity. Biotechnol Lett 42:773–785. https://doi.org/10.1007/s10529-020-02805-4

Li C, Gao S, Li X, Yang X, Lin CSK (2018) Efficient metabolic evolution of engineered *Yarrowia lipolytica* for succinic acid production using a glucose-based medium in an in situ fibrous bioreactor under low-pH condition. Biotechnol Biofuels 11:236–248. https://doi.org/10.1186/s13068-018-1233-6

Li C, Gao S, Yang X, Lin CSK (2017) Green and sustainable succinic acid production from crude glycerol by engineered *Yarrowia lipolytica* via agricultural residue based in situ fibrous bed bioreactor. Bioresour Technol 249:612–619. https://doi.org/10.1016/j.biortech.2017.10.011

Li C, Ong KL, Yang X, Lin CSK (2019) Bio-refinery of waste streams for green and efficient succinic acid production by engineered *Yarrowia lipolytica* without pH control. Chem Eng J 371:804–812. https://doi.org/10.1016/j.cej.2019.04.092

Li C, Yang X, Gao S, Chuh AH, Lin CSK (2018) Hydrolysis of fruit and vegetable waste for efficient succinic acid production with engineered *Yarrowia lipolytica*. J Clean Prod 179:151–159. https://doi.org/10.1016/j.jclepro.2018.01.081

Li C, Yang X, Gao S, Wang H, Lin CSK (2016) High efficiency succinic acid production from glycerol via in situ fibrous bed bioreactor with an engineered *Yarrowia lipolytica*. Bioresour Technol 225:9–16. https://doi.org/10.1016/j.biortech.2016.11.016

Liu X, Lv J, Xu J, Zhang T, Deng Y, He J (2014) Citric acid production in *Yarrowia lipolytica* SWJ-1b yeast when grown on waste cooking oil. Appl Biochem Biotechnol 175:2347–2356. https://doi.org/10.1007/s12010-014-1430-0

Liu XY, Chi Z, Liu GL, Madzak C, Chi ZM (2013) Both decrease in *ACL1* gene expression and increase in ICL1 gene expression in marine-derived yeast *Yarrowia lipolytica* expressing *INU1* gene enhance citric acid production from inulin. Mar Biotechnol (NY) 15:26–36. https://doi.org/10.1007/s10126-012-9452-5

McKinlay JB, Vieille C, Zeikus JG (2007) Prospects for a bio-based succinate industry. Appl Microbiol Biotechnol 76:727–740. https://doi.org/10.1007/s00253-007-1057-y

Mirończuk AM, Biegalska A, Dobrowolski A (2017) Functional overexpression of genes involved in erythritol synthesis in the yeast *Yarrowia lipolytica*. Biotechnol Biofuels 10:77. https://doi.org/10.1186/s13068-017-0772-6

Mironczuk AM, Rzechonek DA, Biegalska A, Rakicka M, Dobrowolski A (2016) A novel strain of *Yarrowia lipolytica* as a platform for value-added product synthesis from glycerol. Biotechnol Biofuels 9:180–192. https://doi.org/10.1186/s13068-016-0593-z

Mirończuk AM, Rzechonek DA, Biegalska A, Rakicka M, Dobrowolski A (2016) A novel strain of *Yarrowia lipolytica* as a platform for value-added product synthesis from glycerol. Biotechnol Biofuels 9. https://doi.org/10.1186/s13068-016-0593-z

Morgunov IG, Kamzolova SV (2015) Physiologo-biochemical characteristics of citrate-producing yeast *Yarrowia lipolytica* grown on glycerol-containing waste of biodiesel industry. Appl Microbiol Biotechnol 99:6443–6450. https://doi.org/10.1007/s00253-015-6558-5

Morgunov IG, Kamzolova SV, Karpukhina OV, Bokieva SB, Inozemtsev AN (2019) Biosynthesis of isocitric acid in repeated-batch culture and testing of its stress-protective activity. Appl Microbiol Biotechnol 103:3549–3558. https://doi.org/10.1007/s00253-019-09729-8

Morgunov IG, Kamzolova SV, Lunina JN (2013) The citric acid production from raw glycerol by *Yarrowia lipolytica* yeast and its regulation. Appl Microbiol Biotechnol 97:7387–7397. https://doi.org/10.1007/s00253-013-5054-z

Niang PM, Arguelles-Arias A, Steels S, Denies O, Nicaud J-M, Fickers P (2020) In *Yarrowia lipolytica* erythritol catabolism ends with erythrose phosphate. Cell Biol Int 44:651–660. https://doi.org/10.1002/cbin.11265

Nicaud J-M (2012) Yarrowia lipolytica. Yeast 29:409–418. https://doi.org/10.1002/yea.2921

Okamoto S, Chin T, Hiratsuka K, Aso Y, Tanaka Y, Takahashi T, Ohara H (2014) Production of itaconic acid using metabolically engineered *Escherichia coli*. J Gen Appl Microbiol 60:191–197. https://doi.org/10.2323/jgam.60.191

Ong KL, Fickers P, Pang KP, Raffel Dharma P, Luk HS, Uisan K, Lin CSK (2020) Fermentation of fruit and vegetable wastes for biobased products. In: Food industry waste: assessment and recuperation of commodities, Kisseva M.R. and Webb C. Academic Press, pp 255–274

Ong KL, Li C, Li X, Zhang Y, Xu J, Lin CSK (2019) Co-fermentation of glucose and xylose from sugarcane bagasse into succinic acid by *Yarrowia lipolytica*. Bioch Eng J 148:108–115. https://doi.org/10.1016/j.bej.2019.05.004

Otten A, Brocker M, Bott M (2015) Metabolic engineering of *Corynebacterium glutamicum* for the production of itaconate. Metab Eng 30:156–165. https://doi.org/10.1016/j.ymben.2015.06.003

Otto C, Yovkova V, Barth G (2011) Overproduction and secretion of alpha-ketoglutaric acid by microorganisms. Appl Microbiol Biotechnol 92:689–695. https://doi.org/10.1007/s00253-011-3597-4

Papanikolaou S, Galiotou-Panayotou M, Fakas S, Komaitis M, Aggelis G (2008) Citric acid production by *Yarrowia lipolytica* cultivated on olive-mill wastewater-based media. Bioresour Technol 99:2419–2428. https://doi.org/10.1016/j.biortech.2007.05.005

Park Y-K, Vandermies M, Soudier P, Telek S, Thomas S, Nicaud J-M, Fickers P (2019) Efficient expression vectors and host strain for the production of recombinant proteins by *Yarrowia lipolytica* in process conditions. Microb Cell Fact 18:167. https://doi.org/10.1186/s12934-019-1218-6

Rakicka M, Biegalska A, Rymowicz W, Dobrowolski A, Mirończuk AM (2017) Polyol production from waste materials by genetically modified *Yarrowia lipolytica*. Bioresour Technol 243:393–399. https://doi.org/10.1016/j.biortech.2017.06.137

Rakicka M, Mirończuk AM, Tomaszewska-Hetman L, Rywińska A, Rymowicz W (2017b) An effective method of continuous production of erythritol from glycerol by *Yarrowia lipolytica* MK1. Food Technol Biotechnol 55:125–130. https://doi.org/10.17113/ftb.55.01.17.4812

Rakicka-Pustułka M, Mirończuk AM, Celińska E, Białas W, Rymowicz W (2020) Scale-up of the erythritol production technology—process simulation and techno-economic analysis. J Clean Prod 257:120533. https://doi.org/10.1016/j.jclepro.2020.120533

Regnat K, Mach RL, Mach-Aigner AR (2018) Erythritol as sweetener—wherefrom and whereto? Appl Microbiol Biotechnol 102:587–595. https://doi.org/10.1007/s00253-017-8654-1

Rice T, Zannini E, Arendt EK, Coffey A (2019) A review of polyols—biotechnological production, food applications, regulation, labeling and health effects. Crit Rev Food Sci Nutr 1–18. https://doi.org/10.1080/10408398.2019.1625859

Rymowicz W, Fatykhova AR, Kamzolova SV, Rywinska A, Morgunov IG (2010) Citric acid production from glycerol-containing waste of biodiesel industry by *Yarrowia lipolytica* in batch, repeated batch, and cell recycle regimes. Appl Microbiol Biotechnol 87:971–979. https://doi.org/10.1007/s00253-010-2561-z

Rzechonek DA, Dobrowolski A, Rymowicz W, Mironczuk AM (2018) Aseptic production of citric and isocitric acid from crude glycerol by genetically modified *Yarrowia lipolytica*. Bioresour Technol 271:340–344. https://doi.org/10.1016/j.biortech.2018.09.118

Rzechonek DA, Dobrowolski A, Rymowicz W, Mirończuk AM (2018) Recent advances in biological production of erythritol. Crit Rev Biotechnol 38:620–633. https://doi.org/10.1080/07388551.2017.1380598

Sarris D, Rapti A, Papafotis N, Koutinas AA, Papanikolaou S (2019) Production of added-value chemical compounds through bioconversions of olive-mill wastewaters blended with crude glycerol by a *Yarrowia lipolytica* strain. Molecules 24:222. https://doi.org/10.3390/molecules24020222

Sassi H, Delvigne F, Kar T, Nicaud J-M, Coq A-MC-L, Steels S, Fickers P (2016) Deciphering how *LIP2* and *POX2* promoters can optimally regulate recombinant protein production in the yeast *Yarrowia lipolytica*. Microb Cell Fact 15:159. https://doi.org/10.1186/s12934-016-0558-8

Sekova VY, Dergacheva DI, Isakova EP, Gessler NN, Tereshina VM, Deryabina YI (2019) Soluble sugar and lipid readjustments in the *Yarrowia lipolytica* yeast at various temperatures and pH. Metabolites 9:307. https://doi.org/10.3390/metabo9120307

Shabbir Hussain M, Gambill L, Smith S, Blenner MA (2016) Engineering promoter architecture in oleaginous yeast *Yarrowia lipolytica*. ACS Synth Biol 5:213–223. https://doi.org/10.1021/acssynbio.5b00100

Tan JP, Md. Jahim J, Wu TY, Harun S, Kim BH, Mohammad AW (2014) Insight into biomass as a renewable carbon source for the production of succinic acid and the factors affecting the metabolic flux toward higher succinate yield. Ind Eng Chem Res 53:16123–16134. https://doi.org/10.1021/ie502178j

Tan Z, Zhu X, Chen J, Li Q, Zhang X (2013) Activating phosphoenolpyruvate carboxylase and phosphoenolpyruvate carboxykinase in combination for improvement of succinate production. App Env Microbiol 79:4838–4844. https://doi.org/10.1128/AEM.00826-13

Tomaszewska L, Rakicka M, Rymowicz W, Rywińska A (2014) A comparative study on glycerol metabolism to erythritol and citric acid in *Yarrowia lipolytica* yeast cells. FEMS Yeast Res 14:966–976. https://doi.org/10.1111/1567-1364.12184

Tomaszewska L, Rywińska A, Gładkowski W (2012) Production of erythritol and mannitol by *Yarrowia lipolytica* yeast in media containing glycerol. J Ind Microbiol Biotechnol 39:1333–1343. https://doi.org/10.1007/s10295-012-1145-6

Trassaert M, Vandermies M, Carly F, Denies O, Thomas S, Fickers P, Nicaud J-M (2017) New inducible promoter for gene expression and synthetic biology in *Yarrowia lipolytica*. Microb Cell Fact 16:141. https://doi.org/10.1186/s12934-017-0755-0

Vandermies M, Denies O, Nicaud JM, celin P (2017) *EYK1* encoding erythrulose kinase as a catabolic selectable marker for genome editing in the non-conventional yeast *Yarrowia lipolytica*. J Microbiol Methods 139

Vandermies M, Fickers P (2019) Bioreactor-scale strategies for the production of recombinant protein in the yeast *Yarrowia lipolytica*. Microorganisms 7:40. https://doi.org/10.3390/microorganisms7020040

Wang N, Chi P, Zou Y, Xu Y, Xu S, Bilal M, Fickers P, Cheng H (2020) Metabolic engineering of *Yarrowia lipolytica* for thermoresistance and enhanced erythritol productivity. Biotechnol Biofuels 13:176. https://doi.org/10.1186/s13068-020-01815-8

Werpy T, Petersen G (2004) Top value-added chemicals from biomass: volume I—results of screening for potential candidates from sugars and synthesis gas United States: N. p., 2004. Web. https://doi.org/10.2172/15008859.

Wojtatowicz M, Rymowicz W, Kautola H (1991) Comparison of different strains of the yeast *Yarrowia lipolytica* for citric acid production from glucose hydrolysate. Appl Biochem Biotechnol 31:165–174

Wong L, Engel J, Jin E, Holdridge B, Xu P (2017) YaliBricks, a versatile genetic toolkit for streamlined and rapid pathway engineering in *Yarrowia lipolytica*. Metab Eng Commun 5:68–77. https://doi.org/10.1016/j.meteno.2017.09.001

Yang X, Wang H, Li C, Lin CSK (2017) Restoring of glucose metabolism of engineered *Yarrowia lipolytica* for succinic acid production via a simple and efficient adaptive evolution strategy. Agri Food Chem 65:4133–4139

Yin X, Madzak C, Du G, Zhou J, Chen J (2012) Enhanced alpha-ketoglutaric acid production in *Yarrowia lipolytica* WSH-Z06 by regulation of the pyruvate carboxylation pathway. Appl Microbiol Biotechnol 96:1527–1537. https://doi.org/10.1007/s00253-012-4192-z

Yovkova V, Otto C, Aurich A, Mauersberger S, Barth G (2013) Engineering the alpha-ketoglutarate overproduction from raw glycerol by overexpression of the genes encoding NADP+-dependent isocitrate dehydrogenase and pyruvate carboxylase in *Yarrowia lipolytica*. Appl Microbiol Biotechnol 98:2003–2013. https://doi.org/10.1007/s00253-013-5369-9

Yuzbashev TV, Yuzbasheva EY, Sobolevskaya TI, Laptev IA, Vybornaya TV, Larina AS, Matsui K, Fukui K, Sineoky SP (2010) Production of succinic acid at low pH by a recombinant strain of the aerobic yeast *Yarrowia lipolytica*. Biotechnol Bioeng 107:673–682. https://doi.org/10.1002/bit.22859

Yuzbasheva EY, Agrimi G, Yuzbashev TV, Scarcia P, Vinogradova EB, Palmieri L, Shutov AV, Kosikhina IM, Palmieri F, Sineoky SP (2019) The mitochondrial citrate carrier in *Yarrowia lipolytica*: Its identification, characterization and functional significance for the production of citric acid. Metab Eng 54:264–274. https://doi.org/10.1016/j.ymben.2019.05.002

Zhang M, Gu L, Cheng C, Ma J, Xin F, Liu J, Wu H, Jiang M (2018) Recent advances in microbial production of mannitol: utilization of low-cost substrates, strain development and regulation strategies. World J Microbiol Biotechnol 34:41. https://doi.org/10.1007/s11274-018-2425-8

Zhao C, Cui Z, Zhao X, Zhang J, Zhang L, Tian Y, Qi Q, Liu J (2019) Enhanced itaconic acid production in *Yarrowia lipolytica* via heterologous expression of a mitochondrial transporter MTT. Appl Microbiol Biotechnol 103:2181–2192. https://doi.org/10.1007/s00253-019-09627-z

Recent Advances in Synthetic Biology Applications of *Pichia* Species

Wan Sun, Yimeng Zuo, Zhanyi Yao, Jucan Gao, Zengyi Shao, and Jiazhang Lian

Abstract

The rapid development of metabolic engineering and synthetic biology has recently attracted great attention to exploring many non-model microbial species. *Pichia pastoris* and *Pichia kudriavzevii* are the two representative *Pichia* species, which possess uniquely desired biochemical, metabolic, and physiological features favoring large-scale industrial production. This review begins with illustrating synthetic biology parts and tools that have been developed for manipulating these two species, followed by summarizing their applications in biomanufacturing as hosts to produce recombinant proteins, bulk

W. Sun, Y. Zuo, Z. Yao and J. Gao contributed equally to this work.

W. Sun · Z. Shao (✉)
Interdepartmental Microbiology Program, Iowa State University, Ames, IA, USA
e-mail: zyshao@iastate.edu

W. Sun · Z. Yao · Z. Shao
NSF Engineering Research Center for Biorenewable Chemicals, Iowa State University, Ames, IA, USA

W. Sun · Z. Shao
The Ames Laboratory, Ames, IA, USA

Y. Zuo · J. Gao · J. Lian (✉)
College of Chemical and Biological Engineering, Zhejiang University, Hangzhou, Zhejiang, China
e-mail: jzlian@zju.edu.cn

Hangzhou Global Scientific and Technological Innovation Center, Zhejiang University, Hangzhou Zhejiang, China

Z. Yao · Z. Shao
Department of Chemical and Biological Engineering, Iowa State University, Ames, IA, USA

Z. Shao
4140 Biorenewables Research Laboratory, Iowa State University, Ames, IA, USA

F. Darvishi Harzevili (ed.), *Synthetic Biology of Yeasts*,
https://doi.org/10.1007/978-3-030-89680-5_10

chemicals, and natural products, and concludes on the challenges and potential strategies for advancing them for broader biotechnological applications.

1 Introduction

With an increasing demand for fuels and petrochemicals, in addition to the growing food shortage and global warming, researchers have been working on engineering microbes to generate desirable products (Choi et al. 2015; Curran and Alper 2012). Model organisms such as *Escherichia coli* and *Saccharomyces cerevisiae* have been studied extensively and engineered as cell factories for the production of value-added chemicals and proteins (Pontrelli et al. 2018; Lian et al. 2018). With the rapid development of next-generation sequencing and synthetic biology tools, non-conventional organisms with unique traits (e.g., thermotolerance (Varela et al. 2017), stress tolerance (Abdel-Mawgoud et al. 2018), special feedstock utilization (Agbogbo and Coward-Kelly 2008; Gong et al. 2015; Yaegashi et al. 2017; Yaguchi et al. 2017), and high protein secretion capacity (Li et al. 2007)) have been explored as well-suited hosts for industrial processes.

The most renowned *Pichia* species is *Pichia pastoris*, a methylotrophic yeast that can utilize methanol as the sole carbon source. It can grow to a very high cell density (>100 g/L dry cell weight) (Wang et al. 2012). The medium for growing *P. pastoris* is simple and inexpensive (containing only methanol or glycerol, biotin, salts, and trace elements), making it ideal for large-scale industrial production (Cereghino et al. 2002). In addition, *P. pastoris* can express heterologous proteins intracellularly or extracellularly at high levels whilst minimizing secretion of endogenous proteins. It possesses the machinery required for proper post-translational modifications (e.g., glycosylation, disulfide bond formation, and proteolytic processing) (Cereghino and Cregg 2000). As a "generally regarded as safe" (GRAS) yeast, it has been engineered to produce industrial enzymes and chemicals (Zhu 2019), therapeutic agents such as vaccines (Wang et al. 2016) and drugs (Yu et al. 2007), and protein-based polymers (Werten et al. 2019). In recent years, it has also been engineered to produce a leghemoglobin protein (LegH) from soy, to create a meat-like flavor in plant-based meat products (Fraser et al. 2018). The well-annotated genome sequence of *P. pastoris* (Love et al. 2016; Sturmberger et al. 2016) and a few genome-scale metabolic models are also available for engineering purposes (Ye et al. 2017; Saitua et al. 2017; Torres 2019).

This review change it to summarizes the current synthetic biology parts and tools available for *P. pastoris*, including promoters, terminators, plasmids, genome-editing tools, and signal peptides, followed by a discussion on the engineering efforts carried out to improve protein secretion and various applications of this yeast. In addition, we have also highlighted *Pichia kudriavzevii*, another *Pichia* species whose acid tolerance is attractive in the production of value-added organic acids such as succinic acid (Xiao et al. 2014), D-lactic acid (Park et al. 2018), and itaconic acid (Sun 2020). This review concludes with future perspectives in establishing *Pichia* species for biotechnological applications.

2 Synthetic Biology Parts and Tools

2.1 Promoters and Terminators

The strength and tunability of promoters are crucial for efficient production (Porro et al. 2005). Strong promoters are usually preferred because of the usually low expression levels of heterologous genes, whereas tunable promoters are desirable when multiple heterologous genes are involved in complex pathways. Commonly used promoters include inducible and constitutive promoters. The former allows genes of interest to be switched on or off at different stages through induction or repression via transcription factors. In practice, applying an inducible promoter can increase the cell density of cultures first and then initiate heterologous protein production. Such a separation of biomass accumulation and protein production provides a great advantage if the accumulated intermediates or products are toxic to the cells (Cereghino and Cregg 2000; Ahmad et al. 2014; Macauley-Patrick et al. 2005). However, using an inducible promoter requires an extra step during cultivation (e.g., carbon source swapping or compound supplementation), which incurs an additional cost for large-scale industrial production. In contrast, strong and steady expression of genes of interest mediated by constitutive promoters contributes to decreasing the operational cost while enhancing the yield, which is a commonly adopted strategy if the constitutive expression does not negatively affect cell growth (Vogl and Glieder 2013).

Inducible promoters are usually identified from unique biochemical pathways, and constitutive promoters generally originate from housekeeping genes. The two most commonly used promoters in *P. pastoris* are the inducible P_{AOX1} and the constitutive P_{GAP} (Rajamanickam et al. 2017). P_{AOX1} is the promoter of the alcohol oxidase gene *AOX1*, which is induced by the inexpensive carbon source, methanol. When methanol is used as a carbon source, the methanol-induced alcohol oxidase expression can reach 30% of the total soluble protein content (Cregg et al. 1993). It is one of the most effective promoters that have been found for protein expression in *P. pastoris* (Cregg et al. 1989; Jahic et al. 2006; Koutz et al. 1989). Heterologous protein synthesis at levels of approximately 20 g/L was achieved two decades ago (Hasslacher et al. 1997; Werten et al. 1999). Extensive efforts have been made to elucidate the regulatory mechanisms of P_{AOX1}. The determination of its *cis*-acting regulatory sequence elements has allowed researchers to engineer the promoter by means of deletion and duplication of putative transcription factor-binding sites, yielding a library of variants with strengths ranging from 6 to 160% of the wild-type P_{AOX1} (Hartner 2008). A dozen transcription factors have been identified to be involved in the induction of P_{AOX1}. This work provided various P_{AOX1} variants with tunable activities by combining *cis*-acting elements with the basal promoter identified based on deletion analysis.

P_{GAP} is the promoter of the glycolytic glyceraldehyde 3-phosphate dehydrogenase gene *GAP*. It remains constitutively expressed, although its activity varies when different carbon sources are used (Waterham et al. 1997). Similar to the engineering strategy applied to P_{GAP}, a library of P_{GAP} variants was constructed using mutagenesis, in which the activity varied from 0.6% to 1960% of the wild-type P_{GAP} (Qin et al. 2011). Several transcription factors have been examined and suggested to play roles in the regulation of P_{GAP}. Most of the available *P. pastoris* promoters are summarized in two review articles, with the strengths benchmarked to those of P_{AOX1} and P_{GAP} (Vogl and Glieder 2013; Turkanoglu Ozcelik et al. 2019). These promoters have been much less extensively studied but are highly desired for controlling multi-gene pathways. Most of their *cis*-acting regulatory sequences and mechanisms remain unclear. Beyond the ones in the two lists, there are also a strong native promoter (P_{CAT1}) from the catalase gene in *P. pastoris* and a strong heterogeneous one (P_{MOX}) originating from the methanol oxidase gene in *Hansenula polymorpha*. The former is induced by methanol with the P_{CAT1} variant P4 even stronger than P_{AOX1} (Nong et al. 2020); the latter was completely inactivated in the presence of xylose and sorbitol but showed strong activities in the glucose, glycerol, and methanol feeds (Mombeni 2020).

In conjunction to a promoter, a terminator (tt) also plays a critical role in regulating the expression level, mainly by influencing the mRNA stability and the subsequent translation process (Shalgi et al. 2005). However, compared with promoter engineering, much less attention has been focused on terminators. In *S. cerevisiae*, hundreds of terminators have been examined and characterized, which have been demonstrated to regulate protein expression levels over a broad range, whereas in *P. pastoris*, only a few recent studies have investigated the impact of terminators.

In *P. pastoris*, both endogenous and heterogeneous terminators have been used for heterologous expression. The commonly used endogenous terminators are mostly from either the methanol utilization pathway (i.e., *AOX1tt*) or housekeeping genes (i.e., *GAPtt*), whereas heterogeneous terminators are from other yeasts such as *S. cerevisiae* (e.g., *CYC1tt* from cytochrome C isoform 1, *PRM9tt* from pheromone-regulated membrane protein 9, and *VPS13tt* from vacuolar protein sorting-associated protein 13), *H. polymorpha* (e.g., *MOXtt* from methanol oxidase), and *Kluyveromyces lactis* (e.g., *LAC4tt* from beta-galactosidase). To date, the impact of different terminators on heterologous expression has been primarily examined by evaluating the expression level under the control of P_{AOX1} and P_{GAP}.

Vogl et al. examined 20 terminators, of which 15 originated from the endogenous methanol assimilation pathway and the remaining five from *S. cerevisiae*. These terminators provided comparable expression levels of green fluorescent protein (GFP) under the control of P_{AOX1}, with the lowest active terminator still reaching 57% of the highest active terminator (Vogl et al. 2016). Prielhofer et al. focused more on the terminators from highly expressed endogenous genes, including many ribosomal terminators. Using *CYC1tt* from *S. cerevisiae* as a reference, all ten terminators yielded similar GFP levels under the control of P_{GAP} (Prielhofer et al. 2017). Interestingly, in both studies, the heterogeneous terminators of genes of interest offered comparable and sometimes even higher activities than the endogenous ones, indicating that the terminators isolated from other yeasts could be effectively recognized in *P. pastoris*. Recently, Ito et al. created a catalog of 72 terminators, including 28 endogenous terminators, 41 heterogeneous terminators from *S. cerevisiae*, and three strong synthetic terminators developed originally for *S. cerevisiae*. Under the control of P_{GAP}, these terminators resulted in a 17-fold degree of tunability in *P. pastoris* (Ito et al. 2020). In these studies, *AOX1tt* seemed to yield the highest activity, regardless of the promoter being used (Karbalaei et al. 2020; Weninger et al. 2016).

A recent study using *Candida antarctica* lipase B (CALB) as a reporter protein showed that the activities of terminators are closely associated with those of promoters (Ramakrishnan et al. 2020). Ten terminators from the endogenous methanol utilization pathway, glycolysis, tricarboxylic acid (TCA) cycle, and other housekeeping genes and five terminators from *S. cerevisiae* were compared. Their activities were estimated by evaluating the corresponding CALB activity under the control of P_{AOX1} and P_{GAP}. Compared to *AOX1tt*, three terminators led to lower lipase activities when paired with P_{AOX1} but higher activities when paired with P_{GAP}, which suggested that the performance of terminators is not insulated from promoter influences and may also be subjected to regulatory mechanisms as seen in promoter studies. However, the mechanism by which individual terminators enhance expression along with different promoters remains unclear in *P. pastoris*. In addition, the terminator of dihydroxyacetone synthase (*DHAStt*) provided a slightly higher CALB expression level than *AOX1tt* under the control of P_{AOX1}, but nearly threefold higher activity under the control of P_{GAP}. Therefore, *DHAStt* can potentially serve as a strong terminator when seeking high heterologous expression in *P. pastoris*. In general, terminators play a critical role in protein expression, but more studies are needed to elucidate terminator-mediated regulatory mechanisms.

2.2 Episomal Plasmids and Integration Plasmids

Most of the protein expression and metabolic engineering tasks in *P. pastoris* were achieved through genome integration, many of which targeted the *AOX1* gene via homologous recombination (HR) (Cereghino and Cregg 2000). However, unlike *S. cerevisiae*, non-homologous end joining (NHEJ) is the dominant mechanism for repairing chromosomal double-stranded breaks (DSBs) in *P. pastoris*. Transformants with targeted integration have to be identified using a laborious screening process because of the uncontrollable random integration events, including large-scale relocation of an integration locus, off-target integration potentially affecting cell growth, and even co-integration of the DNA elements originating from the F-plasmid and the genome of the *E. coli* host used to prepare the shuttle plasmid (Schwarzhans et al. 2016).

An alternative strategy is to use replicative plasmids, which yield higher transformation efficiency and are easier to screen, although stability is sometimes an issue in this case (Lee et al. 2005). An autonomously replicating sequence (ARS) is a key element in replicative plasmids. The first *P. pastoris*-specific ARS, designated *PARS1*, was identified over 35 years ago and enabled the use of replicative plasmids with high transformation efficiency (Cregg et al. 1985). Over the past decade, other elements have been discovered that can serve as alternative ARSs in *P. pastoris*, such as a 452 bp *panARS* identified from *K. lactis* and a 1442 bp mitochondrial DNA fragment from *P. pastoris* itself (Liachko and Dunham 2014; Schwarzhans et al. 2017). Unfortunately, plasmids using these ARSs demonstrate poor stability during mitotic segregation, which is fatal for industrial applications. Recent studies suggest that this inherent instability is caused by the lack of a centromere (CEN), another genomic element that guides stable segregation of chromosomes, and therefore, can be used to improve plasmid stability during cell division (Cao et al. 2017a).

CENs are DNA sequences recognized by kinetochore complexes, which subsequently interact with spindle microtubules and enable equal partitioning of chromosomes to the two dividing cells during mitosis and meiosis. In *S. cerevisiae*, a 125 bp CEN and an ARS have been widely applied as a combination in all low-copy episomal plasmids. Like ARS, a CEN is also species-specific, and recent studies have identified four putative CENs corresponding to the four *P. pastoris* chromosomes (Sturmberger et al. 2016; Coughlan et al. 2016). In a more recent study, a new autonomously replicating plasmid was constructed, harboring an entire putative centromeric region from chromosome 2 (*Cen2*). The plasmid can be replicated and stably distributed in *P. pastoris*. Within this *Cen2*, a ~111 bp sequence was found to enable autonomous replication, which can serve as a new ARS (Nakamura, et al. 2018). Another study confirmed that the entire CENs from chromosomes 1 and 4 (*Cen1* and *Cen4*) could confer replicative stability to plasmids (Piva 2020) although *Cen4* did not support a high number of transformants in a separate study (Nakamura, et al. 2018). Although these new plasmids exhibit relatively high stability and have the potential to expedite cloning and high-throughput screening, harboring the entire *Cen* sequence may lead to other undesired outcomes. First, this kind of plasmid can only be maintained at low copy numbers due to their chromosome-like segregation mechanism, in contrast to those plasmids using *PARS1*. Moreover, the sizes of the plasmids will be much larger since all *Cen* sequences are above 6 kb in length, which does not benefit transformation, especially when large pathways are cloned. Lastly, having the entire centromeric sequence may cause unexpected genomic integration or DNA exchange with chromosomes, which increases the difficulty of screening.

Therefore, genome integration appears to be an alternative for heterologous gene and pathway expression before the functioning mechanisms of ARS and CEN are clearly elucidated and the sequences are optimized in *P. pastoris*. Genome integration is generally achieved via a single crossover or double crossover. A single crossover requires that the circular vector contains a sequence identical to the target locus in the *P. pastoris* genome. After transformation, the linearized vector, including genes of interest, a selectable marker, and backbone, will be inserted into the target site, leaving two copies of the target site flanking the inserted vector. The P_{AOX1} region and the auxotrophic gene *HIS4* have been widely selected as the target sites for single crossover-mediated integration, with an efficiency of 50–80% (Cereghino and Cregg 2000). However, this kind of integration will introduce elements beyond the expression cassette, such as the elements responsible for the replication of the shuttle vector in *E. coli*. In addition, a second single crossover may occur again on the genome between two identical target sequences, especially when the selection pressure is removed, which results in a loss of integrated expression cassette. To achieve a double crossover, a selection marker-containing expression cassette flanked by two homologous arms to the target site is usually transformed, resulting in the direct replacement of the genomic sequence located between the two homologous sequences by the expression cassette. Flanking the desired genes of interest and a selectable marker with the 5' and 3' *AOX1* sequences leads to the disruption of *AOX1*, thereby changing the phenotypic substrate utilization of *P. pastoris*. Lastly, the selection markers commonly used for screening include auxotrophic genes such as *HIS4*, *URA3*, and *ADE1*, as well as the genes encoding resistance to zeocin and G418 sulfate (Daly and Hearn 2005; Papakonstantinou et al. 2009).

2.3 Genome Editing and Integration Loci

In addition to integration vectors or fragments, the Clustered Regularly Interspaced Short Palindromic Repeats (CRISPR) technology has become a revolutionary tool for achieving genomic integration in *P. pastoris* because Cas9, in principle can target any accessible locus containing a 2–6 bp protospacer adjacent motif with the help of a guide RNA (gRNA) molecule. However, it has been reported that a Cas9/gRNA complex may lead to toxic effects due to off-targeting and therefore, the expression level of Cas9 needs to be at an appropriate level to yield a desired efficiency (Weninger et al. 2016). For example, using a weaker promoter for Cas9 expression can lead to higher transformation efficiencies and growth rates (Prielhofer et al. 2017). Cas9 cleaves a genome and introduces a DSB that can be repaired by the NHEJ machinery when donor DNA is not provided, resulting in insertion and deletion (indel) mutations at the target locus. In *P. pastoris*, the inherently strong NHEJ activity enables highly efficient multi-locus disruption or deletion with co-transformation of different gRNAs. For example, Weninger et al. successfully mutated *GUT1* and *AOX1* simultaneously by expressing two gRNAs on a CRISPR/Cas9 plasmid. By testing different combinations of gRNAs, the highest double-editing efficiency of 69% was obtained (Weninger et al. 2016). However, the active NHEJ machinery is a hurdle for achieving precise locus-specific integrations via HR. When a donor DNA is provided, HR must compete with the predominant NHEJ to repair the DSB, and under the selection pressure, the marker can be randomly integrated via NHEJ. The HR activity in the wild-type *P. pastoris* is naturally much lower than that of NHEJ, yielding many false positives (Li et al. 2007). Therefore, it is much more difficult to achieve precise integration in *P. pastoris* than in other yeast species with HR dominancy, such as *S. cerevisiae*. This limitation can be overcome by deleting the *KU70* and *KU80* genes encoding the two proteins comprising a heterodimer threading onto the broken DNA ends. In a recent study, it was reported that upon knock-out of *KU70*,

NHEJ was repressed to a great extent, thereby heightening the role of HR in repairing DSB repair and raising the target integration efficiency to approximately 100% (Weninger et al. 2016, 2018).

In addition to repressing NHEJ, future studies should focus on enhancing HR performance to increase editing efficiency when using CRISPR/Cas9 in *P. pastoris*. For example, the overexpression of RAD family recombinases can be considered. In *S. cerevisiae*, overexpression of Rad51 has been shown to elevate the integration correctness and overexpression of an engineered Rad51 variant, which has a higher affinity to recombinase Rad54, thus significantly increasing the targeting efficiency in *S. cerevisiae* (Liu et al. 2004). Furthermore, expressing Rad52 from *S. cerevisiae* in *Yarrowia lipolytica* has been shown to increase the targeting efficiency from 15 to 95% (Ji 2020). When *KU70* was disrupted, certain chemicals, such as hydroxyurea, were applied to synchronize *Y. lipolytica* cells to the S-phase of the cell cycle, which is when HR has the highest activity (Jang 2018). Similar strategies can be used to enhance the HR performance in *P. pastoris*.

Theoretically, any locus on a genome can serve as a potential target site for integration, except for those related to essential genes. However, several additional features directly affect the integration efficiency. First, a higher accessibility renders a higher chance for the Cas9/gRNA complex, and later, the donor DNA to form a complex with the genome during the cutting and repair process, which is especially important for the integration of large pathways. *AOX1*, *DAS1*, *DAS2*, and *GUT1* have been widely targeted owing to their easy accessibility (Pena et al. 2018). In addition, the nano-environment surrounding the integration locus is key to determining the expression level and dynamics of the integrated gene. Although a high expression level does not necessarily lead to high production, it is usually preferred by the rate-limiting step for the synthesis of a product through a multi-step pathway.

Increasing the copy number of a gene of interest has been found to be an efficient way to increase heterologous protein production. Accordingly, ribosomal RNA (rRNA)-encoding loci, namely ribosomal DNA (rDNA), have been populated because of their highly repeated sequences, where several loci can be targeted simultaneously using a single plasmid containing one gRNA design (Marx et al. 2009). In *P. pastoris*, each repeat consists of 25S, 5.8S, and 18S rRNA genes arranged identically in a head-to-tail tandem array and a non-transcribed spacer (NTS) is located between two rDNA repeats. The sequences and locations of these rDNA repeats on the chromosomes in many yeasts have been specified for decades. For example, Wang et al. successfully integrated ten copies of the resveratrol biosynthetic pathway consisting of genes from *Herpetosiphon aurantiacus*, *Arabidopsis thaliana*, and *Vitis vinifera*, onto the NTS regions of *O. polymorpha* simultaneously via CRISPR/Cas9, without using a selection marker (Wang et al. 2018). In *Y. lipolytica*, Luu et al. obtained an integrant with eight copies of the genes encoding capsid proteins originating from red-spotted grouper nervous necrosis virus at the 26S rRNA loci through HR (Luu et al. 2017). In *P. pastoris*, high copy-number integration of human serum albumin and human superoxide dismutase was achieved by targeting the NTS region of the rDNA locus and repeated selection on increasing the concentration of zeocine™ (Marx et al. 2009). In addition to expressing proteins as products, this strategy has also been applied to enhance the production of D-lactic acid in *P. pastoris* by integrating the D-lactate dehydrogenase gene (*D-LDH*) into the rDNA locus, followed by copy number amplification enabled by gradually increased antibiotic concentration (Yamada et al. 2019).

2.4 Commonly Used Strains

Strains Y-11430 (CBS-7435), GS115, and X-33 are the most commonly used *P. pastoris* strains. Y-11430 is a wild-type strain that was originally isolated from California Black oak and deposited in the United States Department of Agriculture Culture Collection (USDA-NRRL). It is renowned because of its robust growth rate and high activity in the methanol utilization pathway. GS115 is a histidine auxotroph obtained by mutagenizing Y-11430 with nitrosoguanidine. It has become popular for its ease of integration and screening while using *HIS4* as a selectable marker (Cregg et al. 1985; Schutter et al. 2009). X-33 is the revertant of GS115 created by the complementation of *HIS4,* and zeocin or blasticidin can still be used to select X-33 transformants carrying the antibiotic resistance gene (Higgins et al. 1998). X-33 shares a few mutations with GS115 in genes encoding cell wall biosynthesis, which enhance the secretion of membrane-associated proteins and result in a higher transformation efficiency than other species with thick cell walls, making X-33 a popular commercial strain for heterologous protein expression (Brady et al. 2020).

2.5 Signal Peptides for Mediating Protein Secretion

Protein secretion is a complex process that involves multiple steps to generate a mature active protein. A signal peptide is a short sequence that is usually located at the N-terminus of a nascent polypeptide and directs a protein into the secretory pathway (Owji et al. 2018). The most commonly used signal peptide for recombinant protein expression in *P. pastoris* is α-mating factor (α-MF) prepro peptide originating from *S. cerevisiae* (Fig. 1). The α-MF prepropeptide consists of a 19-amino acid presignal sequence and a 66-amino acid pro-sequence. The pre-peptide typically has three domains: a positively charged N-terminal region, a central hydrophobic region, and a polar C-terminal region. There are three main steps in processing an α-MF secretion signal. First, the presignal is cleaved by signal peptidases in the endoplasmic reticulum (ER). Kex2 endopeptidase in the Golgi then cleaves the pro-leader sequence at the dibasic KR site, and finally, Ste13 protein removes the EA repeats (Julius et al. 1984; Brake et al. 1984). α-MF has been used in *P. pastoris* to produce recombinant proteins such as endoglucanase III from *Trichoderma harzianum* (Generoso et al. 2012), human P53 protein

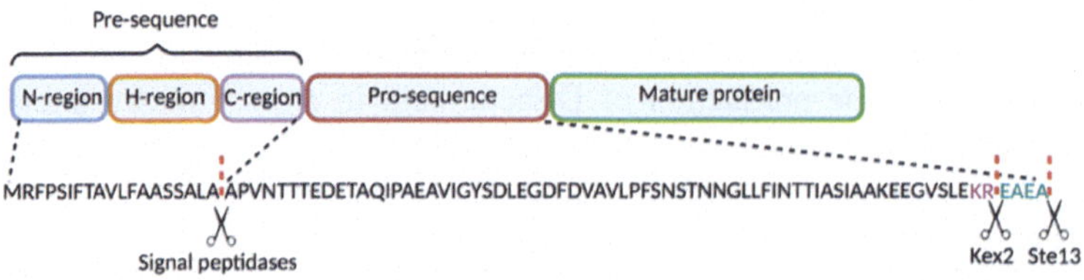

Fig. 1 General organization of the α-mating factor (α-MF) secretion signal originating from *S. cerevisiae* and applied in mediating protein secretion in *P. pastoris*. The pre-sequence consists of three parts and is cleaved by signal peptides in ER. The pro-sequence is cleaved at KR site by endoprotease Kex2 and the two EA repeats are removed by dipeptidyl aminopeptidase Ste13 in the Golgi

(Abdelmoula-Souissi et al. 2013), and manganese superoxide dismutase (PoMn-SOD) from *Pleurotus ostreatus* (Yin et al. 2014), etc. Many recombinant proteins have been successfully expressed in *P. pastoris* with their native signal peptides. For example, the activity of alkaline protease from *Aspergillus oryzae* with its native signal peptide is 1.5-fold higher than that with the α-MF secretion signal peptide (Guo and Ma 2008); the activity of the laccase from white-rot fungus *Polyporus grammocephalus* TR16 with its native signal peptide is threefold higher than that with the α-MF secretion signal peptide (Huang et al. 2011).

To enhance the efficiency of the α-MF signal sequence, various approaches such as codon optimization (Xiong et al. 2005; Ahn et al. 2016), error-prone PCR mutagenesis (Rakestraw et al. 2009), deletion mutagenesis (Aggarwal and Mishra 2020; Chahal et al. 2017; Lin-Cereghino et al. 2013), and synthetic signal peptides (Aza et al. 2021; Obst et al. 2017) have been applied. Upon codon optimization of α-MF, the phytase yield increased approximately sevenfold and CALB production increased by 132–295% compared to the version prior to codon optimization (Xiong et al. 2005; Ahn et al. 2016). Through error-prone PCR and library screening, one α-MF mutant could increase the secretion of a single-chain antibody 4m5.3 up to 16-fold, compared to the wild type. This improvement has also been found in the production of other single-chain antibody fragments and two structurally unrelated proteins, interleukin-2 (IL-2) and horseradish peroxidase (HRP) (Rakestraw et al. 2009). A recent study in *S. cerevisiae* resulted in an optimized α-MF named α_{OPT} with four mutations (Aα9D, Aα20T, Lα42S, and Dα83E), through a bottom-up (i.e., iterations of directed evolution on the native α-MF) and top-down strategy (i.e., examining the evolved signal peptide, namely α_{9H2} leader, and removing potential deleterious or neutral mutations) (Aza et al. 2021). The obtained α_{OPT} could increase the secretion of two laccases, PK2 and ApL, approximately 14- and 26-fold compared to α-MF, respectively. Combinatorial saturation mutagenesis at positions 86 and 87 of the α_{OPT} leader could further enhance laccase secretion. It is appealing to apply this α_{OPT} to protein expression in *P. pastoris*. In addition, deletion of amino acids 57–70 in the propeptide of α-MF enhanced the HRP activity by more than 50% and CALB activity approximately onefold compared to the wild-type α-MF signal sequence (Lin-Cereghino et al. 2013). In a separate study, this strategy led to an increased titer of granulocyte colony-stimulating factor (G-CSF) to 39.4 ± 1.4 mg/L (Aggarwal and Mishra 2020). Structural studies suggested that a specific orientation between both the N- and C-termini of α-MF pro-peptide is required to interact with secretion machinery and therefore facilitate protein secretion. Mutations generated near these termini usually impact secretion negatively, and changes within the interior of the pro-peptide could benefit secretion if these mutations can stabilize the N- and C-termini (Chahal et al. 2017). By combining established leader sequences and α-MF with deletions, Obst et al. designed several synthetic secretion signal peptides and characterized them with a red fluorescent protein (RFP) and yeast-enhanced green fluorescent protein (yEGFP) as reporters under different promoters. However, although these synthetic hybrid peptides yielded a more than tenfold variation in secretion efficiency, all except αMF_no_EAEA with certain promoters were less efficient than α-MF (Obst et al. 2017). The fusion of the *S. cerevisiae* Ost1 signal sequence and α-MF pro-region with two mutations could enhance the secretion of far-red fluorescent protein E2-Crimson by 20-fold and lipase BTL2 by tenfold (Barrero et al. 2018).

In addition to modifications in α-MF, putative secretory signal peptides can be determined by in silico analysis and further confirmed by experiments. Using five computer programs, SignalP4.1, Phobius, WolfPsort0.2, ProP1.0, and NetNGlyc1.0, Massahi and Calik were able to identify eight signal peptides from the sequences of 56 endogenous and exogenous proteins that had higher D-scores than that of *S. cerevisiae* α-MF (Massahi and Çalık 2015). Among the eight signal peptides, five with D-scores higher than 0.8 (SP13, SP23, SP24, SP26, and SP34) were selected for investigation of their efficiency in secreting recombinant human growth hormone. SP23 had the highest secretion efficiency, reaching 70%–80% of the efficiency of α-MF (Massahi and Çalık 2016). There are also eight commercially available signal peptides (The PichiaPink™ Secretion Signal Set) for protein expression in *P. pastoris*. Table 1 summarizes the major signal peptides reported in the literature.

2.6 Co-expression of Chaperones to Facilitate Protein Folding

Secretory proteins enter the ER by translocation in an unfolded state and then undergo chaperone-assisted folding for maturation into their native conformation. Only properly folded proteins are exported from the ER to the Golgi apparatus for further modifications before delivery to intra- or extracellular destinations (Idiris et al. 2010). The folding of secretory proteins is error-prone. When unfolded proteins accumulate in the ER, unfolded protein responses are triggered to decrease the amount of newly unfolded proteins from entering the ER and increasing ER folding capacity. If the ER is overburdened by misfolded proteins, cell apoptosis occurs (Yu et al. 2015; Hetz et al. 2020). Misfolded proteins can also be transported from the ER to the cytosol for ubiquitination and subsequently degraded by

Table 1 Signal peptides used for extracellular protein secretion in *P. pastoris*

Signal peptide	Origin	Target protein	References
α-MF	*S. cerevisiae* α-mating factor	Endoglucanase III (300 mg L^{-1})	Generoso et al. (2012)
W1	Synthetic signal peptide	β-galactosidase (440 mg L^{-1})	Cao et al. (2017b)
nsB	*Candida antarctica* lipase B	Lipase A (220 U mL^{-1}) and lipase B (383 U mL^{-1})	Vadhana et al. (2013)
PHA-E	*Phaseolus vulgaris* agglutinin E	Snowdrop lectin (GNA) and GFP	Raemaekers et al. (1999)
PHOI	*P. pastoris* acid phosphatase 1	Rabies virus glycoprotein (RABV-G)	Ben Azoun et al. (2016)
SUC2	*S. cerevisiae* sucrose invertase	Recombinant human antithrombin (rAT, 324 mg L^{-1})	Kuwae et al. (2005)
HFBII	*Trichoderma reesei* class II hydrophobin HFBII	Endoglucanase (EG27I, 47.7 mg L^{-1})	Su et al. (2017)
nsB-AP2	*Candida antarctica* lipase B + amphipathic peptide	Leech hyaluronidase (LHAase, 1.24×10^6U mL^{-1})	Kang et al. (2016)
SP23	*P. pastoris* protein disulfide isomerase	Human growth hormone (rhGH, 56 mg L^{-1})	Massahi and Çalık (2016)
SPIgG1	Murine IgG1	Anti-HIV antibody (VRC01, 3.05 μg mL^{-1})	Aw et al. (2017)
HSA	*P. pastoris* human serum albumin	Human lysozyme (hLM)	Xiong and Chen (2008)
cSIG	Chicken lysozyme signal peptide	Classical swine fever virus glycoprotein E2	Li (2020)
SP killer protein	*S. cerevisiae* killer protein	Phytase	Helian et al. (2020)
SP invertase	*S. cerevisiae* invertase	Phytase	Helian, et al. (2020)
SP α-amylase	*A. niger* α-amylase	Phytase	Helian, et al. (2020)
SP Inulinase	*Kluyveromyces maxianus* Inulinase	Phytase	Helian, et al. (2020)
SP Lysozyme	*Gallus gallus* lysozyme	Phytase	Helian, et al. (2020)
Scw	*P. pastoris* Scw11p	EGFP, CALB	Liang et al. (2013)
Dse	*P. pastoris* Dse4p	EGFP, CALB	Liang et al. (2013)

(continued)

Table 1 (continued)

Signal peptide	Origin	Target protein	References
Exg	*P. pastoris* Exg1p	EGFP, CALB	Liang et al. (2013)
MF4I	Codon-modified *Sc* α-mating factor	Phytase (6.1 g L^{-1})	Xiong et al. (2005)
HFBI	*Trichoderma reesei* class II hydrophobin HFBI	β-galactosidase	Cao et al. (2017b)
Apre	*S. cerevisiae*, preregion of α-factor	β-galactosidase	Cao et al. (2017b)
SP FAEC	*Talaromyces stipitatus* feruloyl esterase (FAEC)	FAEC (297 mg L^{-1})	Crepin et al. (2003)
SP Fae-1	*Neurospora crassa* Fae-1 feruloyl esterase	FAEC (260 mg L^{-1})	Crepin et al. (2003)
SP Bovine β-casein	Bovine β-casein	Xylanase (2,755.126 IU mL^{-1})	He et al. (2012)
Pptox	Viral K28 preprotoxin	GFP	Eiden-Plach et al. (2004)
SP pGKL	pGKL killer toxin	Mouse α-amylase (~240 mg L^{-1})	Kato et al. (2001)
SP DDDK	*PMT1*-gene-inactivated *P. pastoris* DDDK protein	Porcine carboxypeptidase B (CPB) and Erythrina trypsin inhibitor (ETI)	Govindappa et al. (2014)
Modified signal peptide (MSP)	Mouse salivary α-amylase (S8L) + pro-region of *S. cerevisiae* α-MF	Glucoamylase (GA, 12.619 IU mL^{-1})	Liu et al. (2005)

proteasome. This process is called ER-associated degradation (ERAD) (Römisch 2005). To enhance recombinant protein production in *P. pastoris*, endogenous or exogenous chaperones can be overexpressed to facilitate proper protein folding and secretion (Shen et al. 2012; Navone et al. 2021; Damasceno et al. 2007; Sallada et al. 2019; Jariyachawalid et al. 2012; Summpunn et al. 2018). There are two families of chaperones: molecular chaperones and chaperonins. Molecular chaperones bind to a short segment of substrate proteins and chaperonins form barrel-shaped folding chambers to sequester all or part of the unfolded proteins for proper folding (Evstigneeva et al. 2001).

Protein disulfide isomerase (Pdi) is a commonly used chaperone that is present in the ER lumen. It catalyzes both the formation and isomerization of disulfide bonds (i.e., changing an incorrectly bonded protein to a correct disulfide-bonded protein) and helps with the correct protein folding (Wilkinson and Gilbert 2004). Upon co-expression of Pdi with an IL-1 receptor antagonist and human serum albumin fusion protein (IH) that contains 18 disulfide bonds, there was a significant increase in the yield of IH, as compared to that from the strain expressing only the IH protein at a high copy (Shen et al. 2012). Another study with *E. coli* AppA phytase that contains an extra non-consecutive disulfide bond showed that co-expression of Pdi increased the phytase ApV1 thermostability, and consequently, the production by ~12-fold compared to the expression of ApV1 alone (Navone et al. 2021). Immunoglobulin binding protein (BiP) is another abundant chaperone protein that resides in the ER. Belonging to the heat shock protein Hsp70 family, it facilitates protein folding and plays an important role in the ERAD pathway. Co-expression of BiP with an A33 single-chain antibody fragment (A33scFv) in *P. pastoris* increased the ER folding capacity and resulted in an approximately threefold increase in A33scFv secretion (Damasceno et al. 2007). Co-expression of the chaperon gene *KAR2* with different copies of the gene encoding hydrophobin (HFBI) also resulted in increased HFBI secretion. The highest HFBI secretion with 3-copy *HFBI* was 22 ± 1.6-fold higher than that of the strain overexpressing only single-copy *HFBI* (Sallada et al. 2019).

Apart from molecular chaperones, chaperonins have also been engineered to facilitate protein production in *P. pastoris.* D-phenylglycine aminotransferase (D-PhgAT) from *Pseudomonas stutzeri* ST-201 is an intracellular protein that is difficult to express in the soluble active form. Jariyachawalid et al. overexpressed this enzyme *in P. pastoris* and found that most of the D-PhgAT protein was insoluble. By co-expressing *E. coli* chaperonins GroEL-GroES intracellularly with D-PhgAT, a considerable amount of soluble D-PhgAT was produced, and the activity also increased significantly. Compared to the D-*PhgAT* gene expressed alone, a 14,400-fold higher volumetric activity was achieved when ten copies of chaperonins were co-expressed (Jariyachawalid et al. 2012). In another study, GroEL-GroES residing in the ER was co-expressed with extracellular bacterial phytase or intracellular D-PhgAT in *P. pastoris.* The volumetric activity of extracellular phytase was 1.5–2.3-fold higher than that of phytase expression alone. However, the majority of the D-PhgAT protein was inactive and found in the insoluble protein fraction (Summpunn et al. 2018). These results suggested that the GroEL-GroES chaperone could potentially enhance the production of functional proteins in *P. pastoris* when they are present within the same compartment. Some of the major chaperones overexpressed in *P. pastoris* are summarized in Table 2.

2.7 Cell Surface Display

Cell surface display is a promising method for engineering functional proteins to be expressed on the cell surface through fusing with an anchor protein. Applications include, but are not limited to: whole-cell biocatalysts, bioadsorption and bioremediation, biosensor design, vaccine and antibody development, epitope mapping, library screening, protein engineering (Gai and Wittrup 2007; Kuroda and Ueda 2011; Tanaka et al. 2012; Ueda 2016; Andreu and Olmo 2018). The anchor protein can be fused with a target protein either at either the N-terminus or at the C-terminus (Tanaka et al. 2012). Both the fusion order and the linker between a target protein and an anchor protein can affect the display efficiency and functional properties (Ueda 2016). Commonly used anchor proteins in *P. pastoris* are Aga1 (Wang

Table 2 Commonly used chaperones to increase recombinant protein production in *P. pastoris*

Chaperone	Target protein	References
Glutathione peroxidase 1 (Gpx1)	RABV-G	Ben Azoun et al. (2016)
Protein disulfide Isomerase (Pdi1p)	HFBI, lipase, rhG-CSF, and recombinant *Bombyx mori* acetylcholinesterase 2 (rBmAChE2)	Sallada et al. (2019), Sha et al. (2013), Zhang et al. (2006), Li (2021)
ER-oxidoreductin 1 protein (Ero1p)	HFBI	Sallada et al. (2019)
CNE1p, FAD1p	Δ^9-tetrahydrocannabinolic acid synthase (THCAS)	Zirpel et al. (2018)
Bmh2, Sso2, Ssa4, VHb	*Rhizopus oryzae* lipase (ROL)	Jiao et al. (2018)
Sse1, Bfr2, Cup5, Kin2	Fab fragment of a monoclonal antibody fragment (2F5mAb)	Gasser et al. (2007)
Immunoglobulin binding protein (BiP/Kar2p)	A33scFv, HFBI	Damasceno et al. (2007), Sallada et al. (2019)
Ydj1p, Ssa1p, Sec63p	Lipase	Zhang et al. 2006)
HAC1	Chitinase, lipase	Jiao et al. (2018), Gasser et al. (2007), Song et al. (2020)
Sis1	Porcine growth hormone (pGH)	Deng et al. (2020)
GroEL-GroES	D-phenylglycine aminotransferase (D-PhgAT)	Jariyachawalid et al. (2012), Summpunn et al. (2018)

et al. 2007; Su et al. 2010a; Dong 2013), Sed1 (Su et al. 2010b; Li et al. 2015a), Tip1 (Jo et al. 2011), Aga2 (Jacobs et al. 2008), and Flo1 (Jiang et al. 2007), all of which are from *S. cerevisiae*, as well as, Pir1 (Khasa et al. 2011; Yang et al. 2017) and Pir2 (Khasa et al. 2011) from *P. pastoris*. In another study, 13 endogenous glycosylphosphatidylinositol-modified cell wall proteins were identified upon screening the genome of *P. pastoris* GS115 (Zhang et al. 2013), three of which were chosen as anchor proteins for displaying CALB (Wang et al. 2017). These three anchors (i.e., GCW21, GCW51, and GCW61) have also been applied to display bacterial PETase on the surface of *P. pastoris*, to degrade highly crystallized polyethylene terephthalate (PET). The turnover rate of the whole-cell biocatalyst displaying PETase was approximately 36-fold higher than that of the purified PETase (Chen 2020). Another anchor protein identified in *P. pastoris* is Flo9. The displayed lipase B with Flo9 showed higher thermostability at 45 °C and stability in organic solvents (Moura 2015).

P. pastoris X-33 has also been engineered to assemble protein complexes such as minicellulosomes on the cell surface (Ou and Cao 2014). The truncated CipA, which contains a cellulose-binding module and two cohesin modules from *Clostridium acetobutylicum*, was fused to the C-terminus of the anchor flocculation protein Flo1 from *S. cerevisiae*, whereas a *Nasutitermes takasagoensis* endoglucanase (NtEG) was fused with the dockerin. Fusion proteins were expressed separately in two *P. pastoris* X-33 strains, which were co-cultured for minicellulosome assembly. The surface displayed CipA and assembly of cohesin and dockerin were confirmed using immunofluorescence and western blotting. The hydrolysis efficiencies of NtEG for carboxymethyl cellulose (CMC), microcrystal cellulose (Avicel), and filter paper were enhanced by 1.4-fold, 2.0-fold, and 3.2-fold, respectively, when compared to free NtEG. Another study conducted by Dong et al. utilized an ultra-high-affinity IM7/CL7 protein pair for minicellulosome assembly (Dong et al. 2020). IM7 (including one, two, or three units) was fused to the N-terminus of the anchor protein SED1 from *S. cerevisiae* and expressed in *P. pastoris*. An endoglucanase (EG), an exoglucanase (CBH), a β-glucosidase (BGL), and a carbohydrate-binding module (CBM) from *Thermobifida fusca,* each fused with an N-terminal CL7 tag, were expressed individually in *E. coli*. The secreted proteins from *E. coli* cultures were assembled and displayed on the *P. pastoris* cell surface in vitro. The display system with two or three IM7 showed comparable or even higher efficiency for the hydrolysis of Avicel, phosphoric acid-swollen cellulose (PASC) and CMC, compared to free cellulases. The ethanol titer reached 5.1 g/L when three IM7 units were engaged in CMC fermentation.

In a recent study, Silva et al. displayed specific immunogenic epitopes of ZIKV envelope, NS1 protein, and both on the surface of *P. pastoris* GS115 by fusing these epitopes at the N-terminus of a partial Agα1 (C-terminal portion, nucleotides 970–1950) (Silva et al. 2021). The ability of the recombinant yeast to stimulate immune cells was evaluated in vitro using mouse immunological cells isolated from the spleen. *P. pastoris* displaying EnvNS1 epitopes showed better efficacy in producing IL-6, IL-10, and tumor necrosis factor-alpha (TNF-α) cytokines and an increase in lymphocytes $CD4^+$, $CD8^+$, and $CD16^+$, similar to ZIKV. These epitopes will be beneficial for the development of vaccines against ZIKV infection.

3 Applications of *P. pastoris* in Biomanufacturing

After more than 20 years of development, *P. pastoris* has become one of the most popular protein expression systems that is widely used in protein preparation, structural analysis, and functional characterization. As a GRAS microorganism approved by the United States Food and Drug Administration, thousands of proteins, including medicinal proteins (i.e., insulin, human serum albumin, hepatitis B surface antigen, and epidermal growth factor (Weinacker et al. 2013)) and industrial enzymes (i.e., mannanase, phytase, xylanase, and lipase (Rabert et al. 2013)) have been successfully expressed in *P. pastoris*. In addition, due to the development of pathway assembly and genome editing tools, a growing interest has been seen in establishing *P. pastoris* as a microbial cell factory to produce chemicals and natural products.

3.1 Recombinant Proteins

Recombinant proteins can be produced using bacterial, yeast, mold, insect, plant, and mammalian expression systems. *P. pastoris* is particularly attractive for the large-scale production of recombinant proteins. It naturally contains several highly expressed genes that encode methanol assimilation and dissimilation pathway enzymes, enabling growth with methanol as the sole carbon and energy source (Wegner and Harder 1987). Thus, high-level expression of target proteins can be readily achieved using these methanol-inducible promoters. When compared with the plant and mammalian expression systems, the *P. pastoris* expression system offers the advantages of low cost, fast growth, high cell density fermentation (HCDF), and consequently high expression levels. In contrast to prokaryotes, the biggest advantage of *P. pastoris* is the capability of post-translational modifications (e.g., O- and N-glycosylation and disulfide bond formation. When compared to *S. cerevisiae, P. pastoris* poses little concerns regarding over-glycosylation, and the secretory expression level of recombinant proteins is much higher (Karbalaei et al. 2020). Therefore, *P. pastoris* has been widely used to produce therapeutic glycoproteins. Moreover, *P. pastoris* can efficiently secrete the target proteins into the fermentation broth, making the downstream separation and purification process simpler, which is a paramount variable in designing viable industrial-scale processes.

3.1.1 Medicinal Proteins

Recombinant proteins represent a growing market in medical biotechnology. Many approved biopharmaceuticals are protein-based, such as monoclonal antibodies, growth factors, blood factors, hormones, interleukins, anticoagulants, interferons, and vaccines. Some of the representative medicinal proteins produced by *P. pastoris* are listed in Table 3.

Vaccines represent the largest class of recombinant medicinal proteins produced by *P. pastoris*. Vaccines can be divided into three types: inactivated vaccines, live attenuated vaccines, and recombinant subunit vaccines (Gasser et al. 2006). Recent studies have found that *P. pastoris* is preferred for producing recombinant subunit vaccines compared to other expression systems (Gasser et al. 2007). Compared to the non-glycosylated antigen, the mannose-glycosylated antigen produced by *P. pastoris* has enhanced antigen presentation and T cell activation. Enterovirus 71 (EV71) is the main pathogen that causes hand-foot-mouth disease in children. The establishment of a microbial system for large-scale and safe production of the EV71 vaccine has value in medicinal applications. Yang et al. cloned *P1* and *3C* genes of EV71 and established a microbial system for efficient production of recombinant EV71-VLP (virus-like particle) in *P. pastoris*. Expression levels of as high as 270 mg/L EV71-VLPs antigen have been achieved (Yang et al. 2020). Another example is related to cervical cancer, the fourth most common cancer that threatens the health of women worldwide. Although there is already a market-oriented vaccine, its high price limits its wide application. Recently, Sanchooli et al. inserted the cross-neutralizing epitope of L2-HPV-16 into L1-HPV-16 to form

Table 3 Representative medicinal proteins expressed in *P. pastoris*

Product	Strain	Vector	Titer (mg/L)	Application	References
Bovine IFN-α	GS115	pPIC9K	200	Prevention and therapy of viral diseases	Tu et al. (2016)
hPAB-β	GS115	pPIC9K	241	Antibacterial peptide	Chen et al. (2011)
Bovine lactoferrin	KM71H	pJ902	3500	Transferrin and antibacterial protein	Iglesias-Figueroa, et al. (2016)
Human topoisomerase I	GS115 SMD1168H	pPICZα	11	Antitumor	Chan et al. (2018)
Insulin	SuperMan5	pPICZα	5	Diabetes	Baeshen et al. (2016)
Human gastric lipase	X-33	pGAPZα	7	Diseases of the digestive system	Sams et al. (2017)
Rabies virus glycoprotein	GS115	pHIL-S1	1.23	Rabies vaccine	Ben Azoun et al. (2016)
L1-L2 proteins of HPV virus type 16	GS115	pBLHIS-IX	23.61	Cervical cancer	Bredell et al. (2018)
Human sialyltransferase	KM71H	pPICZαB	130	Pharmacological uses	Luley-Goedl et al. (2016)
Human antiplatelet scFv antibody	X-33	pPICZαA	600	Treatment of atherosclerosis	Vallet-Courbin (2017)
IL-1β	GS115 SMD1168 X-33	pPICZαA	250	Proinflammatory cytokine	Li et al. (2016a)
IL-3	X-33	pPICZαA	2230	Multipotent hematopoietic cytokine	Dagar and Khasa (2018)
IL-11	GS115	pPINKαHC	300	Thrombopoietic growth factor	Yu et al. (2018)
IL-15	X-33	pPICZαA	75	Differentiation and proliferation of T, B, and NK cells	Sun et al. (2016)

an L1/L2-HPV-16 chimeric fragment, which was cloned into the pPICZA plasmid for heterologous expression. The chimeric protein could be positively detected by both L1-HPV-16 and L2-HPV-16 antibodies (Sanchooli et al. 2018). Meanwhile, when Bredell et al. expressed the HPV-16L1/L2 chimeric protein (VLP) in *P. pastoris* KM71 (Mut^S) or GS115 (Mut^+) under a constant dissolved oxygen level

(DO stat) fed-batch culture supplemented with methanol, they achieved a titer of 23.61 mg/L for the chimeric protein (Bredell et al. 2018).

Although a panel of medicinal proteins has been expressed in *P. pastoris*, correct folding of target proteins is a major concern. Due to the lack of sufficient chaperone factors, a considerable number of recombinant proteins cannot fold into their correct configurations. To overcome this limitation and improve the expression level, rational design and reverse engineering strategies can be adopted to improve the protein folding microenvironment. As mentioned above, genes related to protein folding in the ER, such as BiP (an Hsp70 chaperone), can help secretory proteins fold correctly. The secretion of A33scFv fragment was increased approximately threefold upon overexpression of the chaperone protein BiP in *P. pastoris* (Damasceno et al. 2007). Overexpression of Pdi (responsible for the formation of disulfide bonds) in *P. pastoris* can increase the expression of the antibody protein 2F5Fab by 2-fold (Gasser et al. 2006). Zhang et al. tested three factors related to protein transport from *S. cerevisiae* (Sec63p, Ydj1p, and Ssa1p), whose overexpression increased the expression of GCSF by 2.8-, 3.6-, and 6.8-fold, respectively, in *P. pastoris*. Therefore, finding suitable protein mates remains a major challenge in establishing an efficient recombinant protein expression system (Zhang et al. 2006). Gasser et al. identified new protein chaperone genes, including *CUP5*, *SSA4*, *BMH2*, *KIN2*, *SSE1*, and *BFR2* at the transcriptional level. The overexpression of these genes significantly enhanced the secretion of 2F5Fab antibody in *P. pastoris,* with the final titer of 2F5Fab reaching up to 47.27 mg/L (Gasser et al. 2007). Stadlmayr et al. established a cDNA overexpression library in *P. pastoris* and identified three new protein chaperones as the secretion-enhancing factors, which increased the expression level of the model protein by up to 45% (Stadlmayr et al. 2010). Huang et al. identified six significantly upregulated genes related to recombinant protein production using comparative proteomic analysis. In particular, the co-expression of *TPX*, *FBA*, and *PGAM* increased the expression level of the reporter gene by 2.46-, 1.58-, and 1.33-fold, respectively (Huangfu et al. 2015). Noteworthy, owing to the different properties of foreign proteins, there is a dearth of generally applicable engineering approaches. In other words, the optimal protein chaperone or secretory factors can be different case by case and should be evaluated individually. For example, overexpression of Pdi in *P. pastoris* failed to increase the production of A33scFv antibody protein and overexpression of BiP even reduced the yield of glucose oxidase by 10-fold (Heide et al. 2002).

3.1.2 Industrial Enzymes

In recent years, industrial enzymes have been increasingly leveraged in the chemical, food and beverage, pharmaceutical, cosmetic, and textile industries. Owing to the increasing demand for industrial enzymes, the development of production strategies has accelerated. In this regard, the effectiveness of *P. pastoris* as a host for high-level expression of recombinant proteins has attracted increasing attention, because of the presence of strong methanol-inducible promoters (more than 30% of the total proteins) and HCDF (higher than 200 g/L biomass). In addition, *P. pastoris* has a strong ability to secrete target proteins into the fermentation medium, facilitating downstream purification at a much lower cost. Therefore, *P. pastoris* has been regarded as a favorable host for large-scale production of industrial enzymes. Representative industrial enzymes produced in *P. pastoris* using HCDF are listed in Table 4.

As a type of hydrolase, lipase demonstrates high regioselectivity and stereoselectivity and can catalyze ester hydrolysis and transesterification. Therefore, lipases are widely used in food, cosmetics, and pharmaceuticals. Zheng et al. cloned

Table 4 Representative industrial enzymes expressed in *P. pastoris*

Product	Strain	Vector	Scale (L)	Titer (g/L)	Application	References
β-Mannanase	GS115	pPIC9K	5	0.17	Production of manno-oligosaccharides	Li et al. (2018)
Invertase	KM71	pPIC9	7	2.5	Producing inverted sugar	Veana et al. (2014)
Hyaluronidase	GS115	pPIC9K	3	0.42	Degradation of hyaluronic acid	Jin et al. (2014)
Xylanase	X-33	pPICZαA	50	9.88	Bakery, beverages, starch processing	Li et al. (2015b)
Pectinase	GS115	pAO815	–	0.35	Brewing, papermaking	Peng et al. (2019)
Glucose oxidase	GS115	pPIC9	40	0.83	Bakery, eggs, starch processing	Meng et al. (2014)
Formate dehydrogenase	X-33	pPICZαA	3	0.05	Cofactor regeneration, CO_2 reduction	Duman (2020)
Peroxidase	DSMZ 70,382	pUC57	14	0.94	Diagnostics, bioremediation, lignin degradation	Majeke (2020)
L-Glutamate oxidase	GS115	pHBM905A	20	11	Transformation of L-glutamic acid, industrial fermentation, biosensors	YaPing et al. (2017)
Laccase	X-33	pGAPZA	5	0.495	Dye decolorization, beverages	Kittl et al. (2012)
Hydrolases	X-33	pPICZαA	5	0.61	Triacylglycerol lipase	Zheng et al. (2019)
Lipase B	SMD1168	–	10	2	Resolution of chiral compounds	Jahic et al. (2003)
Tannin acyl hydrolase	KM71	pPIC9K	7	0.074	Detannification of food	Zhong et al. (2004)
Phytase	GS115	pPICZαA	10	9.58	Animal feeding	Li et al. (2015c)

the *Aspergillus oryzae* lipase gene (*AOL*) to yield the plasmid pPICZαA-AOL, which was subsequently integrated into the genome of *P. pastoris* X-33. Using the methanol feeding strategy, AOL with a specific activity of 432 U/mg was obtained in a 5 L bioreactor (Zheng et al. 2019). To increase the yield of lipase, Zhang et al. employed a fusion expression strategy by fusing small ubiquitin modifying protein (SUMO) with *Aspergillus niger* lipase (ANL) to obtain SANL. The resultant chimeric gene was cloned into pPIC9K for heterologous expression in *P. pastoris* GS115. The highest activity of SANL was ~960 U/mL in a 3 L fermenter, which was 1.85-fold higher than that of its parent ANL (Zhang et al. 2019a).

Although the expression of recombinant proteins is mainly induced by methanol, a co-substrate culture strategy has been found to increase the yield of industrial enzymes. Berrios et al. cloned the *Rhizopus oryzae* lipase gene (*ROL*) and constructed the plasmid pPICZαA-ROL for heterologous expression in *P. pastoris* X-33. The engineered strain was continuously cultured with methanol and glycerol as co-substrates in a 1.5 L BioStatAplus bioreactor. The results showed that using glycerol as a co-substrate at 22 and 30 °C could increase the volumetric productivity of recombinant lipase and reduce the consumption of methanol (Berrios et al. 2017). In addition, the co-substrate culture could also be applied for the industrial production of phytase. As an animal feed additive, phytase can decompose phytic acid and greatly reduce the input of animal feed. Li et al. engineered phytase production in *P. pastoris* by modifying P_{AOX1} and the α factor signal peptide and increasing gene copy numbers. Phytase activity as high as 2,119 U/mL with a corresponding titer of 0.75 g/L was achieved, which was 4.12-fold higher than that of the parent strain. In a 10 L fermenter, using glycerol and methanol as co-substrates for fed-batch fermentation, the titer and enzyme activity of phytase could be further improved to as high as 9.58 g/L and 35,032 U/mL, respectively (Li et al. 2015c).

Besides the traditional strategies in engineering secretion signals and modifying P_{AOX1} promoter, genome-scale metabolic models can be employed to regulate the metabolic fluxes from a systems perspective, to improve the expression level of recombinant proteins. Saitua et al. employed the dynamic flux balance analysis (dFBA) framework to establish a dynamic genome-scale metabolic model, to simulate recombinant protein expression process in *P. pastoris* (Saitua et al. 2017). Starting with seven state variables including glucose, biomass, and fermentation quantity, they analyzed the kinetics of substrate assumption and distribution of metabolic flux. On this basis, Nocon et al. optimized the dFBA algorithm and predicted gene targets (including both gain- and loss-of-function targets), to enhance the production of recombinant proteins. Overexpression targets were identified to reside in the pentose phosphate pathway and the TCA cycle, whereas knockout targets were found to belong to several branch points of glycolysis (Fig. 2). Five out of the nine predicted targets were found to increase the expression level of cytosolic human superoxide dismutase (hSOD). More importantly, most of the same genetic modifications led to enhanced expression of bacterial β-glucuronidase, indicating the general applicability of the identified metabolic engineering targets (Nocon et al. 2014).

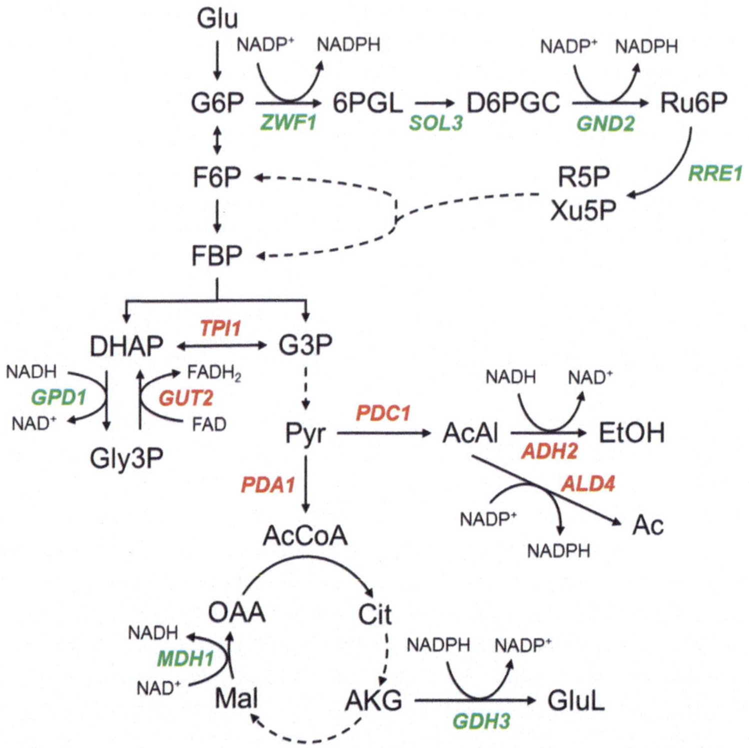

Fig. 2 Implementation of a genome-scale metabolic model to predict gene overexpression and knockout targets of the central metabolism for increased production of recombinant proteins in *P. pastoris* (Nocon et al. 2014). Overexpressed genes are shown in green and deleted genes are shown in red. *ZWF1*: glucose-6-phosphate dehydrogenase; *SOL3*: 6-phosphogluconolactonase; *GND2*: phosphogluconate dehydrogenase; *MDH1*: malate dehydrogenase; *TPI1*: triose-phosphate isomerase; *ADH2*: alcohol dehydrogenase; *ALD4*: aldehyde dehydrogenase; *PDA1*: pyruvate dehydrogenase; *PDC1*: pyruvate decarboxylase; *RPE1*: ribulose 5-phosphate 3-epimerase; *GPD1*: glycerol-3-phosphate dehydrogenase; *GUT2*: glycerol-3-phosphate dehydrogenase; *GDH3*: glutamate dehydrogenase

3.2 Bulk Chemicals

With the rise of synthetic biology, yeast has become an important cell factory to produce fine chemicals. Currently, *S. cerevisiae* is the preferred host to produce a wide range of chemicals, including, but not limited to: glycerol, L-propanediol, lactic acid, succinic acid, and isoprene. Comparative metabolomics indicated that the intermediate metabolites in *P. pastoris* could cover more than 90% of those in *S. cerevisiae*, indicating great potential for chemical production in *P. pastoris* (Carnicer et al. 2012). Currently, the production of S-adenosyl-L-methionine (Chu et al. 2013), xylitol (Louie et al. 2021), hyaluronic acid (Oliveira et al. 2016), gluconic acid (Liu et al. 2016), and lactic acid (Lima et al. 2016) has been reported, confirming the possibility of producing simple and complex chemicals in *P. pastoris*.

To achieve efficient and cost-effective production of chemicals, a combination of metabolic engineering modification modifications and bioprocess optimization is generally employed. Cheng et al. constructed the glucose-D-arabitol-D-xylulose-xylitol pathway to produce xylitol from glucose for the first time. The D-arabitol dehydrogenase gene from *Klebsiella pneumoniae* and the xylitol dehydrogenase gene from *Gluconobacter oxydans* were cloned into pPIC9K for subsequent integration into the genome of the GS225 strain, a derivative strain of *P. pastoris* GS115 strain after adaptive evolution. The yield of xylitol was 0.078 g/g glucose when it was fermented in a 3 L fermenter (Cheng et al. 2014).

Mixed carbon source fermentation may be more beneficial for promoting product production, and this strategy has been used in the production of a variety of chemicals, such as glucaric acid. As an organic acid, glucaric acid is considered to be one of "the most valuable chemicals from biomass" and plays an important role in the synthesis of many biodegradable substances. Much attention has been paid to the production of glucaric acid using a microbial cell factory. Liu et al. overexpressed the inositol oxygenase gene (*MIOX*) and urinate dehydrogenase gene (*UDH*) from *Pseudomonas aeruginosa* KT2440 for glucaric acid production in *P. pastoris*. As MIOX was determined to be rate-limiting, fusion expression of *MIOX* with *UDH* with a flexible linker was employed to improve the conversion efficiency. With glucose and myo-inositol as the co-substrates in fed-batch fermentation, the engineered *P. pastoris* strain was able to produce glucaric acid at a titer as high as 6.61 ± 0.30 g/L (Liu et al. 2016).

2-Phenylethanol (2-PE) is widely used in cosmetics and high-end perfumes because of its rose flavor. In a recent study, Kong et al. overexpressed the 2-ketoacid decarboxylase gene (*ARO10*), aldehyde reductase gene (*ADH6*), and aromatic aminotransferase gene (*ARO8*) from *S. cerevisiae*, together with feedback inhibition-insensitive mutant genes, 3-deoxy-D-arabino-heptulosonate-7-phosphate synthase ($aroG^{fbr}$) and chorismate mutase/prephenate dehydratase ($pheA^{fbr}$) from *E. coli*, under the control of the strong constitutive promoter, P_{GAP}, and achieved de novo biosynthesis of 2-PE from glucose for the first time. Using shake flask fermentation for 36 h, 1,169 mg/L of 2-PE was found to accumulate in the engineered *P. pastoris* strain (Kong 2020). Notably, the titer of 2-PE synthesized by *P. pastoris* was higher than those achieved in *E. coli* and *S. cerevisiae*, indicating that *P. pastoris* has excellent potential as a host strain to produce chemicals.

As mentioned earlier, although *P. pastoris* has been used to synthesize various compounds, most of the engineered strains still use glucose and other fermentable sugars as substrates (Pena et al. 2018). While it is often added as an inducer, methanol has rarely been used as a substrate. Owing to the increasing concerns about sustainability and the abundant availability of methanol, the use of methanol as a substrate to produce chemicals has become a research hotspot. Cai et al. used methanol as the sole carbon and energy source to produce Monachine J and lovastatin in metabolically engineered *P. pastoris,* with titers reaching 60.0 mg/L and 14.4 mg/L, respectively. After bioprocess optimization, including the employment of a co-culture strategy and fed-batch fermentation with glycerol as the co-substrate, the yield of Monachine J and lovastatin reached 594 mg/L and 251 mg/L, respectively. In terms of methanol conversion, Yamada et al. integrated *D-LDH* into the *rDNA* loci of *P. pastoris.* Multicopy integration of the *D-LDH* expression cassette was achieved following post-transformational gene amplification and selection on gradually increasing zeocine concentrations. The optimally engineered *P. pastoris* strain produced D-lactic acid with a titer of 3.48 g/L by means of test-tube fermentation for 96 h, with methanol as the sole carbon source. This is the first report on the establishment of *P. pastoris* as a microbial cell factory for the conversion of methanol to lactic acid (Yamada et al. 2019). This study provides a basis for the application of gene integration strategy at the *rDNA* loci in *P. pastoris* for the construction of microbial cell factories for the production of value-added chemicals. Although research on the synthesis of chemicals from methanol is still in its infancy and the titer is still not high enough for industrial production, the challenges in methanol conversion should be able to be addressed using metabolic engineering and synthetic biology approaches in the future. *P. pastoris* is believed to play an increasingly important role in biorefinery and biomanufacturing in the near future.

3.3 Natural Products

Key enzymes of secondary metabolite (natural product) biosynthetic pathways are often found to have low expression levels and/or limited enzymatic activities, which becomes the bottleneck for efficient biosynthesis of high-value natural products. Considering the advantages of high-level expression and post-translational modifications of complex eukaryotic proteins (i.e., cytochrome P450s, CYPs), *P. pastoris* is a promising microbial cell factory for the synthesis of complex biologically active molecules. Currently, natural products synthesized in *P. pastoris* mainly include terpenoids, polyketides, and flavonoids (Table 5).

Terpenoids are hydrocarbons that are widely found in plants and microorganisms. Many terpenoids have important physiological activities and therefore are important research targets for the development of new drugs. The biosynthetic pathways of a series of terpenoids, such as lycopene, carotene, astaxanthin, (+)-nootkatone, and dammarenediol-II, have been successfully constructed in *P. pastoris.* (+)-Nootkatone is a sesquiterpene compound of great commercial value, with a grapefruit aroma and various biological activities. Although the existing chemical synthesis technology of (+)-nootkatone can meet industrial and commercial needs, heavy metals and flammable compounds are involved in the synthesis process. Therefore, the synthesis of (+)-nootkatone using biological methods is a new trend in the near future (Fig. 3). Through co-expression of the premnaspirodiene oxygenase (HPO) from *Hyoscyamus muticus* and the cytochrome P450 reductase from *A. thaliana*, *trans*-nootkatol was produced by hydroxylation of (+)-valencene, and *trans*-nootkatol was further oxidized to (+)-nootkatone by

Table 5 Production of natural products in *P. pastoris* cell factories

Nature product	Host	Genes	Promoter	Yield	References
Lycopene	X-33	*crtE*, *crtB*, *crtI*	P_{GAP}	1.141 μg/g DCW	Araya-Garay et al. (2012a)
β-Carotene	X-33	*crtE*, *crtB*, *crtI*, *crtL*	P_{GAP}	339 μg/g DCW	Araya-Garay et al. (2012a)
Astaxanthin	X-33	*crtE*, *crtB*, *crtI*, *crtL*, *crtW*, *crtZ*	P_{GAP}	3.7 μg/g DCW	Araya-Garay et al. (2012b)
Nootkatone	CBS7435 *his4 ku70*	*CPR*, *HPO*, *ADH*, truncated *HMG1*	P_{AOX1}	17 mg/L 208 mg/L (fed-batch)	Wriessnegger et al. (2014)
Dammarenediol-II	GS115	*ERG1*, *ERG7*	P_{AOX1}	1.073 mg/L DCW	Liu et al. (2015)
		Repression	P_{THI11}		
		PgDDS	P_{AOX1}		
Menaquinone-4	GS115	*SaGGPPS*	P_{GAP}	0.24 mg/g DCW	Sun et al. (2019)
		HsUBIAD1	P_{AOX1}		
		IDI	P_{GAP}		
6-MSA	GS115	*NpgA*, *ATX*	P_{AOX1}	2.2 g/L (fed-batch)	Gao et al. (2013)
Citrinin	GS115	*pksCT*, *MPL1*, *MPL2*	P_{AOX1}	0.6 ± 0.1 mg/L	Xue et al. (2017)
		MPL4, *MPL6*			
		MPR7			
Lovastatin	GS115	*LovA*, *LovB*	P_{AOX1}	24.6 mg/L 250.8 mg/L (fed-batch)	Liu et al. (2018)
		LovC, *LovD*			
		LovF, *LovG*			

(continued)

Table 5 (continued)

Nature product	Host	Genes	Promoter	Yield	References
		NpgA			
17-Hydroxyprogesterone	GS115	Human CYP17 cDNA	P_{AOX1}	Not reported	Kolar et al. (2007)

Note CrtE: geranylgeranyl diphosphate synthase; *CrtB*: phytoene synthase; *CrtI*: phytoene desaturase; *CrtL*: lycopene β-cyclase; *CrtW*: xanthophylls; *CrtZ*: β-carotene hydroxylase; *CPR*: *A. thaliana* cytochrome P450 reductase; *HPO*: premnaspirodiene oxygenase of *H. muticus*; *ADH*: alcohol dehydrogenase; truncated *HMG1*: truncated hydroxy-methylglutaryl-CoA reductase; *ERG1*: 2,3-oxidosqualene synthase; *ERG7*: responsible for conversion of 2,3-oxidosqualene to lanosterol; *PgDDS*: dammarenediol-II synthase from *Panax ginseng*; *SaGGPPS*: geranylgeranyl diphosphate synthase from the archaebacterium *Sulfolobus acidocaldarius*; *HsUBIAD1*: a human homologue of *E. coli* prenyltransferase menA and mammalian mitochondrial prenyltransferase COQ_2; *IDI*: IPP isomerase; *NpgA*: *A. nidulans* phosphopantetheinyl transferase; *ATX*: *Aspergillus terrus* 6-methylsalicylic acid synthase; *pksCT*: *M. purpureus* citrinin polyketide synthase; *MPL1*: serine hydrolase; *MPL2*: oxygenase. *MPL4*: dehydrogenase. *MPL6*: short-chain dehydrogenase. *MPL7*: oxidoreductase. *MPR1*: a transporter. *LovA*: a cytochrome P450 monooxygenase; *LovB*: lovastatin nonaketide synthase; *LovC*: enoyl-reductase; *LovD*: acyl-transferase; *LovF*: lovastatin diketide synthase; *LovG*: thioesterase; Human CYP17 cDNA: catalyzing the conversion of progesterone to 17-hydroxyprogesterone as well as 16-hydroxyprogesterone

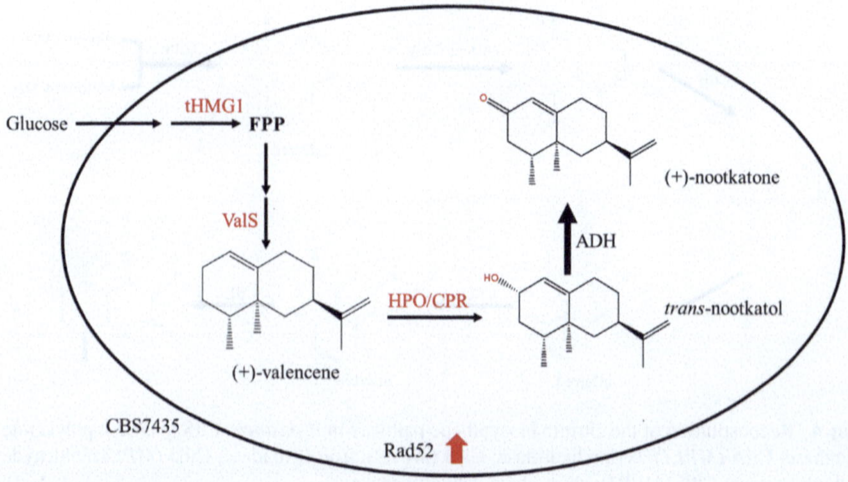

Fig. 3 Synthetic pathway of (+)-nootkatone in *P. pastoris*. (+)-Valencene is synthesized from farnesyl pyrophosphate by valencene synthase (ValS). HPO/CPR converts (+)-valencene into *trans*-nootkatol, which is further converted by the endogenous alcohol dehydrogenase (ADH) to form (+)-nootkatone. Enzymes that can convert *trans*-nootkatol to (+)-nootkatone naturally exist in *P. pastoris*. Overexpression of the endogenous ADH and the truncated hydroxymethylglutaryl coenzyme A reductase (tHMG1) from *S. cerevisiae*, as well as the endogenous Rad52 significantly increased the production of (+)-nootkatone. Exogenous genes are shown in red and bold arrows represent overexpressed endogenous genes. HPO: premnaspirodiene oxygenase from *H. muticus*; CPR: cytochrome P450 reductase from *A. thaliana*

the endogenous dehydrogenase of *P. pastoris*. Further introduction of valencene synthase (ValS) and the truncated *S. cerevisiae* hydroxy-methylglutaryl CoA reductase resulted in the construction of a strain capable of de novo production of (+)-nootkatone from glucose. The production of (+)-nootkatone reached a titer of 17 mg/L in shake flasks and 208 mg/L in the bioreactors (Wriessnegger et al. 2014). Notably, overexpression of *RAD52* and the optimization of the medium have increased the production of *trans*-nootkatone by 5-fold (Wriessnegger, et al. 2016).

Polyketides are a class of secondary metabolites with various structures and biological activities. With 6-methylsalicylic acid (6-MSA) being the first polyketide produced in *P. pastoris*, the biosynthetic pathways of citrinin, monacolin, and other polyketides have also been successfully reconstituted in this host. Synthesized by a relatively small gene cluster (Fig. 4), citrinin serves as a representative for polyketide biosynthesis in *P. pastoris* (Shimizu et al. 2007; Sakai et al. 2008). A total of seven foreign genes was heterologously expressed to construct the citrinin biosynthetic pathway, including the citrinin polyketide synthase gene *PksCT* (*CitS*) from *Monascus purpureus*, the phosphoubiquitin transferase gene *NpgA* from *Aspergillus nidulans*, as well as the cluster genes, including *MPL1* (*CitA*), *MPL2* (*CitB*), *MPL4* (*CitD*), *MPL6* (*CitE*), and *MPL7* (*CitC*) from *M. purpureus*.

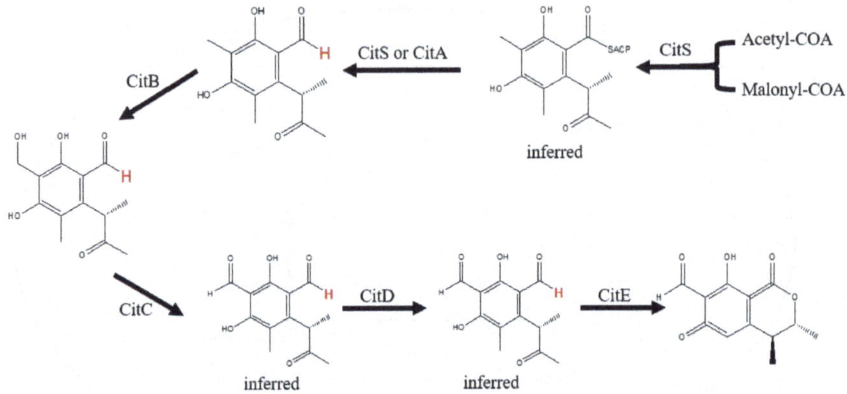

Fig. 4 Reconstitution of the citrinin biosynthetic pathway in *P. pastoris*. CitS (*pksCT*): polyketide synthase; CitA (*MPL1*): serine hydrolase; CitB (*MPL2*): iron II oxidase; CitD (*MPL4*): aldehyde dehydrogenase; CitE (*MPR1*): short-chain dehydrogenase

After 24 h of cultivation with methanol as the sole carbon source, citrinin was produced up to a concentration of 0.6 mg/L (Xue et al. 2017).

4 Recent Advances in Engineering *P. kudriavzeii*

P. kudriavzeii is a non-conventional yeast that can be found in various fermented foods (Choi et al. 2017; Vuyst et al. 2016; Qin et al. 2016), cocoa beans (Delgado-Ospina, et al. 2020), fruits (Park, et al. 2018), wastewater (Pajot et al. 2011), etc. Other names include *Issatchenkia orientalis*, *Candida glycerinogenes*, and *Candida krusei* (Douglass 2018). It is a multistress-tolerant yeast that can grow at low pH (Xiao et al. 2014; Park et al. 2018; Sun et al. 2020; Toivari et al. 2013; Hisamatsu et al. 2006), high temperature (as high as 50 °C) (Park et al. 2018; Yuangsaard et al. 2013; Chamnipa et al. 2018), and high concentrations of salt conditions (Isono et al. 2012); thus, it has been engineered to produce organic acids such as D-xylonic acid (Toivari et al. 2013), succinic acid (Xiao et al. 2014), D-lactic acid (Park et al. 2018), and itaconic acid (Sun et al. 2020). For example, Toivari et al. engineered *P. kudriavzevii* VTT C-79090T to express a D-xylose dehydrogenase gene from *Caulobacter crescentus* at the *PDC1* locus, resulting in 146 g/L D-xylonate production at pH 3.0 (Toivari et al. 2013). Xiao et al. engineered *I. orientalis* SD108 by overexpressing three native genes (i.e., encoding pyruvate carboxylase, malate dehydrogenase, and fumarase) and fumarate reductase that was previously codon-optimized for expression in *S. cerevisiae* via genome integration, enabling production of succinic acid at a titer of 11.63 g/L in shake flask (Xiao et al. 2014). In addition, by replacing the gene encoding pyruvate decarboxylase 1 with the gene encoding D-lactate dehydrogenase from *Lactobacillus plantarum* followed by adaptive evolution, the engineered *P. kudriavzevii* NG7 strain was able to produce D-lactic acid at a titer of 135 g/L (pH 3.6) and 154 g/L (pH 4.7) (Park et al. 2018). Our group also engineered *P. kudriavzevii* YB4010 to produce 1.23 g/L itaconic acid at pH 3.9 in fed batch fermentation by overexpressing a *cis*-aconitic acid decarboxylase gene from *Aspergillus terreus* and a native mitochondrial tricarboxylate transporter in the strain with the isocitrate dehydrogenase gene deleted (Sun et al. 2020). *P. kudriavzevii* can also produce ethanol at a high salt concentration (50 g/L Na_2SO_4) at pH 2.0 or a temperature as high as 43 °C (Isono et al. 2012). Other applications have been demonstrated in wine fermentation (Mónaco et al. 2014, 2016), production of potential probiotics (Greppi et al. 2017), biological control (Bajaj et al. 2013), and bioremediation for heavy metal removal (Li et al. 2016b; Zhang et al. 2019b). To date, the registered *P. kudriavzeii* strains are mainly diploid (Xiao et al. 2014; Xi et al. 2021), although triploid and aneuploid strains have also been reported (Douglass 2018).

Prior to the creation of episomal plasmids, engineering works in *P. kudriavzeii* were usually performed by directly transforming a linear fragment carrying the target gene(s) and a selection marker flanked by homologous arms of the target integration site (Park et al. 2018). *URA3* is often used as a selection marker because of the relatively easy protocol for marker recycling. Considering that the strain is a diploid, a single-round integration will likely create a heterozygous strain, and a second-round integration to the wild-type allele is recommended to improve genetic stability. Recently, Tran and Cao et al. created an episomal plasmid consisting of an ARS and *LEU2* originating from *S. cerevisiae*, *P. kudriavzeii URA3*, and a GFP reporter gene. The percentage of GFP^+ cells in the culture grown from a single colony was approximately 60% after 24 h (Cao et al. 2020). They isolated a centromere-like (CEN-L) sequence from the *P. kudriavzeii* genome with the assistance of an in silico GC_3 analysis to identify the "GC_3 valley" on each chromosome, followed by sequence alignment to identify the conserved regions (Cao et al. 2020). As a CEN is responsible for faithful chromosome segregation and plays a critical role in stabilizing plasmids, the newly constructed plasmid including CEN-L led to an increased percentage of GFP^+ cells, to 81% after cultivation for 24 h and to 67% after cultivation for 120 h.

RNA sequencing is usually implemented to identify strong constitutive promoters and terminators. Cao et al. grew *P. kudriavzeii* under four growth conditions (YNB medium with or without lignocellulosic biomass inhibitors under aerobic or anaerobic conditions) and analyzed the transcriptome. Thirty-five promoters of the most highly expressed genes identified based on RNA-sequencing data were selected and cloned with the *GFP* reporter gene and TEF1 terminator on an episomal plasmid containing ARS. Strong, medium, and weak constitutive promoters were categorized based on flow cytometry. For terminator identification, 14 terminators of the strong promoters identified above were selected for further comparison. Double reporter genes (i.e., *GFP* and *mCherry*) were placed between the TDH3 promoter and the PGK1 terminator. Each of the candidate terminator sequences was cloned between the *GFP* and the *mCherry* open reading frames (ORFs) with a random sequence or no sequence inserted between the two ORFs as controls. Quantitative PCR was used to calculate the transcriptional ratios of *mCherry* and *GFP*. Thirteen of the 14 candidate terminators had ratios below 0.03 and were categorized as strong terminators. Moreover, similar to *S. cerevisiae*, *P. kudriavzeii* has a relatively high HR efficiency. An HR-mediated DNA assembly method was developed to facilitate rapid plasmid construction in a single step. Co-transforming five linear DNA fragments, with 70–80 bp overlaps designed between the adjacent fragments, directly into *I. orientalis* SD108 led to the successful construction of a 14.5 kb plasmid containing the xylose utilization pathway with an assembly efficiency of 100% (Cao et al. 2020).

Genome editing tools have been developed for *P. kudriavzeii.* The promoter used to transcribe the gene encoding sgRNA needs to be an RNA polymerase (RNAP) III promoter because a typical RNAP II promoter used to transcribe proteins will make the gene undergo post-transcriptional modifications such as 5′-end capping and 3′-end polyadenylation, which may inactivate the Cas9/gRNA complex (Gao and Zhao 2014). Tran and Cao et al. chose a series of RNAP III promoters including $tRNA^{Leu}$, $tRNA^{Ser}$, 5S rRNA, RPR1 (the RNA component of RNase P, the 250 bp upstream sequence of RPR1), and fusions of 5S rRNA or RPR1' (the 250 bp upstream sequence of RPR1 and the first 120 bp of RPR1) with $tRNA^{Leu}$. An iCas9 (containing D147Y and P411T) that possesses a higher activity than Cas9 from *Streptococcus pyogenes* was tagged with a nuclear localization sequence and expressed by an episomal plasmid, together with each of the sgRNA cassettes. Targeting *ADE2*, *LEU2*, *HIS3*, and *TRP1* in *I. orientalis* SD108 showed that RPR1'-$tRNA^{Leu}$ led to the highest single-gene disruption efficiency of 97–100% and was therefore used to create double and triple knockouts. Double-gene knockouts of *ADE2*/*TRP1* and *ADE2*/*HIS3* were attained with efficiencies of 72.8% and 89.9%, respectively, whereas triple-gene knockout efficiency for *ADE2*/*HIS3*/*SDH2* was approximately 47% (Tran et al. 2019). In parallel, our group also carried out a similar study by evaluating *ADE2* disruption efficiencies with five versions of Cas9 (including *S. pyogenes* Cas9, iCas9, a codon-optimized Cas9 version for *Homo sapiens*, *Candida albicans*, and *Scheffersomyces stipitis*) and RPR1 (the 311 bp upstream sequence of RPR1) as the sgRNA promoter in *P. kudriavzevii* YB4010. The highest efficiency (42%) was achieved using iCas9 (Sun et al. 2020).

Another interesting area for exploring *P. kudriavzevii* as a production host for organic acids is the identification of their transporters. Previously, our group identified a mitochondrial tricarboxylate transporter, Pk_MttA, which can potentially transport citrate and *cis*-aconitate from the mitochondria to the cytosol, thereby increasing itaconic acid production in *P. kudriavzevii* YB4010 (Sun et al. 2020). A recent genome sequencing and transcriptome analysis of *P. kudriavzevii* CY902 led to the identification of two JEN family carboxylate transporters (PkJEN2-1 and PkJEN2-2), which can import succinate into cell (Xi et al. 2021). Substrate specificity analysis showed that both PkJEN2-1 and PkJEN2-2 are dicarboxylate importers for succinate, L-malate, and fumarate. In addition, PkJEN2-1 can import α-ketoglutarate, whereas PKJEN2-2 can also uptake citrate but not α-ketoglutarate. The structural basis of PkJEN2-2 specificity toward tricarboxylate substrates was studied using model-based structure analysis and rational design. By inactivating both transporters, enhanced extracellular succinate accumulation can be achieved in the late stages of fermentation. This study highlights an important direction for future studies to improve organic acid production in *P. kudriavzevii.*

5 Conclusions and Perspectives

A panel of synthetic biology tools has been developed to establish *P. pastoris* as a microbial cell factory for the efficient production of recombinant proteins, chemicals, and natural products. Although *P. pastoris* has been widely employed for high-level expression of heterologous proteins, its application in the biosynthesis of chemicals and natural products is still limited to a few examples. On the other hand, although *P. pastoris* can utilize methanol as the sole carbon and energy source, the methanol-to-chemical conversion efficiency is still rather low, as most of the methanol is converted to CO_2 via the dissimilatory pathway for energy generation. Thus, the redirection of methanol flux toward the assimilatory pathway rather than the dissimilatory pathway is a prerequisite for establishing *P. pastoris* as a cell factory for chemical production from methanol. In other words, the methanol conversion pathway for the assimilation of methanol and biosynthesis of the desired product should be carefully engineered. Particularly, Lu et al. constructed a three-step synthetic acetyl-CoA (SACA) pathway for the synthesis of acetyl-CoA from formaldehyde by combining an engineered glycolaldehyde synthetase variant (GALS), acetyl phosphate synthetase (ACPS), and phosphate acetyltransferase (PTA). SACA represents the shortest pathway for acetyl-CoA biosynthesis ever reported and is promising for the efficient production of acetyl-CoA-derived compounds from C1 carbon sources (Lu et al. 2019). Considering the capability of formaldehyde formation and the role of acetyl-CoA as an important biosynthesis precursor, the SACA pathway is expected to be employed in *P. pastoris* to produce value-added compounds from methanol in the near future.

The complexity and our insufficient understanding of the cellular metabolic and regulatory networks have largely limited our capability of designing efficient *Pichia* cell factories. For example, the expression of recombinant proteins is affected by central metabolisms in *P. pastoris* (Nocon et al. 2014). One strategy is to establish genome-scale metabolic models, such as PpaMBEL1254 (Sohn et al. 2010), iPP668 (Chung et al. 2010), and iLC915 (Caspeta et al. 2012), to describe the cellular metabolic network from a systems biology perspective. Genome-scale metabolic models have been employed to guide the design of *P. pastoris* cell factories with increased expression of hSOD, human lysozyme, and antibody fragment Fab-3H6 (Cankorur-Cetinkaya et al. 2018). Unfortunately, the complexity of cellular metabolic and regulatory networks is still beyond the reach of the genome-scale metabolic models. For example, the overexpression of a non-intuitive gene, *RAD52*, encoding a protein responsible for DNA recombination was found to be beneficial for the expression of CYPs and correspondingly the production of *trans*-nootkatone (Wriessnegger, et al. 2016). Similar to that of *P. pastoris*, a genome-scale metabolic model, iIsor850, has been developed for *P. kudriavzevii* (i.e., *I. orientalis* SD108). This model contains 850 genes, 1826 reactions, and 1702 metabolites (Suthers 2020). Biomass composition data and the estimated ATP maintenance requirements were collected by cultivating *I. orientalis* SD108 in a chemostat under carbon limitation, to improve the predictive power of the model. The consistency of the model predictions was validated using assessment of substrate utilization and gene knockouts. This model was used to propose engineering strategies for enhanced succinic acid production using the OptKnock framework. This model will be beneficial for the metabolic engineering of *P. kudriavzevii* for other organic acid production.

However, so far, only a small portion of genes is included in the genome-scale metabolic models, and the predictions are not sufficiently accurate and far from perfection. Therefore, an alternative strategy is to employ synthetic biology-based genome-scale engineering, which can perturb many genes in a combinatorial manner, extending our limited knowledge on cellular metabolism and regulation. Such a creation-driven-understanding technology aims to identify non-intuitive engineering targets, to increase the expression of recombinant proteins, as well as the production of desirable compounds. Currently, CRISPR-based genome-scale engineering tools have been developed for *E. coli* (Garst et al. 2017), *S. cerevisiae* (Lian et al. 2017, 2019), and mammalian cells (Gilbert et al. 2014), and are expected to be established in *P. pastoris* and *P. kudriavzevii* in the near future, which can be employed for the construction of efficient cell factories in a high-throughput manner.

Acknowledgements YZ, JG, and JL were supported by the National Key Research and Development Program of China (2018YFA0901800). WS and ZS were supported by the U.S. Department of Energy, Office of Science, Biological and Environmental Research through Ames Laboratory. Ames Laboratory is operated for the U.S. Department of Energy by Iowa State University under Contract No. DE-AC02-07CH11358.

References

Abdel-Mawgoud AM et al (2018) Metabolic engineering in the host *Yarrowia lipolytica*. Metab Eng 50:192–208

Abdelmoula-Souissi S et al (2013) Secreted recombinant P53 protein from *Pichia pastoris* is a useful antigen for detection of serum p53: autoantibody in patients with advanced colorectal adenocarcinoma. Mol Biol Rep 40(5):3865–3872

Agbogbo FK, Coward-Kelly G (2008) Cellulosic ethanol production using the naturally occurring xylose-fermenting yeast *Pichia Stipitis*. Biotechnol Lett 30(9):1515–1524

Aggarwal S, Mishra S (2020) Differential role of segments of α-mating factor secretion signal in *Pichia pastoris* towards granulocyte colony-stimulating factor emerging from a wild type or codon optimized copy of the gene. Microb Cell Fact 19(1):199

Ahmad M et al (2014) Protein expression in Pichia pastoris: recent achievements and perspectives for heterologous protein production. Appl Microbiol Biotechnol 98(12):5301–5317

Ahn J et al (2016) Codon optimization of *Saccharomyces cerevisiae* mating factor alpha prepro-leader to improve recombinant protein production in *Pichia pastoris*. Biotechnol Lett 38(12):2137–2143

Andreu C, Del Olmo ML (2018) Yeast arming systems: pros and cons of different protein anchors and other elements required for display. Appl Microbiol Biotechnol 102(6):2543–2561

Araya-Garay JM et al (2012a) Construction of new *Pichia pastoris* X-33 strains for production of lycopene and beta-carotene. Appl Microbiol Biotechnol 93(6):2483–2492

Araya-Garay JM et al (2012b) Construction of a novel *Pichia pastoris* strain for production of xanthophylls. AMB Express 2(1):24

Aw R et al (2017) Expressing anti-HIV VRC01 antibody using the murine IgG1 secretion signal in Pichia pastoris. AMB Express 7(1):70

Aza P et al (2021) Design of an improved universal signal peptide based on the α-factor mating secretion signal for enzyme production in yeast. Cell Mol Life Sci 78(7):3691–3707

Baeshen MN et al (2016) Expression and purification of C-peptide containing insulin using *Pichia pastoris* expression system. Biomed Res Int 2016:3423685

Bajaj BK, Raina S, Singh S (2013) Killer toxin from a novel killer yeast *Pichia kudriavzevii* RY55 with idiosyncratic antibacterial activity. J Basic Microbiol 53(8):645–656

Barrero JJ et al (2018) An improved secretion signal enhances the secretion of model proteins from *Pichia pastoris*. Microb Cell Fact 17(1):161

Ben Azoun S et al (2016) Molecular optimization of rabies virus glycoprotein expression in *Pichia pastoris*. Microb Biotechnol 9(3):355–368

Berrios J et al (2017) A comparative study of glycerol and sorbitol as co-substrates in methanol-induced cultures of *Pichia pastoris*: temperature effect and scale-up simulation. J Ind Microbiol Biotechnol 44(3):407–411

Brady JR et al (2020) Comparative genome-scale analysis of *Pichia pastoris* variants informs selection of an optimal base strain. Biotechnol Bioeng 117(2):543–555

Brake AJ et al (1984) Alpha-factor-directed synthesis and secretion of mature foreign proteins in *Saccharomyces cerevisiae*. Proc Natl Acad Sci USA 81(15):4642–4646

Bredell H et al (2018) Expression of unique chimeric human papilloma virus type 16 (HPV-16) L1–L2 proteins in *Pichia pastoris* and *Hansenula polymorpha*. Yeast 35(9):519–529

Cankorur-Cetinkaya A et al (2018) Process development for the continuous production of heterologous proteins by the industrial yeast *Komagataella Phaffii*. Biotechnol Bioeng 115(12):2962–2973

Cao M et al (2017a) Centromeric DNA facilitates nonconventional yeast genetic engineering. ACS Synth Biol 6(8):1545–1553

Cao L et al (2017b) Improving the secretion yield of the β-galactosidase Bgal1-3 in *Pichia pastoris* for use as a potential catalyst in the production of prebiotic-enriched milk. J Agric Food Chem 65(49):10757–10766

Cao M et al (2020) A genetic toolbox for metabolic engineering of *Issatchenkia orientalis*. Metab Eng 59:87–97

Carnicer M et al (2012) Development of quantitative metabolomics for *Pichia pastoris*. Metabolomics 8(2):284–298

Caspeta L et al (2012) Genome-scale metabolic reconstructions of *Pichia stipitis* and *Pichia pastoris* and in silico evaluation of their potentials. BMC Syst Biol 6:24

Cereghino JL, Cregg JM (2000) Heterologous protein expression in the methylotrophic yeast *Pichia pastoris*. FEMS Microbiol Rev 24(1):45–66

Cereghino GP et al (2002) Production of recombinant proteins in fermenter cultures of the yeast *Pichia pastoris*. Curr Opin Biotechnol 13(4):329–332

Chahal S et al (2017) Structural characterization of the α-mating factor prepro-peptide for secretion of recombinant proteins in *Pichia pastoris*. Gene 598:50–62

Chamnipa N et al (2018) The potential of the newly isolated thermotolerant yeast *Pichia kudriavzevii* RZ8-1 for high-temperature ethanol production. Braz J Microbiol 49(2):378–391

Chan MK et al (2018) Expression of stable and active human DNA topoisomerase I in *Pichia pastoris*. Protein Expr Purif 141:52–62

Chen Z et al (2011) Recombinant antimicrobial peptide hPAB-beta expressed in *Pichia pastoris*, a potential agent active against methicillin-resistant Staphylococcus aureus. Appl Microbiol Biotechnol 89(2):281–291

Chen Z et al (2020) Efficient biodegradation of highly crystallized polyethylene terephthalate through cell surface display of bacterial PETase. Sci Total Environ 709:136138

Cheng H et al (2014) Genetically engineered *Pichia pastoris* yeast for conversion of glucose to xylitol by a single-fermentation process. Appl Microbiol Biotechnol 98(8):3539–3552

Choi S et al (2015) Biorefineries for the production of top building block chemicals and their derivatives. Metab Eng 28:223–239

Choi DH, Park EH, Kim MD (2017) Isolation of thermotolerant yeast *Pichia kudriavzevii* from nuruk. Food Sci Biotechnol 26(5):1357–1362

Chu J et al (2013) Progress in the research of S-adenosyl-L-methionine production. Appl Microbiol Biotechnol 97(1):41–49

Chung BK et al (2010) Genome-scale metabolic reconstruction and in silico analysis of methylotrophic yeast *Pichia pastoris* for strain improvement. Microb Cell Fact 9:50

Coughlan AY et al (2016) Centromeres of the yeast *Komagataella phaffii* (*Pichia pastoris*) have a simple inverted-repeat structure. Genome Biol Evol 8(8):2482–2492

Cregg JM et al (1985) *Pichia pastoris* as a host system for transformations. Mol Cell Biol 5(12):3376–3385

Cregg JM et al (1989) Functional characterization of the two alcohol oxidase genes from the yeast *Pichia pastoris*. Mol Cell Biol 9(3):1316–1323

Cregg JM, Vedvick TS, Raschke WC (1993) Recent advances in the expression of foreign genes in *Pichia pastoris*. Biotechnology (n y) 11(8):905–910

Crepin VF, Faulds CB, Connerton IF (2003) Production and characterization of the *Talaromyces stipitatus* feruloyl esterase FAEC in *Pichia pastoris*: identification of the nucleophilic serine. Protein Expr Purif 29(2):176–184

Curran KA, Alper HS (2012) Expanding the chemical palate of cells by combining systems biology and metabolic engineering. Metab Eng 14(4):289–297

Dagar VK, Khasa YP (2018) Combined effect of gene dosage and process optimization strategies on high-level production of recombinant human interleukin-3 (hIL-3) in *Pichia pastoris* fed-batch culture. Int J Biol Macromol 108:999–1009

Daly R, Hearn MT (2005) Expression of heterologous proteins in *Pichia pastoris*: a useful experimental tool in protein engineering and production. J Mol Recognit 18(2):119–138

Damasceno LM et al (2007) Cooverexpression of chaperones for enhanced secretion of a single-chain antibody fragment in *Pichia pastoris*. Appl Microbiol Biotechnol 74(2):381–389

de Lima PB et al (2016) Novel homologous lactate transporter improves L-lactic acid production from glycerol in recombinant strains of *Pichia pastoris*. Microb Cell Fact 15(1):158

de Oliveira JD et al (2016) Genetic basis for hyper production of hyaluronic acid in natural and engineered microorganisms. Microb Cell Fact 15(1):119

De Schutter K et al (2009) Genome sequence of the recombinant protein production host *Pichia pastoris*. Nat Biotechnol 27(6):561–566

De Vuyst L et al (2016) Yeast diversity of sourdoughs and associated metabolic properties and functionalities. Int J Food Microbiol 239:26–34

Del Mónaco SM et al (2014) Selection and characterization of a Patagonian *Pichia kudriavzevii* for wine deacidification. J Appl Microbiol 117(2):451–464

Del Mónaco SM, Rodríguez ME, Lopes CA (2016) *Pichia kudriavzevii* as a representative yeast of North Patagonian winemaking terroir. Int J Food Microbiol 230:31–39

Delgado-Ospina J et al (2020) Functional biodiversity of yeasts isolated from Colombian fermented and dry cocoa beans. Microorganisms 8(7)

Deng J et al (2020) Co-expressing GroEL-GroES, Ssa1-Sis1 and Bip-PDI chaperones for enhanced intracellular production and partial-wall breaking improved stability of porcine growth hormone. Microb Cell Fact 19(1):35

Dong JX et al (2013) Surface display and bioactivity of Bombyx mori acetylcholinesterase on Pichia pastoris. PLoS One 8(8):e70451

Dong C et al (2020) Engineering *Pichia pastoris* with surface-display minicellulosomes for carboxymethyl cellulose hydrolysis and ethanol production. Biotechnol Biofuels 13:108

Douglass AP et al (2018) Population genomics shows no distinction between pathogenic *Candida krusei* and environmental *Pichia kudriavzevii*: one species, four names. PLoS Pathog 14(7):e1007138

Duman ZE et al (2020) High-level heterologous expression of active *Chaetomium thermophilum* FDH in *Pichia pastoris*. Enzyme Microb Technol 137:109552
Eiden-Plach A et al (2004) Viral preprotoxin signal sequence allows efficient secretion of green fluorescent protein by *Candida glabrata, Pichia pastoris, Saccharomyces cerevisiae, and Schizosaccharomyces pombe*. Appl Environ Microbiol 70(2):961–966
Evstigneeva ZG, Solov'eva NA, Sidel'nikova LI (2001) Structure and functions of chaperones and chaperonins (Review). Prikl Biokhim Mikrobiol 37(1):5–18
Fraser RZ et al (2018) Safety evaluation of soy leghemoglobin protein preparation derived from *Pichia pastoris*, intended for use as a flavor catalyst in plant-based meat. Int J Toxicol 37(3):241–262
Gai SA, Wittrup KD (2007) Yeast surface display for protein engineering and characterization. Curr Opin Struct Biol 17(4):467–473
Gao Y, Zhao Y (2014) Self-processing of ribozyme-flanked RNAs into guide RNAs in vitro and *in vivo* for CRISPR-mediated genome editing. J Integr Plant Biol 56(4):343–349
Gao L et al (2013) Engineered fungal polyketide biosynthesis in Pichia pastoris: a potential excellent host for polyketide production. Microb Cell Fact 12:77
Garst AD et al (2017) Genome-wide mapping of mutations at single-nucleotide resolution for protein, metabolic and genome engineering. Nat Biotechnol 35(1):48–55
Gasser B et al (2006) Engineering of *Pichia pastoris* for improved production of antibody fragments. Biotechnol Bioeng 94(2):353–361
Gasser B et al (2007) Transcriptomics-based identification of novel factors enhancing heterologous protein secretion in yeasts. Appl Environ Microbiol 73(20):6499–6507
Generoso WC et al (2012) Recombinant expression and characterization of an endoglucanase III (cel12a) from *Trichoderma harzianum* (Hypocreaceae) in the yeast *Pichia pastoris*. Genet Mol Res 11(2):1544–1557
Gilbert LA et al (2014) Genome-scale CRISPR-mediated control of gene repression and activation. Cell 159(3):647–661
Gong Z et al (2015) Efficient conversion of acetate into lipids by the oleaginous yeast *Cryptococcus curvatus*. Biotechnol Biofuels 8:189
Govindappa N et al (2014) A new signal sequence for recombinant protein secretion in *Pichia pastoris*. J Microbiol Biotechnol 24(3):337–345
Greppi A et al (2017) Potential probiotic *Pichia kudriavzevii* strains and their ability to enhance folate content of traditional cereal-based African fermented food. Food Microbiol 62:169–177
Guo JP, Ma Y (2008) High-level expression, purification and characterization of recombinant *Aspergillus oryzae* alkaline protease in *Pichia pastoris*. Protein Expr Purif 58(2):301–308
Hartner FS et al (2008) Promoter library designed for fine-tuned gene expression in *Pichia pastoris*. Nucleic Acids Res 36(12):e76
Hasslacher M et al (1997) High-level intracellular expression of hydroxynitrile lyase from the tropical rubber tree Hevea brasiliensis in microbial hosts. Protein Expr Purif 11(1):61–71
He Z et al (2012) Comparison of alpha-factor preprosequence and a classical mammalian signal peptide for secretion of recombinant xylanase xynB from yeast *Pichia pastoris*. J Microbiol Biotechnol 22(4):479–483
Helian Y et al (2020) A multistrategy approach for improving the expression of *E. coli* phytase in *Pichia pastoris*. J Ind Microbiol Biotechnol 47(12):1161–1172
Hetz C, Zhang K, Kaufman RJ (2020) Mechanisms, regulation and functions of the unfolded protein response. Nat Rev Mol Cell Biol 21(8):421–438
Higgins DR et al (1998) Small vectors for expression based on dominant drug resistance with direct multicopy selection. Methods Mol Biol 103:41–53
Hisamatsu M et al (2006) Isolation and identification of a novel yeast fermenting ethanol under acidic conditions. J Appl Glycosci 53(2):111–113
Huang SJ et al (2011) Molecular cloning and characterization of a novel laccase gene from a white-rot fungus Polyporus grammocephalus TR16 and expression in *Pichia pastoris*. Lett Appl Microbiol 52(3):290–297

Huangfu J et al (2015) Novel helper factors influencing recombinant protein production in *Pichia pastoris* based on proteomic analysis under simulated microgravity. Appl Microbiol Biotechnol 99(2):653–665

Idiris A et al (2010) Engineering of protein secretion in yeast: strategies and impact on protein production. Appl Microbiol Biotechnol 86(2):403–417

Iglesias-Figueroa B et al (2016) High-Level expression of recombinant bovine lactoferrin in *Pichia pastoris* with antimicrobial activity. Int J Mol Sci 17(6)

Isono N et al (2012) A comparative study of ethanol production by *Issatchenkia orientalis* strains under stress conditions. J Biosci Bioeng 113(1):76–78

Ito Y et al (2020) Exchange of endogenous and heterogeneous yeast terminators in *Pichia pastoris* to tune mRNA stability and gene expression. Nucleic Acids Res 48(22):13000–13012

Jacobs PP et al (2008) *Pichia* surface display: display of proteins on the surface of glycoengineered *Pichia pastoris* strains. Biotechnol Lett 30(12):2173–2181

Jahic M et al (2003) Temperature limited fed-batch technique for control of proteolysis in *Pichia pastoris* bioreactor cultures. Microb Cell Fact 2(1):6

Jahic M et al (2006) Process technology for production and recovery of heterologous proteins with *Pichia pastoris*. Biotechnol Prog 22(6):1465–1473

Jang IS et al (2018) Improving the efficiency of homologous recombination by chemical and biological approaches in *Yarrowia lipolytica*. PLoS One 13(3):e0194954

Jariyachawalid K et al (2012) Effective enhancement of Pseudomonas stutzeri D-phenylglycine aminotransferase functional expression in *Pichia pastoris* by co-expressing *Escherichia coli* GroEL-GroES. Microb Cell Fact 11:47

Ji Q et al (2020) Improving the homologous recombination efficiency of *Yarrowia lipolytica* by grafting heterologous component from *Saccharomyces cerevisiae*. Metab Eng Commun 11:e00152

Jiang ZB et al (2007) Cell surface display of functionally active lipases from *Yarrowia lipolytica* in *Pichia pastoris*. Protein Expr Purif 56(1):35–39

Jiao L et al (2018) Efficient heterologous production of *Rhizopus oryzae* lipase via optimization of multiple expression-related helper proteins. Int J Mol Sci 19(11)

Jin P et al (2014) High-yield novel leech hyaluronidase to expedite the preparation of specific hyaluronan oligomers. Sci Rep 4:4471

Jo JH et al (2011) Surface display of human lactoferrin using a glycosylphosphatidylinositol-anchored protein of *Saccharomyces cerevisiae in Pichia pastoris*. Biotechnol Lett 33(6):1113–1120

Julius D et al (1984) Isolation of the putative structural gene for the lysine-arginine-cleaving endopeptidase required for processing of yeast prepro-alpha-factor. Cell 37(3):1075–1089

Kang Z, Zhang N, Zhang Y (2016) Enhanced production of leech hyaluronidase by optimizing secretion and cultivation in *Pichia pastoris*. Appl Microbiol Biotechnol 100(2):707–717

Karbalaei M, Rezaee SA, Farsiani H (2020) *Pichia pastoris*: a highly successful expression system for optimal synthesis of heterologous proteins. J Cell Physiol 235(9):5867–5881

Kato S et al (2001) Efficient expression, purification and characterization of mouse salivary alpha-amylase secreted from methylotrophic yeast *Pichia Pastoris*. Yeast 18(7):643–655

Khasa YP et al (2011) Isolation of *Pichia pastoris* PIR genes and their utilization for cell surface display and recombinant protein secretion. Yeast 28(3):213–226

Kittl R et al (2012) Constitutive expression of *Botrytis aclada* laccase in *Pichia pastoris*. Bioengineered 3(4):232–235

Kolar NW et al (2007) Functional expression and characterisation of human cytochrome P45017alpha in *Pichia pastoris*. J Biotechnol 129(4):635–644

Kong S et al (2020) De novo biosynthesis of 2-phenylethanol in engineered *Pichia pastoris*. Enzyme Microb Technol 133:109459

Koutz P et al (1989) Structural comparison of the *Pichia pastoris* alcohol oxidase genes. Yeast 5(3):167–177

Kuroda K, Ueda M (2011) Cell surface engineering of yeast for applications in white biotechnology. Biotechnol Lett 33(1):1–9

Kuwae S et al (2005) Production of recombinant human antithrombin by *Pichia pastoris*. J Biosci Bioeng 99(3):264–271
Lee CC et al (2005) An episomal expression vector for screening mutant gene libraries in *Pichia pastoris*. Plasmid 54(1):80–85
Li P et al (2007) Expression of recombinant proteins in *Pichia pastoris*. Appl Biochem Biotechnol 142(2):105–124
Li W et al (2015a) Cell surface display and characterization of *Rhizopus oryzae* lipase in *Pichia pastoris* using Sed1p as an anchor protein. Curr Microbiol 71(1):150–155
Li YY et al (2015b) High-level expression and characterization of a thermostable xylanase mutant from *Trichoderma reesei* in *Pichia pastoris*. Protein Expr Purif 108:90–96
Li C et al (2015c) Combined strategies for improving expression of *Citrobacter amalonaticus* phytase in *Pichia pastoris*. BMC Biotechnol 15:88
Li P et al (2016a) High-level secretory expression and purification of recombinant human interleukin 1 beta in *Pichia pastoris*. Protein Pept Lett 23(8):763–769
Li C et al (2016b) Efficient removal of zinc by multi-stress-tolerant yeast *Pichia kudriavzevii* A16. Bioresour Technol 206:43–49
Li YX et al (2018) High-level expression of an engineered beta-mannanase (mRmMan5A) in *Pichia pastoris* for manno-oligosaccharide production using steam explosion pretreated palm kernel cake. Bioresour Technol 256:30–37
Li D et al (2020) Optimized expression of classical swine fever virus E2 protein via combined strategy in *Pichia pastoris*. Protein Expr Purif 167:105527
Li J et al (2021) Preparation of a Bombyx mori acetylcholinesterase enzyme reagent through chaperone protein disulfide isomerase co-expression strategy in *Pichia pastoris* for detection of pesticides. Enzyme Microb Technol 144:109741
Liachko I, Dunham MJ (2014) An autonomously replicating sequence for use in a wide range of budding yeasts. FEMS Yeast Res 14(2):364–367
Lian J et al (2017) Combinatorial metabolic engineering using an orthogonal tri-functional CRISPR system. Nat Commun 8(1):1688
Lian J, Mishra S, Zhao H (2018) Recent advances in metabolic engineering of *Saccharomyces cerevisiae*: new tools and their applications. Metab Eng 50:85–108
Lian J et al (2019) Multi-functional genome-wide CRISPR system for high throughput genotype-phenotype mapping. Nat Commun 10(1):5794
Liang S et al (2013) Endogenous signal peptides efficiently mediate the secretion of recombinant proteins in *Pichia pastoris*. Biotechnol Lett 35(1):97–105
Lin-Cereghino GP et al (2013) The effect of α-mating factor secretion signal mutations on recombinant protein expression in *Pichia pastoris*. Gene 519(2):311–317
Liu L, Maguire KK, Kmiec EB (2004) Genetic re-engineering of *Saccharomyces cerevisiae RAD51* leads to a significant increase in the frequency of gene repair *in vivo*. Nucleic Acids Res 32(7):2093–2101
Liu SH et al (2005) Improved secretory production of glucoamylase in *Pichia pastoris* by combination of genetic manipulations. Biochem Biophys Res Commun 326(4):817–824
Liu XB et al (2015) Metabolic engineering of *Pichia pastoris* for the production of dammarenediol-II. J Biotechnol 216:47–55
Liu Y et al (2016) Production of glucaric acid from myo-inositol in engineered *Pichia pastoris*. Enzyme Microb Technol 91:8–16
Liu Y et al (2018) Engineered monoculture and co-culture of methylotrophic yeast for *de novo* production of monacolin J and lovastatin from methanol. Metab Eng 45:189–199
Louie TM et al (2021) Production of bio-xylitol from D-xylose by an engineered *Pichia pastoris* expressing a recombinant xylose reductase did not require any auxiliary substrate as electron donor. Microb Cell Fact 20(1):50
Love KR et al (2016) Comparative genomics and transcriptomics of *Pichia pastoris*. BMC Genomics 17:550
Lu X et al (2019) Constructing a synthetic pathway for acetyl-coenzyme A from one-carbon through enzyme design. Nat Commun 10(1):1378

Luley-Goedl C et al (2016) Combining expression and process engineering for high-quality production of human sialyltransferase in *Pichia pastoris*. J Biotechnol 235:54–60
Luu VT et al (2017) Development of recombinant *Yarrowia lipolytica* producing virus-like particles of a fish nervous necrosis virus. J Microbiol 55(8):655–664
Macauley-Patrick S et al (2005) Heterologous protein production using the *Pichia pastoris* expression system. Yeast 22(4):249–270
Majeke BM et al (2020) Synergistic codon optimization and bioreactor cultivation toward enhanced secretion of fungal lignin peroxidase in *Pichia pastoris*: enzymatic valorization of technical (industrial) lignins. Enzyme Microb Technol 139:109593
Marx H et al (2009) Directed gene copy number amplification in *Pichia pastoris* by vector integration into the ribosomal DNA locus. FEMS Yeast Res 9(8):1260–1270
Massahi A, Çalık P (2015) *In-silico* determination of *Pichia pastoris* signal peptides for extracellular recombinant protein production. J Theor Biol 364:179–188
Massahi A, Çalık P (2016) Endogenous signal peptides in recombinant protein production by *Pichia pastoris*: from *in-silico* analysis to fermentation. J Theor Biol 408:22–33
Meng Y et al (2014) Production and characterization of recombinant glucose oxidase from *Aspergillus niger* expressed in *Pichia pastoris*. Lett Appl Microbiol 58(4):393–400
Mombeni M et al (2020) pMOX: a new powerful promoter for recombinant protein production in yeast *Pichia pastoris*. Enzyme Microb Technol 139:109582
Moura MV et al (2015) Displaying lipase B from *Candida Antarctica in Pichia pastoris* using the yeast surface display approach: prospection of a new anchor and characterization of the whole cell Biocatalyst. PLoS One 10(10):e0141454
Nakamura Y et al (2018) A stable, autonomously replicating plasmid vector containing *Pichia pastoris* centromeric DNA. Appl Environ Microbiol 84(15)
Navone L et al (2021) Disulfide bond engineering of AppA phytase for increased thermostability requires co-expression of protein disulfide isomerase in *Pichia pastoris*. Biotechnol Biofuels 14(1):80
Nocon J et al (2014) Model based engineering of *Pichia pastoris* central metabolism enhances recombinant protein production. Metab Eng 24:129–138
Nong L et al (2020) Engineering the regulatory site of the catalase promoter for improved heterologous protein production in *Pichia pastoris*. Biotechnol Lett 42(12):2703–2709
Obst U, Lu TK, Sieber V (2017) A modular toolkit for generating *Pichia pastoris* secretion libraries. ACS Synth Biol 6(6):1016–1025
Ou J, Cao Y (2014) Incorporation of *Nasutitermes takasagoensis* endoglucanase into cell surface-displayed minicellulosomes in *Pichia pastoris* X33. J Microbiol Biotechnol 24(9):1178–1188
Owji H et al (2018) A comprehensive review of signal peptides: structure, roles, and applications. Eur J Cell Biol 97(6):422–441
Pajot HF et al (2011) Unraveling the decolourizing ability of yeast isolates from dye-polluted and virgin environments: an ecological and taxonomical overview. Antonie Van Leeuwenhoek 99(3):443–456
Papakonstantinou T, Harris S, Hearn MT (2009) Expression of GFP using *Pichia pastoris* vectors with zeocin or G-418 sulphate as the primary selectable marker. Yeast 26(6):311–321
Park HJ et al (2018) Low-pH production of d-lactic acid using newly isolated acid tolerant yeast *Pichia kudriavzevii* NG7. Biotechnol Bioeng 115(9):2232–2242
Park HJ et al (2018) Draft genome sequence of a multistress-tolerant yeast, *Pichia kudriavzevii* NG7. Genome Announ 6(3)
Pena DA et al (2018) Metabolic engineering of *Pichia pastoris*. Metab Eng 50:2–15
Peng XB et al (2019) High-level secretive expression of a novel achieved Talaromyces cellulolyticus endo-polygalacturonase in *Pichia pastoris* by improving gene dosage for hydrolysis of natural pectin. World J Microbiol Biotechnol 35(6):84
Piva LC et al (2020) Construction and characterization of centromeric plasmids for *Komagataella phaffii* using a color-based plasmid stability assay. PLoS One 15(7):e0235532
Pontrelli S et al (2018) *Escherichia coli* as a host for metabolic engineering. Metab Eng 50:16–46
Porro D et al (2005) Recombinant protein production in yeasts. Mol Biotechnol 31(3):245–259

Prielhofer R et al (2017) GoldenPiCS: a Golden Gate-derived modular cloning system for applied synthetic biology in the yeast *Pichia pastoris*. BMC Syst Biol 11(1):123

Qin X et al (2011) GAP promoter library for fine-tuning of gene expression in *Pichia pastoris*. Appl Environ Microbiol 77(11):3600–3608

Qin H et al (2016) Microbial diversity and biochemical analysis of suanzhou: a traditional Chinese fermented cereal gruel. Front Microbiol 7:1311

Rabert C et al (2013) Recombinants proteins for industrial uses: utilization of *Pichia pastoris* expression system. Braz J Microbiol 44(2):351–356

Raemaekers RJ et al (1999) Functional phytohemagglutinin (PHA) and *Galanthus nivalis* agglutinin (GNA) expressed in *Pichia pastoris* correct N-terminal processing and secretion of heterologous proteins expressed using the PHA-E signal peptide. Eur J Biochem 265(1):394–403

Rajamanickam V et al (2017) A novel bi-directional promoter system allows tunable recombinant protein production in *Pichia pastoris*. Microb Cell Fact 16(1):152

Rakestraw JA et al (2009) Directed evolution of a secretory leader for the improved expression of heterologous proteins and full-length antibodies in *Saccharomyces cerevisiae*. Biotechnol Bioeng 103(6):1192–1201

Ramakrishnan K et al (2020) Transcriptional control of gene expression in *Pichia pastoris* by manipulation of terminators. Appl Microbiol Biotechnol 104(18):7841–7851

Römisch K (2005) Endoplasmic reticulum-associated degradation. Annu Rev Cell Dev Biol 21:435–456

Saitua F et al (2017) Dynamic genome-scale metabolic modeling of the yeast *Pichia pastoris*. BMC Syst Biol 11(1):27

Sakai K et al (2008) Construction of a citrinin gene cluster expression system in heterologous *Aspergillus oryzae*. J Biosci Bioeng 106(5):466–472

Sallada ND, Harkins LE, Berger BW (2019) Effect of gene copy number and chaperone coexpression on recombinant hydrophobin HFBI biosurfactant production in *Pichia pastoris*. Biotechnol Bioeng 116(8):2029–2040

Sams L et al (2017) Constitutive expression of human gastric lipase in *Pichia pastoris* and site-directed mutagenesis of key lid-stabilizing residues. Biochim Biophys Acta Mol Cell Biol Lipids 1862(10 Pt A):1025–1034

Sanchooli A et al (2018) VLP production from recombinant L1/L2 HPV-16 protein expressed in *Pichia Pastoris*. Protein Pept Lett 25(8):783–790

Schwarzhans JP et al (2016) Non-canonical integration events in *Pichia pastoris* encountered during standard transformation analysed with genome sequencing. Sci Rep 6:38952

Schwarzhans JP et al (2017) A mitochondrial autonomously replicating sequence from *Pichia pastoris* for uniform high level recombinant protein production. Front Microbiol 8:780

Sha C et al (2013) Enhancement of lipase r27RCL production in *Pichia pastoris* by regulating gene dosage and co-expression with chaperone protein disulfide isomerase. Enzyme Microb Technol 53(6–7):438–443

Shalgi R et al (2005) A catalog of stability-associated sequence elements in 3' UTRs of yeast mRNAs. Genome Biol 6(10):R86

Shen Q et al (2012) The effect of gene copy number and co-expression of chaperone on production of albumin fusion proteins in *Pichia pastoris*. Appl Microbiol Biotechnol 96(3):763–772

Shimizu T, Kinoshita H, Nihira T (2007) Identification and in vivo functional analysis by gene disruption of ctnA, an activator gene involved in citrinin biosynthesis in *Monascus purpureus*. Appl Environ Microbiol 73(16):5097–5103

Silva AJD et al (2021) *Pichia pastoris* displaying ZIKV protein epitopes from the envelope and NS1 induce in vitro immune activation. Vaccine 39(18):2545–2554

Sohn SB et al (2010) Genome-scale metabolic model of methylotrophic yeast *Pichia pastoris* and its use for in silico analysis of heterologous protein production. Biotechnol J 5(7):705–715

Song W et al (2020) Multiple strategies to improve the yield of chitinase a from *Bacillus licheniformis in Pichia pastoris* to obtain plant growth enhancer and GlcNAc. Microb Cell Fact 19(1):181

Stadlmayr G et al (2010) Genome-scale analysis of library sorting (GALibSo): Isolation of secretion enhancing factors for recombinant protein production in *Pichia pastoris*. Biotechnol Bioeng 105(3):543–555

Sturmberger L et al (2016) Refined Pichia pastoris reference genome sequence. J Biotechnol 235:121–131

Su GD et al (2010a) Display of *Candida antarctica* lipase B on *Pichia pastoris* and its application to flavor ester synthesis. Appl Microbiol Biotechnol 86(5):1493–1501

Su GD, Zhang X, Lin Y (2010b) Surface display of active lipase in *Pichia pastoris* using Sed1 as an anchor protein. Biotechnol Lett 32(8):1131–1136

Su X et al (2017) High-level expression and purification of a molluskan endoglucanase from *Ampullaria crossean in Pichia pastoris*. Protein Expr Purif 139:8–13

Summpunn P, Jomrit J, Panbangred W (2018) Improvement of extracellular bacterial protein production in *Pichia pastoris* by co-expression of endoplasmic reticulum residing GroEL-GroES. J Biosci Bioeng 125(3):268–274

Sun W et al (2016) High level expression and purification of active recombinant human interleukin-15 in *Pichia pastoris*. J Immunol Methods 428:50–57

Sun X et al (2019) Construction of a novel MK-4 biosynthetic pathway in *Pichia pastoris* through heterologous expression of HsUBIAD1. Microb Cell Fact 18(1):169

Sun W et al (2020) Metabolic engineering of an acid-tolerant yeast strain *Pichia kudriavzevii* for itaconic acid production. Metab Eng Commun 10:e00124

Suthers PF et al (2020) Genome-scale metabolic reconstruction of the non-model yeast *Issatchenkia orientalis* SD108 and its application to organic acids production. Metab Eng Commun 11:e00148

Tanaka T et al (2012) Recent developments in yeast cell surface display toward extended applications in biotechnology. Appl Microbiol Biotechnol 95(3):577–591

Toivari M et al (2013) Low pH D-xylonate production with *Pichia kudriavzevii*. Bioresour Technol 133:555–562

Torres P et al (2019) Contextualized genome-scale model unveils high-order metabolic effects of the specific growth rate and oxygenation level in recombinant *Pichia pastoris*. Metab Eng Commun 9:e00103

Tran VG et al (2019) Development of a CRISPR/Cas9-based tool for gene deletion in *Issatchenkia orientalis*. mSphere 4(3)

Tu Y et al (2016) Extracellular expression and antiviral activity of a bovine interferon-alpha through codon optimization in *Pichia pastoris*. Microbiol Res 191:12–18

Turkanoglu Ozcelik A, Yilmaz S, Inan M (2019) *Pichia pastoris promoters*. Methods Mol Biol 1923:97–112

Ueda M (2016) Establishment of cell surface engineering and its development. Biosci Biotechnol Biochem 80(7):1243–1253

Vadhana AK et al (2013) Improved secretion of *Candida antarctica* lipase B with its native signal peptide in *Pichia pastoris*. Enzyme Microb Technol 52(3):177–183

Vallet-Courbin A et al (2017) A Recombinant human anti-platelet scFv antibody produced in *Pichia pastoris* for atheroma targeting. PLoS One 12(1):e0170305

van der Heide M et al (2002) Overproduction of BiP negatively affects the secretion of *Aspergillus* niger glucose oxidase by the yeast *Hansenula polymorpha*. Appl Microbiol Biotechnol 58(4):487–494

Varela JA et al (2017) Applications of *Kluyveromyces marxianus* in biotechnology. In: Satyanarayana T, Kunze G (eds) Yeast diversity in human welfare. Springer Singapore, Singapore, pp 439–453

Veana F et al (2014) Gene encoding a novel invertase from a xerophilic *Aspergillus niger* strain and production of the enzyme in *Pichia pastoris*. Enzyme Microb Technol 63:28–33

Vogl T, Glieder A (2013) Regulation of *Pichia pastoris* promoters and its consequences for protein production. N Biotechnol 30(4):385–404

Vogl T et al (2016) A toolbox of diverse promoters related to methanol utilization: functionally verified parts for heterologous pathway expression in *Pichia pastoris*. ACS Synth Biol 5(2):172–186

Wang Q et al (2007) Construction of a novel system for cell surface display of heterologous proteins on *Pichia pastoris*. Biotechnol Lett 29(10):1561–1566

Wang X et al (2012) Constitutive expression of *Yarrowia lipolytica* lipase LIP2 in *Pichia pastoris* using GAP as promoter. Appl Biochem Biotechnol 166(5):1355–1367

Wang M, Jiang S, Wang Y (2016) Recent advances in the production of recombinant subunit vaccines in *Pichia pastoris*. Bioengineered 7(3):155–165

Wang P et al (2017) Accurate analysis of fusion expression of *Pichia pastoris* glycosylphosphatidylinositol-modified cell wall proteins. J Ind Microbiol Biotechnol 44(9):1355–1365

Wang L et al (2018) Efficient CRISPR-Cas9 mediated multiplex genome editing in yeasts. Biotechnol Biofuels 11:277

Waterham HR et al (1997) Isolation of the *Pichia pastoris* glyceraldehyde-3-phosphate dehydrogenase gene and regulation and use of its promoter. Gene 186(1):37–44

Wegner GH, Harder W (1987) Methylotrophic yeasts-1986. Antonie Van Leeuwenhoek 53(1):29–36

Weinacker D et al (2013) Applications of recombinant *Pichia pastoris* in the healthcare industry. Braz J Microbiol 44(4):1043–1048

Weninger A et al (2016) Combinatorial optimization of CRISPR/Cas9 expression enables precision genome engineering in the methylotrophic yeast *Pichia pastoris*. J Biotechnol 235:139–149

Weninger A et al (2018) Expanding the CRISPR/Cas9 toolkit for Pichia pastoris with efficient donor integration and alternative resistance markers. J Cell Biochem 119(4):3183–3198

Werten MW et al (1999) High-yield secretion of recombinant gelatins by *Pichia pastoris*. Yeast 15(11):1087–1096

Werten MWT et al (2019) Production of protein-based polymers in *Pichia pastoris*. Biotechnol Adv 37(5):642–666

Wilkinson B, Gilbert HF (2004) Protein disulfide isomerase. Biochim Biophys Acta 1699(1–2):35–44

Wriessnegger T et al (2014) Production of the sesquiterpenoid (+)-nootkatone by metabolic engineering of *Pichia pastoris*. Metab Eng 24:18–29

Wriessnegger T et al (2016) Enhancing cytochrome P450-mediated conversions in *P. pastoris* through RAD52 over-expression and optimizing the cultivation conditions. Fungal Genet Biol 89;114–125

Xi Y et al (2021) Characterization of JEN family carboxylate transporters from the acid-tolerant yeast *Pichia kudriavzevii* and their applications in succinic acid production. Microb Biotechnol 14(3):1130–1147

Xiao H et al (2014) Exploiting *Issatchenkia orientalis* SD108 for succinic acid production. Microb Cell Fact 13:121

Xiong R, Chen J (2008) Secreted expression of human lysozyme in the yeast *Pichia pastoris* under the direction of the signal peptide from human serum albumin. Biotechnol Appl Biochem 51(Pt 3):129–134

Xiong AS et al (2005) High level expression of a recombinant acid phytase gene in *Pichia pastoris*. J Appl Microbiol 98(2):418–428

Xue Y et al (2017) Methylotrophic yeast *Pichia pastoris* as a chassis organism for polyketide synthesis via the full citrinin biosynthetic pathway. J Biotechnol 242:64–72

Yaegashi J et al (2017) Rhodosporidium toruloides: a new platform organism for conversion of lignocellulose into terpene biofuels and bioproducts. Biotechnol Biofuels 10:241

Yaguchi A et al (2017) Metabolism of aromatics by *Trichosporon oleaginosus* while remaining oleaginous. Microb Cell Fact 16(1):206

Yamada R et al (2019) Toward the construction of a technology platform for chemicals production from methanol: D-lactic acid production from methanol by an engineered yeast *Pichia pastoris*. World J Microbiol Biotechnol 35(2):37

Yang S et al (2017) Cell-surface displayed expression of trehalose synthase from *Pseudomonas putida* ATCC 47054 in Pichia pastoris using Pir1p as an anchor protein. Front Microbiol 8:2583

Yang Z et al (2020) Development and characterization of an enterovirus 71 (EV71) virus-like particles (VLPs) vaccine produced in *Pichia pastoris*. Hum Vaccin Immunother 16(7):1602–1610

YaPing W et al (2017) High-level expression of l-glutamate oxidase in *Pichia pastoris* using multi-copy expression strains and high cell density cultivation. Protein Expr Purif 129:108–114

Ye R et al (2017) Comprehensive reconstruction and evaluation of *Pichia pastoris* genome-scale metabolic model that accounts for 1243 ORFs. Bioresour Bioprocess 4(1):22

Yin C et al (2014) High-level expression of a manganese superoxide dismutase (PoMn-SOD) from *Pleurotus ostreatus in Pichia pastoris*. Appl Biochem Biotechnol 174(1):259–269

Yu X et al (2007) Expression and purification of ancrod, an anticoagulant drug *Pichia Pastoris*. Protein Expr Purif 55(2):257–261

Yu P et al (2015) Improving the secretory production of the heterologous protein in *Pichia pastoris* by focusing on protein folding. Appl Biochem Biotechnol 175(1):535–548

Yu KM et al (2018) Efficient expression and isolation of recombinant human interleukin-11 (rhIL-11) in *Pichia pastoris*. Protein Expr Purif 146:69–77

Yuangsaard N et al (2013) Selection and characterization of a newly isolated thermotolerant *Pichia kudriavzevii* strain for ethanol production at high temperature from cassava starch hydrolysate. Antonie Van Leeuwenhoek 103(3):577–588

Zhang W et al (2006) Enhanced secretion of heterologous proteins in *Pichia pastoris* following overexpression of *Saccharomyces cerevisiae* chaperone proteins. Biotechnol Prog 22(4):1090–1095

Zhang L et al (2013) Screening for glycosylphosphatidylinositol-modified cell wall proteins in *Pichia pastoris* and their recombinant expression on the cell surface. Appl Environ Microbiol 79(18):5519–5526

Zhang XF et al (2019a) High-level expression of *Aspergillus niger* lipase in *Pichia pastoris*: characterization and gastric digestion in vitro. Food Chem 274:305–313

Zhang D et al (2019b) Improved cadmium resistance and removal capacity in *Pichia kudriavzevii* A16 by sucrose preincubation. J Basic Microbiol 59(9):867–878

Zheng JY et al (2019) High-level expression and characterization of a stereoselective lipase from *Aspergillus oryzae* in *Pichia pastoris*. Protein Expr Purif 155:1–7

Zhong X et al (2004) Secretion, purification, and characterization of a recombinant *Aspergillus oryzae* tannase in *Pichia pastoris*. Protein Expr Purif 36(2):165–169

Zhu T et al (2019) *Pichia pastoris* as a versatile cell factory for the production of industrial enzymes and chemicals: current status and future perspectives. Biotechnol J 14(6):e1800694

Zirpel B et al (2018) Optimization of Δ(9)-tetrahydrocannabinolic acid synthase production in *Komagataella phaffii* via post-translational bottleneck identification. J Biotechnol 272–273:40–47

Kluyveromyces marxianus as a Platform in Synthetic Biology for the Production of Useful Materials

Noppon Lertwattanasakul, Mochamad Nurcholis, Nadchanok Rodrussamee, Tomoyuki Kosaka, Masayuki Murata, and Mamoru Yamada

Abstract

In *Kluyveromyces marxianus*, a thermotolerant yeast that has already been utilized for producing various useful materials, recent advances including complete genome sequencing and transcriptome analysis have provided valuable information on its physiological and metabolic characteristics including its capacity for assimilation of various sugars and tolerance to high temperatures, which are distinct from *Saccharomyces cerevisiae.* These prominent properties enable the development of high-temperature fermentation (HTF) technology for industrial applications. In addition, the yeast is able to integrate DNA fragments into its genome, which makes it easy to introduce foreign genes or gene sets

N. Lertwattanasakul
Department of Microbiology, Faculty of Science, Kasetsart University, Bangkok 10900, Thailand

M. Nurcholis
Department of Food Science and Technology, Faculty of Agricultural Technology, Brawijaya University, Malang 65145, Indonesia

N. Rodrussamee
Department of Biology, Faculty of Science, Chiang Mai University, Chiang Mai 50200, Thailand

T. Kosaka · M. Yamada (✉)
Department of Biological Chemistry, Faculty of Agriculture, Yamaguchi University, Yamaguchi 753-8515, Japan
e-mail: m-yamada@yamaguchi-u.ac.jp

T. Kosaka · M. Murata · M. Yamada
Graduate School of Science and Technology for Innovation, Yamaguchi University, Yamaguchi 753-8515, Japan

T. Kosaka · M. Yamada
Research Center for Thermotolerant Microbial Resources, Yamaguchi University, Yamaguchi 753-8515, Japan

F. Darvishi Harzevili (ed.), *Synthetic Biology of Yeasts*,
https://doi.org/10.1007/978-3-030-89680-5_11

into the genome. Moreover, DNA editing technology and simulation models are in place. Therefore, *K. marxianus* is a new platform in synthetic biology for producing useful materials.

1 Introduction

Microbial production of fuels and chemicals from biomass and other renewable carbon sources is an attractive alternative to petroleum-derived production. *Saccharomyces cerevisiae* is the organism of choice because of its high rate of production and tolerance to ethanol titers upwards of 120 g L^{-1} (Qiu and Jiang 2017; Nielsen et al. 2013). These outstanding phenotypes, among others, may have led to the widespread study of *S. cerevisiae* and its development as a model eukaryotic host for chemical biosynthesis.

Kluyveromyces marxianus is a Crabtree-negative yeast and favors respiration over fermentation. This yeast is also industrially relevant because of its wide substrate spectrum including pentose sugars, fast-growth characteristics, and thermotolerance to ~50 °C (Varela et al. 2017; Löbs et al. 2016), making it a promising host for industrial biotechnology to produce renewable chemicals from agricultural biomass feedstocks. Native strains of *K. marxianus* are also known to synthesize ethyl acetate at rates above 2 g L^{-1} h^{-1} in aerated bioreactors (Löser et al. 2013, 2015). In recent years, increasing interest has been shown in several new applications including production of biomolecules (Hughes et al. 2017; Lin et al. 2017), biocatalysts (Wang et al. 2017; Simoness et al. 2017), and heterologous protein expression (Lee et al. 2017; Gombert et al. 2016). However, major genetic engineering limitations have kept this yeast from replacing the commonly used yeast *S. cerevisiae* in industrial applications.

2 Assimilation Capacity of Various Substrates

One of the features of *K. marxianus* that are suitable for industrial applications is its ability to utilize a broad range of substrates. The major common feature of sugar utilization by *K. marxianus* is the ability to assimilate lactose and inulin as a carbon source, a feature that is absent in *S. cerevisiae* (Lane and Morrissey 2010). The ability has evolved by acquiring *LAC* genes, *LAC12* and *LAC4*, that are accountable for the uptake of lactose and subsequent cleavage into galactose and glucose, respectively, and the inulinase gene *INU1*, which liberates fructose molecules from oligo- or poly-sugars with β-(2,1)-linked fructose units at the terminal (Rouwenhorst et al. 1988).

Consistent with the capacity of this yeast to utilize a wide variety of substrates, there are a number of sugar transporters, including 27 putative sugar transporters, encoded by the genome of *K. marxianus* DMKU 3-1042 (Lertwattanasakul et al. 2015). The ability for utilization of a wide variety of carbon sources by various strains of *K. marxianus* compared to *S. cerevisiae* has been shown (Nonklang et al.

2008) (Fig. 1). These carbon sources include monosaccharides of six-carbon sugars (galactose, fructose, and mannose), monosaccharides of five-carbon sugars (xylose, xylitol, and arabinose), disaccharides (sucrose, lactose, and cellobiose), trisaccharide (raffinose), polysaccharide (inulin), and a non-fermentable carbon source of glycerol. Besides, *K. marxianus* DMB1 can utilize sorbitol, which is not generally utilized by *K. marxianus* strains, as a non-fermentable carbon source (Goshima et al. 2013b). *K. marxianus* can utilize these carbon sources, while *S. cerevisiae* cannot utilize cellobiose, xylose, xylitol, arabinose, glycerol, sorbitol, and lactose.

Ethanol is one of the classic bulk chemical products of yeasts. Much interest has been shown in *K. marxianus* for the possibility of it being a platform for ethanol production because of not only its capability of utilizing a broad range of substrates but also its fermentation ability at high temperatures. These properties lead us to develop a stable HTF technology with several benefits as described below and to utilize not only first-generation substrates but also second- or third-generation substrates.

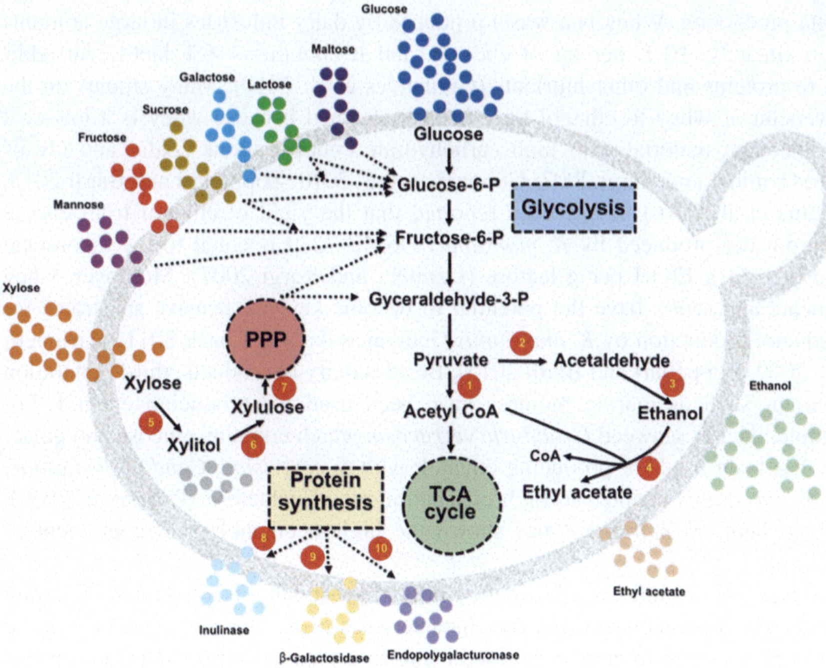

Fig. 1 Metabolic pathways for synthesis of high-value chemicals in *K. marxianus*. Abbreviated versions of how these processes that lead to the formation of classes of metabolite are shown, along with examples cited in the main text. Reactions are depicted as unidirectional, although many are reversible. Important enzymatic steps are indicated by the key, as follows: 1. pyruvate dehydrogenase; 2. pyruvate decarboxylase; 3. alcohol dehydrogenase; 4. alcohol acyl transferase; 5. xylose reductase; 6. xylitol dehydrogenase; 7. xylulokinase; 8. inulinase; 9. β-galactosidase; 10. endopolygalacturonase

Related to conversion from lactose to ethanol, which has been received much attention, expression of genes for β-galactosidases in *K. marxianus* is induced by its natural inducers, galactose, and lactose. However, the production of β-galactosidases seems to be dependent on the substrate concentration. In *K. marxianus* CBS6556, the maximum activity of β-galactosidase is obtained at low concentrations of the inducer carbohydrates, in the range between 0.5 and 15 mM. When shifted to a high concentration of D-galactose or lactose, a repressive mechanism is superimposed to the inducing effect of the substrate (Martins et al. 2002). *K. marxianus* exhibits glucose repression in galactose utilization, and galactose is utilized after diminishing glucose in a mixed-sugar medium (Rodrussamee et al. 2011). Moreover, mutation of *MIG1*, which plays key roles as a regulator complex in glucose repression, increases the activity of lactose hydrolysis (Zoppellari and Bardi 2013).

In dairy industries, dairy effluents as high strength wastewater contains a high concentration of lactose, which is responsible for the high chemical oxygen demand (COD), causing an environmental problem (Karim et al. 2020). Whey and permeate are effluents from cheese processing, while scotta is effluent from ricotta processing. Whey is a waste produced by dairy industries in large amounts (approximately, 10 L per kg of cheese), and it contains 4–6% lactose in addition to proteins and other nutrients (Guimarães et al. 2010). Many studies on the conversion of whey to ethanol have been performed because whey is a low-cost and abundant material with high carbohydrate concentrations (Zafar and Owais 2006; Ozmihci and Kargi 2007; Guimarães et al. 2010; Zoppellari and Bardi 2013; Roohina et al. 2016). It has been reported that the yield of ethanol from cheese whey powder produced by *K. marxianus* DSMZ-7239 is equal to the theoretical yield of 0.54 g EtOH per g lactose (Ozmihci and Kargi 2007). Moreover, whey permeate and scotta have the potential to become key inexpensive substrates for bioethanol production by *K. marxianus* (Jedrzejewska and Kozak 2011; Sansonetti et al. 2009; Zoppellari and Bardi 2013). Besides dairy byproducts, third-generation substrates such as marine biomass have been used for producing ethanol. For example, the red seaweed *Gracilaria verrucosa*, which contains glucose and galactose, has been used for producing ethanol, with *S. cerevisiae*, *Candida lusitaniae,* and *K. marxianus* adapted to high concentrations of galactose (Park et al. 2020). Among them, *K. marxianus* has shown the highest ethanol yield coefficient of 0.47 g g^{-1}.

Conversion of inulin as a D-fructose polymer to ethanol is a notable conversion process that maximally utilizes the characteristics of *K. marxianus,* and extensive studies on inulin have been performed using Jerusalem artichoke (JA), a crop that contains nearly 20% of carbohydrates, 70–90% of which is inulin (Hu et al. 2012; Yuan et al. 2012; Kim et al. 2013; Kim and Kim 2014; Charoensopharat et al. 2015). Inulinase encoded by *INU1* is very effective in hydrolyzing inulin to D-fructose monomers, which are then imported into cells by the fructose transporter Frt1p and catabolized through the glycolytic pathway. Regulation of the expression of *INU1* in *K. marxianus* has been studied using various strains, and it has been shown that the production of inulinase depends on the carbon source (Sokolenko and Karpechenko 2015; Lertwattanasakul et al. 2011; Hoshida et al. 2018). In *K. marxianus* NCIM 3231, sucrose and fructose acted as inducers for inulinase production, though their actions were weaker than that of inulin, whereas in *K. marxianus* CDBBL278, no such inulinase induction was found (Cruz-Guerrero et al. 1995). The expression of *INU1* also seems to be under the control of glucose repression (Gupta et al. 1994). *K. marxianus* is one of the non-conventional yeast candidates as ideal consolidated bioprocessing (CBP). Ethanol production from inulin by CBP in *K. marxianus* Y179 was found to be dependent on inulin concentrations and aeration levels, with the yeast producing ethanol up to 98 g $L^{-1}$from inulin with an ethanol yield of 0.43 g g^{-1} at 30 °C (Gao et al. 2015). Direct ethanol fermentation from fresh JA tubers without inulin hydrolysis by CBP using *K. marxianus* DBKKUY-102 achieved a maximum ethanol concentration of 97.46 g L^{-1} at 40 °C (Charoensopharat et al. 2015).

Lignocellulosic biomass is an attractive source, which does not compete with land for food production, because it is the most abundant renewable and inexpensive biomass (Hahn-Hägerdal et al. 2006). The primary components in lignocellulosic biomass are cellulose (35–50 wt.%, dry basis), hemicellulose (15–30%), pectin (2–5%), and lignin (12–35%). Cellulose and hemicellulose, which account for more than 50% of the total mass, can be converted to sugars for their conversion to ethanol (Kumar et al. 2016). In contrast to cellulose and starch, lignocellulose is composed of a mixture of hexose (glucose, mannose, and galactose) and pentose (xylose and arabinose) sugars, which can be released by pretreatment and enzymatic hydrolysis steps (Margeot et al. 2009). Cost-competitive ethanol yields from lignocellulose, require fermentation of both hexose and pentose constituents (Galbe and Zacchi 2007). In order to realize economical ethanol production from lignocellulose, one of the prerequisites for making lignocellulosic ethanol processes economically competitive is a robust-fermenting microorganism that can effectively convert all sugars released from lignocellulose to ethanol with a high yield and high productivity. Unlike *S. cerevisiae*, which is unable to utilize pentose sugars as a sole carbon source for growth and fermentation (Kuhn et al. 1995), *K. marxianus* can effectively uptake and catabolize the pentoses, xylose, and arabinose, via the aldose reductase pathway and produce ethanol from all sugar components in lignocelluloses except for arabinose (Rodrussamee et al. 2011). However, the yeast inefficiently produces ethanol from xylose compared to hexose and co-produces xylitol and/or acetic acid as byproducts, and the consumption of xylose is exposed to glucose repression (Harner et al. 2015; Nitiyon et al. 2016; Rodrussamee et al. 2011). Xylose fermentation of this strain also requires microaerophilic conditions for relieving the cofactor imbalance due to different preferred coenzymes of xylose reductase (XR) and xylitol dehydrogenase (XDH), having a higher affinity for NADPH and NAD^+, respectively (Harner et al. 2015). Some strains are capable of fermenting xylose, though inefficiently, even at temperatures of more than 40 °C (Suryawati et al. 2008; Kumar et al. 2009; Rodrussamee et al. 2011). Metabolic engineering strategies may thus be required to improve the capability for fermentation of xylose, like those adopted for *S. cerevisiae* (Matsushika et al. 2009). Many attempts have been made to improve the coenzyme specificities by replacing *KmXR* and/or *KmXDH* with the corresponding genes from other microorganisms with or without modification by site-directed mutagenesis (Table 1) such as by using XR of *Scheffersomyces stipitis* (*SsXR*) (Zhang et al. 2013), XR and XDH of *S. stipitis* (*SsXR* and *SsXDH*) together with xylulokinase (XK) of *S. cerevisiae* (*ScXK*) (Goshima et al. 2013a), and XR of *Neurospora crassa* (*NcXR*) and *SsXDH* together with overexpression of several downstream *K. marxianus* genes (Zhang et al. 2015a, b). Recently, Suzuki et al. (2019) developed a recombinant *K. marxianus* DMB13 strain by multiple site-directed mutagenesis to overexpress an $NADP^+$-dependent *KmXDH* mutant gene and by overexpression of the wild-type *KmXR* and *KmXK* genes. The recombinant strain rapidly converted xylose to ethanol after depletion of glucose and achieved the maximum ethanol yield of 0.402 g g^{-1} in xylose/glucose co-fermentation at 40 °C. An alternative approach of expressing a xylose isomerase (XI), which directly converts xylose to xylulose, has been performed in *K. marxianus.* Wang et al. (2013) transformed the xylose isomerase gene *XYLA* from the fungus *Orpinomyces* under control of the *GAPDH* promoter in *K. marxianus* YHJ010 in which *KmXR* and *KmXDH* had been disrupted and adapted the transformant to a xylose medium by repetitive cultivation. The resultant strain achieved ethanol yields of 0.38 g g^{-1} and 0.31 g g^{-1} at 42 °C and 45 °C, respectively, and produced 8.25 g L^{-1} ethanol from the corn cob hydrolysate containing 20.04 g L^{-1} xylose. Similar evolutionary adaptation to xylose at 45 °C improved ethanol and xylitol yields in the xylose medium (Sharma et al. 2016, 2017). A trial of an evolutionary adaptation approach together with random mutagenesis using ethyl methanesulfonate also improved the ethanol yield from xylose, being 2.31-fold higher than that of the parental strain (Kwon et al. 2019). On the other hand, a random *kan*MX4-insertion mutagenesis revealed essential factors for pentose metabolism in *K. marxianus*, including a cytochrome oxidase assembly factor (singleton) (*KmCOX15*), a transcription factor required for assembly of the Atp9p subunit of mitochondrial ATP synthase (*KmATP25*) and cytochrome *c* heme lyase (*KmCYC3*), suggest that respiratory activity is essential for utilization of pentose in (Lertwattanasakul et al. 2013).

Consolidated bioprocessing (CBP) is an alternative and interesting technology for conversion of lignocellulose into desired products in one step without exogenous enzymes because of its simplicity and potentially low cost (Olson et al. 2012). Breakdown of lignocellulose into fermentable sugars requires three major types of enzymes, endo-1,4-β-glucanase (EC 3.2.1.4), cellobiohydrolase (EC 3.2.1.91), and β-glucosidase (EC 3.2.1.21), and several hemicellulases such as endo-1,4-β-xylanase (EC 3.2.1.8), endo-1,4-β-mannanase (EC 3.2.1.78), β-xylosidase (EC 3.2.1.27), and hemicellulolytic esterases (Shallom and Shoham 2003; Li et al. 2017). However, no natural microorganisms with both efficient enzyme production capability for lignocellulose saccharification and ethanol production ability are currently available (Hasunuma and Kondo 2012). Since, fortunately, *K. marxianus* has some capability to utilize cellobiose and xylose, heterologous expression of other cellulolytic and/or xylanolytic enzymes may expand its capability for the conversion of lignocellulose to biorefinery products including ethanol.

Table 1 Examples of native and non-native chemicals produced by *K. marxianus*

Aimed product	Modification	Source	Parental strain	Cultivation condition	Titer	References
L-Lactic acid	↑*PfLDH*	*Plasmodium falciparum*	YZB058	Glucose/xylose, 42 °C	103 g/L	Kong et al. (2019)
	↑*BmLDH*	*Bacillus megaterium*				
	↑*ScJEN1*	*Saccharomyces cerevisiae*				
	↑*KmPFK* Δ*KmDLD1*	*Kluyveromyces marxianus*				
	↑*Bmldh*	*B. megaterium*	KM1	–	24 g/L	Pecota et al. (2007)
Xylitol	↑*ScGAL2* (N376F)	*S. cerevisiae*	KCTC 17,555	Glucose/xylose, 30 °C	95.58 g/L	Kwon et al. (2020)
	↑*NcXYL1*	*Neurospora crassa*	NBRC1777	Xylose, 45 °C	60.03 g/L	Zhang et al. (2014)
2-Phenylethanol	↑*KmARO4* (K221L) ↑*KmPHA2* ↑*KmARO7* (G141S) ↑*KmARO10* Δ*KmEAT1*	*K. marxianus*	CBS 6556	Glucose, 30 °C	1943 ± 63 mg/L	Li et al. (2021)
	↑*KmTKL1* ↑*KmTAL1* ↑*KmARO3* (feedback resistant) Δ*KmARO8* *KmTYR1pr*Δ::*REV1pr*	*K. marxianus*	NBRC1777	Glucose, 30 °C	850 mg/L	Rajkumar and Morrissey (2020)
	↑*Bbxfpk*	*Bifidobacterium breve*				
	↑*Septa*	*Salmonella enterica*				

(continued)

Table 1 (continued)

Aimed product	Modification	Source	Parental strain	Cultivation condition	Titer	References
	↑*EcppsA*	*Escherichia coli*				
	↑*ScARO10* ↑*ScADH2*	*S. cerevisiae*	DMKU 3-1042	Glucose, 30 °C	1.0 g/L	Kim et al. (2014)
Hexanoic acid	↑*atoB*	*E. coli*	ATCC17555	Galactose, 37 °C	154 mg/L	Cheon et al. (2014)
	↑*bktB*	*Ralstonia eutropha*				
	↑*crt* ↑*hbd*	*Clostridium acetobutylicum*				
	↑*ter*	*Treponema denticola*				
	↑*MCT1*	*S. cerevisiae*				
Pyruvate	Δ*KmPDC1* Δ*KmGPD1* ↑*KmMTH1-ΔT*	*K. marxianus*	YZJ051	Glucose/xylose, 42 °C	29.21 g/L	Zhang et al. (2017a, b)
	↑*SsXYL2*	*Scheffersomyces stipitis*				
	↑*ScGAL2* (N376F)	*S. cerevisaie*				
β-Carotene	↑*HpCHYb*	*Haematococcus pluvialis*	KY3	Galactose, 30 °C	224.4 ± 9.9 mg/g DCW	Chang et al. (2015)
Triacetic acid lactone	↑*G2PS1*	*Gerbera hybrida*	CBS 6556	Xylose, 37 °C	1.24 g/L	McTaggart et al. (2019)
D-Allulose	↑*dpe*	*Agrobacterium tumefaciens*	CICC 1911	Fructose, 55 °C	190 g/L	Yang et al. (2018)
Fatty acid ethyl esters	↑*KmATF1*	*K. marxianus*	UMPe-1	Glucose, 30 °C	1.53%	Campos-García et al. (2018)
Cellulases	↑*EGI* ↑*EGIA*	*Aspergillus niger*	KY3	Glucose	–	Chang et al. (2017)

(continued)

Table 1 (continued)

Aimed product	Modification	Source	Parental strain	Cultivation condition	Titer	References
	↑*EGIII* ↑*CBHI*	*Trichoderma reesei*				
	↑*CBHII*	Chemically synthesized and optimized for *K. marxianus*				
	↑*NpaBGS*	*Neocallimastix patriciarum*				
Tannase	–	–	NRRL Y-8281	Olive pomace, 45 °C	1026.12 U/mg	Mahmoud et al. (2018)
α-Galactosidase	–	*Cyamopsis tetragonoloba*	CBS 6556	Starch	153 mg/L	Bergkamp et al. (1993)
α-Amylase	↑*TAA*	*A. oryzae*	DMKU 3-1042	–	–	Hoshida et al. (2014)
β-Galactosidase	↑*KlLAC4*	*Kluyveromyces lactis*	NBRC1777	Glucose, 45 °C	20 U/mg	Yang et al. (2015)
β-Glucuronidase	↑*gusA*	*E. coli*	NBRC1777	Glucose, 45 °C	0.8 U/mg	Yang et al. (2015)
Superoxide dismutase	↑*KlSOD1*	*K. lactis*	L3	Glucose, 30 °C	24 kU/L	Raimondi et al. (2010)
β-Glucuronidase	↑*gusA*	*E. coli*	KM1	Glucose, 37 °C	50 nmole 4-MU/min /mL	Pecota and Da silva (2005)
Glucose oxidase	↑*GOX*	*A. niger*	CBS 6556	Glucose, 30 °C	1722 U/g DCW	Rocha et al. (2010)

(continued)

Table 1 (continued)

Aimed product	Modification	Source	Parental strain	Cultivation condition	Titer	References
Esterase	↑*est*	*Thermus thermophilus*	CBS 6556	Sucrose	56.9 U/g DCW	Rocha et al. (2011)
Glucoamylase	↑*GAA*	*Arxula adeninivorans*	L3	Glucose, 40 °C	80 U/mL	Raimondi et al. (2013)
Dengue virus type I non-structural protein 1	–	Dengue virus	UFV-3	Galactose, 37 °C	1.2 mg/mL	Bragança et al. (2015)
Cytochrome P450 monooxygenase	↑*CYP505A1*	*Fusarium oxysporum*	UOFS Y1185	Glucose, 28 °C	–	Theron et al. (2014)
	↑*CYP102A1*	*B. megaterium*				
Xylose isomerase	↑*XYLA*	*Orpinomyces*	NBRC1777	Xylose	–	Wang et al. (2013)
Porcine circovirus type 2 virus-like particles	–	Porcine circovirus type 2	FIM-1	Glucose, 30 °C	1.91 g/L	Duan et al. (2019)
Porcine parvovirus virus-like particles	–	Parvovirinae virus	FIM-1	Glucose, 30 °C	2.5 g/L	Yang et al. (2021)
Single-chain antibody	↑scFv	HyHEL-10 scFv	NBRC1777	Xylose, 30 °C	–	Nambu-Nishida et al. (2018)
Ethanol	↑*KmXYL2* ($NADP^+$-dependent)	*K. marxianus*	DMB13	Xylose/glucose, 40 °C	–	Suzuki et al. (2019)
	↑*PsXYL1* (N272D)	*Pichia stipitis*	NBRC1777	Xylose, 42 °C	3.55 g/L	Zhang et al. (2013)
	↑*NcXYL1*	*N. crassa*	NBRC1777	Xylose/glucose, 42 °C	6.22 g/L	Zhang et al. (2015a, b)

(continued)

Table 1 (continued)

Aimed product	Modification	Source	Parental strain	Cultivation condition	Titer	References
	↑*PsXYL2*	*P. stipitis*				
	↑*PsXYL1* ↑*PsXYL2*	*P. stipitis*	DMB1	Xylose, 45 °C	3.3 g/L	Goshima et al. (2013a)
	↑*ScXYL3*	*S. cerevisiae*				
	↑*AMY*	*A. oryzae*	NBRC1777	Starch, 48 °C	36.88 g/L	Wang et al. (2014)
	↑*AMY* ↑*GAM1*	*Debaryomyces occidentalis*				
	↑*AMY*	*A. oryzae*	DMKU 3-1042	–	–	Nonklang et al. (2008)
	↑*EGIII* ↑*CBHI*	*T. reesei*	KY3	Cellulose, 30 °C	0.9 g/L	Chang et al. (2012)
	↑*EG* ↑*CBH*	*T. reesei*	KY3	Cellulose, 40 °C	0.6 g/L	Chang et al. (2013)
	↑*EG* ↑*CBH*	*A. niger*				
	↑*BGL*	*N. patriciarum*				
	↑*CDT*	*N. crassa*				
	↑*EG*	*A. niger*	NBRC1777	Cellobiose, 45 °C	43.4 g/L	Hong et al. (2007)
	↑*CBH* ↑*BGL*	*T. reesei*				
	↑*EG*	*T. reesei*	NBRC1777	Cellulosic β-glucan, 48 °C	4.24 g/L	Yanase et al. (2010)
	↑*BGL*	*A. aculeatus*				
	↑*FLO1* ↑*FLO5* ↑*FLO9* or ↑*FLO10*	*S. cerevisiae*	DMKU 3-1042	Glucose, 40 °C	48 g/L	Nonklang et al. (2009)

(continued)

Table 1 (continued)

Aimed product	Modification	Source	Parental strain	Cultivation condition	Titer	References
	↑*BGL*	*Thermoascus aurantiacus*	NBRC1777	Cellobiose, 45 °C	29.5 g/L	Matsuzaki et al. (2012)
	↑*ScGAL2* ↑*ScGAL2* (N376F)	*S. cerevisiae*	KCTC 17555	Glucose/galactose, 30 °C	36.14 g/L	Kwon et al. (2020)
	↑*olpB-Sc*GPI ↑*cipA*	*C. thermocellum*	4G5	Avicel, 30 °C	3.09 g/L	Anandharaj et al. (2020)
		S. cerevisiae		Phosphoric acid-swollen cellulose, 30 °C	8.61 g/L	

LDH lactate dehydrogenase, *JEN1* proton-coupled monocarboxylate transporter, *PFK* 6-phosphofructokinase, *dld1* putative d-lactate dehydrogenase, *GAL2* galactose permease, *XYL1* xylose reductase, *XYL2* xylitol dehydrogenase, *XYL3* xylulokinase, *XYLA* xylose isomerase, *ARO3/ARO4* DAHP synthase, *ARO7* chorismate mutase, *ARO8* aromatic aminotransferase, *ARO10* phenylpyruvate decarboxylase, *PHA2* prephenate dehydratase, *eat1* ethanol acetyltransferase, *TKL1* transketolase, *TAL1* transaldolase, *xfpk* phosphoketolase, *pta* phosphotransacetylase, *ppsA* phosphoenolpyruvate synthase, *ADH2* alcohol dehydrogenase 2, *atoB* acetyl-CoA acetyltransferase, *bktB* β-ketothiolase, *crt* crotonase, *hbd* 3-hydroxybutyryl-CoA dehydrogenase, *ter* trans-enoyl-CoA reductase, *MCT1* malonyl CoA-acyl carrier protein transacylase, *PDC1* pyruvate decarboxylase, *GPD1* glycerol-3-phosphate dehydrogenase, *MTH1* negative regulator of the glucose-sensing signal transduction pathway, *CHYb* β-carotene hydroxylase, *G2PS1* 2-pyrone synthase, *dpe* D-psicose-3-epimerase, *ATF1* alcohol acetyl-transferase, *EG* endoglucanase, *CBH* cellobiohydrolase *BGL* β-glucosidase, *LAC4* β-galactosidase, *gusA* β-glucuronidase, *SOD1* superoxide dismutase, *GOX* glucose oxidase, *est* esterase, *GAA* glucoamylase, *CYP* cytochrome P450 monooxygenase, *AMY* amylase, *GAM1* glucoamylase, *CDT* cellodextrin transporter, *FLO* flocculating, *GPI* glycosylphosphatidylinositol, *cipA* scaffoldin, *olpB* anchoring protein

A recombinant of *K. marxianus* NBRC1777 expressing endoglucanase from *Trichoderma reesei* and β-glucosidase from *Aspergillus aculeatus* on its cell surface produced 43.4 g L^{-1} ethanol, corresponding to 90% of the theoretical yield from 10% cellobiose (Yanase et al. 2010). Chang et al. (2012) developed a technique named Promoter-based Gene Assembly and Simultaneous Overexpression (PGASO), which uses overlapping oligonucleotides for recombinatorial assembly of gene cassettes with individual promoters to engineer *K. marxianus* KY3 for co-expressing three types of cellulase genes for exoglucanase and endoglucanase from *Trichoderma reesei* and β-glucosidase from a cow rumen fungus. The engineered strain was found to utilize β-glycan, cellobiose, or carboxymethyl cellulose as the sole carbon source for growth and directly convert cellobiose and β-glycan to ethanol. In addition, the co-expression of five cellulase genes for two cellobiohydrolases, two endo-β-1,4-glucanases from *Trichoderma reesei*, and β-glucosidase from the cow rumen fungus *Neocallimastix patriciarum*, and a gene for the cellodextrin transporter from *N. crassa* resulted in direct conversion of cellulose to ethanol, producing 0.6 g L^{-1} ethanol from 10% avicel (Chang et al. 2013). Zhou et al. (2018) attempted heterologous expression and secretion of lignocellulolytic enzymes, endo-1,4-β-mannanase, endo-1,4-β-endoxylanase, endo-1,4-β-glucanase, and feruloyl esterase in *K. marxianus* using the *INU1* promoter and the signal sequence of inulinase.

3 Thermotolerance and High-Temperature Fermentation Ability

K. marxianus has the interesting feature of producing several useful materials such as ethanol as its main product (Table 2). The application of HTF at a temperature of about 40 °C or higher is expected to provide many advantages over general fermentation at a temperature of about 30 °C or less including reducing operating costs of cooling systems, minimizing risk of contamination, reducing viscosity of the fermentation broth, efficiently achieving simultaneous saccharification and fermentation, robustness process against accidental temperature rise even in tropical countries, operating continuous ethanol removal, and increasing enzyme activity for biomass hydrolysis (Banat et al. 1998; Fonseca et al. 2008; Hoshida and Akada 2017; Kosaka et al. 2018). Fermenting microbes for HTF must be thermotolerant, that is, they must grow sufficiently and ferment efficiently at high temperatures.

Yeast strains that have often been isolated as ethanol-producing thermotolerant yeasts are strains of *K. marxianus* and *S. cerevisiae* because they are highly thermotolerant yeasts and efficient ethanol-producing yeasts, respectively (Hoshida and Akada 2017). *K. marxianus* DMKU 3-1042 efficiently produced ethanol from sugar cane juice at a temperature of 40 °C, and the maximal ethanol concentration was 67.8 g L^{-1} and the yield was 60.4% of the theoretical yield (Limtong et al. 2007). Five *K. marxianus* strains isolated in Laos exhibited strong fermentation abilities in a 16% sugars-containing medium of glucose, sucrose, sugarcane, or molasses at 40 °C (Keo-oudone et al. 2016). The evolutionary adapted KM-100d

Table.2 Comparison of ethanol production levels and yields among various strains of *K. marxianus*

Strains	Temp (°C)	Carbon sources	Time (h)	Ethanol production (g L^{-1})	Ethanol yield (g g^{-1})	References
K. marxianus DMKU 3-1042	30	Xylose	48	1.71 ± 0.45	0.09 ± 0.03	Nitiyon et al. (2016)
	30	Xylose	72	~2.60	0.13	Rodrussamee et al. (2011)
	37	Xylose	36	1.29 ± 0.23	0.07 ± 0.01	Nitiyon et al. (2016)
	40	Xylose	72	~2.20	0.11	Rodrussamee et al. (2011)
	45	Xylose	48	~0.96	0.06	Rodrussamee et al. (2011)
	40	Sugar cane juice	60	67.8	0.6	Limtong et al. (2007)
K. marxianus BUNL-21	30	Xylose	48	2.91 ± 0.40	0.15 ± 0.02	Nitiyon et al. (2016)
	37	Xylose	36	2.58 ± 0.05	0.14 ± 0.00	Nitiyon et al. (2016)
	40	Glucose	24	68.7	0.43	Keo-oudone et al. (2016)
K. marxianus IMB3	45	Xylose		0.8–1.2	0.08–0.12	Banat et al. (1998)
K. marxianus FIM1	30	Glucose	48	110	0.55	Mo et al. (2019)
	45	Glucose	48	50	0.25	Mo et al. (2019)
K. marxianus	40	Mannose	24	8.52	0.426	Rouhollah et al. (2007)
K. marxianus SBK1	40	Glucose and xylose	72	23.82	0.35	Kim et al. (2019)
K. marxianus DMB1	42	Lignocellulosic hydrolysates		25.98		Goshima et al. (2013b)
K. marxianus CICC 1727–5	40	Lignocellulosic hydrolysates		42.6		Du et al. (2019)
K. marxianus DBKKU Y-102	37	Inulin jerusalem artichoke	72	~104.83	0.47	Charoensopharat et al. (2015)
	40	Inulin jerusalem artichoke	72	~97.46	0.45	Charoensopharat et al. (2015)
K. marxianus Y179	30	Inulin	36	98	0.43	Gao et al. (2015)

mutant improved ethanol productivity to produce about 50 g L^{-1} in a glucose-containing medium at 45 °C (Mo et al. 2019). Rouhollah et al. (2007) co-cultured *P. stipitis* and *K. marxianus* to convert mixed sugars of glucose and xylose to ethanol and achieved a high ethanol yield (0.42 g g^{-1}) and high maximum ethanol (31.87 g L^{-1}) at 40 °C. Goshima et al. (2013b) reported fermentation of ethanol from lignocellulosic hydrolysates of Japanese cedar and eucalyptus with ethanol concentrations of 25.98 g L^{-1} and 13.4 g L^{-1} at 42 and 45 °C, respectively. Other yeast strains could also produce ethanol at high temperatures including *Saccharomyces uvarum* (Hacking et al. 1984), *Saccharomyces carlsbergensis* (Szczodrak and Targoński 1988), *Ogataea polymorpha* (Ryabova et al. 2003), *Candida pseudotropicalis* (Hacking et al. 1984), *Candida acidothermophilum* (Kadam and Schmidt 1997), and *Pichia kudriavzevii* (Chamnipa et al. 2018; Koutinas et al. 2016; Oberoi et al. 2012; Pongcharoen et al. 2018; Yuangsaard et al. 2013). Among these yeasts, some *K. marxianus* strains are able to produce ethanol at temperatures above 40 °C and to have maximum growth temperatures of 47 °C (Anderson et al. 1986), 49 °C (Hughes et al. 1984; Nonklang et al. 2008), and even 52 °C (Banat et al. 1992; Lane and Morrissey 2010). These abilities of *K. marxianus* are superior to those of other species that have an optimum fermentation temperature of about 40 °C, suggesting that *K. marxianus* is a better species for HTF (Hoshida and Akada 2017) and for simultaneous saccharification and fermentation (SSF) (Choudhary et al. 2016).

There have been many studies on SSF using *K. marxianus* as shown in Table 2. However, the major problem with ethanol production using *K. marxianus* strains is the low production levels of lignocellulolytic enzymes. Heterologous expression of lignocellulolytic enzymes needs to be improved by optimizing the inulinase promoter and signal sequence of *K. marxianus* through mutagenesis. A mutation improved the secretory expression of lignocellulolytic enzymes, including endo-1,4-β-glucanase, endo-1,4-β-endoxylanase, and endo-1,4-β-mannanase by up to threefold (Zhou et al. 2018). On the other hand, a thermotolerant mutant derived from *K. marxianus* SBK1 that is capable of simultaneous co-fermentation of glucose and xylose produced 23.82 g L^{-1} ethanol at 40 °C. A mutant strain that alleviated catabolite repression fermented mixed sugars simultaneously (Kim et al. 2019). Co-culture of *K. marxianus* CICC-1727–5 with *Spathaspora passalidarum* ATCC MYA-4345 was performed to convert lignocellulosic biomass to ethanol (Du et al. 2019).

4 Findings from Complete Genome Sequencing and Transcriptome Analysis

Genomic and transcriptomic studies have shed light on *K. marxianus*, and an increasing number of genome sequences of *K. marxianus* strains are becoming available. Genomic studies have been performed on KCTC 17555 (Jeong et al. 2012), DMB1 (Suzuki et al. 2014), CCT 7735 (Silveira et al. 2014), NBRC1777 (Inokuma et al. 2015), DMKU 3-1042 (Lertwattanasakul et al. 2015), B0399

Table.3 Using *K. marxianus* for ethanol production in SSF at high temperatures

Strains of *K. marxianus*	Biomass	Temp. (°C)	Pretreatment	Enzymes used in saccharification	EtOH production (g L^{-1})	Yield ($Y_{E/S}$)	References
L. G	10% Solka-floc (microcrystalline cellulose)	42	–	15 FPU/g (cellulase)	37.6	0.50 g g^{-1}	Ballesteros et al. (1991)
Y01070	5% Spruce chip	42	Pretreatment by impregnation with 2.5% SO_2 and then steam at 215 °C for 5 min	37 FPU/g (cellulase) and 38 IU/g (β-glucosidase)	7.5	31.5 g/100 g	Bollók et al. (2000)
NCIM 3358	10% Solka-floc (microcrystalline cellulose)	43	–	40 FPU/g (cellulase) and 50 U/g (β-glucosidase)	35.0	2.5–3.5% w/v	Krishna et al. (2001)
Y01070	6% Solka-floc 6% OCC 6% paper sludge	40	–	15 FPU/g (cellulase) and 15 IU (β-glucosidase)	17.8 14.1 8.8	0.34 0.31 0.33	Kádár et al. (2004)
IMB4	Switchgrass (4.1% glucan)	45	Pretreatment by hydrothermolysis at 200 °C for 10 min	15 FPU/g (cellulase)	16.6	NR (78% theoretical yield)[a]	Suryawati et al. (2008)
IMB3	8% Switchgrass	45	Pretreatment by hydrothermolysis at 200 °C for 10 min	82.2 FPU/mL (cellulase)	22.5	NR (86% theoretical yield)[a]	Pessani et al. (2011)

(continued)

Table.3 (continued)

Strains of *K. marxianus*	Biomass	Temp. (°C)	Pretreatment	Enzymes used in saccharification	EtOH production (g L^{-1})	Yield ($Y_{E/S}$)	References
ATCC 36907	8% Sunflower meal	38	Pretreatment by 6% H_2SO_4 (w/v), at 121 °C, for 20 min and then 1% NaOH	20 FPU/g (cellulase) and 13.3 CBU/g (β-glucosidase)	27.88	0.47 g g^{-1}	Camargo et al. (2014)
NRRLY-6860	8% Rice straw cellulignin	45	Pretreatment by dilute acid (100 mg H_2SO_4) at 120 °C for 30 min	25 FPU/g (cellulase) and 25 UI/g (β-glucosidase)	11.5	0.24 g g^{-1}	Castro and Roberto (2014)
K21	8.05% Taro waste (=40 g L^{-1} glucose)	40	–	α-amylase (500–1500 units/mg)	48.98	NR (94.2% theoretical yield)[a]	Wu et al. (2016)
MTCC 1389	5% woody stem of *Prosopis juliflora*	41	Pretreatment by 3% (v/v) nitric acid, and sonication (40 kHz)	12 FPU/g (cellulase)	21.45	0.67	Sivarathnakumar et al. (2019)

NR not reported; [a]reported value; *OCC* old corrugated cardboard

(Quarella et al. 2016), UFS-Y2791 (Schabort et al. 2016), and other nine strains: L01, L02, L03, L04, L05, CBS397, NBRC0272, NBRC0288, and NBRC0617 (Ortiz-Merino et al. 2018). However, complete genome sequences of only two strains, DMKU 3-1042 and NBRC1777, are currently available. The former genome of 11.0 Mb is composed of 8 chromosomes in total including mitochondrial DNA. Annotation of the genome revealed a total of 4952 genes. A total of 202 tRNAs and 8 rDNAs were identified.

A major potential future application of *K. marxianus* may be ethanol production from lignocellulosic biomass, which is an anaerobic or oxygen-limited process in which both glucose and xylose are present. Detailed transcription start site sequencing (TSS Seq) to explore the response of *K. marxianus* DMKU 3-1042 was thus performed under four different conditions: shaking condition in a rich medium at 30 °C (30D) or 45 °C (45D), a static condition in a rich medium at 30 °C (30DS), and shaking condition in a xylose-containing rich medium at 30 °C (30X) (Lertwattanasakul et al. 2015).

Under the 30DS condition, *K. marxianus* may increase the turnover of RNAs and proteins in addition to suppression of transporters that depend on the mitochondrial respiratory activity. Most of the genes for several oxygen-dependent biosynthetic pathways, such as those for heme, sterols, unsaturated fatty acids, pyrimidine, and deoxyribonucleotides (Ishtar Snoek and Yde Steensma 2006), are crucial for cellular metabolism under a static condition. Under the 45D condition, *K. marxianus* seems to drastically change metabolic pathways, that is, enhancement of the pentose phosphate pathway (PPP) and attenuation of the TCA cycle after the fumarate-producing step. Several genes involved in both the DNA repair pathways of homologous recombination (HR) and non-homologous end-joining (NHEJ) are upregulated. Heat shock proteins and chaperones, such as Hsp26, Hsp60, Hsp78, Hsp82, Ssa3, and Cpr6, are crucial for survival at high temperatures. The thermotolerance of *K. marxianus* is thus likely to be achieved by systematic mechanisms consisting of various strategies. The yeast prevents the generation of reactive oxygen species (ROS) by minimizing mitochondrial activity and mainly acquires ATP from glycolysis rather than from the TCA cycle at high temperatures. Fu et al. (2019) also reported that excess ROS generated during a high temperature fermentation condition could be neutralized by NADPH in *K. marxianus*. The degree of fatty acid unsaturation may be reduced to adapt to high temperatures. Genes associated with DNA repair or lipid composition of the plasma membrane are upregulated. The yeast also produces more ergosterol to deal with ethanol stress. Under the 30X condition, degradation of lipids in the peroxisome seems to be stimulated and amino acid synthesis is kept at a low level, indicating the possibility that fatty acids could be a subsidiary intracellular carbon source in the xylose medium. Consistently, Schabort et al. (2016) also reported that peroxisomal fatty acid catabolism is dramatically upregulated in a defined xylose mineral medium without fatty acids. They also described mechanisms by which fatty acids are activated and products of β-oxidation are transferred to the mitochondria.

Notably, oxidative stress-response genes were highly induced under the three conditions tested, indicating that ROS accumulated in the cytoplasm, mitochondria, and peroxisome under the 30DS and 30X conditions and in the cytoplasm and mitochondria under the 45D condition. In conclusion, *K. marxianus* likely adapts to the three different growth conditions by distinctive metabolic pathways from the control condition. Interestingly, the yeast appears to overcome the problem of ROS, which tend to accumulate under all three conditions. Nicotinamide adenine dinucleotide phosphate (NADPH) synthesis from several reactions is the key for cells to cope with ROS.

Marcišauskas et al. (2019) reported the first genome-scale metabolic model of *K. marxianus*, iSM996, using TSS data reported by Lertwattanasakul et al. (2015). The model includes 1913 reactions associated with 996 genes and 1531 metabolites. The iSM996 was used to construct three condition-specific models in YPD medium while considering the low oxygen and high temperature conditions. The results suggest that at a high temperature, the cell turns off more genes, thereby introducing new auxotrophies and utilizing as many resources as possible from the medium. These findings may be used in the design of growth media at low levels of oxygen and/or high temperatures.

The global transcriptional response of *K. marxianus* to multiple inhibitors including acetic acid, phenols, furfural, and HMF at 42 °C was also studied via RNA-seq technology (Wang et al. 2018). Genes involved in the glycolysis pathway, fatty acid metabolism, ergosterol metabolism, and vitamin B6 and B1 metabolic process were enriched in the downregulated gene set, while genes involved in the TCA cycle, respiratory chain and detoxification of ROS and transporter coding genes were enriched in the upregulated gene set in response to the stress with multiple inhibitors. Redox balance and $NAD(P)^+/NAD(P)H$ homeostasis play an important role in tolerance to lignocellulose-derived inhibitors.

5 Possible Mechanism of Thermotolerance

K. marxianus can grow well at temperatures over 45 °C, unlike *K. lactis*, which belongs to the same genus, or *S. cerevisiae*, which is a closely related yeast in hemiascomycetous yeasts. *K. marxianus* may thus have an intrinsic mechanism to survive at high temperatures. *K. marxianus* strains are relatively resistant against hydrogen peroxide, furfural, and hydroxymethyl furfural (Nitiyon et al. 2016). Thermotolerance of the yeast may thus overlap with other stress tolerances, assuming a common mechanism of robustness against stressors. *K. marxianus*, which is one of the mesophilic yeasts, may have evolved into yeast with thermal resistance due to natural selection pressure in a high-temperature environment, and the acquisition of thermotolerance might have allowed it to withstand other stresses.

A clue for the mechanism of thermotolerance might be provided by transcriptome analysis, flux analysis, or comparison with a non-thermotolerant yeast. Transcriptome analysis (Lertwattanasakul et al. 2015) revealed that there are a tremendous number of significantly upregulated and downregulated genes in *K.*

marxianus at high temperatures, suggesting a drastic change of metabolism from that at low temperatures. The metabolic flow inside cells at high temperatures seems to be different from that at low temperatures. Compared to low temperatures, repression of glycolysis and enhancement of PPP activity are thought to occur at high temperatures and attenuation of the TCA cycle after the fumarate-producing step is also thought to occur at high temperatures. These changes and the fact that the intracellular level of ROS increases at high temperatures (Zhang et al. 2015a, b) lead to the speculation that the former provides NADPH for scavenging ROS and that the latter deals with H_2O_2 via electron transfer from succinate dehydrogenase to cytochrome *c* peroxidase. Consistent with these conjectures, a higher temperature generates more ROS, which causes DNA damage (Hori et al. 2009). In fact, a change in temperature from low to high temperatures causes a drastic increase in the level of ROS and similar transcriptional change, suggesting a shift of the metabolic flow from glycolysis to the PPP (unpublished data). Notably, genes for DNA double-strand break repair and removal of uracil in DNA molecules are upregulated, suggesting enhancement of double-strand breaks or deamination of cytosines in DNA at high temperatures. Results of transcriptome analysis have suggested additional strategies for survival at high temperatures: alteration of ribosome biogenesis including pre-rRNA processing presumably for stable and efficient protein synthesis, reduction of mitochondrial ribosome biogenesis probably for saving energy, minimization of electron leakage in the respiratory chain by reduction of its components, and enhanced expression of heat shock proteins and chaperones.

Lehnen et al. (2019) reported distinct metabolic responses of thermotolerant *Ogataea* and *Kluyveromyces* strains to high temperatures, which were investigated by ^{13}C-metabolic flux and physiology analyses, suggesting that there are no highly conserved metabolic traits among thermotolerant yeasts. However, compared to *S. cerevisiae*, both thermotolerant species exhibited high PPP and TCA cycle activities under all temperature conditions. While the maximum growth temperatures are similar for the thermotolerant strains, the metabolic network response to high temperatures is not conserved among the different species. Metabolic flux distributions in *O. polymorpha* are irresponsive to high temperatures, while *K. marxianus* strains exhibit flux rerouting at elevated temperatures. Mejía-Barajas et al. (2017) compared two *K. marxianus* strains and one *S. cerevisiae* strain isolated from hot environments with a laboratory yeast strain. One of the *K. marxianus* strains exhibited strong thermotolerant traits with a high specific growth rate and biomass productivity together with shorter duplication time, lower ROS production level, and lower lipid peroxidation level and also increase in the activity of catalase and amount of saturated fatty acids in membranes after elevation of temperature.

Although there is still insufficient evidence for concluding the mechanisms of thermotolerance in yeasts, thermotolerant yeasts may share strategies to maintain a low level of ROS, which generally increase at high temperatures (Zhang et al. 2015a, b) and are mainly generated by leakage of electrons from the respiratory chain in mitochondria (Pan 2011), which is enhanced at high temperatures due to structural instability of the mitochondrial membrane (Tarrío et al. 2008) or by an increase in respiration rate (Abbott et al. 2009). Low levels of ROS are scavenged by non-enzymatic and enzymatic anti-oxidizing agents such as glutathione (GSH), thioredoxin (TRX), superoxide dismutase, catalase, and peroxidases, but high levels of ROS cause oxidation of intracellular components, such as DNA, protein, and lipid and induce apoptosis (Madeo et al. 1999; Scherz-Shouval and Elazar 2007). Increased activity of the PPP supports the high demand of NADPH at high temperatures for glutathione reductase-mediated protection against oxidative stress (Konings 1988; Grant 2001; Sugiyama et al. 2000). However, it is not likely that a higher level of NADPH production is sufficient for thermotolerance because increased PPP activity (Celton et al. 2012; Frick and Wittmann 2005) in *S. cerevisiae* is not capable of supporting growth at elevated temperatures, and thermosensitive *K. lactis* (Tarrío et al. 2006) and *Pichia pastoris* (Jorda et al. 2014) have relatively high PPP activity. Notably, genome-wide analysis of thermotolerant genes supporting cell survival at a critical high temperature (CHT), an upper limit of temperature, in three bacteria, *Escherichia coli* (Murata et al. 2011, 2018), thermotolerant *Acetobacter tropicalis* (Soemphol et al. 2011), and thermotolerant *Zymomonas mobilis* (Charoensuk et al. 2017), revealed that they share thermotolerant genes for membrane stabilization, protection against oxidative stress, and repair of damage of DNA or proteins, which may contribute to minimization of ROS generation, avoidance of ROS damage, and recovery from ROS damage, respectively.

As for thermotolerance, microbes seem to have repair mechanisms for damage of proteins, lipid, or DNA caused by ROS (Piper 1993) and structural specialties of the membranes and enzymes, which may be crucial for metabolic activity at high temperatures (Mejía-Barajas et al. 2018; Fields 2001). Heat shock proteins including proteinases participate in the renaturation or degradation of unfolded or damaged proteins. Enhanced expression of genes for heat shock proteins as well as genes for ROS-scavenging enzymes improves thermotolerance in *Z. mobilis* (Anggarini et al. 2016). The susceptibility of lipids to oxidation depends on the lipid composition and degree of unsaturation (Catalá 2012). The extent of cellular damage under heat shock conditions is correlated positively with increasing unsaturation of fatty acids (Suryawati et al. 2008). Consistently, Steels et al. (1994) found that the most stress-resistant yeast membranes were enriched in saturated fatty acids. Membrane fluidity is related to the ratio of saturated to unsaturated fatty acids (Los and Murata 2004).

To further understand the mechanism of thermotolerance in yeasts, thermal adaptation followed by detailed analysis of causative mutations or enhanced thermotolerance of a thermosensitive yeast by heterologous expression of genes from a thermotolerant yeast may be useful. Heterologous expression of heat shock genes from thermophiles has been shown to improve thermotolerance of *S. cerevisiae* (Liu et al. 2014). The transcription factors *Km*Hsf1 and *Km*Msn2 of *K. marxianus* can promote both cell growth and ethanol fermentation of *S. cerevisiae* at high temperatures (Li et al. 2017). These two transcription factors might increase ethanol production by different mechanisms. In addition, *Km*Msn2 might also help to cope with a high temperature by regulating genes associated with lipid metabolism to change the membrane fluidity.

The thermal adaptation may lead to a decrease in the generation of ROS in cells that produce higher levels of ROS at higher temperatures, suggesting that the thermally adapted cells could become robust and resistant to many stressors (Matsushita et al. 2016). Kosaka et al. (2019) performed a thermal adaptation of two species and three strains of mesophilic microbes for the improvement of their CHTs. The results of experiments including analysis of the characteristics of mutants suggested that these microbes have a genomic potential to endure a 2–3 °C rise in temperature but possess a limited variety of strategies for thermal adaptation. The thermoadapted mutants bear 2–15 mutations in protein-coding regions of genes, which are categorized into 6 groups. Most of them are overlapping with the common classification for thermotolerant genes of *E. coli*, *A. tropicalis*, and *Z. mobilis* (see above), suggesting that mutations generated by thermal adaptation enhance physiological functions of the products of thermotolerant genes or their closely related genes (Kosaka et al. 2019).

In conclusion, the common and basic mechanisms of thermotolerance of mesophilic microbes including yeasts that have been tested at least are avoidance of ROS generation and damage by ROS and repair systems of ROS-directed cellular damage. To maintain ROS at a low level, they may have developed common or unique strategies. The thermotolerance trait may thus be achieved by an integrated mechanism consisting of various strategies. The thermotolerance of *K. marxianus* may also have been achieved by a systematic mechanism including enhanced production of NADPH by the PPP. Especially, the yeast would mainly acquire ATP from glycolysis rather than the TCA cycle at high temperatures, which could prevent the generation of ROS by minimization of mitochondrial activity. The mechanisms against ROS under a high-temperature condition may allow cells to endure other stresses because most of the stresses seem to generate ROS. Transcriptome analysis suggests that ROS accumulate at low temperatures under a static condition compared to a shaking condition or in xylose medium compared to glucose medium (Lertwattanasakul et al. 2015).

6 High Protein Production Ability

Escherichia coli has so far been primarily used as a host for the production of useful proteins by genetic engineering, probably due to the accumulation of advanced methodologies including the development of various vectors that are superior to other microbes (Rosano and Ceccarelli 2014). Although various kinds of cytokines or growth factors are produced with *E. coli* and are already commercially available, *E. coli* has some drawbacks (Nausch et al. 2013; Feng et al. 2015). For example, lipopolysaccharide as a constituent of the membrane in *E. coli* is harmful to the human body, and great care must therefore be taken, especially in the purification of pharmaceutical products (Storeng et al. 1987). The surface antigen protein of hepatitis B virus (HBsAg) or tissue plasminogen activators (TPA) cannot be produced as active proteins, and small substances such as peptide hormones are decomposed in *E. coli* cells (Pumpen et al. 1984; Elghanam et al. 2012; Xu et al. 2017). Therefore, attempts have been made to use other hosts to compensate for the disadvantage of *E. coli,* and yeast is drawing attention as one of the candidates. The utilization of *S. cerevisiae*, which is generally used as a host for DNA recombination research (Mattanovich et al. 2012; Porro et al. 2005), has several advantages: 1. a wealth of genetic knowledge has accumulated, 2. the fundamental mechanisms of replication, transcription and translation have been elucidated, making it a suitable model of analysis of biological phenomena in higher organisms, 3. the yeast has a long history of being used industrially as a useful microorganism for food, feed, and pharmaceutical raw materials, and 4. there is abundant information on fermentation engineering and culture engineering. However, *S. cerevisiae* also has drawbacks for protein production. *S. cerevisiae* as a Crabtree-positive yeast consumes oxygen in a limited range in the presence of glucose above a certain density, despite high levels of dissolved O_2 (van Urk et al. 1990). As a result, *S. cerevisiae* specializes in ethanol production, reducing the yield of the target product per sugar. Therefore, the Crabtree effect may be one of the causes of the small applicable range of material production of *S. cerevisiae.* In addition, since the yeast growth temperature is around 30 °C, temperature control during fermentation is indispensable and costly.

Thermotolerant *K. marxianus*, which is generally accepted as safe (GRAS), is a Crabtree-negative yeast that acts on the metabolism from a TCA cycle with a priority independent of glucose concentration and performs aerobic alcohol fermentation (Blank et al. 2005; Vandijken et al. 1993). Additionally, *K. marxianus* is the fastest-growing yeast known so far (Groeneveld et al. 2009), reported growth rates of *K. marxianus, K. lactis, S. cerevisiae* and *P. pastoris* being 0.80 h^{-1}, 0.50 h^{-1}, 0.37 h^{-1}, and 0.18 h^{-1}, respectively (Lane and Morrissey 2010; Gao et al. 2012), and has prominent features including efficient fermentation at high temperatures, assimilation of various sugars, and advanced genetic tools as mentioned above. Therefore, *K. marxianus* is one of the alternative yeasts that have great potential in technology to produce various useful proteins under aerobic conditions.

K. marxianus has promising properties for producing various enzymes such as β-glucosidase, β-galactosidase, inulinase, endopolygalacturonase, and xylosidase (Fonseca et al. 2008; Lertwattanasakul et al. 2015) (Fig. 1). β-glucosidase of *Kluyveromyces fragilis* ATCC 12424 has been cloned and expressed in *S. cerevisiae*, and it has been applied for hydrolysis of cellulosic materials (Raynal et al. 1987). The ability to hydrolyze cellulose is useful for further application in ethanol production from cellulosic biomass. β-Glucosidase is one of three major cellulase enzymes that are responsible for the regulation of the whole cellulolytic process to produce fermentable sugars (Zhou et al. 2018). β-Galactosidase can be produced from a *MIG1* mutant of *K. marxianus* KM-15 for hydrolysis of lactose in foods to become glucose and galactose (Zhou et al. 2013). This property is very attractive because it produces lactose-free or low-lactose foods and reduces health issues of people who suffered from lactose intolerance (Wolf et al. 2018). Inulinase can be produced from the disrupted *MIG1* gene and the overexpressed *INU1* gene (Zhou et al. 2014). The enzyme is used to hydrolyze inulin by degrading the β-2,1-fructosyl bond to become fructose (Hoshida et al. 2018). The enzyme can be applied in food, fermentation, pharmaceutical, and chemical industries (Chi et al. 2011). Endopolygalacturonase produced by *K. marxianus* BKM Y-719 is used for the reduction of viscosity in fruit processing products (Šiekštele et al. 1999). Xylosidase is related to catalysis of the reducing site of xylooligosaccharides and liberates xylose. Efficient liberation of xylose is related to the ability of a non-conventional yeast to convert xylose to ethanol. Xylosidase activity of thermotolerant yeast *K. marxianus* strains NIRE-K1 and NIRE-K3 was found to be higher than that of *Candida tropicalis*. An increased amount of xylosidase was found in adapted *K. marxianus* cells (Behera et al. 2016). Other possible enzymes related to biomass utilization have been predicted from the whole genome sequence (Lertwattanasakul et al. 2015), and they include β-glucosidase (*LAC4*), endo-1,3(4)-β-glucanase 1 (*DSE4*), and endo-1,3(4)-β-glucanase 2 (*ACF2*) in addition to three lactose permeases (3 copies of *LAC12*). Phosphatase, aminopeptidase, and carboxypeptidase have also been reported as possible enzymes related to biomass utilization (Nurcholis et al. 2020).

Production of proteins with *K. marxianus* that have been investigated so far include enzymes (β-galactosidase, β-glucosidase, inulinase, polygalacturonase, and others) (Topete et al. 1997; Su et al. 2021; Yarimizu et al. 2015; Hoshida et al. 2018), single-cell proteins (Rajkumar and Morrissey 2020), and antibodies (Duan et al. 2019; Yang et al. 2021). However, there is a problem in producing a protein using *K. marxianus*. *K. marxianus* cells have a strong cell wall, and it takes a lot of labor to purify the target protein produced inside cells, and thus the production amount is limited. Therefore, if the yeast can secrete the protein outside the cell body, application of highly efficient production and continuous in vitro cultivation as well as simplification of the industrial process by the ease of purification would be possible. In order to release the produced protein extracellularly, the N-terminal signal sequence of a known secretory protein must be connected to the protein of interest. Zhou et al. (2018) reported that a P10L substitution in the signal sequence of the *INU1* gene increased the secretory expression of lignocellulolytic enzymes in *K. marxianus*. The P10L substitution extended the hydrophobic core of the signal sequence and promoted secretion of mature proteins. Raimondi et al. (2010) succeeded in extracellular secretion of superoxide dismutase (SOD) by fusion with an appropriate signal sequence with SOD. In addition to enzymes, the production of a single-chain Fv antibody (Nambu-Nishida et al. 2018) and virus-like particles of porcine parvovirus (Yang et al. 2021) and porcine circovirus (Duan et al. 2019) has been reported.

7 High Ability of Non-homologous End-Joining for Genetic Engineering

There are two major mechanisms by which cells can repair DNA double-strand breaks (DSBs), which are intimately related to the classes of genetic recombination, homologous and non-homologous recombinations (Daley et al. 2005). The core components of NHEJ required for rejoining of any DSB can be experimentally defined as the proteins needed for simple religation. In the NHEJ pathway, binding of the Ku heterodimer (Ku70/Ku80) to both ends of the DNA DSB with the participation of additional proteins, such as Lig4, Nej1, and Lif1, promotes repair of the DSB (Kooistra et al. 2004; Palmbos et al. 2005; Kegel et al. 2006; Maassen et al. 2008). Ku, which is conserved from bacteria to humans, is an indispensable protein for NHEJ (Doherty et al. 2001), and disruption of *KU70* or *KU80*, results in efficient gene targeting in *K. lactis* (Kooistra et al. 2004), *P. stipitis* (Maassen et al. 2008), and *K. marxianus* (Abdel-Banat et al. 2010). In yeast lacking *KU70*, a high frequency of non-homologous gene integration was abolished and *KmKU70* mutants showed 82–95% homologous gene targeting efficiency using homologous sequences of 40–1000 bp, indicating that the highly efficient NHEJ pathway can be used in random gene disruption techniques such as transposon mutagenesis and plasmid-free gene manipulations in *K. marxianus* (Abdel-Banat et al. 2010). In addition, the linear DNA integrative technique can eliminate the burden of plasmid construction for targeted gene manipulation in *K. marxianus*.

8 Genome Editing Tools

Synthetic biology leverages the metabolic capacity of microorganisms for the biosynthesis of simple and complex compounds that are sourced unsustainably from fossil fuels or that are too expensive to use chemical synthesis on an industrial scale (Cernak et al. 2018). The yeast *S. cerevisiae* functions as a major eukaryote for synthetic biology but lacks the metabolic potential available in many of the more than one thousand yeast species identified so far. However, these yeasts remain difficult to use due to the lack of synthetic biology tools to access the underlying metabolic networks and physiology (Löbs et al. 2017).

The ability of *K. marxianus* to grow at high temperatures and utilize a wide variety of industry-related substrates reflects its potential for biotechnological applications (Foseca et al. 2008, Lane and Morrissey 2010). However, metabolic engineering of *K. marxianus* is limited by the lack of sophisticated genome editing tools and an incomplete understanding of its genetics, metabolism, and cellular physiology. A basic requirement of metabolic engineering is the ability to express native or heterologous genes from expression cassettes. In *K. marxianus*, stable plasmid options are limited to the centromeric region and autonomous replicating sequence of the host genome and a selectable auxotrophic or drug resistance marker (Hoshida et al. 2014). Transformation with a linear DNA fragment containing an expression cassette and a selectable marker result in genomic integration in one of two ways: heterologous DNA is either incorporated into the genome at a random locus (Kegel et al. 2006) by NHEJ, or the cassette is targeted to a specific site on the genome by homologous recombination (HR) to the site of interest (Lieber 2010; Löbs et al. 2017). In *S. cerevisiae*, HR is the primary DNA repair pathway, and its high capabilities have made genomic engineering relatively efficient and have facilitated the development of a wide range of in vivo DNA assembly tools (Shao et al. 2009; Horwitz et al. 2015). However, in most other yeasts, NHEJ is the preferred DNA repair pathway and genome engineering with HR is inefficient. The high functionality of *K. marxianus* NHEJ can limit genome editing in many applications. However, NHEJ-mediated functional marker selection as a novel DNA cloning method has been developed in the yeast (Hoshida et al. 2014). Some researchers have taken advantage of this ability for multiplexed gene integration (Cheon et al. 2014), but successful transformants differ significantly in hexanoic acid-producing ability, probably due to gene insertion at crucial genomic loci. Based on available genomic and transcriptomic data, Rajkumar et al. (2019) characterized a set of native sequences in *K. marxianus*, including constitutive and inducible promoters and terminators for expression of multiple genes, for metabolic engineering and synthetic biology. Several known centromeres and autonomous replication sequences (ARS) are also included in their collection. These tools will serve as the basis for efficiently building next-generation cell factories from this alternative yeast. A widely used strategy for enhancing HR in non-conventional yeasts is the disruption of genes essential for the NHEJ pathway such as *KU70* or *KU80*. In *K. marxianus*, the disruption of *KU70* and *KU80* increased HR rates to 95% and 70%, respectively (Abdel-Banat et al. 2010; Choo et al. 2014).

An alternative strategy for achieving efficient HR is the introduction of a genomic DSB using a programmable endonuclease in the presence of a homologous repair template (Liu et al. 2017). Several programmable tools exist for a targeted DSB, including dimeric meganucleases, zinc finger nucleases (ZFNs), transcription activator-like effector nucleases (TALENs), and clustered regularly interspaced short palindromic repeats (CRISPR) and CRISPR-associated 9 (CRISPR-Cas9) (Liu et al. 2017). The CRISPR-Cas9 gene-editing system has been widely used in many yeasts including *S. cerevisiae*, *Schizosaccharomyces pombe*, *Yarrowia lipolytica*, and *Kluyveromyces lactis* and recently in *K. marxianus* (Nambu-Nishida et al. 2017; Löbs et al. 2017; Lee et al. 2018; Juergens et al. 2018). The CRISPR-Cas9 system has been developed for use with *K. marxianus*, enabling both NHEJ-based and HR-based genome editing (Cernak et al. 2018; Rajkumar et al. 2019). The genetic loci, *ALPHA3* and *KAT1*, responsible for mating-type switching in *K. marxianus* were identified and heterothallic haploid strains were constructed using the CRISPR-Cas9 system. This is indispensable for genetic manipulation of the most desired traits, which are likely to depend on multiple unlinked genetic loci, and remain difficult to identify without the ability to carry out genetic crosses. Three complex traits, the ability to take up exogenous DNA, thermotolerance, and high lipid production, have been successfully combined into a single *K. marxianus* isolate (Cernak et al. 2018).

9 Examples of Production of Other Useful Materials by *K. marxianus*

9.1 Flavor Metabolites

A non-conventional yeast can increase the variety and complexity of aroma profiles of bakery products such as nuts and fruity aroma (Aslankoohi et al. 2016) and alcoholic beverages such as wine and beer (Gamero et al. 2020). In addition to producing ethanol, *K. marxianus* is capable of producing a variety of volatile molecules or aromatic esters used as fragrances or flavors (Table 1). The ability of the yeast to produce acetate esters such as 2-phenyl ethyl acetate (2-PEA) and isoamyl acetate from 2-phenyl ethanol (2-PE) and isoamyl alcohol is due to the presence of alcohol acetyltransferase or AATase (Gethins et al. 2015). The ability to produce 2-phenylethanol from glucose without an additional L-phenylalanine supplement can be achieved in *K. marxianus* by genetic engineering via overexpression of *ARO10* for phenylpyruvate decarboxylase and *ADH2* for alcohol dehydrogenase II from *S. cerevisiae* (Kim et al. 2014). Genes involved in ethyl acetate biosynthesis in *K. marxianus* were identified (Löbs et al. 2017). KmAdh2 was found to be critical for aerobic and anaerobic ethanol production. Aerobically produced ethanol is supplied for the biosynthesis of ethyl acetate catalyzed by KmAtf. KmAdh7 was found to exhibit activity toward the oxidation of hemiacetal, a possible alternative route for the synthesis of ethyl acetate.

9.2 Fructose

Fructose is a saccharide used for sweeteners, seasonings, humectant, color and flavor development, freezing-point depression, and osmotic stability in foods and beverages such as confectionary, baby food, and high-fructose syrup (Hanover and White 1993). A high concentration of glucose-free fructose can be produced by overexpressing *GLK1* in an *HXK1*-knockout mutant or by a *RAG5*-knockout mutant of *K. marxianus* (Zhang et al. 2017a, b; Nurcholis et al. 2019). These findings suggest that disruption of *RAG5* results in reduced kinase activity for glucose and fructose, reduced glucose uptake due to prevention of the expression of *RAG1* for a glucose transporter, and increased inulinase activity, which in turn lead to decreased glucose utilization and fructose accumulation (Nurcholis et al. 2019).

9.3 Xylitol Formation

Xylitol has gained increasing attention in recent years due to its use in several industries such as food, dental products, and pharmaceuticals (Ravella et al. 2012). The ability of *K. marxianus* to ferment xylose under oxygen-limited conditions is weak due to its redox imbalance. Xylose consumption and fermentation can be enhanced by genetically engineered strains of *K. marxianus*, in which xylose reductase (XR) is one of the targeted genes. Replacing the native XR of *K. marxianus* with the XR from *Pichia stipitis* significantly improved xylose assimilation to ethanol and xylitol at high temperatures (Zhang et al. 2013). There are several reports of xylitol production (Table 1). Hua et al. (2019) obtained a high concentration of xylitol, 82.85 g L^{-1}, from detoxified corncob hydrolysate at 42 °C, and Du et al. (2019) reported xylitol production from a xylose medium with a yield of 0.58 g g^{-1} at 40 °C. Notably, the *MIG1*-disrupted mutant of *K. marxinaus* exhibited increased accumulation of xylitol in a xylose medium (Nurcholis et al. 2019). It is assumed that Mig1 represses xylose utilization but that dysfunction of Mig1 causes increased xylose utilization, leading to a limitation of NAD^+ required for conversion of xylitol to xylulose. Most xylose-fermenting yeasts, including *K. marxianus,* have a xylose metabolic pathway, in which XR converts xylose to xylitol and further to xylulose under cofactors-balanced conditions, but xylitol oxidation is prevented if NAD^+ is insufficient. The XR of *K. marxianus* NBRC1777 has sole coenzyme specificity, which is activated only by NADPH (Zhang et al. 2011), while the xylitol dehydrogenase (XDH) of *K. marxianus* strains prefers to use NAD^+ (Lulu et al. 2013).

10 Conclusions

To be a platform of synthetic biology, there are at least three necessary conditions: excellent potential, sufficient available information, and genetic engineering tools. In this chapter, with a focus on those three points for *K. marxianus*, the prominent properties of *K. marxianus* including thermotolerance, broad substrate specificity and protein productivity were introduced in Sects. 2–3, 5–6, and 9, genome and transcription information was presented in Sect. 4, and genetic engineering tools including modeling were introduced in Sects. 7 and 8. It was shown that this yeast has the necessary conditions for use as a platform of synthetic biology, and the development of synthetic biology that utilizes *K. marxianus* for enterprises is expected. Furthermore, in order to accelerate synthetic biology, it is desirable to enhance metabolic modeling in addition to these three points, but, as with other organisms, comprehensive metabolic modeling has not yet been established.

The degree of interest of *K. marxianus* is reflected in the difference in the number of publications for *K. marxianus* compared to that for *K. lactis* (Fig. 2). The number of publications sharply increased from 2005 to 2014 and the number of publications has been maintained each year since 2014. On the other hand, there were many publications for *K. lactis* from around 1992, but the number has gradually decreased since 2011. The trends in publications for the two species may imply a shift of trends towards thermotolerance, which is beneficial for industrial applications. On the other hand, although basic research on *K. marxianus* is far behind that on *S. cerevisiae*, several milestones have been achieved so far for supporting further basic research on *K. marxianus* (Nurcholis et al. 2020), including complete genome sequencing of two strains (Lertwattanasakul et al. 2015;

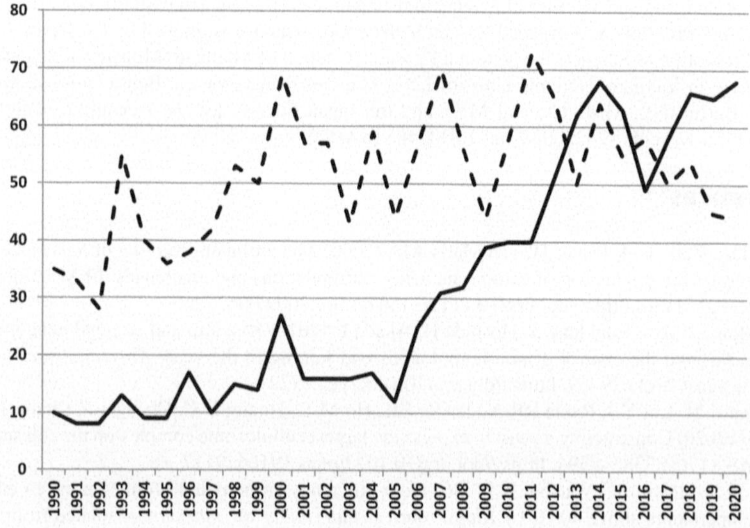

Fig. 2 Changes in the number of publications related to *K. marxianus* (straight lines) and the number of publications related to *K. lactis* (broken lines). Analysis was performed using PubMed

Inokuma et al. 2015), transcriptome analysis under various conditions (Lertwattanasakul et al. 2015; Gao et al. 2015; Schabort et al. 2016; Mo et al. 2019; Fu et al. 2019; Rollero et al. 2019), transformation with a linear DNA fragment (Nonklang et al. 2008), genome editing (Nambu-Nishida et al. 2017; Rajkumar et al. 2019), and modeling (Pentjuss et al. 2017; Marcišauskas et al. 2019).

S. cerevisiae has been utilized in food production in human society for a long time and has recently become a major industrial and model microorganism due to the many genetic and genomic tools that have become available to understand its biology. However, it has been difficult to expand the capabilities of *S. cerevisiae* for the use of various carbon sources, production of various products, and tolerance to stresses for industrial applications. Other non-conventional yeasts need to be developed to solve many of these problems. The beneficial characteristics and potentials of *K. marxianus* were summarized in this review. Due to such intrinsically outstanding features in addition to the development of biotechnological tools, *K. marxianus* has become one of the most interesting non-conventional yeasts, comparable to *S. cerevisiae*, at least for industrial applications. The recent advances as mentioned above may enable the integration of comprehensive data and biotechnological tools to promote industrial applications as well as further basic research.

Acknowledgements This study was supported by the Advanced Low Carbon Technology Research and Development Program, which was granted by the Japan Science and Technology Agency (JPMJAL1106) (MM, TK, and MY) and e-ASIA Joint Research Program, which was granted by Japan Science (JPMJSC16E5) (MM, TK, and MY) and Technology Agency, Ministry of Research, Technology and Higher Education of the Republic of Indonesia, Agricultural Research Development

Agency of Thailand and Ministry of Science and Technology of Laos, and partially supported by the Core to Core Program A. Advanced Research Networks, which was granted by the Japan Society for the Promotion of Science, the National Research Council of Thailand, Ministry of Science and Technology in Vietnam, National Univ. of Laos, Univ. of Brawijaya and Beuth Univ. of Applied Science Berlin (NL, MM, TK, and MY), and the Japan Society for the Promotion of Science, MEXT/JSPS Kakenhi (25250028 and 16H02485 to MY).

References

Abbott DA, Zelle RM, Pronk JT, van Maris AJA (2009) Metabolic engineering of *Saccharomyces cerevisiae* for production of carboxylic acids: current status and challenges. FEMS Yeast Res 9(8):1123–1136. https://doi.org/10.1111/j.1567-1364.2009.00537.x

Abdel-Banat BMA, Nonklang S, Hoshida H, Akada R (2010) Random and targeted gene integrations through the control of non-homologous end joining in the yeast *Kluyveromyces marxianus*. Yeast 27(1):29–39. https://doi.org/10.1002/yea.1729

Anandharaj M, Lin Y-J, Rania RP, Nadendla EK, Ho M-C, Huang C-C, Cheng J-F, Chang J-J, Li W-H (2020) Constructing a yeast to express the largest cellulosome complex on the cell surface. PNAS 117(5):2385–2394. https://doi.org/10.1073/pnas.1916529117

Anderson P, McNeil K, Watson, K (1986) High-efficiency carbohydrate fermentation to ethanol at temperatures above 40 °C by *Kluyveromyces marxianus* var. *marxianus* isolated from sugar mills. Appl Environ Microbiol (USA) 51(6):1314–1320

Anggarini S, Murata M, Kido K, Kosaka T, Sootsuwan K, Thanonkeo P, Yamada M (2019) Improvement of thermotolerance of *Zymomonas mobilis* by genes for reactive oxygen species-scavenging enzymes and heat shock proteins. Front Microbiol 10:3073. https://doi.org/10.3389/fmicb.2019.03073

Aslankoohi E, Herrera-Malaver B, Rezaei MN, Steensels J, Courtin CM, Verstrepen KJ (2016) Non-conventional yeast strains increase the aroma complexity of bread. PLoS ONE 11(10):1–18. https://doi.org/10.1371/journal.pone.0165126

Ballesteros I, Ballesteros M, Cabanas A, Carrasco J, Martin C, Negro M, Saez R (1991) Selection of thermotolerant yeasts for simultaneous saccharification and fermentation (SSF) of cellulose to ethanol. Appl Biochem Biotechnol (USA) 28:307–315

Banat I, Nigam P, Singh D, Marchant R, McHale A (1998) Ethanol production at elevated temperatures and alcohol concentrations: part I—yeasts in general. World J Microbiol Biotechnol (united Kingdom) 14(6):809–821

Banat IM, Nigam P, Marchant R (1992) Isolation of thermotolerant, fermentative yeasts growing at 52 °C and producing ethanol at 45 and 50 °C. World J Microbiol Biotechnol 8(3):259–263. https://doi.org/10.1007/BF01201874

Behera S, Sharma N, Arora R, Kumar S (2016) Effect of evolutionary adaption on xylosidase activity in thermotolerant yeast isolates *Kluyveromyces marxianus* NIRE-K1 and NIRE-K3. Appl Biochem Biotechnol 179(7):1143–1154. https://doi.org/10.1007/s12010-016-2055-2

Bergkamp RJM, Bootsman TC, Toschka HY, Mooren ATA, Kox L, Verbakel JMA, Geerse RH, Planta RJ (1993) Expression of an alpha-galactosidase gene under control of the homologous inulinase promoter in *Kluyveromyces marxianus*. Appl Microbiol Biotechnol 40:309–317. https://doi.org/10.1007/BF00170386

Blank LM, Lehmbeck F, Sauer U (2005) Metabolic-flux and network analysis in fourteen hemiascomycetous yeasts. FEMS Yeast Res 5:545–558

Bollók M, Réczey K, Zacchi G (2000) Simultaneous saccharification and fermentation of steam-pretreated spruce to ethanol. Appl Biochem Biotechnol 84–6:69–80. https://doi.org/10.1385/ABAB:84-86:1-9:69

Bragança CRS, Colombo LT, Roberti AS, Alvim MCT, Cardoso SA, Reis KCP, de Paula SO, da Silveira WB, Passos FML (2015) Construction of recombinant *Kluyveromyces marxianus* UFV-3 to express dengue virus type 1 nonstructural protein 1 (NS1). Appl Microbiol Biotechnol 99:1191–1203. https://doi.org/10.1007/s00253-014-5963-5

Camargo D, Gomes S, Sene L (2014) Ethanol production from sunflower meal biomass by simultaneous saccharification and fermentation (SSF) with *Kluyveromyces marxianus* ATCC 36907. Bioprocess Biosyst Eng 37(11):2235–2242. https://doi.org/10.1007/s00449-014-1201-x

Campos-García J, Vargas A, Farías-Rosales L, Miranda AL, Meza-Carmen V, Díaz-Pérez AL (2018) Increase in fatty acid ethyl ester content through *ATF1* expression in an engineered *Kluyveromyces marxianus* UMPe-1 yeast improves the organoleptic properties of a craft Mezcal beverage. J Agric Food Chem 66(17):4469–4480. https://doi.org/10.1021/acs.jafc.8b00730

Castro R, Roberto I (2014) Selection of a thermotolerant *Kluyveromyces marxianus* strain with potential application for cellulosic ethanol production by simultaneous saccharification and fermentation. Appl Biochem Biotechnol 172(3):1553–1564. https://doi.org/10.1007/s12010-013-0612-5

Catalá A (2012) Lipid peroxidation modifies the picture of membranes from the "Fluid Mosaic Model" to the "Lipid Whisker Model." Biochimie 94(1):101–109. https://doi.org/10.1016/j.biochi.2011.09.025

Celton M, Sanchez I, Goelzer A, Fromion V, Camarasa C, Dequin S (2012) A comparative transcriptomic, fluxomic and metabolomic analysis of the response of *Saccharomyces cerevisiae* to increases in NADPH oxidation. BMC Genomics 13(1):317–330. https://doi.org/10.1186/1471-2164-13-317

Cernak P, Estrela R, Poddar S, Skerker JM, Cheng Y-F, Carlson AK, Chen B, Glynn VM, Furlan M, Ryan OW, Donnelly MK, Arkin AP, Taylor JW, Cate JHD (2018) Engineering *Kluyveromyces marxianus* as a robust synthetic biology platform host. mBio 9(5):e01410– e01418.https://doi.org/10.1128/mBio.01410-18

Chamnipa N, Thanonkeo S, Klanrit P, Thanonkeo P (2018) The potential of the newly isolated thermotolerant yeast *Pichia kudriavzevii* RZ8-1 for high-temperature ethanol production. Braz J Microbiol 49(2):378–391. https://doi.org/10.1016/j.bjm.2017.09.002

Chang J-J, Ho C-Y, Ho F-J, Tsai T-Y, Ke H-M, Wang CH, Li W-H (2012) PGASO: A synthetic biology tool for engineering a cellulolytic yeast. Biotechnol Biofuels 5(1):53–64. https://doi.org/10.1186/1754-6834-5-53

Chang J-J, Ho F-J, Ho C-Y, Wu Y-C, Hou Y-H, Huang C-C, Shih M-C, Li W-H (2013) Assembling a cellulase cocktail and a cellodextrin transporter into a yeast host for CBP ethanol production. Biotechnol Biofuels 6(1):1–13. https://doi.org/10.1186/1754-6834-6-19

Chang J-J, Lin Y-J, Lay C-H, Thia C, Wu Y-C, Hou Y-H, Huang C-C, Li W-H (2017) Constructing a cellulosic yeast host with an efficient cellulase cocktail. Biotechnol Bioeng 1–11. https://doi.org/10.1002/bit.26507

Chang J-J, Thia C, Lin H-Y, Liu H-L, Ho F-J, Wu J-T, Shih M-C, Li W-H, Huang C-C (2015) Integrating an algal β-carotene hydroxylase gene into a designed carotenoid-biosynthesis pathway increases carotenoid production in yeast. Biores Technol 184:2–8. https://doi.org/10.1016/j.biortech.2014.11.097

Charoensopharat K, Thanonkeo P, Thanonkeo S, Yamada M (2015) Ethanol production from Jerusalem artichoke tubers at high temperature by newly isolated thermotolerant inulin-utilizing yeast *Kluyveromyces marxianus* using consolidated bioprocessing. Antonie Van Leeuwenhoek 108(1):173–190. https://doi.org/10.1007/s10482-015-0476-5

Charoensuk K, Sakurada T, Tokiyama A, Murata M, Kosaka T, Thanonkeo P, Yamada M (2017) Thermotolerant genes essential for survival at a critical high temperature in thermotolerant ethanologenic *Zymomonas mobilis* TISTR 548. Biotechnol Biofuels 10(1):1–11. https://doi.org/10.1186/s13068-017-0891-0

Cheon Y, Kim J-S, Park J-B, Heo P, Lim JH, Jung GY, Seo J-H, Park JH, Koo HM, Cho KM, Park J-B, Ha S-J, Kweon D-H (2014) A biosynthetic pathway for hexanoic acid production in *Kluyveromyces marxianus*. J Biotechnol 182–183:30–36. https://doi.org/10.1016/j.jbiotec.2014.04.010

Chi Z-M, Zhang T, Cao T-S, Liu X-Y, Cui W, Zhao C-H (2011) Biotechnological potential of inulin for bioprocesses. Biores Technol 102(6):4295–4303. https://doi.org/10.1016/j.biortech.2010.12.086

Choo JH, Han C, Kim J-Y, Kang HA (2014) Deletion of a KU80 homolog enhances homologous recombination in the thermotolerant yeast *Kluyveromyces marxianus*. Biotechnol Lett 36(10):2059–2067. https://doi.org/10.1007/s10529-014-1576-4

Choudhary J, Singh S, Nain L (2016) Thermotolerant fermenting yeasts for simultaneous saccharification fermentation of lignocellulosic biomass. Electron J Biotechnol 21:82–92. https://doi.org/10.1016/j.ejbt.2016.02.007

Cruz-Guerrero A, Garcia-Peña I, Barzana E, Garcia-Garibay M, Gomez-Ruiz L (1995) *Kluyveromyces marxianus* CDBB-L-278: a wild inulinase hyperproducing strain. J Fermentation Bioeng (japan) 80(2):159–163

Daley JM, Palmbos PL, Wu D, Wilson TE (2005) Nonhomologous end joining in yeast. Annu Rev Genet 39:431–451

Doherty AJ, Jackson SP, Weller GR (2001) Identification of bacterial homologues of the Ku DNA repair proteins. FEBS Lett 500:186–188

Du C, Li Y, Zhao X, Pei X, Yuan W, Bai F, Jiang Y (2019) The production of ethanol from lignocellulosic biomass by *Kluyveromyces marxianus* CICC 1727–5 and *Spathaspora passalidarum* ATCC MYA-4345. Appl Microbiol Biotechnol 103(6):2845–2855. https://doi.org/10.1007/s00253-019-09625-1

Duan J, Yang D, Chen L, Yu Y, Zhou J, Lu H (2019) Efficient production of porcine circovirus virus-like particles using the nonconventional yeast *Kluyveromyces marxianus*. Appl Environ Microbiol 103:833–842

Elghanam S-M, Attia S-A, Shoeb A-H, Hashem MA-E (2012) Expression and purification of hepatitis B surface antigen S from *Escherichia coli*; a new simple method. BMC Res Notes 5:125–133

Feng J, Wan R, Yi Q, He L, Yang L, Tang L (2015) Examination of alternate codon bias solutions for expression and purification of recombinant mechano-growth factor in *Escherichia coli*. Biotechnol Appl Biochem 62(5):690–698. https://doi.org/10.1002/bab.1312

Fields PA (2001) Review: protein function at thermal extremes: balancing stability and flexibility. Comparative Biochem Physiol Part A 129(2):417–431. https://doi.org/10.1016/S1095-6433(00)00359-7

Fonseca GG, Heinzle E, Wittmann C, Gombert AK (2008) The yeast *Kluyveromyces marxianus* and its biotechnological potential. Appl Microbiol Biotechnol 79(3):339–354. https://doi.org/10.1007/s00253-008-1458-6

Frick O, Wittmann C (2005) Characterization of the metabolic shift between oxidative and fermentative growth in *Saccharomyces cerevisiae* by comparative ^{13}C flux analysis. Microb Cell Fact 4:1–16. https://doi.org/10.1186/1475-2859-4-30

Fu X, Li P, Zhang L, Li S (2019) Understanding the stress responses of *Kluyveromyces marxianus* after an arrest during high-temperature ethanol fermentation based on integration of RNA-Seq and metabolite data. Appl Microbiol Biotechnol 103(6):2715–2729. https://doi.org/10.1007/s00253-019-09637-x

Galbe M, Zacchi G (2007) Pretreatment of lignocellulosic materials for efficient bioethanol production. Adv Biochem Eng Biotechnol 108:41–65. https://doi.org/10.1007/10_2007_070

Gamero A, Dijkstra A, Smit B, de Jong C (2020) Aromatic potential of diverse non-conventional yeast species for winemaking and brewing. Fermentation 6(2):50. https://doi.org/10.3390/fermentation6020050

Gao J, Yuan W, Li Y, Xiang R, Hou S, Zhong S, Bai F (2015) Transcriptional analysis of *Kluyveromyces marxianus* for ethanol production from inulin using consolidated bioprocessing technology. Biotechnol Biofuels 8(1):1–17. https://doi.org/10.1186/s13068-015-0295-y

Gao M-J, Zheng Z-Y, Wu J-R, Dong S-J, Li Z, Jin H, Zhan X-B, Lin C-C (2012) Improvement of specific growth rate of *Pichia pastoris* for effective porcine interferon-α production with an on-line model-based glycerol feeding strategy. Appl Microbiol Biotechnol 93(4):1437–1445. https://doi.org/10.1007/s00253-011-3605-8

Gethins L, Guneser O, Demirkol A, Rea MC, Stanton C, Ross RP, Yuceer Y, Momissey JP (2015) Influence of carbon and nitrogen sources on production of volatile fragrance and flavour metabolites by the yeast *Kluyveromyces marxianus*. Yeast 32(1):67–76. https://doi.org/10.1002/yea.3047

Gombert AK, Madeira JV, Cerdán ME, González-Siso MI (2016) *Kluyveromyces marxianus* as a host for heterologous protein synthesis. Appl Microbiol Biotechnol 100(14):6193–6208. https://doi.org/10.1007/s00253-016-7645-y

Goshima T, Tsuji M, Inoue H, Yano S, Hoshino T, Matsushika A (2013a) Bioethanol production from lignocellulosic biomass by a novel *Kluyveromyces marxianus* strain. Biosci Biotechnol Biochem 77(7):1505–1510

Goshima T, Negi K, Tsuji M, Inoue H, Yano S, Hoshino T, Matsushika A (2013b) Ethanol fermentation from xylose by metabolically engineered strains of *Kluyveromyces marxianus*. J Biosci Bioeng 116(5):551–554. https://doi.org/10.1016/j.jbiosc.2013.05.010

Grant CM (2001) Role of the glutathione/glutaredoxin and thioredoxin systems in yeast growth and response to stress conditions. Mol Microbiol 39(3):533–541

Groeneveld P, Stouthamer AH, Westerhoff HV (2009) Super life—how and why 'cell selection' leads to the fastest-growing eukaryote. FEBS J 276(1):254–270. https://doi.org/10.1111/j.1742-4658.2008.06778.x

Guimarães PMR, Teixeira JA, Domingues L (2010) Fermentation of lactose to bio-ethanol by yeasts as part of integrated solutions for the valorisation of cheese whey. Biotechnol Adv 28(3):375–384. https://doi.org/10.1016/j.biotechadv.2010.02.002

Gupta AK, Singh DP, Kaur N, Singh R (1994) Production, purification and immobilisation of inulinase from *Kluyveromyces fragilis*. J Chem Technol Biotechnol 54(4):377–385

Hacking AJ, Taylor IWF, Hanas CM (1984) Selection of yeast able to produce ethanol from glucose at 40 °C. Appl Microbiol Biotechnol 19(5):361–363. https://doi.org/10.1007/BF00253786

Hahn-Hägerdal B, Galbe M, Gorwa-Grauslund MF, Lidén G, Zacchi G (2006) Bio-ethanol—the fuel of tomorrow from the residues of today. Trends Biotechnol 24(12):549–556. https://doi.org/10.1016/j.tibtech.2006.10.004

Hanover LM, White JS (1993) Manufacturing, composition, and applications of fructose. Am J Clin Nutrition 724S–732S

Harner NK, Wen X, Bajwa PK, Austin GD, Ho C-Y, Habash MB, Trevors JT, Lee H (2015) Genetic improvement of native xylose-fermenting yeasts for ethanol production. J Ind Microbiol Biotechnol 42(1):1–20. https://doi.org/10.1007/s10295-014-1535-z

Hasunuma T, Kondo A (2012) Development of yeast cell factories for consolidated bioprocessing of lignocellulose to bioethanol through cell surface engineering. Biotechnol Adv 30(6):1207–1218. https://doi.org/10.1016/j.biotechadv.2011.10.011

Hong J, Wang Y, Kumagai H, Tamaki H (2007) Construction of thermotolerant yeast expressing thermostable cellulase genes. J Biotechnol 130:114–123. https://doi.org/10.1016/j.jbiotec.2007.03.008

Hori A, Yoshida M, Shibata T, Ling F (2009) Reactive oxygen species regulate DNA copy number in isolated yeast mitochondria by triggering recombination-mediated replication. Nucleic Acids Res 37(3):749–761

Horwitz AA, Walter JM, Schubert MG, Kung SH, Hawkins K, Platt DM, Hernday AD, Mahatdejkul-Meadows T, Szeto W, Chandran SS, Newman JD (2015) Efficient multiplexed integration of synergistic alleles and metabolic pathways in yeasts via CRISPR-Cas. Cell Syst 1(1):88–96. https://doi.org/10.1016/j.cels.2015.02.001

Hoshida H, Akada R (2017) High-temperature bioethanol fermentation by conventional and non-conventional yeasts. In: Biotechnology of yeasts and filamentous fungi. Springer, pp 39–61

Hoshida H, Kidera K, Takishita R, Fujioka N, Fukagawa T, Akada R (2018) Enhanced production of extracellular inulinase by the yeast *Kluyveromyces marxianus* in xylose catabolic state. J Biosci Bioeng 125(6):676–681. https://doi.org/10.1016/j.jbiosc.2017.12.024

Hoshida H, Murakami N, Suzuki A, Tamura R, Asakawa J, Abdel-Banat BMA, Nonklang S, Nakamura M, Akada R (2014) Non-homologous end joining-mediated functional marker selection for DNA cloning in the yeast *Kluyveromyces marxianus*. Yeast 31(1):29–46. https://doi.org/10.1002/yea.2993

Hu N, Yuan B, Sun J, Wang S-A, Li F-L (2012) Thermotolerant *Kluyveromyces marxianus* and *Saccharomyces cerevisiae* strains representing potentials for bioethanol production from Jerusalem artichoke by consolidated bioprocessing. Appl Microbiol Biotechnol 95(5):1359–1368. https://doi.org/10.1007/s00253-012-4240-8

Hua Y, Wang J, Zhu Y, Zhang B, Kong X, Li W, Wang D, Hong J (2019) Release of glucose repression on xylose utilization in *Kluyveromyces marxianus* to enhance glucose-xylose co-utilization and xylitol production from corncob hydrolysate. Microb Cell Fact 18:24. https://doi.org/10.1186/s12934-019-1068-2

Hughes DB, Tudroszen NJ, Moye CJ (1984) Effect of temperature on the kinetics of ethanol production by a thermotolerant strain of *Kluyveromyces marxianus*. Biotech Lett 6(1):1–6

Hughes SR, Qureshi N, Lopez-Nunez JC, Jones MA, Jarodsky JM, Galindo-Leva LA, Lindquist MR (2017) Utilization of inulin-containing waste in industrial fermentations to produce biofuels and bio-based chemicals. World J Microbiol Biotechnol (4):1.https://doi.org/10.1007/s11274-017-2241-6

Inokuma K, Ishii J, Hara KY, Mochizuki M, Hasunuma T, Kondo A (2015) Complete genome sequence of *Kluyveromyces marxianus* NBRC1777, a nonconventional thermotolerant yeast. Genome Announcements, 3. https://doi.org/10.1128/genomeA.00389-15

Ishtar Snoek IS, Yde Steensma H (2006) Why does *Kluyveromyces lactis* not grow under anaerobic conditions? Comparison of essential anaerobic genes of *Saccharomyces cerevisiae* with the *Kluyveromyces lactis* genome. FEMS Yeast Res 6:393–403

Jedrzejewska M, Kozak K (2011) Ethanol production from whey permeate in a continuous anaerobic bioreactor by *Kluyveromyces marxianus*. Environ Technol 32(1):37–42. https://doi.org/10.1080/09593331003616805

Jeong H, Lee D-H, Kim SH, Kim H-J, Lee K, Song JY, Kim BK, Sung BH, Park JC, Sohn JH, Koo HM, Kim JF (2012) Genome sequence of the thermotolerant yeast *Kluyveromyces marxianus* var. *marxianus* KCTC 17555. Eukaryot Cell 11:1584–1585. https://doi.org/10.1128/ec.00260-12

Jorda J, Rojas HC, Carnicer M, Wahl A, Ferrer P, Albiol J (2014) Quantitative metabolomics and in stationary ^{13}C-metabolic flux analysis reveals impact of recombinant protein production on trehalose and energy metabolism in *Pichia pastoris*. Metabolites 4(2):281–299. https://doi.org/10.3390/metabo4020281

Juergens H, Varela JA, Gorter de Vries AR, Perli T, Gast VJM, Gyurchev NY, Rajkumar AS, Mans R, Pronk JT, Morrissey JP, Daran J-MG (2018) Genome editing in *Kluyveromyces* and *Ogataea* yeasts using a broad-host-range Cas9/gRNA co-expression plasmid. FEMS Yeast Res 18(3):1–16. https://doi.org/10.1093/femsyr/foy012

Kadam KL, Schmidt SL (1997) Evaluation of *Candida acidothermophilum* in ethanol production from lignocellulosic biomass. Appl Microbiol Biotechnol 48(6):709–713. https://doi.org/10.1007/s002530051120

Kádár Z, Szengyel Z, Réczey K (2004) Simultaneous saccharification and fermentation (SSF) of industrial wastes for the production of ethanol. Ind Crops Prod 20(1):103–110. https://doi.org/10.1016/j.indcrop.2003.12.015

Karim A, Gerliani N, Aïder M (2020) *Kluyveromyces marxianus*: an emerging yeast cell factory for applications in food and biotechnology. Int J Food Microbiol 333. https://doi.org/10.1016/j.ijfoodmicro.2020.108818

Kegel A, Martinez P, Carter SD, Åström SU (2006) Genome wide distribution of illegitimate recombination events in *Kluyveromyces lactis*. Nucleic Acids Res 34(5):1633–1645. https://doi.org/10.1093/nar/gkl064

Keo-oudone C, Nitiyon S, Sotitham P, Tani A, Lertwattanasakul N, Yuangsaard N, Bounphanmy S, Lintong S, Yamada M (2016) Isolation and characterization of thermotolerant ethanol-fermenting yeasts from Laos and application of whole-cell matrix-assisted laser desorption/ionization time-of-flight mass spectrometry (MALDI-TOF/MS) analysis for their quick identification. African J Biotechnol 15(6):153–164. https://doi.org/10.5897/AJB2015.14984

Kim S-B, Kwon D-H, Park J-B, Ha S-J (2019) Alleviation of catabolite repression in *Kluyveromyces marxianus*: the thermotolerant SBK1 mutant simultaneously coferments glucose and xylose. Biotechnol Biofuels 12:90. https://doi.org/10.1186/s/s13068-019-1431-x

Kim S, Kim CH (2014) Evaluation of whole Jerusalem artichoke (*Helianthus tuberosus* L.) for a consolidated bioprocessing ethanol production. Renewable Energy 65:83–91

Kim S, Park JM, Kim CH (2013) Ethanol production using whole plant biomass of jerusalem artichoke by *Kluyveromyces marxianus* CBS1555. Appl Biochem Biotechnol 169(5):1531–1545. https://doi.org/10.1007/s12010-013-0094-5

Kim T-Y, Lee S-W, Oh M-K (2014) Biosynthesis of 2-phenylethanol from glucose with genetically engineered *Kluyveromyces marxianus*. Enzyme Microb Technol 61–62:44–47. https://doi.org/10.1016/j.enzmictec.2014.04.011

Kong X, Zhang B, Hua Y, Zhu Y, Li W, Wang D, Hong J (2019) Efficient L-lactic acid production from corncob residue using metabolically engineered thermo-tolerant yeast. Biores Technol 273:220–230. https://doi.org/10.1016/j.biortech.2018.11.018

Konings AW (1988) Importance of the glutathione level and the activity of the pentose phosphate pathway in cellular heat sensitivity. Recent Results in Cancer Research Fortschritte Der Krebsforschung Progres Dans Les Recherches Sur Le Cancer 109:109–125. https://doi.org/10.1007/978-3-642-83263-5_14

Kooistra R, Hooykaas PJJ, Steensma HY (2004) Efficient gene targeting in *Kluyveromyces lactis*. Yeast 21:781–792

Kosaka T, Lertwattanasakul N, Rodrussamee N, Nurcholis M, Dung NTP, Keo-Oudone C, Murata M, Götz P, Theodoropoulos C, Suprayogi; Maligan JM, Limtong S, Yamada M (2018) Potential of thermotolerant ethanologenic yeasts isolated from ASEAN countries and their application in high-temperature fermentation. Book Chapter IntechOpen. https://doi.org/10.5772/intechopen.79144

Kosaka T, Nakajima Y, Ishii A, Yamashita M, Yoshida S, Murata M, Kato K, Shiromaru Y, Kato S, Kanasaki Y, Yoshikawa H, Matsutani M, Thanonkeo P, Yamada M (2019) Capacity for survival in global warming: adaptation of mesophiles to the temperature upper limit. PLoS ONE 14(5):1–15. https://doi.org/10.1371/journal.pone.0215614

Koutinas M, Patsalou M, Stavrinou S, Vyrides I (2016) High temperature alcoholic fermentation of orange peel by the newly isolated thermotolerant *Pichia kudriavzevii* KVMP10. Lett Appl Microbiol 62(1):75–83. https://doi.org/10.1111/lam.12514

Krishna SH, Reddy TJ, Chowdary G (2001) Simultaneous saccharification and fermentation of lignocellulosic wastes to ethanol using a thermotolerant yeast. Biores Technol 77(2):193–196

Kuhn A, Van Zyl C, Van Tonder A, Prior BA (1995) Purification and partial characterization of an aldo-keto reductase from *Saccharomyces cerevisiae*. Appl Environ Microbiol 61(4):1580–1585. https://doi.org/10.1128/aem.61.4.1580-1585.1995

Kumar R, Tabatabaei M, Karimi K, Sárvári Horváth I (2016) Recent updates on lignocellulosic biomass derived ethanol—a review. Biofuel Res J 3(1):347–356

Kumar S, Singh SP, Mishra IM, Adhikari DK (2009) Ethanol and xylitol production from glucose and xylose at high temperature by *Kluyveromyces* sp. IIPE453. J Industr Microbiol Biotechnol 36(12):1483–1489. https://doi.org/10.1007/s10295-009-0636-6

Kwon D-H, Kim S-B, Park J-B, Ha S-J (2020) Overexpression of mutant galactose permease (ScGal2_N376F) effective for utilization of glucose/xylose or glucose/galactose mixture by engineered *Kluyveromyces marxianus*. J Microbiol Biotechnol 30(12):1944–1949. https://doi.org/10.4014/jmb.2008.08035

Kwon D-H, Park J-B, Hong E, Ha S-J (2019) Ethanol production from xylose is highly increased by the *Kluyveromyces marxianus* mutant 17694-DH1. Bioprocess Biosyst Eng 42(1):63–70. https://doi.org/10.1007/s00449-018-2014-0

Lane MM, Morrissey JP (2010) *Kluyveromyces marxianus*: a yeast emerging from its sister's shadow. Fungal Biol Rev 24(1):17–26. https://doi.org/10.1016/j.fbr.2010.01.001

Lee JW, In JH, Park J-B, Shin J, Park JH, Sung BH, Sohn J-H, Seo J-H, Park J-B, Kim SR, Kweon D-H (2017) Co-expression of two heterologous lactate dehydrogenases genes in *Kluyveromyces marxianus* for L-lactic acid production. J Biotechnol 241:81–86. https://doi.org/10.1016/j.jbiotec.2016.11.015

Lee MH, Lin JJ, Ke HM, Wang TY, Li WH, Lin YJ, Chang JJ, Fan WL (2018) Genome-wide prediction of CRISPR/Cas9 targets in *Kluyveromyces marxianus* and its application to obtain a stable haploid strain. Sci Rep 8(1). https://doi.org/10.1038/s41598-018-25366-z

Lehnen M, Ebert BE, Blank LM (2019) Elevated temperatures do not trigger a conserved metabolic network response among thermotolerant yeasts. BMC Microbiol 19(1):100. https://doi.org/10.1186/s12866-019-1453-3

Lertwattanasakul N, Kosaka T, Hosoyama A, Suzuki Y, Rodrussamee N, Matsutani M, Murata M, Fujimoto N, Suprayogi TK, Limtong S, Fujita N, Yamada M (2015) Genetic basis of the highly efficient yeast *Kluyveromyces marxianus*: complete genome sequence and transcriptome analyses. Biotechnol Biofuels 8(1):47. https://doi.org/10.1186/s13068-015-0227-x

Lertwattanasakul N, Rodrussamee N, Suprayogi LS, Thanonkeo P, Kosaka T, Yamada M (2011) Utilization capability of sucrose, raffinose and inulin and its less-sensitiveness to glucose repression in thermotolerant yeast *Kluyveromyces marxianus* DMKU 3-1042. AMB Express 1:20–31

Lertwattanasakul N, Suprayogi MM, Rodrussamee N, Limtong S, Kosaka T, Yamada M (2013) Essentiality of respiratory activity for pentose utilization in thermotolerant yeast *Kluyveromyces marxianus* DMKU 3-1042. Antonie Van Leeuwenhoek 103:933–945. https://doi.org/10.1007/s10482-012-9874-0

Li M, Lang X, Cabrera MM, De Keyser S, Sun X, Da Silva N, Wheeldon I (2021) CRISPR-mediated multigene integration enables Shikimate pathway refactoring for enhanced 2-phenylethanol biosynthesis in *Kluyveromyces marxianus*. Biotechnol Biofuels 14:3. https://doi.org/10.1186/s13068-020-01852-3

Li P, Fu X, Zhang L, Zhang Z, Li J, Li S (2017) The transcription factors Hsf1 and Msn2 of thermotolerant *Kluyveromyces marxianus* promote cell growth and ethanol fermentation of *Saccharomyces cerevisiae* at high temperatures. Biotechnol Biofuels 10(1). https://doi.org/10.1186/s13068-017-0984-9

Li Z, Liu G, Qu Y (2017) Improvement of cellulolytic enzyme production and performance by rational designing expression regulatory network and enzyme system composition. Biores Technol 245(Part B):1718–1726. https://doi.org/10.1016/j.biortech.2017.06.120

Lieber MR (2010) The mechanism of double-strand DNA break repair by the nonhomologous DNA end-joining pathway. Annu Rev Biochem 79:181e211

Limtong S, Sringiew C, Yongmanitchai W (2007) Production of fuel ethanol at high temperature from sugar cane juice by a newly isolated *Kluyveromyces marxianus*. Biores Technol 98(17):3367–3374. https://doi.org/10.1016/j.biortech.2006.10.044

Lin Y-J, Chang J-J, Lin H-Y, Thia C, Kao Y-Y, Huang C-C, Li W-H (2017) Metabolic engineering a yeast to produce astaxanthin. Biores Technol 245(Part A):899–905. https://doi.org/10.1016/j.biortech.2017.07.116

Liu Y, Li C, Zhang G, Sun H, Sun X, Jiang N, Rasool A, Lin Z (2014) Enhanced pathway efficiency of *Saccharomyces cerevisiae* by introducing thermo-tolerant devices. Biores Technol 170:38–44. https://doi.org/10.1016/j.biortech.2014.07.063

Liu Z, Liang Y, Ang EL, Zhao H (2017) A new era of genome integration-simply cut and paste! ACS Synth Biol 6(4):601e9

Löbs AK, Lin JL, Cook M, Wheeldon I (2016) High throughput, colorimetric screening of microbial ester biosynthesis reveals high ethyl acetate production from *Kluyveromyces marxianus* on C5, C6, and C12 carbon sources. Biotechnol J 11(10):1274–1281. https://doi.org/10.1002/biot.201600060

Löbs A-K, Engel R, Schwartz C, Flores A, Wheeldon I (2017) CRISPR-Cas9-enabled genetic disruptions for understanding ethanol and ethyl acetate biosynthesis in *Kluyveromyces marxianus*. Biotechnol Biofuels 10:1–14. https://doi.org/10.1186/s13068-017-0854-5

Los DA, Murata N (2004) Membrane fluidity and its roles in the perception of environmental signals. BBA Biomembranes 1666(1):142–157. https://doi.org/10.1016/j.bbamem.2004.08.002

Löser C, Urit T, Keil P, Bley T (2015) Studies on the mechanism of synthesis of ethyl acetate in *Kluyveromyces marxianus* DSM 5422. Appl Microbiol Biotechnol 99(3):1131–1144. https://doi.org/10.1007/s00253-014-6098-4

Löser C, Urit T, Stukert A, Bley T (2013) Formation of ethyl acetate from whey by *Kluyveromyces marxianus* on a pilot scale. J Biotechnol 163(1):17–23. https://doi.org/10.1016/j.jbiotec.2012.10.009

Lulu L, Dongmei W, Xiaolian G, Jiong H, Ling Z, Hisanori T, Hidehiko K (2013) Identification of a xylitol dehydrogenase gene from *Kluyveromyces marxianus* NBRC1777. Mol Biotechnol 53(2):159–169. https://doi.org/10.1007/s12033-012-9508-9

Maassen N, Freese S, Schruff B, Passoth V, Klinner U (2008) Nonhomologous end joining and homologous recombination DNA repair pathways in integration mutagenesis in xylose-fermenting yeast *Pichia stipitis*. FEMS Yeast Res 8:735–743

Madeo F, Fröhlich KU, Fröhlich E, Ligr M, Wolf DH, Grey M, Sigrist SJ (1999) Oxygen stress: a regulator of apoptosis in yeast. J Cell Biol 145(4):757–767. https://doi.org/10.1083/jcb.145.4.757

Mahmoud AE, Fathy SA, Rashad MM, Ezz MK, Mohammed AT (2018) Purification and characterization of a novel tannase produced by *Kluyveromyces marxianus* using olive pomace as solid support, and its promising role in gallic acid production. Int J Biol Macromol 107:2342–2350. https://doi.org/10.1016/j.ijbiomac.2017.10.117

Marcišauskas S, Ji B, Nielsen J (2019) Reconstruction and analysis of a *Kluyveromyces marxianus* genome-scale metabolic model. BMC Bioinform 20(1). https://doi.org/10.1186/s12859-019-3134-5

Margeot A, Hahn-Hagerdal B, Edlund M, Slade R, Monot F (2009) New improvements for lignocellulosic ethanol. Curr Opin Biotechnol 20(3):372–380. https://doi.org/10.1016/j.copbio.2009.05.009

Martins DBG, de Souza Jr CG, Simões DA, de Morais Jr MA (2002) The β-galactosidase activity in *Kluyveromyces marxianus* CBS6556 decreases by high concentrations of galactose. Curr Microbiol 44(5):379–382. https://doi.org/10.1007/s00284-001-0052-2

Matsushika A, Inoue H, Kodaki T, Sawayama S (2009) Ethanol production from xylose in engineered *Saccharomyces cerevisiae* strains: current state and perspectives. Appl Microbiol Biotechnol 84(1):37–53. https://doi.org/10.1007/s00253-009-2101-x

Matsushita K, Azuma Y, Kosaka T, Yakushi T, Hoshida H, Akada R, Yamada M (2016) Genomic analyses of thermotolerant microorganisms used for high-temperature fermentations. Biosci Biotechnol Biochem 80(4):655–668. https://doi.org/10.1080/09168451.2015.1104235

Matsuzaki C, Nakagawa A, Koyanagi T, Tanaka K, Minami H, Tamaki H, Katayama T, Yamamoto K, Kumagai H (2012) *Kluyveromyces marxianus*-based platformfor direct ethanol fermentation and recovery from cellulosic materials under air-ventilated conditions. J Biosci Bioeng 113:604–607. https://doi.org/10.1016/j.jbiosc.2011.12.007

Mattanovich D, Branduardi P, Dato L, Gasser B, Sauer M, Porro D (2012) Recombinant protein production in yeasts. Methods Mol Biol 824:329–358. https://doi.org/10.1007/978-1-61779-433-9_17

McTaggart TL, Bever D, Bassett S, Da Silva NA (2019) Synthesis of polyketides from low cost substrates by the thermotolerant yeast *Kluyveromyces marxianus*. Biotechnol Bioeng 116:1721–1730. https://doi.org/10.1002/bit.26976

Mejía-Barajas J, Montoya-Pérez R, Manzo-Avalos S, Cortés-Rojo C, Riveros-Rosas H, Cervantes C, Saavedra-Molina A (2018) Fatty acid addition and thermotolerance of *Kluyveromyces marxianus*. FEMS Microbiol Lett 365(7):1–5. https://doi.org/10.1093/femsle/fny043

Mejía-Barajas JA, Montoya-Pérez R, Salgado-Garciglia R, Aguilera-Aguirre L, Cortés-Rojo C, Mejía-Zepeda R, Arellano-Plaza M, Saavedra-Molina A (2017) Oxidative stress and antioxidant response in a thermotolerant yeast. Braz J Microbiol 48(2):326–332. https://doi.org/10.1016/j.bjm.2016.11.005

Mo W, Wang M, Zhan R, Yu Y, Lu H, He Y (2019) *Kluyveromyces marxianus* developing ethanol tolerance during adaptive evolution with significant improvements of multiple pathways. Biotechnol Biofuels 12(1). https://doi.org/10.1186/s13068-019-1393-z

Murata M, Fujimoto H, Nishimura K, Charoensuk K, Nagamitsu H, Raina S, Kosaka T, Oshima T, Ogasawara N, Yamada M (2011) Molecular strategy for survival at a critical high temperature in *Eschierichia coli*. PLoS ONE 6(6):1–9. https://doi.org/10.1371/journal.pone.0020063

Murata M, Ishii A, Fujimoto H, Nishimura K, Kosaka T, Mori H, Yamada M (2018) Update of thermotolerant genes essential for survival at a critical high temperature in *Escherichia coli*. PLoS ONE 13(2):1–10. https://doi.org/10.1371/journal.pone.0189487

Nambu-Nishida Y, Kondo A, Nishida K, Hasunuma T (2017) Development of a comprehensive set of tools for genome engineering in a cold- and thermo-tolerant *Kluyveromyces marxianus* yeast strain. Sci Rep 7(1). https://doi.org/10.1038/s41598-017-08356-5

Nambu-Nishida Y, Kondo A, Nishida K, Hasunuma T (2018) Genetic and physiological basis for antibody production by *Kluyveromyces marxianus*. AMB Express 8(1). https://doi.org/10.1186/s13568-018-0588-1

Nausch H, Huckauf J,Koslowski R, Meyer U, Broer I, Mikschofsky H (2013) Recombinant production of human interleukin 6 in *Escherichia coli*. PLoS One 8:e54933

Nielsen J, Larsson C, van Maris A, Pronk J (2013) Metabolic engineering of yeast for production of fuels and chemicals. Curr Opin Biotechnol 24(3):398–404. https://doi.org/10.1016/j.copbio.2013.03.023

Nitiyon S, Keo-oudone C, Murata M, Lertwattanasakul N, Limtong S, Kosaka T, Yamada M (2016) Efficient conversion of xylose to ethanol by stress-tolerant *Kluyveromyces marxianus* BUNL-21. Springerplus 5(1):1–12. https://doi.org/10.1186/s40064-016-1881-6

Nonklang S, Abdel-Banat BMA, Cha-aim K, Hoshida H, Yamada M, Akada R, Moonjai N, Limtong S (2008) High-temperature ethanol fermentation and transformation with linear DNA in the thermotolerant yeast *Kluyveromyces marxianus* DMKU3-1042. Appl Environ Microbiol 74(24):7514–7521. https://doi.org/10.1128/AEM.01854-08

Nonklang S, Ano A, Abdel-Banat BMA, Saito Y, Hoshida H, Akada R (2009) Construction of flocculent *Kluyveromyces marxianus* strains suitable for high-temperature ethanol fermentation. Biosci Biotechnol Biochem 73:1090–1095. https://doi.org/10.1271/bbb.80853

Nurcholis M, Lertwattanasakul N, Rodrussamee N, Kosaka T, Murata M, Yamada M (2020) Integration of comprehensive data and biotechnological tools for industrial applications of *Kluyveromyces marxianus*. Appl Microbiol Biotechnol 104:475–488. https://doi.org/10.1007/s00253-019-10224-3

Nurcholis M, Nitiyon S, Suprayogi RN, Lertwattanasakul N, Limtong S, Kosaka T, Yamada M (2019) Functional analysis of Mig1 and Rag5 as expressional regulators in thermotolerant yeast *Kluyveromyces marxianus*. Appl Microbiol Biotechnol 103(1):395–410. https://doi.org/10.1007/s00253-018-9462-y

Oberoi H, Babbar N, Sandhu S, Dhaliwal S, Kaur U, Chadha B, Bhargav V (2012) Ethanol production from alkali-treated rice straw via simultaneous saccharification and fermentation using newly isolated thermotolerant *Pichia kudriavzevii* HOP-1. J Ind Microbiol Biotechnol 39(4):557–566. https://doi.org/10.1007/s10295-011-1060-2

Olson DG, McBride JE, Joe Shaw A, Lynd LR (2012) Recent progress in consolidated bioprocessing. Curr Opin Biotechnol 23(3):396–405. https://doi.org/10.1016/j.copbio.2011.11.026

Ortiz-Merino RA, Varela JA, Coughlan AY, Hoshida H, da Silveira WB, Wilde C, Kuijpers NGA, Geertman J-M, Wolfe KH, Morrissey JP (2018) Ploidy variation in *Kluyveromyces marxianus* separates dairy and non-dairy isolates. Frontiers Genetics 9. https://doi.org/10.3389/fgene.2018.00094

Ozmihci S, Kargi F (2007) Comparison of yeast strains for batch ethanol fermentation of cheese-whey powder (CWP) solution. Lett Appl Microbiol 44(6):602–606. https://doi.org/10.1111/j.1472-765X.2007.02132.x

Palmbos PL, Daley JM, Wilson TE (2005) Mutation of the Yku80 C terminus and Xrs2 FHA domain specifically block yeast nonhomologous end joining. Mol Cell Biol 25:10782–10790

Pan Y (2011) Mitochondria, reactive oxygen species, and chronological aging: a message from yeast. Exp Gerontol 46(11):847–852. https://doi.org/10.1016/j.exger.2011.08.007

Park Y, Sunwoo IY, Yang J, Jeong G-T, Kim S-K (2020) Comparison of ethanol yield coefficients using *Saccharomyces cerevisiae*, *Candida lusitaniae*, and *Kluyveromyces marxianus* adapted to high concentrations of galactose with *Gracilaria verrucosa* as substrate. J Microbiol Biotechnol 30(6):930–936. https://doi.org/10.4014/jmb.2002.02014

Pecota DC, Da Silva NA (2005) Evaluation of the tetracycline promoter system for regulated gene expression in *Kluyveromyces marxianus*. Biotechnol Bioeng 92:117–123. https://doi.org/10.1002/bit.20584

Pecota DC, Rajgarhia V, Da Silva NA (2007) Sequential gene integration for the engineering of *Kluyveromyces marxianus*. J Biotechnol 127:408–416. https://doi.org/10.1016/j.jbiotec.2006.07.031

Pentjuss A, Stalidzans E, Liepins J, Kokina A, Martynova J, Zikmanis P, Mozga I, Scherbaka R, Vigants A, Hartman H, Poolman M, Fell D (2017) Model-based biotechnological potential analysis of *Kluyveromyces marxianus* central metabolism. J Ind Microbiol Biotechnol 44(8):1177–1190. https://doi.org/10.1007/s10295-017-1946-8

Pessani NK, Atiyeh HK, Wilkins MR, Bellmer DD, Banat IM (2011) Simultaneous saccharification and fermentation of Kanlow switchgrass by thermotolerant *Kluyveromyces marxianus* IMB3: The effect of enzyme loading, temperature and higher solid loadings. Biores Technol 102(22):10618–10624. https://doi.org/10.1016/j.biortech.2011.09.011

Piper PW (1993) Molecular events associated with acquisition of heat tolerance by the yeast *Saccharomyces cerevisiae*. FEMS Microbiol Rev 11(4):339–355. https://doi.org/10.1111/j.1574-6976.1993.tb00005.x

Pongcharoen P, Chawneua J, Tawong W (2018) High temperature alcoholic fermentation by new thermotolerant yeast strains *Pichia kudriavzevii* isolated from sugarcane field soil. Agric Nat Res 52(6):511–518. https://doi.org/10.1016/j.anres.2018.11.017

Porro D, Sauer M, Branduardi P, Mattanovich D (2005) Recombinant protein production in yeasts. Mol Biotechnol 31(3):245–259. https://doi.org/10.1385/MB:31:3:245

Pumpen P, Kozlovskaya TM, Borisova GP, Bichko VV, Dishler A, Kalis J, Kukaine RA, Gren EJ (1984) Expression of hepatitis B virus surface antigen gene in *Escherichia coli*. Gene 30:201–210

Qiu Z, Jiang R (2017) Improving *Saccharomyces cerevisiae* ethanol production and tolerance via RNA polymerase II subunit Rpb7. Biotechnol Biofuels 10(1).https://doi.org/10.1186/s13068-017-0806-0

Quarella S, Lovrovich P, Scalabrin S, Campedelli I, Backovic A, Gatto V, Cattonaro F, Turello A, Torriani S, Felis GE (2016) Draft genome sequence of the probiotic yeast Kluyveromyces marxianus fragilis B0399. Genome Announcements 4. https://doi.org/10.1128/genomeA.00923-16

Raimondi S, Uccelletti D, Amaretti A, Leonardi A, Palleschi C, Rossi M (2010) Secretion of *Kluyveromyces lactis* Cu/Zn SOD: strategies for enhanced production. Appl Microbiol Biotechnol 86:871–878

Raimondi S, Zanni E, Amaretti A, Palleschi C, Uccelletti D, Rossi M (2013) Thermal adaptability of *Kluyveromyces marxianus* in recombinant protein production. Microb Cell Factories 12:1–7. https://doi.org/10.1186/1475-2859-12-34

Rajkumar AS, Varela JA, Juergens H, Daran J-MG, Morrissey JP (2019) Biological parts for *Kluyveromyces marxianus* synthetic biology. Front Bioeng Biotechnol. https://doi.org/10.3389/fbioe.2019.00097

Rajkumar SA, Morrissey PJ (2020) Rational engineering of *Kluyveromyces marxianus* to create a chassis for the production of aromatic products. Microb Cell Fact 19:207. https://doi.org/10.1186/s12934-020-01461-7

Ravella SR, Gallagher J, Fish S, Prakasham RS (2012) Overview on commercial production of xylitol, economic analysis and market trends. Springer, Berlin Heidelberg

Raynal A, Gerbaud C, Francingues MC, Guerineau M (1987) Sequence and transcription of the β-glucosidase gene of *Kluyveromyces fragilis* cloned in *Saccharomyces cerevisiae*. Curr Genet 12(3):175–184. https://doi.org/10.1007/BF00436876

Rocha SN, Abrahão-Neto J, Cerdán ME, Gombert AK, González-Siso MI (2011) Heterologous expression of a thermophilic esterase in *Kluyveromyces* yeasts. Appl Microbiol Biotechnol 89:375–385. https://doi.org/10.1007/s00253-010-2869-8

Rocha SN, Abrahão-Neto J, Cerdán ME, González-Siso MI, Gombert AK (2010) Heterologous expression of glucose oxidase in the yeast *Kluyveromyces marxianus*. Microb Cell Factories 9:1–12. https://doi.org/10.1186/1475-2859-9-4

Rodrussamee N, Lertwattanasakul N, Hirata K, Limtong S, Kosaka T, Yamada M (2011) Growth and ethanol fermentation ability on hexose and pentose sugars and glucose effect under various conditions in thermotolerant yeast *Kluyveromyces marxianus*. Appl Microbiol Biotechnol 90(4):1573–1586. https://doi.org/10.1007/s00253-011-3218-2

Rollero S, Bauer FF, Divol B, Bloem A, Camarasa C, Ortiz-Julien A (2019) A comparison of the nitrogen metabolic networks of *Kluyveromyces marxianus* and *Saccharomyces cerevisiae*. Environ Microbiol 21(11):4076–4091. https://doi.org/10.1111/1462-2920.14756

Roohina F, Mohammadi M, Najafpour GD (2016) Immobilized *Kluyveromyces marxianus* cells in carboxymethyl cellulose for production of ethanol from cheese whey: experimental and kinetic studies. Bioprocess Biosyst Eng 39(9):1341–1349. https://doi.org/10.1007/s00449-016-1610-0

Rosano GL, Ceccarelli EA (2014) Recombinant protein expression in *Escherichia coli*: advances and challenges. Front Microbiol 5:172. https://doi.org/10.3389/fmicb.2014.00172

Rouhollah H, Iraj N, Giti E, Sorah A (2007) Mixed sugar fermentation by *Pichia stipitis*, *Saccharomyces cerevisiae*, and an isolated xylose-fermenting *Kluyveromyces marxianus* and their cocultures. Afr J Biotech 6(9):1110–1114

Rouwenhorst RJ, Visser LE, Van Der Baan AA, Scheffers WA, Van Dijken JP (1988) Production, distribution, and kinetic properties of inulinase in continuous cultures of *Kluyveromyces marxianus* CBS 6556. Appl Environ Microbiol 54(5):1131–1137. https://doi.org/10.1128/AEM.54.5.1131-1137.1988

Ryabova OB, Chmil OM, Sibirny AA (2003) Xylose and cellobiose fermentation to ethanol by the thermotolerant methylotrophic yeast *Hansenula polymorpha*. FEMS Yeast Res 4(2):157–164. https://doi.org/10.1016/S1567-1356(03)00146-6

Saet-Byeol K, Deok-Ho K, Jae-Bum P, Suk-Jin H (2019) Alleviation of catabolite repression in *Kluyveromyces marxianus*: the thermotolerant SBK1 mutant simultaneously coferments glucose and xylose. Biotechnol Biofuels 12(1):1–9. https://doi.org/10.1186/s13068-019-1431-x

Sansonetti S, Curcio S, Calabrò V, Iorio G (2009) Bio-ethanol production by fermentation of ricotta cheese whey as an effective alternative non-vegetable source. Biomass Bioenerg 33(12):1687–1692. https://doi.org/10.1016/j.biombioe.2009.09.002

Schabort DTW, Letebele PK, Steyn L, Kilian SG, du Preez JC (2016) Differential RNA-seq, multinetwork analysis and metabolic regulation analysis of *Kluyveromyces marxianus* reveals a compartmentalised response to xylose. PLoS ONE 11(6):1–31. https://doi.org/10.1371/journal.pone.0156242

Scherz-Shouval R, Elazar Z (2007) ROS, mitochondria and the regulation of autophagy. Trends Cell Biol 17(9):422–427. https://doi.org/10.1016/j.tcb.2007.07.009

Shallom D, Shoham Y (2003) Microbial hemicellulases. Curr Opin Microbiol 6(3):219–228. https://doi.org/10.1016/S1369-5274(03)00056-0

Shao Z, Zhao H, Zhao H. (2009) DNA assembler, an in vivo genetic method for rapid construction of biochemical pathways. Nucleic Acids Res 37(2):e16

Sharma NK, Behera S, Arora R, Kumar S (2016) Enhancement in xylose utilization using *Kluyveromyces marxianus* NIRE-K1 through evolutionary adaptation approach. Bioprocess Biosyst Eng 39(5):835–843. https://doi.org/10.1007/s00449-016-1563-3

Sharma NK, Behera S, Arora R, Kumar S (2017) Evolutionary adaptation of *Kluyveromyces marxianus* NIRE-K3 for enhanced xylose utilization. Front Energy Res 5(DEC). https://doi.org/10.3389/fenrg.2017.00032

Šiekštele R, Bartkevičiute D, Sasnauskas K (1999) Cloning, targeted disruption and heterologous expression of the *Kluyveromyces marxianus* endopolygalacturonase gene (*EPG1*). Yeast 15(4):311–322. https://doi.org/10.1002/(SICI)1097-0061(19990315)15:4%3c311::AID-YEA379%3e3.0.CO;2-9

Silveira WB, Diniz RH, Cerdán ME, González-Siso MI, Souza RA, Vidigal PM, Brustolini OJ, de Almeida Prata ER, Medeiros AC, Paiva LC, Nascimento M, Ferreira EG, Dos Santos VC, Bragança CR, Fernandes TA, Colombo LT, Passos FM (2014) Genomic sequence of the yeast

Kluyveromyces marxianus CCT 7735 (UFV-3), a highly lactose-fermenting yeast isolated from the Brazilian dairy industry. Genome Announcements 2. https://doi.org/10.1128/genomeA.01136-14

Simoness O, Murilol B, Carlosr R, Paulalde A, Mariackv R, Franciscorde A-N, Soreleb F, Luizars D (2017) Asymmetric bioreduction of β-ketoesters derivatives by *Kluyveromyces marxianus*: Influence of molecular structure on the conversion and enantiomeric excess. An Acad Bras Ciênc 89:1403–1415. https://doi.org/10.1590/0001-3765201720170118

Sivarathnakumar S, Jayamuthunagai J, Baskar G, Praveenkumar R, Selvakumari IAE, Bharathiraja B (2019) Bioethanol production from woody stem *Prosopis juliflora* using thermo tolerant yeast *Kluyveromyces marxianus* and its kinetics studies. Biores Technol 293. https://doi.org/10.1016/j.biortech.2019.122060

Soemphol W, Deeraksa A, Matsutani M, Yakushi T, Toyama H, Adachi O, Yamada M, Matsushita K (2011) Global analysis of the genes involved in the thermotolerance mechanism of thermotolerant *Acetobacter tropicalis* SKU1100. Biosci Biotechnol Biochem 75(10):1921–1928

Sokolenko G, Karpechenko N (2015) Expression of inulinase genes in the yeasts *Saccharomyces cerevisiae* and *Kluyveromyces marxianus*. Microbiology (00262617) 84(1):23–27.https://doi.org/10.1134/S0026261715010142

Steels EL, Learmonth RP, Watson K (1994) Stress tolerance and membrane lipid unsaturation in *Saccharomyces cerevisiae* grown aerobically or anaerobically. Microbiology 140(3):569–576

Storeng R, Johne B (1987) Toxic effects of lipopolysaccharide from *Bacteroides intermedius* and *Escherichia coli* assessed in the pre-implantation mouse embryo culture system. Acta Pathol Microbiol Immunol Scand B 95:135–139

Su M, Hu Y, Cui Y, Wang Y, Yu H, Liu J, Dai W, Piao C (2021) Production of b-glucosidase from okara fermentation using *Kluyveromyces marxianus*. J Food Sci Technol 58:366–376. https://doi.org/10.1007/s13197-020-04550-y

Sugiyama KI, Izawa S, Inoue Y (2000) The Yap1p-dependent induction of glutathione synthesis in heat shock response of *Saccharomyces cerevisiae*. J Biol Chem 275(20):15535–15540. https://doi.org/10.1074/jbc.275.20.15535

Suryawati L, Wilkins MR, Bellmer DD, Huhnke RL, Maness NO, Banat IM (2008) Simultaneous saccharification and fermentation of Kanlow switchgrass pretreated by hydrothermolysis using *Kluyveromyces marxianus* IMB4. Biotechnol Bioeng 101(5):894–902. https://doi.org/10.1002/bit.21965

Suzuki T, Hoshino T, Matsushika A (2014) Draft genome sequence of *Kluyveromyces marxianus* strain DMB1, isolated from sugarcane bagasse hydrolysate. Genome Announcements 2. https://doi.org/10.1128/genomeA.00733-14

Suzuki T, Hoshino T, Matsushika A (2019) High-temperature ethanol production by a series of recombinant xylose-fermenting *Kluyveromyces marxianus* strains. Enzyme Microbial Technol 129. https://doi.org/10.1016/j.enzmictec.2019.109359

Szczodrak J, Targonski Z (1988) Selection of thermotolerant yeast strains for simultaneous saccharification and fermentation of cellulose. Biotechnol Bioeng (USA) 31(4):300–303

Tarrío N, Becerra M, Cerdan ME, González-Siso MI (2006) Reoxidation of cytosolic NADPH in *Kluyveromyces lactis*. FEMS Yeast Res 6(3):371–380

Tarrío N, García-Leiro A, Cerdán ME, González-Siso MI (2008) The role of glutathione reductase in the interplay between oxidative stress response and turnover of cytosolic NADPH in *Kluyveromyces lactis*. FEMS Yeast Res 8(4):597–606. https://doi.org/10.1111/j.1567-1364.2008.00366.x

Theron CW, Labuschagné M, Gudiminchi R, Albertyn J, SmitMS, (2014) A broad-range yeast expression system reveals *Arxula adeninivorans* expressing a fungal self-sufficient cytochrome P450 monooxygenase as an excellent whole-cell biocatalyst. FEMS Yeast Res 14:556–566. https://doi.org/10.1111/1567-1364.12142

Topete M, Casas TL, Galindo E (1997) β-Galactosidase production by *Kluyveromyces marxianus* cultured in shake flasks. Rev Latinoam Microbiol 39:101–107

Van Urk H, Voll WS, Scheffers WA, Van Dijken JP (1990) Transient-state analysis of metabolic fluxes in crabtree-positive and crabtree-negative yeasts. Appl Environ Microbiol 56(1):281–287. https://doi.org/10.1128/AEM.56.1.281-287.1990

Vandijken JP, Weusthuis RA, Pronk JT (1993) Kinetics of growth and sugar consumption in yeasts. Antonie Van Leeuwenhoek Int J Gen Mol Microbiol 63:343–352

Varela JA, Gethins L, Stanton C, Ross P, Morrissey JP (2017) Applications of *Kluyveromyces marxianus* in biotechnology. In: Satyanarayana T, Kunze G (eds) Yeast diversity in human welfare. Springer, p 439e53

Wang D, Wu D, Yang X, Hong J (2018) Transcriptomic analysis of thermotolerant yeast: *Kluyveromyces marxianus* in multiple inhibitors tolerance. RSC Adv 8(26):14177–14192. https://doi.org/10.1039/c8ra00335a

Wang R, Li L, Zhang B, Gao X, Wang D, Hong J (2013) Improved xylose fermentation of *Kluyveromyces marxianus* at elevated temperature through construction of a xylose isomerase pathway. J Ind Microbiol Biotechnol 40(8):841–854. https://doi.org/10.1007/s10295-013-1282-6

Wang R, Wang D, Gao X, Hong J (2014) Direct fermentation of raw starch using a *Kluyveromyces marxianus* strain that express glucoamylase and alpha-amylase to produce ethanol. Biotechnol Prog 30:338–347. https://doi.org/10.1002/btpr.1877

Wang YJ, Ying BB, Shen W, Zheng RC, Zheng YG (2017) Rational design of *Kluyveromyces marxianus* ZJB14056 aldo–keto reductase KmAKR to enhance diastereoselectivity and activity. Enzyme Microb Technol 107:32–40. https://doi.org/10.1016/j.enzmictec.2017.07.012

Wolf M, Gasparin BC, Paulino AT (2018) Hydrolysis of lactose using β-d-galactosidase immobilized in a modified Arabic gum-based hydrogel for the production of lactose-free/low-lactose milk. Int J Biol Macromol 115:157–164. https://doi.org/10.1016/j.ijbiomac.2018.04.058

Wu WH, Cheng KC, Hung WC, Wan HP, Lo KY, Chen YH (2016) Bioethanol production from taro waste using thermo-tolerant yeast *Kluyveromyces marxianus* K21. Biores Technol 201:27–32. https://doi.org/10.1016/j.biortech.2015.11.015

Xu F, Wang KY, Wang N, Li G, Liu D (2017) Modified human glucagon-like peptide-1 (GLP-1) produced in *E. coli* has a long-acting therapeutic effect in type 2 diabetic mice. PLoS One 12(7):e0181939. https://doi.org/10.1371/journal.pone.0181939

Yanase S, Yamada R, Ogino C, Kondo A, Hasunuma T, Tanaka T, Fukuda H (2010) Direct ethanol production from cellulosic materials at high temperature using the thermotolerant yeast *Kluyveromyces marxianus* displaying cellulolytic enzymes. Appl Microbiol Biotechnol 88(1):381–388. https://doi.org/10.1007/s00253-010-2784-z

Yang C, Hu S, Zhu S, Wang D, Gao X, Hong J (2015) Characterizing yeast promoters used in *Kluyveromyces marxianus*. World J Microbiol Biotechnol 31:1641–1646. https://doi.org/10.1007/s11274-015-1899-x

Yang D, Chen L, Duan J, Yu Y, Zhou J, Lu H (2021) Investigation of *Kluyveromyces marxianus* as a novel host for large-scale production of porcine parvovirus-like particles. Microb Cell Fact 20:24. https://doi.org/10.1186/s12934-021-01514-5

Yang P, Zhu X, Zheng Z, Mu D, Jiang S, Luo S, Wu Y, Du M (2018) Cell regeneration and cyclic catalysis of engineered *Kluyveromyces marxianus* of a D-psicose-3-epimerase gene from *Agrobacterium tumefaciens* for D-allulose production. World J Microbiol Biotechnol 34:65. https://doi.org/10.1007/s11274-018-2451-6

Yarimizu T, Nakamura M, Hoshida H, Akada R (2015) Synthetic signal sequences that enable efficient secretory protein production in the yeast *Kluyveromyces marxianus*. Microb Cell Fact 14:20–34

Yuan WJ, Chang BL, Ren JG, Bai FW, Li YY, Liu JP (2012) Consolidated bioprocessing strategy for ethanol production from Jerusalem artichoke tubers by *Kluyveromyces marxianus* under high gravity conditions. J Appl Microbiol 112(1):38–44. https://doi.org/10.1111/j.1365-2672.2011.05171.x

Yuangsaard N, Yongmanitchai W, Limtong S, Yamada M (2013) Selection and characterization of a newly isolated thermotolerant *Pichia kudriavzevii* strain for ethanol production at high temperature from cassava starch hydrolysate. Antonie Van Leeuwenhoek Int J General Mol Microbiol 103(3):577–588. https://doi.org/10.1007/s10482-012-9842-8

Zafar S, Owais M (2006) Ethanol production from crude whey by *Kluyveromyces marxianus*. Biochem Eng J 27(3):295–298

Zhang B, Li L, Zhang J, Gao X, Wang D, Hong J (2013) Improving ethanol and xylitol fermentation at elevated temperature through substitution of xylose reductase in *Kluyveromyces marxianus*. J Industr Microbiol Biotechnol 40(3–4):305–316. https://doi.org/10.1007/s10295-013-1230-5

Zhang B, Zhang L, Wang D, Gao X, Hong J (2011) Identification of a xylose reductase gene in the xylose metabolic pathway of *Kluyveromyces marxianus* NBRC1777. J Ind Microbiol Biotechnol 38(12):2001–2010. https://doi.org/10.1007/s10295-011-0990-z

Zhang B, Zhu Y, Zhang J, Wang D, Sun L, Hong J (2017a) Engineered *Kluyveromyces marxianus* for pyruvate production at elevated temperature with simultaneous consumption of xylose and glucose. Biores Technol 224:553–562. https://doi.org/10.1016/j.biortech.2016.11.110

Zhang G, Lu M, Wang J, Wang D, Gao X, Hong J (2017b) Identification of hexose kinase genes in *Kluyveromyces marxianus* and thermo-tolerant one step producing glucose-free fructose strain construction. Scientific Reports 7. https://doi.org/10.1038/srep45104

Zhang J, Zhang B, Wang D, Gao X, Hong J (2014) Xylitol production at high temperature by engineering *Kluyveromyces marxianus*. Bioresour Technol 152:192–201. https://doi.org/10.1016/j.biortech.2013.10.109

Zhang J, Zhang B, Wang D, Gao X, Sun L, Hong J (2015a) Rapid ethanol production at elevated temperatures by engineered thermotolerant *Kluyveromyces marxianus* via the NADP(H)-preferring xylose reductase-xylitol dehydrogenase pathway. Metab Eng 31:140–152. https://doi.org/10.1016/j.ymben.2015.07.008

Zhang M, Jiang L, Shi J (2015b) Modulation of mitochondrial membrane integrity and ROS formation by high temperature in *Saccharomyces cerevisiae*. Electron J Biotechnol 18(3):202–209. https://doi.org/10.1016/j.ejbt.2015.03.008

Zhou HX, Xin FH, Chi Z, Liu GL, Chi ZM (2014) Inulinase production by the yeast *Kluyveromyces marxianus* with the disrupted *MIG1* gene and the over-expressed inulinase gene. Process Biochem 49(11):1867–1874. https://doi.org/10.1016/j.procbio.2014.08.001

Zhou HX, Xu JL, Chi Z, Liu GL, Chi ZM (2013) β-Galactosidase over-production by a *mig1* mutant of *Kluyveromyces marxianus* KM for efficient hydrolysis of lactose. Biochem Eng J 76:17–24. https://doi.org/10.1016/j.bej.2013.04.010

Zhou J, Zhu P, Hu X, Lu H, Yu Y (2018) Improved secretory expression of lignocellulolytic enzymes in *Kluyveromyces marxianus* by promoter and signal sequence engineering. Biotechnol Biofuels 11(1). https://doi.org/10.1186/s13068-018-1232-7

Zoppellari F, Bardi L (2013) Production of bioethanol from effluents of the dairy industry by *Kluyveromyces marxianus*. New Biotechnol 30(6):607–613. https://doi.org/10.1016/j.snbt.2012.11.017

Synthetic Biology in the *Candida* (CTG) Clade

Dalal Kasir, Sébastien Besseau, Marc Clastre, Audrey Oudin, Monzer Hamze, Vincent Courdavault, Marwan Osman, and Nicolas Papon

Abstract

The continuous bio-race to find an ideal microbial chassis opens the path toward the *Candida* CTG clade to enter the competition. Unexpectedly, this unique CUG-serine coding clade successfully mastered the production of various nutraceutical, commercial, and pharmaceutical valuable compounds. The following chapter aims to snapshot the *Candida* CTG clade's bioengineering properties and their biotechnological applications in synthetic biology.

1 Introduction

The ever-increasing progress in technologies and genetic engineering has enabled this discipline to book a place in the list of revolutionizing scientific fields. Generally, the application of synthetic biology on "microbial chassis" aims to draw a detailed roadmap to successfully predict, guide, and control the functionality of cellular behavior (Leonard et al. 2008). Briefly, the manipulation of the microbial genomic repertoire by knockin/knockout of certain DNA sequences with known

D. Kasir · M. Hamze · M. Osman
Laboratoire Microbiologie Santé et Environnement (LMSE), Doctoral School of Sciences and Technology, Faculty of Public Health, Lebanese University, Tripoli, Lebanon

S. Besseau · M. Clastre · A. Oudin · V. Courdavault (✉)
Université de Tours, BBV EA 2106, 31 avenue Monge, 37200 Tours, France
e-mail: vincent.courdavault@univ-tours.fr

M. Osman
Department of Population Medicine and Diagnostic Sciences, College of Veterinary Medicine, Cornell University, Ithaca, NY 14850, USA

N. Papon (✉)
Université d'Angers, GEIHP, E3142, SFR 4208 ICAT, Angers, France
e-mail: nicolas.papon@univ-angers.fr

F. Darvishi Harzevili (ed.), *Synthetic Biology of Yeasts*,
https://doi.org/10.1007/978-3-030-89680-5_12

behaviors can generate synthetic bio-factories with new desirable properties (Lam et al. 2010). The decrease in DNA synthesis and sequencing cost, the improvement of the gene-to-protein relationship decoding, the standardization of DNA assembly modalities, and the implementation of detector–reporter systems are considered key players in this modern sector (Jensen and Keasling 2015). Various microbial factories are considered promising candidates in several biotechnological applications. Among these, the mycotic CTG clade corresponds to a subdivision of restricted ascomycetous yeast group that displays a distinctive genetic code (Papon et al. 2014). In these species, a mistranslation reassigned the universal canonical leucine CUG codon to serine predominantly (Papon et al. 2014). Despite the presence of numerous *Candida* species that pose a significant health challenge (due to their ability to develop muco-cutaneous and/or systemic infections), the CTG clade encompasses several species with strong biotechnological application prepotency (Defosse et al. 2018a; Turner and Butler 2014). Currently, these mycotic microorganisms are considered as suitable bio-producer candidates for industrial enzymes, single-cell protein, bioethanol, vitamins, sweeteners, lipids, and metabolites with pharmaceutical and nutritional values (Johnson 2013). Displaying extraordinary potentials is one of the reasons that bring these yeasts into the discipline of synthetic biology, namely tolerate niche osmolarity changes, grow on a variety of carbon sources (including pentoses, n-alkanes, fatty acids, and phenols), metabolize inexpensive substrate, and stand for extreme environmental stresses (Papon et al. 2014; Johnson 2013). The genetic toolbox of CTG clade yeasts is progressively prepared to generate a stand-alone bio-factory via recipient strain construction, selectable marker adaptation, control (inducible or repressible) gene expression system formation, and transformation protocol optimization (Papon et al. 2012). Such efforts with the recent advancements in omics studies and metabolic modeling improve the understanding of CTG clade metabolic behavior, bringing it to the list of microbial competitors in the field of synthetic biology (Papon et al. 2014). The following chapter will cover the taxonomy, the main applied classical genetic tools, omics resources, and the major biotechnological potentials in *Candida* CTG clade.

2 Taxonomy

Similar to all microbial taxonomic relationships, the phylogeny of *Candida* species was reformulated after the introduction of advanced molecular techniques. Recently, whole genome analysis-based studies steered the subdivision of Saccharomycotina into 8 clades including in particular: (i) the Saccharomycetaceae clade including *Candida glabrata* and yeasts of *Saccharomyces* genus and (ii) CTG clade comprising *Candida albicans*, *Candida dubliniensis*, *Candida tropicalis*, *Candida parapsilosis*, *Candida maltosa*, *Candida famata*, *Candida lusitaniae*, *Candida oleophila*, *Candida auris*, *Diutina rugosa*, *Lodderomyces elongisporus*, *Metschnikowia fructicola*, *Meyerozyma guilliermondii*, *Millerozyma farinosa*, *Spathaspora passalidarum*, *Yamadazyma tenuis*, and *Scheffersomyces*

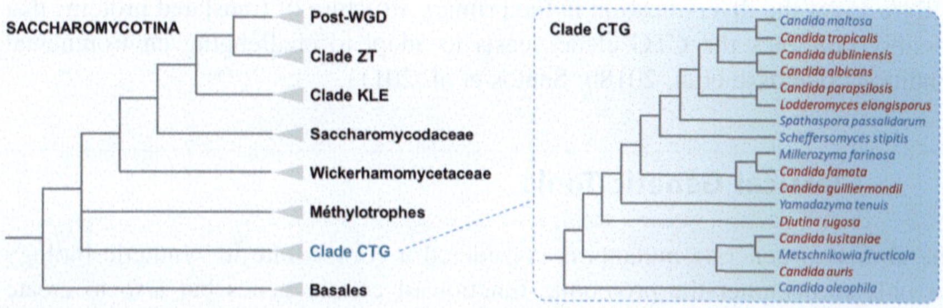

Fig. 1 Phylogeny of Saccharomycotina and CTG clade yeasts. The topology represents the current vision of the relations between clades and other species [according to (Dujon and Louis 2017; Shen et al. 2016; Chen et al. 2000; Tsui et al. 2008)]. The human opportunistic species are shown in red. The phylogeny of Saccharomycotina presented here in 8 clades is not definitive nor used by all authors (Dujon and Louis 2017). Only a few species of the CTG clade are indicated in the phylogenetic tree on the right. Adapted from Defosse et al. (2018a)

stipitis (Fig. 1) (Fitzpatrick et al. 2006; Butler et al. 2009; Nosek et al. 2009; Defosse et al. 2018b). The unique genetic decoding ability of CTG clade was characterized for the first time by Kawaguchi et al. (1989) in *D. rugosa*. Generally, the deviation of this clade's coding system greatly supports the "ambiguous intermediate theory," which hypothesized that codons are reassigned through enigmatic decoding in which a mutation in tRNA can expand its decoding array capacity (Santos et al. 2004). Mühlhausen et al. (2014) described the cause for the translation codon shift in CTG clade by demonstrating the polyphyletic relationship among the *Candida* genus using phylogenomic analyses of 26 motor and cytoskeletal proteins. The evolutionary driving force mainly led to the emergence of two groups with codon bias; one used the standard CUG translation, while the other is an alternative yeast codon usage (AYCU) (Mühlhausen and Kollmar 2014). Comparative and molecular phylogenomics studies showed that $tRNA_{CAG}^{Ser}$ emerged (272 ± 25 million years ago) prior to the separation between the *Saccharomyces* and *Candida* genera (170 ± 27 million years ago); such appearance drives the common ancestor of CTG clade to lose the cognate $tRNA_{CAG}^{leu}$ to the newly mutant $tRNA_{CAG}^{Ser}$ (Defosse et al. 2018b; Santos et al. 2011).

Comparative genomics analyses of CTG-Ser clade species revealed that CTG codons are most frequently aligned with Ser rather than Leu amino acid (Riley et al. 2016). Furthermore, their predicted $tRNA_{CAG}^{Ser}$ mainly has the three serylation features: (i) a guanine at position 33 (G33) in the anticodon loop (may lessen the rates of leucylation), (ii) Ser identity element (positioned in the variable loop), and (iii) G discriminator base at position 73 (Riley et al. 2016). The genomic comparison of ascomycete yeasts gives a snapshot of their biotechnological exploitations, which reveals a heterogeneous distribution of metabolic traits, with restriction to a single clade (such as methylotrophy), and patches distribution for others (such as D-xylose utilization) (Riley et al. 2016). Functionally speaking, the incorporation of serine in polypeptides originating from CUG-loaded mRNA drives the diversification in the primary structure of translated proteins that possibly promotes the CTG clade yeasts to adapt to challenging environmental conditions (Defosse et al. 2018b; Santos et al. 2011).

3 Classical Genetic Tools

The ability to generate mutants is considered a cornerstone in synthetic biology not only to decipher the proteomic function of certain genes but also to create sophisticated bio-systems with the potency to produce beneficial compounds. Early researchers introduced mutations into the genomes of tested strains by exposing them to either UV or chemical mutagens, causing inaccurate and undesirable mutations. The innovation of genome modification techniques over the past decade revolutionized the propensity to accurately regulate a vast array of genes at the genome level (Ren et al. 2020). Genome-scale engineering represents the keystone methodology to precisely manipulate the microbiological factories' operation, generating a reproducible bio-system (Ren et al. 2020). The following section will cover the main standard and advanced genetic tools applied in *Candida* CTG clade yeasts.

3.1 Selection Markers

Selection markers are considered an integral part of the construction of transformation vectors since they allow the screening for successful transformants. The choice of selection markers is dependent on the genomic repertoire of the transformant host strains (Hinchliffe et al. 1993). Löbs et al. (2017) recently described the procedure of generation and utilization of auxotrophic markers for engineered yeast. Typically, random mutagenesis or homologous recombination facilitates the generation of stable auxotrophic strains, which allows further advanced genome editing (random/targeted integration, Cre-lox, HisG/lacZ, and CRISPR–Cas9) to be applied in the targeted yeast species (Löbs et al. 2017). Generally, in *Candida,* the standard dominant selection genes are divided into two main groups: metabolic (nutritional) markers and drug-resistant cassettes (Papon et al. 2012). In the metabolic-marker-mediated yeast transformation, the formation of selection markers undergoes three main steps: (i) generation and isolation of an auxotrophic recipient strain, (ii) identification of the interrupted metabolic pathway by "medium complementation" with the appropriate amino acid or base, and (iii) restoration of the metabolically recessive mutation by using the gene encoding the functional "repairing" enzyme (Papon et al. 2012). Currently, numerous metabolic genes are used as selectable markers in *Candida* CTG clade, such as *HIS1, ARG4, URA3, ADE2, MET2, LYS4,* and *ARO4* (Samaranayake and Hanes 2011; Vyas et al. 2015; Kosa et al. 2007; Dmytruk et al. 2011).

However, the extensive investigation and optimization of drug-resistant cassettes facilitate the functionality of such markers in the CTG clade (Papon et al. 2012). Four main codon-optimized genes constitute the drug-resistant markers in *Candida* spp.: (i) hygromycin phosphotransferase (*HPH*; confers resistance to hygromycin B), (ii) streptothricin acetyltransferase (*SAT-1*; confers resistance to nourseothricin), (iii) glyoxalase (*ble*; confers resistance to bleomycin/phleomycin/zeocin), and iv) inosine 5-monophosphate dehydrogenase (*IMH*; confers resistance to mycophenolic acid) (Papon et al. 2012). Furthermore, new dominant selectable markers for *C. famata* were recently developed by Bratiichuk et al. (2020) termed *BSD* (*Aspergillus terreus* origin) encoding for blasticidin S deaminase, thus conferring resistance against blasticidin (Table 1).

3.2 Reporter Genes

Reporter genes are considered a robust tool in genetic analysis, reporting successful transformation and monitoring protein interaction and localization (Papon et al. 2012; Expression et al. 2002; Jones and Thomas 2003). Furthermore, apart from reporting successful recipients of an episomal system, various reporter genes facilitate the dual-labeling approach in different *Candida* CTG clade yeasts which aids in studying protein co-expression and co-localization (Courdavault et al. 2011; Reijnst et al. 2011). Among the reporter genes used in *Candida* genetic manipulation are:

3.2.1 Fluorescent Reporters

Generally, they are implicated in tracking protein localization/interaction and monitoring promoter activity in a broad range of fungal species (Papon et al. 2012). Numerous reengineered fluorescent proteins have been developed, aiming for codon optimization and brightness improvement (Cormack et al. 1996, 1997). Initially, the green fluorescent protein (GFP; *Aequorea victoria* origin) was successfully integrated into *C. albicans*-manipulating cassettes after integration of chromophore mutations and exhibited strong fluorescence emission in comparison with the wild-type strains (Morschhäuser et al. 1998). Later, certain genetic manipulations (site-directed mutagenesis) of the yeast-enhanced GFP synthetic gene yEGFP3 facilitated the production of two variants, yellow fluorescent protein (YFP) and cyan fluorescent protein (CFP) (Papon et al. 2012). Both GFP variants were reengineered to enhance brightness, provide further emission wavelengths, and promote multicolor imaging of differential gene expression and protein localization in *C. albicans* (Gerami-Nejad et al. 2001).

3.2.2 Enzymatic Reporters

In *Candida* CTG clade, two enzymatic systems are mainly applied, Luciferase and β-Galactosidase: Luciferase, encoded from certain bioluminescent genes to produce different varieties of luciferase enzyme (Roda et al. 2004). For a light emission, a substrate called luciferin is added to the manipulation platform (Roda et al. 2004). The luciferase system was notably recognized in *Candida* genetic

Table 1 Molecular toolbox for *Candida* CTG clade engineering

Candida CTG clade spp.	Selectable marker	Reporter genes	Genetic editing tool				References
			Plasmid	Recombinase	Tet-system	CRISPR-Cas	
C. albicans	*HIS1* *ARG4* *URA3* *ADE2* *CaHygB* *caSAT-1* *caSAT1* *CaNAT1* *IMH3*	*yEGFP3* *GFP* *CaGFPc* *YFP* *CFP* *Venus* *YFP* *yEmRFP* *DsRFP* *RLUC* *FLUC* *gLUC59* *LAC4* *lacZ*	pAYCU267 pAYCU268 pAYCU228	*caFLP* *ecaFLP* *cre*	*tTA* *cartTA*	*CaCas9* *CRIME*	Defosse et al. (2018a), Samaranayake and Hanes (2011), Reijnst et al. (2011), Cormack et al. (1997), Morschhäuser et al. (1998), Gerami-Nejad et al. (2001, 2009), Enjalbert et al. (2009), Uhl and Johnson (2001), Staib et al. (1999, 2000a), Morschhäuser et al. (1999), Reuss et al. (2004), Park and Morschhäuser (2005), Dennison et al. (2005), Nakayama et al. (2000), Roemer et al. (2003), Vyas et al. (2015), Huang and Mitchell (2017), Shen et al. (2005), Basso et al. (2010), Zhang and Konopka (2010), Leuker et al. (1992), Köhler et al. (1997), Srikantha et al. (1996), Doyle et al. (2006a, b), Keppler-Ross et al. (2008)
C. dubliniensis	*HIS1* *ARG4* *LEU2* MPA^{R}	*GFP*	pAYCU267 pAYCU268 pAYCU228	*FLP*	–	–	Defosse et al. (2018a), Staib et al. (2000b), Mancera et al. (2019), Wirsching et al. (2001))
C. tropicalis	*URA3* *HYG#* $IMH3^{r}$	*LAC4* *lacZ*	pAYCU267 pAYCU268 pAYCU228	*FLP*	–	*CRISPR/Cas9*	Defosse et al. (2018a) Lombardi et al. (2019), Mancera et al. (2019), Hara et al. (2001), Xiang et al. (2014), Beckerman et al. (2001)

(continued)

Table 1 (continued)

Candida CTG clade spp.	Selectable marker	Reporter genes	Genetic editing tool				References
			Plasmid	Recombinase	Tet-system	CRISPR-Cas	
C. parapsilosis	*URA3* *MET2* *LYS4* *IMH3*^{r} *CpSAT1* *CaNAT1*	*LAC4*	pAYCU257 pAYCU281 pAYCU230 pAYCU210 pAYCU212	*ecaFLP* *CpFLP*	–	*CRISPR/Cas9*	Defosse et al. (2018a), Kosa et al. (2007), Lombardi et al. (2019), Ding and Butler (2007), Lombardi et al. (2017) Shen et al. (2005), Gácser et al. (2007), Nguyen et al. (2009)
C. maltosa	*URA3*	*LAC4*	pAYCU267 pAYCU268 pAYCU228	–	–	–	Defosse et al. (2018a), Papon et al. (2012), Masuda et al. (1994)
C. famata	*BSD* *Dh_IMH3* *Sa_ble* *ARO4*	*LAC4*	pAYCU272 pAYCU273 pAYCU244	–	–	*CRISPR/Cas9*	Defosse et al. (2018a), Dmytruk et al. (2011), Bratiichuk et al. (2020), Spasskaya et al. (2021), Dmytruk et al. (2014), Ishchuk et al. (2008)
C. lusitaniae	*URA3* *NAT* *IMH3.2*	*GFP* *YFP*	pAYCU257 pAYCU211 pAYCU256 pAYCU281 pAYCU230 pAYCU210 pAYCU212	–	–	*CRISPR/Cas9* (RNPs)	Defosse et al. (2018a), Grahl et al. (2017), Norton et al. (2017), Lin et al. (2011), Gabriel et al. (2014), Defosse et al. (2018b)
C. oleophila	*HYG#*	*lacZ*	–	–	–	–	Segal et al. (2002), Yehuda et al. (2002)
C. auris	*NAT1*	*mCh*	pAYCU257	–	–	*CRISPR/Cas9* (RNPs)	Defosse et al. (2018a), Grahl et al. (2017)
D. rugosa	*zeo-n*	ND	pAYCU267 pAYCU268 pAYCU228	–	–	–	Defosse et al. (2018a), Tang et al. (2003)

(continued)

Table 1 (continued)

Candida CTG clade spp.	Selectable marker	Reporter genes	Genetic editing tool				References
			Plasmid	Recombinase	Tet-system	CRISPR-Cas	
L. elongisporus	*IMH3.2*	*YFP*	pAYCU254	–	–	–	Defosse et al. (2018a, b)
M. guilliermondii	*MET2* *HYG#* *SAT1* *URA5*	*yEGFP3* *YFP* *CFP* *yEmRFP*	pAYCU211 pAYCU256 pAYCU281 pAYCU230 pAYCU210 pAYCU212	–	–	–	Defosse et al. (2018a), Courdavault et al. (2011), Millerioux et al. (2011), Obando Montoya et al. (2014), Foureau et al. (2013)
M. farinosa	*ble*	ND	pAYCU230 pAYCU210 pAYCU212	–	–	–	Defosse et al. (2018a), Wang et al. (2006)
M. fructicola	ND	*CFP*	pAYCU211 pAYCU256 pAYCU230 pAYCU210 pAYCU212	–	–	–	Defosse et al. (2018a)
S. stipitis	*ble* *IMH3.2*	*yEGFP3*	pAYCU257 pAYCU211 pAYCU256 pAYCU230 pAYCU210 pAYCU212	*Cre*	–	*CRISPR/Cas9*	Defosse et al. (2018a, b), Laplaza et al. (2006), Cao et al. (2017b), Passoth et al. (2003)
S. passalidarum	*IMH3.2*	*GFP*	pAYCU270 pAYCU271 pAYCU229	–	–	–	Defosse et al. (2018a, b)
Y. tenuis	ND	ND	pCtAKR	–	–	–	Wohlbach et al. (2011)

engineering especially after developing a codon-optimized luciferase gene (*Gaussia princeps* origin) fused to the gene encoding glycosylphosphatidylinositol-linked cell wall protein (Pga59p) (Moreno-Ruiz et al. 2009; Enjalbert et al. 2009). This construction overcomes the low cell permeability to luciferin by expressing Pga59p product to the cell surface of the engineered strain (Papon et al. 2012).

The enzyme β-Galactosidase encoded by *lacZ* gene is used to generate high versatile reporting approach in several bioengineered models (Papon et al. 2012; Smale 2010). Two main detector substrates are used in β-Galactosidase system, *o*-nitrophenyl galactoside (ONPG) and 5-bromo-4-chloro-3-indolyl β-D-galactoside (X-gal) triggering the appearance of yellow (soluble) and blue (insoluble) products, respectively (Möckli and Auerbach 2004; Juers et al. 2012; Clark et al. 2019). Notably, in contrast to the previous optimization process applied in *Candida* toolbox, the codon-optimized version *lacZ* gene (*Streptococcus thermophiles* origin) developed by Uhl et al. (2001) showed an equivalent activity to the wild-type sequence, which justifies the preferable application of *S. thermophiles lacZ* gene in numerous studies targeting *C. albicans.*

3.3 Auto-replicating Sequence

The discovery of the microbial system's transformation modality (firstly reported by Frederick Griffith in 1928) revolutionized the vision, the orientation, and the plasticity of scientific research (Griffith 1928; Schindler 2020). The transformation phenomenon enables yeast to acquire new genetic traits by either the integration of non-replicative plasmids or the acquisition of an independent replicative episome (Orr-Weaver et al. 1981). Moreover, the random gene integration modality demonstrated its efficacy in generating CUG-Ser yeast successful transformants, mediated by electroporation and lithium acetate approaches (Gordon et al. 2019). In addition to plasmid's capacity to transform traditional genes (selectable markers, reporter genes, and gene of interest), they are also capable of transforming a full gene manipulating system into various recipient strains (Lombardi et al. 2019). In the CTG clade, the episomal transformation was prone to certain difficulties due to their alternative translation behavior. Optimization of selectable markers and reporter genes represents the key players in building a successful transformation construct. The recent identification of centromere sequences in non-conventional yeasts allowed the creation of a building stable episome construct for heterologous expression in *S. stipitis* (Cao et al. 2017a). Additionally, Defosse et al. (2018a) developed an episomal module to genetically manipulate CTG yeast clade with DNA-interchangeable capacity (customization property; as an action of commonly used and rare-cutter restriction enzymes) called pAYCU plasmids (Fig. 2), consisting of two main gene cassettes: 1) drug-resistant cassette and 2) reporter gene expression cassette. The main advantage of this flexible construct is the numerous possibilities offered for both, the selection marker and the reporter gene cassettes (Defosse et al. 2018a). *SAT1*, *HPH*#, and *MgIMH3.2* are examples of ectopically expressed selectable markers in the designed episomal construct (Defosse et al.

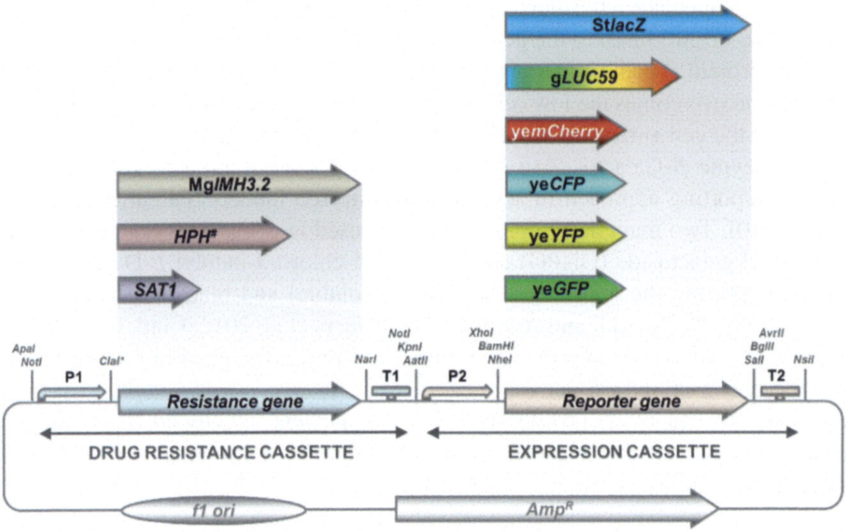

Fig. 2 Schematic illustration of the standard pAYCU plasmid constructs (Defosse et al. 2018a)

2018a). Regarding reporter genes, the synthetic construct offers a broad spectrum of ORF-adapted fluorescent protein alternatives including yeast-enhanced green (*yeGFP*), yellow (*yeYFP*), cyan (*yeCFP*), monomeric cherry (*yemCherry*), the surface-exposed Gaussia princeps luciferase (*gLUC59*), and *lacZ* gene (*StlacZ*) (Courdavault et al. 2011; Enjalbert et al. 2009; Uhl and Johnson 2001) (Fig. 2). Various combinations of pAYCU episome construct are offered to adapt the genetic transformation and manipulation in *Candida* spp. (Defosse et al. 2018a) (Table 1). Nowadays, the plasmid transformation is no longer restricted to the conventional ectopic gene manipulation but rather is extended to markerless gene editing modalities (Lombardi et al. 2019; Wang et al. 2018). Wang et al. (2018) recently reported the construction of a "suicide plasmid" named pPICPJ-*mazF*, facilitating scarless metabolic manipulation in *C. tropicalis* to increase the conversion rate of oils into long-chain dicarboxylic acids (DCAs). The future perspective for *Candida* gene manipulation will be mainly shifting from standardized conventional plasmid transformation to genetic circuit design with directed functional behavior (Papon et al. 2014; Røkke et al. 2014).

3.4 Recombinases

The emergence of the site-specific recombinases upgraded genetic engineering to the next level in numerous prokaryotic and eukaryotic genomes (Papon et al. 2012). Their ability to promote safe targeted genetic modifications (without the addition of toxic compounds such as 5-fluoroorotate applied in URA3 blaster systems) grants it access to the list of adopted powerful molecular tools (Papon et al.

2012). The intensive efforts over the last two decades facilitated the application of two main recombinase systems in the CTG clade, the flippase (FLP) recognition target (FRT) and the Cre recombinase (Cre) locus of crossing over (x), P1 (loxP) site-specific recombinases (Papon et al. 2012; Nunes-Düby et al. 1998; Austin et al. 1981; Sternberg and Hamilton 1981).

3.4.1 FLP-FRT System

Generally, the system is composed of two functional elements, FLP and FRT sites (Holkers et al. 2006). Its mode of action is primed by FLP that catalyzes high-fidelity recombination between two FRT sites (Holkers et al. 2006). Moreover, FLP can mediate DNA-*FRT*-flanked segment excision and inversion in direct and inverted repeat configurations, respectively (Holkers et al. 2006). Since the FLP-FRT modality is considered as one of the first engineered recombinase strategies in *Candida,* several codon optimization processes and transcription regulation exchanges were carried out to optimize its expression in *Candida* spp. (Papon et al. 2012*).* For instance, Staib et al. (1999) and Morschhäuser et al. (1999) successfully developed *caFLP* (*Saccharomyces cerevisiae* origin) under the control of *SAP2* promoter (inducible, expressing genes according to the stage of mycotic infection) to guide the excision of *FRT*-flanked *IMH3, CDR4,* and *MDR1* (the last two genes were sequentially disrupted) genes in *C. albicans* strains. Notably, an improved enzymatic activity was recognized in *C. albicans* for the new codon-optimized FLP recombinase, called *ecaFLP*, which was expressed under *SAP2* and *MAL2* promoters, and in a tetracycline-inducible system (Staib et al. 2000a; Michel et al. 2002; Sánchez-Martínez and Pérez-Martín 2002; Reuss et al. 2004; Park and Morschhäuser 2005). A more interesting capability was spotted on *FRT*-flanked-*MAL2* promoter-driven *ecaFLP* coupled with the *caSAT1* marker cassette, facilitating the selection of nourseothricin-resistant strains. The FLP-mediated manipulation modality could offer an excision property for the ectopic cassette driven by culturing the transformants in a maltose-enriched medium (selection marker recycling) (Reuss et al. 2004). Worth mentioning, the *MAL2* promoter-driven *FLP* manipulation strategy was successfully integrated into *C. parapsilosis* upon replacing the promoter with a species-specific *MAL2* transcription regulatory region (Ding and Butler 2007).

3.4.2 Cre-LoxP System

Typically, this system (originating from P1 bacteriophage) is composed of two main elements, causing recombination (Cre) and *loxP* flanked (floxed) DNA segment (McLellan et al. 2017). It was recently engineered to direct an efficient genetic manipulation in CTG clade (Papon et al. 2012). Dennison et al. (2005) reported successful sequential disruption of the *ADE2* and *MET15* genes in *C. albicans* mediated by the codon-optimized version of *cre* gene (driven by *MET3* promoter). Furthermore, Laplaza et al. (2006) reported the construction of another effective version of mutagenized *cre* gene that allows functional genomics and metabolic engineering in *Candida* CTG clade. More recently, Shahana et al. (2014) constructed a new *Clox* system with an efficacy-improved *cre* gene (containing synthetic intron). Interestingly, this system facilitates multi-marker recycling, with open choices for selection markers (*NAT1*, *URA3*, *HIS1*, *ARG4*) (Shahana et al. 2014).

3.5 Tetracycline-Regulatable Dual System

The myriad potency of certain bacteria to resist tetracycline antibiotics (as an action of their special molecular toolbox) opens the path toward developing a powerful system, allowing the control of gene expression in prokaryotes and eukaryotes (Papon et al. 2012; Berens and Hillen 2004). The two key players in this system are the TetR transactivator and the tetO operator (from the promoter of tetracycline-resistant genes) (Papon et al. 2012; Berens and Hillen 2004). Remarkably, this system can control the genetic switches via two modalities, Tet-off and Tet-on (Sprengel and Hasan 2007). In CTG clade, the Tet-off system consists of the tTA, hybrid codon-modified *TetR* (*E. coli* origin) fused to *ScHPA4-TA* (*S. cerevisiae* origin; encoding Hap4 activation domain), and the TR promoter (Papon et al. 2012; Nakayama et al. 2000). The tetracycline-deficient medium enables the binding of tTA transactivator on the TR promoter, allowing an active expression of the gene of interest (Papon et al. 2012), whereas tetracycline-enriched medium acts as a negative regulator for gene expression (Papon et al. 2012). On the contrary, the Tet-on system behaves oppositely, where the addition of tetracycline triggers the expression of the gene of interest (Park and Morschhäuser 2005). To perform such modality in CTG clade yeast, a reverse tetracycline-controlled transactivator (cartTA) and the P_{tet} promoter are needed (Papon et al. 2012; Park and Morschhäuser 2005). Typically, the cartTA is the result of fusing *carTetR* nucleotide sequence to the codon-optimized sequence, *ScGAL4-TA* (*S. cerevisiae* origin, encoding Gal4 activation domain) (Papon et al. 2012; Park and Morschhäuser 2005). Notably, the modification performed in the *carTetR* sequence reversed the effect of tetracycline on the expressed transactivator *cartTA* (Papon et al. 2012) (Fig. 3). In fact, the emergence of this regulatable system facilitates the development of gene replacement and conditional expression (GRACE) strategy in *C. albicans* (Roemer et al. 2003). Additionally, it demonstrated its efficacy in regulating the gene expression of *C. tropicalis* toolbox (Bijlani et al. 2018).

3.6 CRISPR–Cas9 System

The recent breakthrough discovery of clustered regularly interspaced short palindromic repeats (CRISPR)/CRISPR-associated system (Cas) greatly facilitated the birth of a new genome-editing era (Jinek et al. 2012). The natural implication of CRISPR–Cas9 in bacterial adaptive immunity against invading bacteriophages supports its theoretical potency to target any "suspected DNA" (DiCarlo et al. 2013). The recent advancement in bioengineering succeeded in translating the extraordinary potentials (accuracy, precision, and flexibility) of CRISPR–Cas9 in the area of

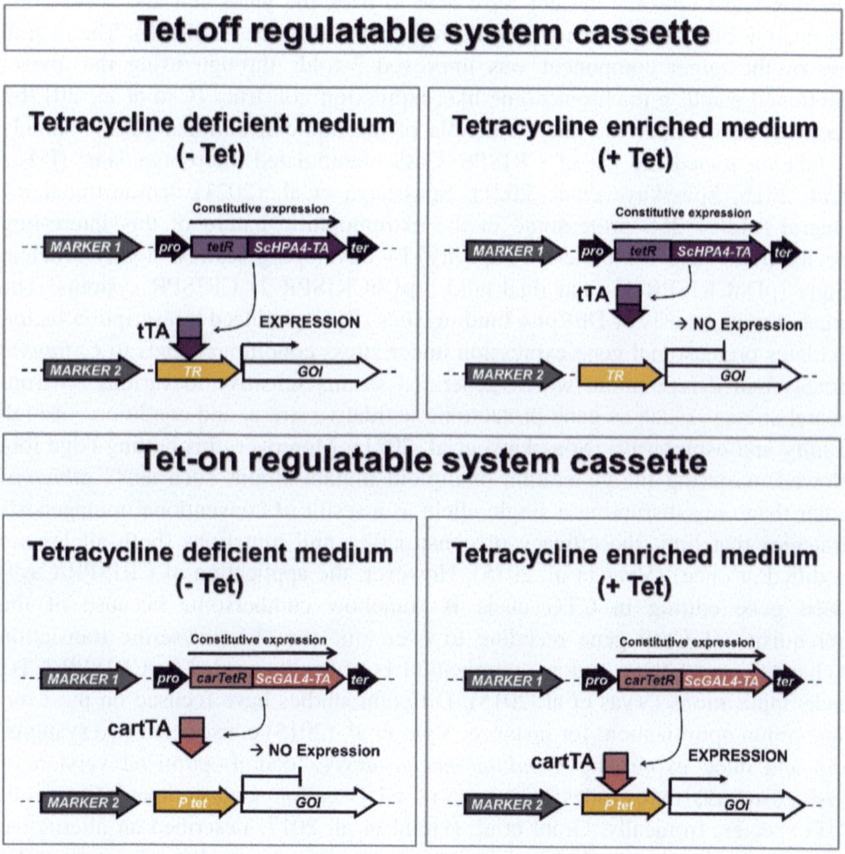

Fig. 3 Tet-off-/Tet-on-regulatable dual system in CTG clade (Nakayama et al. 2000)

genome editing (Heidari et al. 2017; Stoneman et al. 2020). Typically, the selection of a single-guide RNA (sgRNA) target site is considered as the prime step in designing each CRISPR–Cas9 experiment (Stoneman et al. 2020). Upon the assembly of the sgRNA and Cas9 (endonuclease) enzyme complex, the guide RNA directs Cas9 for manipulating its complementary region in the genome (Stoneman et al. 2020).

Experimentally speaking, this novel tool demonstrates its ability to meticulously reduce the transformation steps and minimize the off-target effects and the overall number of "unwanted" genomic changes, especially hyperploidy events (Marton et al. 2020; Chandrasegaran and Carroll 2016). Cao et al. (2017b) demonstrated the potency of using CRISPR technology in non-conventional yeasts, namely *S. stipitis* to overcome the genome editing's limited efficacy, where the rate of *ade2*-knockout and *trp1*-knockout was boosted up from<1% to 80%. In addition, both centromere (CEN) and autonomously replicating sequence (ARS) of the

CRISPR–Cas9 episome system were able to trick the yeast cell and solve both, segregation bias and directing plasmid replication (Cao et al. 2017b). The overall titer of the target component was improved 3-folds through using the above-mentioned stable minichromosome-like expression construct (Cao et al. 2017b). Recently, a new species, "the Cinderella of the non-conventional yeast," namely C. *famata*, joined the list of CRISPR–Cas9-manipulated microorganisms (Prista et al. 2016; Spasskaya et al. 2021). Spasskaya et al. (2021) demonstrated the integral role of 26S proteasome in the extremophilic nature of this interesting species (halo- and osmotolerant capacity) by developing scarless highly efficient single (pDhCRISPR-1)- and dual-guide (pDhCRISPR-2) CRISPR systems. The targeted mutagenesis of DhRpn4-binding sites (Rpn4: induced transcription factor, regulates proteasomal gene expression under stress conditions) leads to *C. famata* proteasomal deregulation, which generated strains sensitive to various environmental stresses (such as gene-proteotoxic, oxidative stress, and conditions of high salinity and osmolarity) (Spasskaya et al. 2021). Moreover, this cutting-edge tool allowed mastering the generation of diploid mutant strains such as *C. albicans*, rather than only disrupting a single allele as a result of conventional mutagenesis strategies that limit the efficacy of constructing null mutations (both alleles are modified at once) (Vyas et al. 2015). However, the application of CRISPR/Cas9-based gene editing in CTG clade is somehow cumbersome because of the prerequisite of Cas9 gene recoding to overcome the "leucine/serine translation exchange" issue; thus, codon optimization is crucially needed in CRISPR-CTG clade applications (Vyas et al. 2015). Different studies have focused on the *Candida* codon optimization; for instance, Vyas et al. (2015) constructed two systems: solo and duet, expressing *Candida/Saccharomyces* codon-optimized version of *Cas9 (CaCas9)* that evades the usage of CUG codon with compatibility to all CTG species. Ironically, Grahl et al. (Grahl et al. 2017) described an alternative approach to circumvent the need for species-specific manipulation constructs by applying expression-free CRISPR genome editing in *C. lusitaniae*, *C. glabrata*, and *C. auris*. The alternative construct is composed of purified Cas9 protein and gene-specific and scaffold RNAs referred as RNA–protein complexes (RNPs) (Grahl et al. 2017).

The transient system is typically used as a safeguard to avoid the negative feedback (cell toxicity and strain fitness drop) of CRISPR/Cas9 constitutive expression (Min et al. 2016). In 2017, Norton et al. (2017) developed the first transient CRISPR–Cas9 expression with gene deletion efficiencies up to 81% in *Ku70* and *LIG4* (non-homologous end joining (NHEJ) factors) in *C. lusitaniae*-deficient strains. Huang and Mitchell (2017) generated a marker recycling model called CRISPR–Cas9-induced marker excision (CRIME) in *C. albicans* to avoid the paucity of drug-resistant makers. Methodologically speaking, the selection marker that is used to delete the target gene is flanked by directly repeated sequences. The CRISPR–Cas9 system is used dually to delete the gene of interest subsequently to the deletion of the marker itself by double-strand break (DSB) induction, leading to marker excision by recombination between the direct repeats (Huang and Mitchell 2017). Furthermore, Nguyen et al. (2017) developed two recyclable CRISPR-mediated approaches, namely the LEUpOUT system (applicable to LEU2/leu2Δ parental strain) and the HIS–FLP system (applicable theoretically to any nourseothricin-sensitive *C. albicans* strain), to allow scarless genome manipulation of *C. albicans* without the integration of permanent markers or the application of a cloning step. Furthermore, Lombardi et al. (2017) applied an episome-based CRISPR–Cas9 system in *C. parapsilosis*; the adopted plasmid system called pRIBO theoretically offered the ability to manipulate any genetic background with the possibility of sequential genome editing in the same background since the plasmid is escaped easily in the absence of selection. Afterward, the former system was further improved into markerless-gene manipulation system adapting various *Candida* spp., namely pCP-tRNA and pCT-tRNA applied on *C. parapsilosis* (and their sister species, *Candida orthopsilosis* and *Candida metapsilosis*) and *C. tropicalis*, respectively (Lombardi et al. 2019).

Interestingly, since bioinformatics is the keystone in scientific speaking languages, a breakthrough in CRISPR–Cas9 application area was made by Stoneman et al. (2020) who developed recently CRISpy-pop, a flexible Python-based Web application with a user-friendly graphical interface. The substantial aim of this application is to theoretically design CRISPR/Cas9-driven genetic manipulations on individual strain or population of strains to predict the most efficient combination (sgRNA and strain) (Stoneman et al. 2020). Currently, this powerful tool can cover both bacteria (*Zymomonas mobilis*) and yeasts (*S. cerevisiae*), with a future perspective to integrate new biological models (Stoneman et al. 2020).

4 Omics Resources

The complexity of the living cell hindered the scientists' ability to see the whole picture behind any given phenotype for a long period of time, since they mainly adopted the "one–one analysis" (one gene/protein at a time) in different biomedical researches, neglecting the myriad interactions between genes, proteins, carbohydrates, lipids and metabolites that are eventually translated into a certain phenotype (Stagljar 2016). The innovations of technological techniques have significantly facilitated the conjugation of the "omics" suffix to both, the genomic and post-genomic levels (transcriptomics, proteomics, and metabolomics) that decode holistically the physiological and biochemical framework of numerous organisms (Hasin et al. 2017; Babar et al. 2018; Narad et al. 2018; Patra et al. 2021). Since omics experiments (high-dimensional biology) are hypothesis-generating rather than hypothesis-driven studies (traditional experiment approach), it could aid in opening the path toward building a nonbiased picture of the biological system (Horgan and Kenny 2011). Generally, systemic biology represents an interdisciplinary platform for mathematical analysis and computational modeling approaches that translate the complexity of a defined biological system via either top-down or bottom-up modalities (Patra et al. 2021).

The application of the state-of-the-art multi-omics approaches is a cornerstone in synthetic biology, not only for reducing the time needed for bio-product candidates to be commercially available but also by allowing the utilization of numerous strategies for process optimization via metabolic engineering modality (Roldão et al. 2012). The significance of the implantation of omics system in microbial cell factory optimization relies basically on the ability of such multi-directory approach to decode the underlined intra- and inter-cellular network communication, aiding in a greater understanding of principal cellular elements' properties and functions and thus retrofitting and controlling the optimum niche for high quality and yield production (Roldão et al. 2012). Generally speaking, the adopted omics approaches differ according to the application field. In industrial biotechnology and bioprocess engineering, five main omics modalities are applied, including the following.

4.1 Genomics

It represents a snapshot of the production repertoire of a defined microbe. In general, genomics approach offers the potency to analyze and study comprehensively microorganisms' genomes to understand the function/interaction of genes within the entire genome network (Roldão et al. 2012). The whole-genome sequencing is undoubtedly crucial in screening and identifying the gene pool that leads to specific phenotype expression. Additionally, such a platform can allow predicting the ideal growth conditions for targeted bio-factories (Herrgård and Panagiotou 2012).

4.2 Transcriptomics

An integrative approach provides a precise view of all the microorganism's possible expression profiles under specific environmental conditions. Additionally, it directs the spotlight on important gene interactome, possible gene targets for metabolic engineering, and functional diversity among yeast strains (Roldão et al. 2012). The continuous development in this field facilitated the accurate prediction of microbe's expression profile under heterogeneous environmental conditions. The recent construction of Reaction Inclusion by Parsimony and Transcript Distribution (RIPTiDe) platform aids greatly in implementing the transcriptome analysis in identifying the ideal metabolic pathways that lead to producing a desirable compound regardless of the external niche (Jenior et al. 2020).

4.3 Proteomics

This modality represents the third stage in the holistic understanding of a cell's biological processes. It offers in-depth characterization of microorganisms' protein profiles (including their location, structure, and function) since the transcript and protein levels are weakly related due to the specific regulatory mechanisms associated with the translational control (posttranslational modifications and translation efficiency) (Maier et al. 2009; Olivares-Hernández et al. 2011, 2010). The variable nature of proteomics in response to environmental conditions imposes the need for a constant tracking system for such a dynamic platform (Raghavachari 2011).

4.4 Metabolomics

These approaches constitute the fourth stage of the "biological puzzle-solving" that provide a snapshot of metabolite concentrations (Pickford et al. 2019). Generally, the metabolomics approach aims to study quantitatively the microorganism's cellular metabolite profile under specific culture conditions (Roldão et al. 2012). The construction of mathematical models is considered one of the efficient tools in plotting and understanding the metabolic profile in a defined organism by integrating data science platforms such KBase, which involves semi-automated reconstruction approaches (such as RAVEN, Merlin, and ModelSEED) (Arkin et al. 2018). It displays a detailed map for the underlying molecular interaction that can evaluate the efficacy of future network response under specific genetic/environmental interventions (Roldão et al. 2012). A corrected meticulously comprehensive view of microbe's metabolic trait could be achieved by adding species-specific software databases such as BRENDA, KEGG, and MetaCyc (Patra et al. 2021).

4.5 Fluxomics

Since metabolomics neglects the control and the functional regulation of metabolic networks, the fluxomics platform acts as a complementary approach to intensively understand the microorganism's phenotype by determining the rates of metabolic reactions (Pickford et al. 2019; Ratcliffe and Shachar-Hill 2005). Thereby improving our ability to predict the effect of genetic manipulation on the entity of the microorganism's physiological behavior, which leads to the formation of microbial systems with improved metabolic capacity (Yu et al. 2012). The offered flux balance analysis (FBA) in the fluxomics discipline greatly facilitates the reading of metabolite flow through a defined metabolic model (Orth et al. 2010; Price et al. 2004). The FBA analysis involves several advantages, such as the analysis of phenotypic behaviors and flux coupling, the evolution of metabolic status with environmental and genetic changes, and the prediction of metabolic engineering foci (Patra et al. 2021).

The high-throughput potency of omics to comprehensively study the microbial system is directly dependent on the available pool of technologies to scientifically decode such huge data. The genomics platform constituted various technologies to systematically analyze the *Candida* genome such as *Candida* Gene Order Browser (CGOB), multiplex PCR, whole-genome sequencing, flow cytometry, slot blot hybridization, microarray, Avadis, FuncExpression, High-Throughput GoMiner, and GREAT (Roldão et al. 2012; Sampaio et al. 2005; Asadzadeh et al. 2020;

Maldonado et al. 2018; Chang et al. 2015; Loeffler et al. 2000; Campa et al. 2008; Maguire et al. 2013).

Regarding transcriptomics, numerous techniques and sites are established to facilitate integrative analysis of *Candida* transcripts, e.g., dual RNA sequencing, CLASSIFY, FunCluster, GARBAN, GOdist, dChip, ErmineJ, GoMiner, Matrix2png, TM4, MAGIC, GeneSpring, Array-Pro Analyzer, and ArrayStar (Roldão et al. 2012; Schulze et al. 2016).

For the *Candida* proteomics and metabolomics profiling, different methodological approaches are developed to perform a holistic analysis of yeast proteomic and metabolomic repertoire, e.g., sodium dodecyl sulfate polyacrylamide gel electrophoresis (SDS-PAGE), two-dimensional gel electrophoresis (2DE), differential in-gel electrophoresis (DIGE), electrospray ionization-mass spectrometry (ESI–MS), matrix-assisted laser desorption/ionization-MS (MALDI-MS), MALDI-time-of-flight (MALDI-TOF), liquid chromatography-MS (LC–MS), high-performance liquid chromatography (HPLC), gas chromatography-MS (GC–MS), ultra-performance liquid chromatography-MS (UPLC-MS), AMDIS software, Pathway Activity Profiling (PAPi) algorithm, UPLC-QTOF-MS Perseus software, SIMCA, and FiatFlux (Wang et al. 2009, 2020; Larbi and Jefferies 2009; Gräslund et al. 2008; Kim et al. 2005; Awad et al. 2018; Karkowska-Kuleta et al. 2020; Cabrera et al. 2010; Christen and Sauer 2011; Han et al. 2012).

The fluxomics yeast profiling is still restricted due to the disciplines nascent state, leaving behind some constructed tools to examine the microbial phenotypic expression such as ^{13}C metabolic flux analysis (^{13}C MFA), OptKnock, OptReg, OptStrain, optimal metabolic network identification (OMNI), OptGene, thermodynamics-based MFA, COBRA toolbox, and OpenFLUX (Pharkya and Maranas 2006; Henry et al. 2007; Feng et al. 2010; Pharkya et al. 2004; Herrgård et al. 2006; Blazeck and Alper 2010; Schellenberger et al. 2011; Veras et al. 2019).

5 Candida Metabolic Engineering

From adapting modern gene-editing tools to the implication in different industrial and medicinal approaches, *Candida* CTG clade yeasts continuously demonstrate their potency to update and track the latest scientific trends. The recent intensive efforts in optimizing/advancing the CTG clade molecular toolbox and providing a series of synthetic constructs facilitate the rewiring of various *Candida* spp. in several biotechnological applications such as bioremediation, biocontrol, biofuel, vitamins, and industrial enzyme production (Defosse et al. 2018a, b; Papon et al. 2012). The current deviation of bioengineering orientation toward the notion of microbial consortia (rather than sufficing only with the monoculture application) modernizes the vision of culturing several microbial factories into bio-consortia (offers an ideal niche for efficient production), the *Candida* CTG clade makes no exception for this modern application (McCarty and Ledesma-Amaro 2019; Kouzuma et al. 2015; Brenner et al. 2008; Bhatia et al. 2018).

The microbial community takes advantage of "the power of the group" which offers multiple boosters for a higher production rate such as having a robust bio-community, tolerating various environmental challenges, reducing the "metabolic load," allowing the exchange of resources, activating bio-communication (between species), and expanding the metabolic fitness (Stenuit and Agathos 2015; Tsoi et al. 2018; Bassler and Losick 2006). Interestingly, *C. famata* was recently reported as a potential candidate to enter the microbial consortia construction for the industrial production of riboflavin (Bhatia et al. 2018). The application of such bio-consortium is expected to increase in the near future especially with the successful integration of CRISPR/Cas9 system in the consortium building process (Wang et al. 2016a). The following section will cover the current main metabolic engineering applications and biotechnological potentials of *Candida* CTG clade yeasts (Table 2).

5.1 Ethanol Production

The world's modern and technological lifestyle posed many challenges in sustaining energy sources especially with the existence of ever-increasing concerns about economic and environmental consequences (Pereira et al. 2015). This stimulates collaborative research efforts aiming to produce renewable, eco-friendly biofuel sources using microbial factories (Du et al. 2019). Lignocellulose's (consists of cellulose, hemicellulose, and lignin) bioconversion into ethanol is currently being considered as an alternative source to petroleum-based fuels (Rodrussamee et al. 2018). The efficient fermentation of lignocellulose sugars represents a limiting factor in the production process, especially xylose sugar (most abundant in hemicellulose) (Rodrussamee et al. 2018). However, since CTG clade species are considered as xylose growers (utilized as their sole carbon source) with some having the ability to ferment xylose naturally into ethanol, they represent an attractive superior microbial platform in such an industrial section (Papon et al. 2014). The comparative genomic studies for xylose fermenters revealed the presence of amplification in sugar transporters and cell surface proteins, which may explain their exceptional sugar environment (Wohlbach et al. 2011). Interestingly, Wohlbach et al. reported an increase in the flux of the xylose assimilation pathway with the absence of xylitol accumulation in strains engineered with pCtAKR (*Candida tenuis* aldo/keto reductase), suggesting their involvement in stimulating NADH recycling and glycerol production (Wohlbach et al. 2011). The insertion of optimized (to frequently used codons in *S. cerevisiae*) *exo*-inulinase gene *INU1* from *M. guilliermondii* (*INU1Y*) in *S. cerevisiae* resulted in Y13 recombinant strains with inulinase activity up to 43.84 U/mL and 126.30 mg/mL ethanol production from 300.0 g/L inulin (Liu et al. 2014). In addition, the fusion of *S. stipitis* with *S. cerevisiae* (the microbe of choice in the ethanol industry) protoplasts generates a hybrid (unstable) strain with enhanced xylose-based ethanol production in comparison with *S. stipitis* wild strain (Ruchala et al. 2020; Yoon et al. 1996). Furthermore, the generation of *S. stipitis Δhxk1* and 2-deoxyglucose-resistant mutants has resulted in the derepression of xylose utilization in both engineered strains

Table 2 Current biotechnological potentials of the *Candida* CTG clade yeasts

Produced compound	*Candida* CTG clade spp. (parental strain)	Modification	Titer	References
Adipic acid	*C. tropicalis* KCTC 7212	*ΔAOX4:AOX5*	12.1 g/L	Ju et al. (2020)
Citric acid	*C. oleophila* ATCC 20177	Submerged fermentation	20.7 g/L	Kim et al. (2015)
	C. oleophila ATCC 20373		60.1 g/L	
	C. tropicalis ATCC 20115		45.0 g/L	
	C. oleophila ATCC 20177	Nitrogen limitation condition	74.2 g/L	Anastassiadis et al. (2005)
Dicarboxylic acids	*C. tropicalis* 1798	*ΔCART*	32.84 g/L	Wang et al. (2018)
	M. guilliermondii Δku70 KU141F1	↑*CYP52A12*	ND	Werner et al. (2017)
Ethanol	*S. passalidarum* ATCC MYA-4345	Fermentation at 30 °C	31.9 g/L	Du et al. (2019)
Fumaric acid	*S. stipitis* CBS6054	*ΔPsfum1 ΔPsfum2* ↑*YMAE1* Pyruvate carboxylase, malate dehydrogenase, and fumarase codon optimization	4.67 g/L	Wei et al. (2015)
Lipid	*C. tropicalis* SY005	↑*CtRAP1*	0.37 g/g	Chattopadhyay et al. (2020)
L-lactic acid	*S. stipitis* CBS6054	↑*LDH*	58 g/L	Ilmén et al. (2007)
2-phenylethanol	*M. guilliermondii* YLG18	Optimization fermentation condition In situ product recovery (ISPR)	3.20 g/L	Yan et al. (2021)
Riboflavin	*C. famata* VKMY-9	↑*SEF1* ↑*IMH3* ↑*RIB1* ↑*RIB7*	1.026 g/L	Dmytruk et al. (2011)
	C. famata AF-4	↑*RIB1* ↑*RIB7*	16.4 g/L	Dmytruk et al. (2014)
	C. famata VKMY-9	Δ*Sef1 C. famata*: p*SEF1 C. albicans* /*C. tropicalis* + *C. famata SEF1* ORF	0.028–0.040 g/L	Andreieva et al. (2020)
S-adenosyl-l-methionine	*S. stipitis* BS 5776	*Δ Erg6p*	0.052 g/g	Križanović et al. (2015)

(continued)

Table 2 (continued)

Produced compound	*Candida* CTG clade spp. (parental strain)	Modification	Titer	References
Sophorolipids	*C. albicans* O-13–1	Semi-continuous fermentation by using a novel bioreactor with DVDSB	477 g/L	Zhang et al. (2018)
Shikimate	*S. stipitis* FPL-UC7	↑*aro4K220L* ↑*aro1D900A* ↑*tkt1*	3.11 g/L	Gao et al. (2017)
Xylitol	*C. tropicalis* LXU1-NXRG	↑*At5g17010*	1.14 g/L	Jeon et al. (2013)
	C. guilliermondii FTI 20,037 (ATCC 201,935)	Permeabilized with Triton X-100	2.65 g/L	Cortez et al. (2016)

ND not determined

(Sreenath and Jeffries 1999; Dashtban et al. 2015). Remarkably, comparative metabolite profiling of xylose-fermenting yeasts showed a significant capability of *S. stipitis* to produce ethanol more effectively than *S. cerevisiae* SR8 at xylose (as the sole carbon source) concentration of 40 g/L with negligible by-product (xylitol and glycerol) formation (Shin et al. 2019). Genome shuffling in *S. stipitis* was recognized as a competent approach in xylose-based ethanol production, which facilitates the design of TJ2-3 strain that produced 21.9 g/L of ethanol (Shi et al. 2014).

Papon et al. (2014) summarized three modalities to augment the ethanol production using engineered CTG clade yeasts: (i) comparative physiological analysis of xylose growers and xylose fermenters, (ii) omics studies and metabolic modeling, and (iii) construction of molecular tools. Yeast metabolic engineering in xylose-based ethanol production is mainly focused on improving sugar uptake and the initial assimilation steps (Vleet and Jeffries 2009). For instance, the heterologous expression of xylose transporter gene *At5g17010* (*Arabidopsis thaliana* origin) in *C. tropicalis* strain (LXU1-NXRG) blocked the glucose suppression effect on xylose assimilation, with an increase in the xylose uptake rate by 37–73% (depending on the glucose/xylose mixture ratio) (Jeon et al. 2013). The potency of *C. tropicalis* in producing a high yield of ethanol is also associated with its intrinsic tolerance toward generated biomass pretreatment inhibitors such as furfural (Wang et al. 2016b). The underlined furfural detoxification mechanism is correlated to the regulated expression of alcohol dehydrogenase 1 (*ADH1*) gene; such capability was demonstrated after the heterologous expression of *ctADH1* in *Escherichia coli* BL21, resulting in a furfural degradation rate up to 1.59-fold (Wang et al. 2016b). Rodrussamee et al. (2018) reported the efficient activity of newly isolated thermotolerant *S. passalidarum* CMUWF1–2 in xylose–ethanol bioconversion. The strain reached 0.43 g, 0.40 g, and 0.20 g ethanol/g xylose productivity rate at 30 °C, 37 °C and 40 °C, respectively. Worth mentioning, this strain shows remarkable resistance ability to glucose repression phenomenon (Rodrussamee et al. 2018). Also, *S. passalidarum* ATCC MYA-4345 strain succeeded lately in conferment glucose and xylose with peak ethanol production up to 31.9 g/L (Du et al. 2019).

5.2 Lipid Production

The microbial lipid production has revolutionized both the industrial sector that relies intensively on non-renewable, environmentally hazardous resources of fossil fuel and the bio-medicinal area that recently integrated the lipo-molecules in numerous scientific applications (Li et al. 2020; Younes et al. 2020). Naturally speaking, the lipid accumulation in oleaginous microorganisms can take two pathways, the "de novo" (fermentation on hydrophilic substrates such as sugars and related substrates) and the "*ex novo*" lipid synthesis (fermentation on hydrophobic substrates such as oils and alkane; greatly depends on the carbon source) (Yan et al. 2020; Huang et al. 2017). For example, sophorolipids, and low-toxic, biodegradable, and ecofriendly biosurfactants have gained interest in various therapeutic biological applications since they exhibit antineoplastic, anti-bacterial, and anti-inflammatory properties (Li et al. 2020). The application spectrum of sophorolipids was extended to additional fields such as petroleum, soil remediation, food, cosmetics, detergents, and pharmacy (Li et al. 2020; Wang et al. 2019; Hirata et al. 2009). Interestingly, *C. albicans* O-13-1 was reported to produce sophorolipids up to 477 g/L with a productivity rate equal to 1.59 g/L by applying a semi-continuous fermentation strategy and the usage of a novel bioreactor with dual ventilation pipes and dual sieve plates (DVDSB) (Zhang et al. 2018). Another example of *Candida* CTG clade contribution in lipo-synthesis was recently reported by Chattopadhyay et al. (2020) who constructed a novel strategy to boost lipid production in *C. tropicalis* SY005. The procedure highlighted the usage of two approaches, FLP/FRT-based recombination system and pCtINT1 integrative episome that facilitate the ectopic expression of *CtRAP1* transcription factor (under the control of GAL1p promoter), which upregulated the expression of *FAS1*, *FAS2*, *PAH1*, and *DGAT* (play a role in fatty acid and triacylglycerol biosynthetic pathways) (Chattopadhyay et al. 2020). The prescribed approach aids in increasing lipid accumulation up to 60% (0.37 g/g dry cell weight) in comparison with the wild-type strain (Chattopadhyay et al. 2020).

Additionally, the CTG clade yeasts are also presented as competent candidates in the adipic acid industry, which is mainly involved in manufacturing nylons, pH adjustment, adhesives, resins, and synthetic lubricants (Polen et al. 2013; Castellan et al. 1991). Ju et al. (2020) developed metabolically engineered *C. tropicalis* ΔAOX4:AOX5 (encoding acyl-CoA oxidases) mutant strain with adipic acid maximum production, 12.1 g/L and 0.1 g/L productivity rate. The analysis of a genome-scale metabolic model of *C. tropicalis* (iCT646) deciphers the presence of 52 unique reactions (some belong to the lipid metabolic pathways) in this oleaginous yeast relative to the genome-scale models of *S. cerevisiae*, *Yarrowia lipolytica*, and *C. glabrata* (Mishra et al. 2016). Such profile may explain the capability of *C. tropicalis* in catalyzing the oxidation pathway, thus facilitating the synthesis of DCA (Mishra et al. 2016). Wang et al. (2018) reported the construction of *C. tropicalis ΔCART* strain mutant with sophisticated metabolic behavior to effectively produce long-chain DCA (intermediate, involved in the chemical industry). The application of pPICPJ-*mazF*-based markerless editing system facilitates the *CART* gene (encoding carnitine acetyltransferase) knockout and the replacement of the NADPH–cytochrome P450 reductase gene promoter with the constitutively expressed promoter pGAP (Wang et al. 2018). Such manipulation improved the yield of long-chain DCA up to 32.84 g/L (11.4-fold relative to the parental strain) (Wang et al. 2018). Werner et al. (Werner et al. 2017) reported *M. guilliermondii Δku70* KU141F1 as a promising biocatalyst in DCA production. The overexpression of *CYP52A12* (encoding cytochrome P450 monooxygenase, catalyzing the ω-oxidation of fatty acids and alkanes) showed a significant increase in DCA production after 72 h (Werner et al. 2017). However, the introduction of *Δpox1Δpox2* double mutant (*POX1* and *POX2* encoding for the main acyl-CoA oxidases, catalyzing the oleic acid degradation in the β-oxidation pathway) in the strain

KU141F1 showed higher DCA degradation (relative to the parental strain), suggesting the coexistence of non-peroxisomal fatty acid catabolism pathway (Werner et al. 2017).

5.3 Vitamin Production

The current diet–vitamin insufficiency posed an urgent demand for compensatory production of these valuable nutraceuticals compounds that shape good health and mental strength (Revuelta et al. 2016; Ledesma-Amaro et al. 2013). To overcome this drawback, a determinant orientation toward scaling up vitamin productivity via adopting specialized microbial factories (especially yeast) dominated in synthetic biology field recently (Ledesma-Amaro et al. 2013; Varga and Maraz 2002). For instance, vitamin B2 (riboflavin) is a metabolic precursor of flavin mononucleotide (FMN) and flavin adenine dinucleotide (FAD) (both act as coenzymes in various enzymatic redox reactions), firstly discovered in 1879 as a yellow pigment from milk (named lactoflavin) (Dmytruk et al. 2011; Liu et al. 2020). The production of this vitamin is stimulated in the bioengineering area mainly due to its nutritional and pharmaceutical values (Abbas and Sibirny 2011). There is a specialized group of yeast species with an ability to produce riboflavin, termed "flavogenic yeast" including, *C. famata*, *C. albicans*, *M. guilliermondii*, *Schwanniomyces occidentalis*, and others (Papon et al. 2013). For instance, *C. famata* which is considered a potential riboflavin producer demonstrates a noticeable enhancement in riboflavin synthesis especially with *RIB1* + (GTP cyclohydrolase II), *RIB7* + (RF synthase) expression profile (Dmytruk et al. 2014). The two former metabolic enzymes (under iron starvation) aid in 62-fold productivity enhancement compared to other tested candidates that give only a 3.1- or 16-fold productivity rate (with different expression profiles) (Liu et al. 2020; Dmytruk et al. 2014). Furthermore, the metabolic manipulation of *C. famata* dep8-conventionally mutated strain via the introduction of additional copies of *SEF1*, *IMH3*, *RIB1*, and *RIB7* aids in a 4.1-fold increase in riboflavin production (Dmytruk et al. 2011). Similarly, *M. guilliermondii* which were firstly described by Tanner et al. (1945) in 1945 for their ability to overproduce vitamin B2 under iron deficiency were found to upregulate riboflavin production under iron-enriched medium using *rib80*, *rib81*, *hit1*, and *red1–red6* mutant strains (Papon et al. 2013). Andreieva et al. (2020) recently generated a species–hybrid vitamin B2 inducing construct that restores riboflavin overproduce ability in *sef1*Δ mutant strain, involving the fusion of *C. albicans* (flavinogenic) or *C. tropicalis* (non-flavinogenic) promoter *SEF1* with *SEF1* ORF of *C. famata*. In addition, they demonstrated the positive feedback of *Sfu1* (GATA-type transcription factor) and/or vacuolar ATPase subunit A (*VMA1*) repression on vitamin B2 upregulation (Andreieva et al. 2020).

5.4 Other Valuable Compound Bio-production

The biotechnological potentials of *Candida* CTG clade are not constrained only to the former cited applications; rather, numerous species show additional capabilities in producing valuable bio-molecules such as industrial enzymes and therapeutic compounds (Papon et al. 2013; Roa Engel et al. 2008; Križanović et al. 2015).

Various *Candida* CTG clade yeasts demonstrate a remarkable capacity in producing citric acid, an organic acid commonly used in food and pharmaceutical industries (Singh Dhillon et al. 2011). Anastassiadis et al. (2005) reported a successful continuous citric acid fermentation by *Candida oleophila* ATCC 20177 strain under nitrogen limitation condition with synthesis rate up to 74.2 g/L. Furthermore, Kim et al. described considerable citric acid productivity by submerged fermentation of *C. oleophila* ATCC 20177, *C. tropicalis* ATCC 20115, and *C. oleophila* ATCC 20373 strains up to 20.7 g/L, 45.0 g/L, and 60.1 g/L, respectively (Kim et al. 2015).

M. guilliermondii appears to be one of the most efficient candidates in the xylose-xylitol conversion process with considerable variation in xylitol yield based on the accredited strain (Papon et al. 2013). The generated xylitol is beneficial in various nutraceuticals (chewing gum, candies, wafer fillings, and chocolate) and pharmaceuticals (protein extraction stabilizer and antineoplastic properties) industries (Papon et al. 2013; Ur-Rehman et al. 2015). Cortez et al. (2016) reported an effective xylitol production from D-xylose using *M. guilliermondii* (permeabilized with Triton X-100) with a yield up to 0.80 g/g and 2.65 g/L volumetric productivity. Furthermore, *M. guilliermondii* showed high potentials in producing 2-phenylethanol (2-PE), aromatic compound (rose scent) used in cosmetics, perfume, and food industries (Yan et al. 2020). Yan et al. (2021) recently reported newly identified *M. guilliermondii* strain YLG18 as a promising *candida*te for 2-PE high production from L-phe. Upon optimization process, this strain was able to synthesize 2-PE up to 3.20 g/L (Yan et al. 2021).

S. stipitis shows promising potentials in producing various valuable compounds among which is fumaric acid, a C4-dicarboxylic acid applied in nutraceutical, pharmaceutical, and chemical industries (Roa Engel et al. 2008). Wei et al. (2015) described the development of an optimized strain PSYPMFfS with the following properties, overexpression of heterologous reductive fumaric acid synthetic pathway (*Rhizopus oryzae* FM19), codon modification, blockage of the conversion of fumaric acid (double-deletion of fumarase genes (*Psfum1* and *Psfum2*), and overexpression of fumaric acid heterologous transporter (*YMAE1*). The obtained strain produced fumaric acid titer up to 4.67 g/L from xylose (Wei et al. 2015). Additionally, Ilmén et al. (2007) described an efficient production of l-lactic acid by an engineered *S. stipitis* strain; the random integration of heterologous *LDH* gene enhanced the yield of lactate up to 58 g/L from 100 g/L xylose. The produced lactic acid can be applied in various industrial applications such as biodegradable plastics and textile fiber manufacturers (Ilmén et al. 2007). The systematic metabolic engineering driven by flux facilitates the design of *S. stipitis*-mutant strains with improved biomass yield up to 44% (in ZWF1 overexpressed mutant) (Unrean et al.

2016). Further application of metabolic evolution results in maximum biomass of 9.81 g/L (Unrean et al. 2016). Remarkably, *S. stipitis* was reported to be an effective producer of S-adenosyl-l-methionine (SAM), a metabolic intermediate in enzymatic reactions which participates in several biological processes, commonly applied in medicine as a chemotherapeutic agent in the treatment of numerous pathologies, including depression, liver disease, Lesch-Nyhan disease, Alzheimer's disease, and diarrhea (Križanović et al. 2015; Chen et al. 2016). Križanović et al. (2015) showed that *S. stipitis* strain M12 with disrupted Erg6p (responsible for C-24 methylation in ergosterol biosynthesis) resulted in higher accumulation of SAM up to 52.48 mg/g CDW. The production potency of *S. stipitis* is not limited to the former molecules; rather, it demonstrates a remarkable capability in producing shikimate, a building block for several industrial, pharmaceutical, and nutraceutical applications (Gao et al. 2017). Gao et al. (2017) reported the successful generation of a recombinant *S. stipitis* FPL-UC7 platform, carrying *aro4K220L*, *aro1D900A*, and *tkt1* (encoding enzymes involved in the shikimate pathway) and strong promoters and terminators, synthesizing shikimate up to 3.11 g/L (7-folds higher relative to *S. cerevisiae* shikimate yield).

5.5 Bioremediation

Nowadays, environmental pollution is considered one of the most challenging worldwide health-related problems especially with the excessive application of pesticides and chemical fertilizers in various agricultural practices (Dangi et al. 2019). The ever-increasing of industrial development is significantly contributing to the liberation of toxic organic and inorganic compounds and heavy metals that pose a disastrous effect on the whole ecosystem (Cheng 2016; Larsson 2014). Bioremediation, a promising newly developed eco-friendly practice, aims to break the pollution cycle through the usage of "bio-cleaners," which are certain microbes (bacteria, fungi, and algae) with specialized ability to remove or neutralize contaminants present in the environment (Dangi et al. 2019). Remarkably, the fungi have many advantages over other adopted microbes in bioremediation process (Defosse et al. 2018b). Higher resistance toward toxins, natural self-regulation of accumulated pollutants, quick nutrients capturing, and biomass development are among the advantages offered by the mycoremediation approach (Defosse et al. 2018b; Treu and Falandysz 2017). The *Candida* CTG clade is no exception for such potencies (Defosse et al. 2018b). For instance, soil contaminated with chromium (a toxic metal) was efficiently restored after the implication of *C. maltosa* that partially reduced the two forms of chromium metal (trivalent Cr(III) and hexavalent Cr(VI) (Ramírez-Ramírez et al. 2004). Furthermore, the live and dead biomasses of *C. tropicalis* were capable to limit the phytoavailability of the mutagenic form of chromium Cr(VI) (Bahafid et al. 2013). Coimbra et al. (2009) demonstrated the capability of both *M. guilliermondii* and *C. tropicalis* to effectively remove hydrophobic contaminants (petroleum and motor oil) with the implementation of the substrates, soybean oil, and soybean oil + glucose. Recently, García-Béjar et al. (2020) demonstrated that the bioremediatory activity of *C. albicans* and *C.*

parapsilosis ranged from 50 to 64% in removing aflatoxin B1, and the combination of the aforementioned species with *C. tropicalis* displayed a removal capacity ranging from 56 to 68% for zinc metal.

5.6 Biocontrol

The spread of phytopathogens and crop pests in agricultural industries poses a major challenge to food security and the balance of the ecosystem since it represents one of the main causes of worldwide crop losses (Babbal and Khasa 2017; Marian and Shimizu 2019). Despite the periodical application of chemical pesticides, the estimated loss of the annual global crop yields is up to 40% (Messing and Brodeur 2018). Microbial biocontrol, a fast-growing practice, aims to use microbes (fungi, bacteria, yeast, and viruses) with the ability to control the spreading of such pests in a safe and eco-friendly manner (Babbal and Khasa 2017). Interestingly, different *Candida* CTG spp. are being considered as potential candidates for the application of microbial biocontrol approaches (Defosse et al. 2018b). Various *Candida* spp. are granted the access to the list of registered biocontrol yeasts for their adaptation properties such as secretion of hydrolytic enzymes (proteases, chitinases, glucanases) and volatile compounds (implicated in antimycotic activity), biofilm formation, high osmotolerance, induction of resistance in the plant/fruit (to control post-harvest decay), and guide of hyphal parasitism (Freimoser et al. 2019).

For instance, Guerrero et al. (2014) reported a significant suppression of *Penicillium expansum* spore germination on the post-harvest apple fruit after the application of three strains of *C. oleophila* (L06, L07 smooth, and L07 rough) with an inhibitory rate up to 97%. The suggested mode of action for the efficient bio-controllability of *C. oleophila* is mainly associated with the production of exo-β1,3-glucanase enzyme (Guerrero et al. 2014). Interestingly, numerous antagonistic yeast-based commercial products are being developed for the control of post-harvest pathogens, among which are *C. oleophila*-based products, namely Aspire (controls decay on stone fruit, pome, citrus, strawberry) and Nexy (controls decay on pome, banana, citrus) (Zhang et al. 2020). The biocontrol potency is not exclusive to *C. oleophila*. Recently, Sun et al. (Sun et al. 2021) described the synergistic bio-controllability action of *M. guilliermondii* Y-1 (1×10^8 cells/mL) and melatonin (100 μmol/L; possibly implicated in improving the strain Y-1 colonization ability) against apple gray mold (caused by *Botrytis cinerea*). A significant reduction in disease incidence and lesion diameter is achieved with the aforementioned combination. Furthermore, the co-culture of *M. guilliermondii* KL3 and *Aureobasidium pullulans* GE17 was able to suppress the spore germination of both, *Penicillium digitatum* (green mold) and *P. expansum* (blue mold) by the efficacy of 86–95% (Agirman and Erten 2020).

6 Concluding Remarks and Perspectives

The importance of CTG clade yeasts in the field of synthetic biology is not doubtable anymore; all the identified capabilities in this unique clade are just presenting the tip of submerged potentials. The future vision of *Candida* metabolic engineering will mainly deviate to the *Candida* consortium rather than the single strain notion with the application of more species-directed (species–hybrid) and synthetic standardized gene rewiring modules. The unlimited efforts to standardize the advanced gene-editing tools are the keystone in deciphering the competent species. Thus, the continuous optimization and integration of both, gene-engineering strategies and omics platforms, will facilitate building specialized potentially unexplored CTG clade yeast strains with desirable properties in additional biotechnological fields.

References

Abbas CA, Sibirny AA (2011) Genetic control of biosynthesis and transport of riboflavin and flavin nucleotides and construction of robust biotechnological producers. Microbiol Mol Biol Rev 75(2):321–360

Agirman B, Erten H (2020) Biocontrol ability and action mechanisms of *Aureobasidium pullulans* GE17 and *Meyerozyma guilliermondii* KL3 against *Penicillium digitatum* DSM2750 and *Penicillium expansum* DSM62841 causing postharvest diseases. Yeast 37(9–10):437–448

Anastassiadis S, Wandrey C, Rehm HJ (2005) Continuous citric acid fermentation by *Candida oleophila* under nitrogen limitation at constant C/N ratio. World J Microbiol Biotechnol 21(5):695–705

Andreieva Y, Petrovska Y, Lyzak O, Liu W, Kang Y, Dmytruk K et al (2020) Role of the regulatory genes *SEF1*, *VMA1* and *SFU1* in riboflavin synthesis in the flavinogenic yeast *Candida famata* (*Candida flareri*). Yeast (chichester, England) 37(9–10):497–504

Arkin AP, Cottingham RW, Henry CS, Harris NL, Stevens RL, Maslov S et al (2018) KBase: the United States department of energy systems biology knowledgebase. Nat Biotechnol 36(7):566–569

Asadzadeh M, Dashti M, Ahmad S (2020) Whole genome and targeted-amplicon sequencing of fluconazole-susceptible and -resistant *Candida parapsilosis* isolates from Kuwait reveals a previously undescribed N1132D polymorphism in CDR1. Antimicrob Agents Chemother

Austin S, Ziese M, Sternberg N (1981) A novel role for site-specific recombination in maintenance of bacterial replicons. Cell 25(3):729–736

Awad A, El Khoury P, Wex B, Khalaf RA (2018) Proteomic analysis of a *Candida albicans* pga1 Null Strain. EuPA Open Proteom 18:1–6

Babar MM, Afzaal H, Pothineni VR, Zaidi N-u-SS, Ali Z, Zahid MA et al (2018) Omics approaches in industrial biotechnology and bioprocess engineering ((Chap 14)). In: Barh D, Azevedo V (eds) Omics technologies and bio-engineering. Academic Press, pp 251–69

Babbal, Adivitiya, Khasa YP (2017) Microbes as biocontrol Agents. In: Kumar V, Kumar M, Sharma S, Prasad R (eds) Probiotics and plant health. Springer, Singapore, pp 507–552

Bahafid W, Tahri Joutey N, Sayel H, Boularab I, Ghachtouli N (2013) Bioaugmentation of chromium-polluted soil microcosms with *Candida tropicalis* diminishes phytoavailable chromium. J Appl Microbiol 115(3):727–734

Bassler BL, Losick R (2006) Bacterially speaking. Cell 125(2):237–246

Basso LR Jr, Bartiss A, Mao Y, Gast CE, Coelho PSR, Snyder M et al (2010) Transformation of *Candida albicans* with a synthetic hygromycin B resistance gene. Yeast (chichester, England) 27(12):1039–1048

Beckerman J, Chibana H, Turner J, Magee PT (2001) Single-copy IMH3 allele is sufficient to confer resistance to mycophenolic acid in *Candida albicans* and to mediate transformation of clinical *Candida* species. Infect Immun 69(1):108–114

Berens C, Hillen W (2004) Gene regulation by tetracyclines. Genet Eng (NY) 26:255–277

Bhatia S, Bhatia R, Choi Y-K, Kane E, Kim Y-G, Yanga Y-H (2018) Biotechnological potential of microbial consortia and future perspective. Crit Rev Biotechnol 38

Bijlani S, Nahar AS, Ganesan K (2018) Improved Tet-On and Tet-Off systems for tetracycline-regulated expression of genes in *Candida*. Curr Genet 64(1):303–316

Blazeck J, Alper H (2010) Systems metabolic engineering: genome-scale models and beyond. Biotechnol J 5(7):647–659

Bratiichuk D, Kurylenko O, Vasylyshyn R, Zuo M, Kang Y, Dmytruk K et al (2020) Development of new dominant selectable markers for the nonconventional yeasts Ogataea polymorpha and *Candida famata*. Yeast 37(9–10):505–513

Brenner K, You L, Arnold FH (2008) Engineering microbial consortia: a new frontier in synthetic biology. Trends Biotechnol 26(9):483–489

Butler G, Rasmussen MD, Lin MF, Santos MA, Sakthikumar S, Munro CA et al (2009) Evolution of pathogenicity and sexual reproduction in eight *Candida* genomes. Nature 459(7247):657–662

Cabrera Z, Gutarra MLE, Guisan JM, Palomo JM (2010) Highly enantioselective biocatalysts by coating immobilized lipases with polyethyleneimine. Catal Commun 11(11):964–967

Campa D, Tavanti A, Gemignani F, Mogavero CS, Bellini I, Bottari F et al (2008) DNA microarray based on arrayed-primer extension technique for identification of pathogenic fungi responsible for invasive and superficial mycoses. J Clin Microbiol 46(3):909–915

Cao M, Seetharam AS, Severin AJ, Shao Z (2017a) Rapid isolation of centromeres from *Scheffersomyces stipitis*. ACS Synth Biol 6(11):2028–2034

Cao M, Gao M, Lopez-Garcia CL, Wu Y, Seetharam AS, Severin AJ et al (2017b) Centromeric DNA facilitates nonconventional yeast genetic engineering. ACS Synth Biol 6(8):1545–1553

Castellan A, Bart JCJ, Cavallaro S (1991) Industrial production and use of adipic acid. Catal Today 9(3):237–254

Chandrasegaran S, Carroll D (2016) Origins of programmable nucleases for genome engineering. J Mol Biol 428(5 Pt B):963–89

Chang W, Zhang M, Li Y, Lou H (2015) Flow cytometry-based method to detect persisters in *Candida albicans*. Antimicrob Agents Chemother 59(8):5044–5048

Chattopadhyay A, Gupta A, Maiti MK (2020) Engineering an oleaginous yeast *Candida tropicalis* SY005 for enhanced lipid production. Appl Microbiol Biotechnol 104(19):8399–8411

Chen YC, Eisner JD, Kattar MM, Rassoulian-Barrett SL, LaFe K, Yarfitz SL et al (2000) Identification of medically important yeasts using PCR-based detection of DNA sequence polymorphisms in the internal transcribed spacer 2 region of the rRNA genes. J Clin Microbiol 38(6):2302–2310

Chen H, Wang Z, Cai H, Zhou C (2016) Progress in the microbial production of S-adenosyl-L-methionine. World J Microbiol Biotechnol 32(9):153

Cheng Z (2016) The spatial correlation and interaction between manufacturing agglomeration and environmental pollution. Ecol Ind 61:1024–1032

Christen S, Sauer U (2011) Intracellular characterization of aerobic glucose metabolism in seven yeast species by 13C flux analysis and metabolomics. FEMS Yeast Res 11(3):263–272

Clark DP, Pazdernik NJ, McGehee MR (2019) Analysis of gene expression (Chap 21). In: Clark DP, Pazdernik NJ, McGehee MR (eds) Molecular biology, 3rd ed. Academic Cell, pp 654–90

Coimbra CD, Rufino RD, Luna JM, Sarubbo LA (2009) Studies of the cell surface properties of *Candida* species and relation to the production of biosurfactants for environmental applications. Curr Microbiol 58(3):245–251

Cormack BP, Valdivia RH, Falkow S (1996) FACS-optimized mutants of the green fluorescent protein (GFP). Gene 173(1):33–38

Cormack BP, Bertram G, Egerton M, Gow NA, Falkow S, Brown AJ (1997) Yeast-enhanced green fluorescent protein (yEGFP): a reporter of gene expression in Candida albicans. Microbiology 143(2):303–311

Cortez DV, Mussatto SI, Roberto IC (2016) Improvement on D-xylose to xylitol biotransformation by *Candida guilliermondii* using cells permeabilized with triton X-100 and selected process conditions. Appl Biochem Biotechnol 180(5):969–979

Courdavault V, Millerioux Y, Clastre M, Simkin AJ, Marais E, Crèche J et al (2011) Fluorescent protein fusions in *Candida guilliermondii*. Fungal Genet Biol 48(11):1004–1011

Dangi AK, Sharma B, Hill RT, Shukla P (2019) Bioremediation through microbes: systems biology and metabolic engineering approach. Crit Rev Biotechnol 39(1):79–98

Dashtban M, Wen X, Bajwa PK, Ho CY, Lee H (2015) Deletion of *hxk1* gene results in derepression of xylose utilization in *Scheffersomyces stipitis*. J Ind Microbiol Biotechnol 42(6):889–896

Defosse TA, Courdavault V, Coste AT, Clastre M, de Bernonville TD, Godon C et al (2018a) A standardized toolkit for genetic engineering of CTG clade yeasts. J Microbiol Methods 144:152–156

Defosse TA, Le Govic Y, Courdavault V, Clastre M, Vandeputte P, Chabasse D et al (2018b) Les levures du clade CTG (clade Candida): biologie, incidence en santé humaine et applications en biotechnologie. Med Mycol J 28(2):257–268

Defosse TA, Mélin C, Clastre M, Besseau S, Lanoue A, Glévarec G et al (2016) An additional *Meyerozyma guilliermondii* IMH3 gene confers mycophenolic acid resistance in fungal CTG clade species. FEMS Yeast Res 16(6)

Dennison PM, Ramsdale M, Manson CL, Brown AJ (2005) Gene disruption in *Candida albicans* using a synthetic, codon-optimised Cre-loxP system. Fungal Genet Biol 42(9):737–748

DiCarlo JE, Norville JE, Mali P, Rios X, Aach J, Church GM (2013) Genome engineering in *Saccharomyces cerevisiae* using CRISPR-Cas systems. Nucleic Acids Res 41(7):4336–4343

Ding C, Butler G (2007) Development of a gene knockout system in *Candida parapsilosis* reveals a conserved role for BCR1 in biofilm formation. Eukaryot Cell 6(8):1310–1319

Dmytruk KV, Yatsyshyn VY, Sybirna NO, Fedorovych DV, Sibirny AA (2011) Metabolic engineering and classic selection of the yeast *Candida famata* (*Candida flareri*) for construction of strains with enhanced riboflavin production. Metab Eng 13(1):82–88

Dmytruk K, Lyzak O, Yatsyshyn V, Kluz M, Sibirny V, Puchalski C et al (2014) Construction and fed-batch cultivation of *Candida famata* with enhanced riboflavin production. J Biotechnol 172:11–17

Doyle TC, Nawotka KA, Purchio AF, Akin AR, Francis KP, Contag PR (2006a) Expression of firefly luciferase in *Candida albicans* and its use in the selection of stable transformants. Microb Pathog 40(2):69–81

Doyle TC, Nawotka KA, Kawahara CB, Francis KP, Contag PR (2006b) Visualizing fungal infections in living mice using bioluminescent pathogenic *Candida albicans* strains transformed with the firefly luciferase gene. Microb Pathog 40(2):82–90

Du C, Li Y, Zhao X, Pei X, Yuan W, Bai F et al (2019) The production of ethanol from lignocellulosic biomass by *Kluyveromyces marxianus* CICC 1727-5 and *Spathaspora passalidarum* ATCC MYA-4345. Appl Microbiol Biotechnol 103(6):2845–2855

Dujon BA, Louis EJ (2017) Genome diversity and evolution in the budding yeasts (Saccharomycotina). Genetics 206(2):717–750

Enjalbert B, Rachini A, Vediyappan G, Pietrella D, Spaccapelo R, Vecchiarelli A et al (2009) A multifunctional, synthetic *Gaussia princeps* luciferase reporter for live imaging of *Candida albicans* infections. Infect Immun 77(11):4847–4858

Expression G, Sundaresan G, Gambhir SS (2002) Radionuclide imaging of reporter gene expression (Chap 29). In: Toga AW, Mazziotta JC (eds) Brain mapping: the methods, 2nd edn. Academic Press, San Diego, pp 799–818

Feng X, Page L, Rubens J, Chircus L, Colletti P, Pakrasi H et al (2010) Bridging the gap between fluxomics and industrial biotechnology. J Biomed Biotechnol 2010:460717

Fitzpatrick DA, Logue ME, Stajich JE, Butler G (2006) A fungal phylogeny based on 42 complete genomes derived from supertree and combined gene analysis. BMC Evol Biol 6(1):99

Foureau E, Courdavault V, Navarro Gallón SM, Besseau S, Simkin AJ, Crèche J et al (2013) Characterization of an autonomously replicating sequence in *Candida guilliermondii*. Microbiol Res 168(9):580–588

Freimoser FM, Rueda-Mejia MP, Tilocca B, Migheli Q (2019) Biocontrol yeasts: mechanisms and applications. World J Microbiol Biotechnol 35(10):154

Gabriel F, Accoceberry I, Bessoule JJ, Salin B, Lucas-Guérin M, Manon S et al (2014) A Fox2-dependent fatty acid ß-oxidation pathway coexists both in peroxisomes and mitochondria of the ascomycete yeast *Candida lusitaniae*. PLoS One 9(12):e114531

Gácser A, Trofa D, Schäfer W, Nosanchuk JD (2007) Targeted gene deletion in *Candida parapsilosis* demonstrates the role of secreted lipase in virulence. J Clin Invest 117(10):3049–3058

Gao M, Cao M, Suástegui M, Walker J, Rodriguez Quiroz N, Wu Y et al (2017) Innovating a nonconventional yeast platform for producing shikimate as the building block of high-value aromatics. ACS Synth Biol 6(1):29–38

García-Béjar B, Arévalo-Villena M, Guisantes-Batan E, Rodríguez-Flores J, Briones A (2020) Study of the bioremediatory capacity of wild yeasts. Sci Rep 10(1):11265

Gerami-Nejad M, Berman J, Gale CA (2001) Cassettes for PCR-mediated construction of green, yellow, and cyan fluorescent protein fusions in *Candida albicans*. Yeast 18(9):859–864

Gerami-Nejad M, Dulmage K, Berman J (2009) Additional cassettes for epitope and fluorescent fusion proteins in *Candida albicans*. Yeast 26(7):399–406

Gordon ZB, Soltysiak MPM, Leichthammer C, Therrien JA, Meaney RS, Lauzon C et al (2019) Development of a transformation method for *Metschnikowia borealis* and other CUG-Serine Yeasts. Genes (Basel) 10(2)

Grahl N, Demers EG, Crocker AW, Hogan DA (2017) Use of RNA-protein complexes for genome editing in non-*albicans Candida* species. mSphere 2(3)

Gräslund S, Sagemark J, Berglund H, Dahlgren LG, Flores A, Hammarström M et al (2008) The use of systematic N- and C-terminal deletions to promote production and structural studies of recombinant proteins. Protein Expr Purif 58(2):210–221

Griffith F (1928) The significance of pneumococcal types. J Hyg 27(2):113–159

Guerrero V, Guigón-López C, Berlanga D, Ojeda-Barrios D (2014) Complete control of *Penicillium expansum* on apple fruit using a combination of antagonistic yeast *Candida oleophila*. Chilean J Agric Res 74:427–431

Han T-l, Cannon RD, Villas-Bôas SG (2012) Metabolome analysis during the morphological transition of *Candida albicans*. Metabolomics 8(6):1204–1217

Hara A, Arie M, Kanai T, Matsui T, Matsuda H, Furuhashi K et al (2001) Novel and convenient methods for *Candida tropicalis* gene disruption using a mutated hygromycin B resistance gene. Arch Microbiol 176(5):364–369

Hasin Y, Seldin M, Lusis A (2017) Multi-omics approaches to disease. Genome Biol 18(1):83

Heidari R, Shaw DM, Elger BS (2017) CRISPR and the rebirth of synthetic biology. Sci Eng Ethics 23(2):351–363

Henry CS, Broadbelt LJ, Hatzimanikatis V (2007) Thermodynamics-based metabolic flux analysis. Biophys J 92(5):1792–1805

Herrgård M, Panagiotou G (2012) Analyzing the genomic variation of microbial cell factories in the era of "New Biotechnology". Comput Struct Biotechnol J 3(4):e201210012

Herrgård M, Fong S, Palsson B (2006) Identification of genome-scale metabolic network models using experimentally measured flux profiles. PLoS Comput Biol 2:e72

Hinchliffe E, Kenny E (1993) Yeast as a vehicle for the expression of heterologous genes (Chap 9). In: Rose AH, Stuart Harrison J (eds) The yeasts, 2nd edn. Academic Press, San Diego, pp 325–356

Hirata Y, Ryu M, Oda Y, Igarashi K, Nagatsuka A, Furuta T et al (2009) Novel characteristics of sophorolipids, yeast glycolipid biosurfactants, as biodegradable low-foaming surfactants. J Biosci Bioeng 108(2):142–146

Holkers M, Vries AAFd, Gonçalves MAFV (2006) Modular and excisable molecular switch for the induction of gene expression by the yeast FLP recombinase. BioTech 41(6):711–713

Horgan RP, Kenny LC (2011) 'Omic' technologies: genomics, transcriptomics, proteomics and metabolomics. Obstet Gynaecol 13(3):189–195

Huang C, Luo M-T, Chen X-F, Qi G-X, Xiong L, Lin X-Q et al (2017) Combined "de novo" and "ex novo" lipid fermentation in a mix-medium of corncob acid hydrolysate and soybean oil by *Trichosporon dermatis*. Biotechnol Biofuels 10(1):147

Huang MY, Mitchell AP (2017) Marker recycling in *Candida albicans* through CRISPR-Cas9-induced marker excision. mSphere 2(2):e00050-17

Ilmén M, Koivuranta K, Ruohonen L, Suominen P, Penttilä M (2007) Efficient production of lactic acid from Xylose by *Pichia stipitis*. Appl Environ Microbiol 73(1):117–123

Ishchuk O, Dmytruk K, Rohulya O, Voronovsky A, Abbas C, Sibirny A (2008) Development of a promoter assay system for the flavinogenic yeast *Candida famata* based on the *Kluyveromyces lactis* β-galactosidase LAC4 reporter gene. Enzyme Microb Technol 42:208–215

Jenior ML, Moutinho TJ Jr, Dougherty BV, Papin JA (2020) Transcriptome-guided parsimonious flux analysis improves predictions with metabolic networks in complex environments. PLoS Comput Biol 16(4):e1007099-e

Jensen MK, Keasling JD (2015) Recent applications of synthetic biology tools for yeast metabolic engineering. FEMS Yeast Res 15(1):1–10

Jeon WY, Shim WY, Lee SH, Choi JH, Kim JH (2013) Effect of heterologous xylose transporter expression in *Candida tropicalis* on xylitol production rate. Bioprocess Biosyst Eng 36(6):809–817

Jinek M, Chylinski K, Fonfara I, Hauer M, Doudna JA, Charpentier E (2012) A Programmable Dual-RNA–Guided DNA Endonuclease in Adaptive Bacterial Immunity. Science 337(6096):816–821

Johnson EA (2013) Biotechnology of non-*Saccharomyces* yeasts-the basidiomycetes. Appl Microbiol Biotechnol 97(17):7563–7577

Jones HD (2003) Genetic modification | transformation, general principles. In: Thomas B (ed) Encyclopedia of applied plant sciences. Elsevier, Oxford, pp 377–382

Ju JH, Oh BR, Heo SY, Lee YU, Shon JH, Kim CH et al (2020) Production of adipic acid by short- and long-chain fatty acid acyl-CoA oxidase engineered in yeast *Candida tropicalis*. Bioprocess Biosyst Eng 43(1):33–43

Juers DH, Matthews BW, Huber RE (2012) LacZ β-galactosidase: structure and function of an enzyme of historical and molecular biological importance. Protein Sci 21(12):1792–1807

Karkowska-Kuleta J, Kulig K, Karnas E, Zuba-Surma E, Woznicka O, Pyza E et al (2020) Characteristics of extracellular vesicles released by the pathogenic yeast-like fungi *Candida glabrata*, *Candida parapsilosis* and *Candida tropicalis*. Cells 9(7):1722

Kawaguchi Y, Honda H, Taniguchi-Morimura J, Iwasaki S (1989) The codon CUG is read as serine in an asporogenic yeast *Candida cylindracea*. Nature 341(6238):164–166

Keppler-Ross S, Noffz C, Dean N (2008) A new purple fluorescent color marker for genetic studies in *Saccharomyces cerevisiae* and *Candida albicans*. Genetics 179(1):705–710

Kim SH, Shin DH, Liu J, Oganesyan V, Chen S, Xu QS et al (2005) Structural genomics of minimal organisms and protein fold space. J Struct Funct Genomics 6(2–3):63–70

Kim KH, Lee H-Y, Lee CY (2015) Pretreatment of sugarcane molasses and citric acid production by *Candida zeylanoides*. Microbiol Biotechnol Lett 43(2):164–168

Köhler GA, White TC, Agabian N (1997) Overexpression of a cloned IMP dehydrogenase gene of *Candida albicans* confers resistance to the specific inhibitor mycophenolic acid. J Bacteriol 179(7):2331–2338

Kosa P, Gavenciakova B, Nosek J (2007) Development of a set of plasmid vectors for genetic manipulations of the pathogenic yeast *Candida parapsilosis*. Gene 396(2):338–345

Kouzuma A, Kato S, Watanabe K (2015) Microbial interspecies interactions: recent findings in syntrophic consortia. Front Microbiol 6(477)

Križanović S, Butorac A, Mrvčić J, Krpan M, Cindrić M, Bačun-Družina V et al (2015) Characterization of a S-adenosyl-l-methionine (SAM)-accumulating strain of Scheffersomyces stipitis. Int Microbiol 18(2):117–125

Lam CMC, Godinho M, dos Santos VAPM (2010) An introduction to synthetic biology. In: Schmidt M, Kelle A, Ganguli-Mitra A, Vriend H (eds) Synthetic biology: the technoscience and its societal consequences. Springer, Netherlands, Dordrecht, pp 23–48

Laplaza JM, Torres BR, Jin Y-S, Jeffries TW (2006) Sh ble and Cre adapted for functional genomics and metabolic engineering of *Pichia stipitis*. Enzyme Microb Technol 38(6):741–747

Larbi NB, Jefferies C (2009) 2D-DIGE: comparative proteomics of cellular signalling pathways. Methods in Molecular Biology (clifton, NJ) 517:105–132

Larsson DGJ (2014) Pollution from drug manufacturing: review and perspectives. Philos Trans R Soc Lond B Biol Sci 369(1656):20130571

Ledesma-Amaro R, Santos MA, Jiménez A, Revuelta JL (2013) Microbial production of vitamins (Chap 21). In: McNeil B, Archer D, Giavasis I, Harvey L (eds) Microbial production of food ingredients, enzymes and nutraceuticals. Woodhead Publishing, pp 571–594

Leonard E, Nielsen D, Solomon K, Prather KJ (2008) Engineering microbes with synthetic biology frameworks. Trends Biotechnol 26(12):674–681

Leuker CE, Hahn AM, Ernst JF (1992) beta-Galactosidase of *Kluyveromyces lactis* (Lac4p) as reporter of gene expression in *Candida albicans* and *C. tropicalis*. Mol Gen Genet 235(2–3):235–241

Li Y, Chen Y, Tian X, Chu J (2020) Advances in sophorolipid-producing strain performance improvement and fermentation optimization technology. Appl Microbiol Biotechnol 104:10325–10337

Lin C-H, Choi A, Bennett RJ (2011) Defining pheromone-receptor signaling in *Candida albicans* and related asexual *Candida* species. Mol Biol Cell 22(24):4918–4930

Liu S, Hu W, Wang Z, Chen T (2020) Production of riboflavin and related cofactors by biotechnological processes. Microb Cell Fact 19(1):31

Liu GL, Fu GY, Chi Z, Chi ZM (2014) Enhanced expression of the codon-optimized exo-inulinase gene from the yeast *Meyerozyma guilliermondii* in *Saccharomyces* sp. W0 and bioethanol production from inulin. Appl Microbiol Biotechnol 98(21):9129–9138

Löbs AK, Schwartz C, Wheeldon I (2017) Genome and metabolic engineering in non-conventional yeasts: current advances and applications. Synth Syst Biotechnol 2(3):198–207

Loeffler J, Hebart H, Magga S, Schmidt D, Klingspor L, Tollemar J et al (2000) Identification of rare *Candida* species and other yeasts by polymerase chain reaction and slot blot hybridization. Diagn Microbiol Infect Dis 38(4):207–212

Lombardi L, Turner SA, Zhao F, Butler G (2017) Gene editing in clinical isolates of *Candida parapsilosis* using CRISPR/Cas9. Sci Rep 7(1):8051

Lombardi L, Oliveira-Pacheco J, Butler G (2019) Plasmid-based CRISPR-Cas9 gene editing in multiple *Candida* species. mSphere 4(2)

Maguire SL, ÓhÉigeartaigh SS, Byrne KP, Schröder MS, O'Gaora P, Wolfe KH et al (2013) Comparative genome analysis and gene finding in *Candida* Species using CGOB. Mol Biol Evol 30(6):1281–1291

Maier T, Güell M, Serrano L (2009) Correlation of mRNA and protein in complex biological samples. FEBS Lett 583(24):3966–3973

Maldonado I, Cataldi S, Garbasz C, Relloso S, Striebeck P, Guelfand L et al (2018) Identification of *Candida* yeasts: conventional methods and MALDI-TOF MS. Rev Iberoam Micol 35(3):151–154

Mancera E, Frazer C, Porman AM, Ruiz-Castro S, Johnson AD, Bennett RJ (2019) Genetic modification of closely related *Candida* species. Front Microbiol 10(357)

Marian M, Shimizu M (2019) Improving performance of microbial biocontrol agents against plant diseases. J Gen Plant Pathol 85(5):329–336

Marton T, Maufrais C, d'Enfert C, Legrand M (2020) Use of CRISPR-Cas9 to target homologous recombination limits transformation-induced genomic changes in *Candida albicans*. mSphere 5(5)

Masuda Y, Park SM, Ohkuma M, Ohta A, Takagi M (1994) Expression of an endogenous and a heterologous gene in *Candida maltosa* by using a promoter of a newly-isolated phosphoglycerate kinase (PGK) gene. Curr Genet 25(5):412–417

McCarty NS, Ledesma-Amaro R (2019) Synthetic biology tools to Engineer microbial communities for biotechnology. Trends Biotechnol 37(2):181–197
McLellan MA, Rosenthal NA, Pinto AR (2017) Cre-loxP-mediated recombination: general principles and experimental considerations. Curr Protoc Mouse Biol 7(1):1–12
Messing R, Brodeur J (2018) Current challenges to the implementation of classical biological control. Biocontrol 63(1):1–9
Michel S, Ushinsky S, Klebl B, Leberer E, Thomas D, Whiteway M et al (2002) Generation of conditional lethal *Candida albicans* mutants by inducible deletion of essential genes. Mol Microbiol 46(1):269–280
Millerioux Y, Clastre M, Simkin AJ, Courdavault V, Marais E, Sibirny AA et al (2011) Drug-resistant cassettes for the efficient transformation of *Candida guilliermondii* wild-type strains. FEMS Yeast Res 11(6):457–463
Min K, Ichikawa Y, Woolford CA, Mitchell AP (2016) *Candida albicans* gene deletion with a transient CRISPR-Cas9 system. mSphere 1(3)
Min et al (2016) *Candida albicans* gene deletion with a transient CRISPR-Cas9 system. https://doi.org/10.1128/mSphere.00130-16
Mishra P, Park GY, Lakshmanan M, Lee HS, Lee H, Chang MW et al (2016) Genome-scale metabolic modeling and in silico analysis of lipid accumulating yeast *Candida tropicalis* for dicarboxylic acid production. Biotechnol Bioeng 113(9):1993–2004
Möckli N, Auerbach D (2004) Quantitative β-galactosidase assay suitable for high-throughput applications in the yeast two-hybrid system. Biotechniques 36(5):872–876
Moreno-Ruiz E, Ortu G, de Groot PWJ, Cottier F, Loussert C, Prévost M-C et al (2009) The GPI-modified proteins Pga59 and Pga62 of *Candida albicans* are required for cell wall integrity. Microbiology 155(6):2004–2020
Morschhäuser J, Michel S, Hacker J (1998) Expression of a chromosomally integrated, single-copy GFP gene in *Candida albicans*, and its use as a reporter of gene regulation. Mol Gen Genet MGG 257(4):412–420
Morschhäuser J, Michel S, Staib P (1999) Sequential gene disruption in *Candida albicans* by FLP-mediated site-specific recombination. Mol Microbiol 32(3):547–556
Mühlhausen S, Kollmar M (2014) Molecular phylogeny of sequenced *Saccharomycetes* reveals polyphyly of the alternative yeast codon usage. Genome Biol Evol 6(12):3222–3237
Nakayama H, Mio T, Nagahashi S, Kokado M, Arisawa M, Aoki Y (2000) Tetracycline-regulatable system to tightly control gene expression in the pathogenic fungus *Candida albicans*. Infect Immun 68(12):6712–6719
Narad P, Kirthanashri SV (2018) Introduction to omics. In: Arivaradarajan P, Misra G (eds) Omics approaches, technologies and applications: integrative approaches for understanding OMICS data. Springer Singapore, Singapore, pp 1–10
Nguyen LN, Trofa D, Nosanchuk JD (2009) Fatty acid synthase impacts the pathobiology of *Candida parapsilosis* in vitro and during mammalian infection. PLoS One 4(12):e8421
Nguyen N, Quail MMF, Hernday AD (2017) An efficient, rapid, and recyclable system for CRISPR-mediated genome editing in *Candida albicans*. mSphere 2(2)
Norton EL, Sherwood RK, Bennett RJ (2017) Development of a CRISPR-Cas9 system for efficient genome editing of *Candida lusitaniae*. mSphere 2(3)
Nosek J, Holesova Z, Kosa P, Gacser A, Tomaska L (2009) Biology and genetics of the pathogenic yeast *Candida parapsilosis*. Curr Genet 55(5):497–509
Nunes-Düby SE, Kwon HJ, Tirumalai RS, Ellenberger T, Landy A (1998) Similarities and differences among 105 members of the Int family of site-specific recombinases. Nucleic Acids Res 26(2):391–406
Obando Montoya EJ, Mélin C, Blanc N, Lanoue A, Foureau E, Boudesocque L et al (2014) Disrupting the methionine biosynthetic pathway in *Candida guilliermondii*: characterization of the MET2 gene as counter-selectable marker. Yeast 31(7):243–251
Olivares-Hernández R, Usaite R, Nielsen J (2010) Integrative analysis using proteome and transcriptome data from yeast to unravel regulatory patterns at post-transcriptional level. Biotechnol Bioeng 107(5):865–875

Olivares-Hernández R, Bordel S, Nielsen J (2011) Codon usage variability determines the correlation between proteome and transcriptome fold changes. BMC Syst Biol 5(1):33
Orr-Weaver TL, Szostak JW, Rothstein RJ (1981) Yeast transformation: a model system for the study of recombination. Proc Natl Acad Sci U S A 78(10):6354–6358
Orth JD, Thiele I, Palsson BØ (2010) What is flux balance analysis? Nat Biotechnol 28(3):245–248
Papon N, Courdavault V, Clastre M, Simkin AJ, Crèche J, Giglioli-Guivarc'h N (2012) Deus ex *Candida* genetics: overcoming the hurdles for the development of a molecular toolbox in the CTG clade. Microbiology 158(Pt 3):585–600
Papon N, Savini V, Lanoue A, Simkin AJ, Crèche J, Giglioli-Guivarc'h N et al (2013) *Candida guilliermondii*: biotechnological applications, perspectives for biological control, emerging clinical importance and recent advances in genetics. Curr Genet 59(3):73–90
Papon N, Courdavault V, Clastre M (2014) Biotechnological potential of the fungal CTG clade species in the synthetic biology era. Trends Biotechnol 32(4):167–168
Park Y-N, Morschhäuser J (2005) Tetracycline-inducible gene expression and gene deletion in *Candida albicans*. Eukaryot Cell 4(8):1328–1342
Passoth V, Cohn M, Schäfer B, Hahn-Hägerdal B, Klinner U (2003) Analysis of the hypoxia-induced ADH2 promoter of the respiratory yeast *Pichia stipitis* reveals a new mechanism for sensing of oxygen limitation in yeast. Yeast 20(1):39–51
Patra P, Das M, Kundu P, Ghosh A (2021) Recent advances in systems and synthetic biology approaches for developing novel cell-factories in non-conventional yeasts. Biotechnol Adv 47:107695
Pereira SC, Maehara L, Machado CMM, Farinas CS (2015) 2G ethanol from the whole sugarcane lignocellulosic biomass. Biotechnol Biofuels 8(1):44
Pharkya P, Maranas C (2006) An optimization framework for identifying reaction activation/inhibition or elimination candidates for overproduction in microbial systems. Metab Eng 8:1–13
Pharkya P, Burgard A, Maranas C (2004) OptStrain: A computational framework for redesign of microbial production systems. Genome Res 14:2367–2376
Pickford R (2019) Mass spectrometry-based metabolomic analysis. In: Ranganathan S, Gribskov M, Nakai K, Schönbach C (eds) Encyclopedia of bioinformatics and computational biology. Academic Press, Oxford, pp 410–425
Polen T, Spelberg M, Bott M (2013) Toward biotechnological production of adipic acid and precursors from biorenewables. J Biotechnol 167(2):75–84
Price ND, Reed JL, Palsson B (2004) Genome-scale models of microbial cells: evaluating the consequences of constraints. Nat Rev Microbiol 2(11):886–897
Prista C, Michán C, Miranda IM, Ramos J (2016) The halotolerant *Debaryomyces hansenii*, the Cinderella of non-conventional yeasts. Yeast 33(10):523–533
Raghavachari N (2011) Overview of omics pp 1–20
Ramírez-Ramírez R, Calvo-Méndez C, Avila-Rodriguez M, Lappe-Oliveras P, Ulloa M, Vázques-Juárez R et al (2004) Cr(VI) reduction in chromate-resistant strain of *Candida maltosa* isolated from the leather industry. Antonie Van Leeuwenhoek 85:63–68
Ratcliffe RG, Shachar-Hill Y (2005) Revealing metabolic phenotypes in plants: inputs from NMR analysis. Biol Rev Camb Philos Soc 80(1):27–43
Reijnst P, Walther A, Wendland J (2011) Dual-colour fluorescence microscopy using yEmCherry-/GFP-tagging of eisosome components Pil1 and Lsp1 in *Candida albicans*. Yeast 28(4):331–338
Ren J, Lee J, Na D (2020) Recent advances in genetic engineering tools based on synthetic biology. Microbiology 58(1):1–10
Reuss O, Vik A, Kolter R, Morschhäuser J (2004) The SAT1 flipper, an optimized tool for gene disruption in *Candida albicans*. Gene 341:119–127
Revuelta JL, Buey RM, Ledesma-Amaro R, Vandamme EJ (2016) Microbial biotechnology for the synthesis of (pro)vitamins, biopigments and antioxidants: challenges and opportunities. Microb Biotechnol 9(5):564–567

Riley R, Haridas S, Wolfe KH, Lopes MR, Hittinger CT, Göker M et al (2016) Comparative genomics of biotechnologically important yeasts. Proc Natl Acad Sci U S A 113(35):9882–9887

Roa Engel CA, Straathof AJJ, Zijlmans TW, van Gulik WM, van der Wielen LAM (2008) Fumaric acid production by fermentation. Appl Microbiol Biotechnol 78(3):379–389

Roda A, Pasini P, Mirasoli M, Michelini E, Guardigli M (2004) Biotechnological applications of bioluminescence and chemiluminescence. Trends Biotechnol 22(6):295–303

Rodrussamee N, Sattayawat P, Yamada M (2018) Highly efficient conversion of xylose to ethanol without glucose repression by newly isolated thermotolerant *Spathaspora passalidarum* CMUWF1–2. BMC Microbiol 18(1):73

Roemer T, Jiang B, Davison J, Ketela T, Veillette K, Breton A et al (2003) Large-scale essential gene identification in *Candida albicans* and applications to antifungal drug discovery. Mol Microbiol 50(1):167–181

Røkke G, Korvald E, Pahr J, Øyås O, Lale R (2014) BioBrick assembly standards and techniques and associated software tools. In: Valla S, Lale R (eds) DNA cloning and assembly methods. Humana Press, Totowa, NJ, pp 1–24

Roldão A, Kim I-K, Nielsen J (2012) Bridging omics technologies with synthetic biology in yeast industrial biotechnology. In: Wittmann C, Lee SY (eds) Systems metabolic engineering. Springer, Netherlands, Dordrecht, pp 271–327

Ruchala J, Kurylenko OO, Dmytruk KV, Sibirny AA (2020) Construction of advanced producers of first- and second-generation ethanol in *Saccharomyces cerevisiae* and selected species of non-conventional yeasts (*Scheffersomyces stipitis*, *Ogataea polymorpha*). J Ind Microbiol Biotechnol 47(1):109–132

Samaranayake DP, Hanes SD (2011) Milestones in *Candida albicans* gene manipulation. Fungal Genet Biol: FG & B 48(9):858–865

Sampaio P, Gusmão L, Correia A, Alves C, Rodrigues AG, Pina-Vaz C et al (2005) New microsatellite multiplex PCR for *Candida albicans* strain typing reveals microevolutionary changes. J Clin Microbiol 43(8):3869–3876

Sánchez-Martínez C, Pérez-Martín J (2002) Site-specific targeting of exogenous DNA into the genome of *Candida albicans* using the FLP recombinase. Mol Genet Genomics 268(3):418–424

Santos MAS, Moura G, Massey SE, Tuite MF (2004) Driving change: the evolution of alternative genetic codes. Trends Genet: TIG 20(2):95–102

Santos MA, Gomes AC, Santos MC, Carreto LC, Moura GR (2011) The genetic code of the fungal CTG clade. C R Biol 334(8–9):607–611

Schellenberger J, Que R, Fleming RMT, Thiele I, Orth JD, Feist AM et al (2011) Quantitative prediction of cellular metabolism with constraint-based models: the COBRA Toolbox v2.0. Nature Protocols 6(9):1290–1307

Schindler D (2020) Genetic engineering and synthetic genomics in yeast to understand life and boost biotechnology. Bioengineering 7(4)

Schulze S, Schleicher J, Guthke R, Linde J (2016) How to predict molecular interactions between species? Front Microbiol 7:442

Segal E, Yehuda H, Droby S, Wisniewski M, Goldway M (2002) Cloning and analysis of CoEXG1, a secreted 1,3-β-glucanase of the yeast biocontrol agent *Candida oleophila*. Yeast 19(13):1171–1182

Shahana S, Childers DS, Ballou ER, Bohovych I, Odds FC, Gow NA et al (2014) New clox systems for rapid and efficient gene disruption in *Candida albicans*. PLoS One 9(6):e100390

Shen J, Guo W, Köhler JR (2005) CaNAT1, a heterologous dominant selectable marker for transformation of *Candida albicans* and other pathogenic *Candida* species. Infect Immun 73(2):1239–1242

Shen X-X, Zhou X, Kominek J, Kurtzman CP, Hittinger CT, Rokas A (2016) Reconstructing the backbone of the saccharomycotina yeast phylogeny using genome-scale data. G3 Genes Genomes Genet 6(12):3927–3939

Shi J, Zhang M, Zhang L, Wang P, Jiang L, Deng H (2014) Xylose-fermenting *Pichia stipitis* by genome shuffling for improved ethanol production. Microb Biotechnol 7(2):90–99

Shin M, Kim JW, Ye S, Kim S, Jeong D, Lee DY et al (2019) Comparative global metabolite profiling of xylose-fermenting *Saccharomyces cerevisiae* SR8 and *Scheffersomyces stipitis*. Appl Microbiol Biotechnol 103(13):5435–5446

Singh Dhillon G, Kaur Brar S, Verma M, Tyagi RD (2011) Recent advances in citric acid bioproduction and recovery. Food Bioprocess Technol 4(4):505–529

Smale ST (2010) Beta-galactosidase assay. Cold Spring Harb Protoc 2010(5):pdb.prot5423

Spasskaya DS, Kotlov MI, Lekanov DS, Tutyaeva VV, Snezhkina AV, Kudryavtseva AV et al (2021) CRISPR/Cas9-mediated genome engineering reveals the contribution of the 26S proteasome to the extremophilic nature of the yeast *Debaryomyces hansenii*. ACS Synthetic Biol

Sprengel R, Hasan MT (2007) Tetracycline-controlled genetic switches. Handb Exp Pharmacol 178:49–72

Sreenath HK, Jeffries TW (1999) 2-Deoxyglucose as a selective agent for derepressed mutants of *Pichia stipitis*. Appl Biochem Biotechnol 77(1):211–222

Srikantha T, Klapach A, Lorenz WW, Tsai LK, Laughlin LA, Gorman JA et al (1996) The sea pansy *Renilla reniformis* luciferase serves as a sensitive bioluminescent reporter for differential gene expression in *Candida albicans*. J Bacteriol 178(1):121–129

Stagljar I (2016) The power of OMICs. Biochem Biophys Res Commun 479(4):607–609

Staib P, Kretschmar M, Nichterlein T, Köhler G, Michel S, Hof H et al (1999) Host-induced, stage-specific virulence gene activation in *Candida albicans* during infection. Mol Microbiol 32(3):533–546

Staib P, Kretschmar M, Nichterlein T, Hof H, Morschhäuser J (2000a) Differential activation of a *Candida albicans* virulence gene family during infection. PNAS 97(11):6102–6107

Staib P, Michel S, Köhler G, Morschhäuser J (2000b) A molecular genetic system for the pathogenic yeast *Candida dubliniensis*. Gene 242(1–2):393–398

Stenuit B, Agathos SN (2015) Deciphering microbial community robustness through synthetic ecology and molecular systems synecology. Curr Opin Biotechnol 33:305–317

Sternberg N, Hamilton D (1981) Bacteriophage P1 site-specific recombination. I. Recombination between loxP sites. J Mol Biol 150(4):467–486

Stoneman HR, Wrobel RL, Place M, Graham M, Krause DJ, De Chiara M et al (2020) CRISpy-pop: a web tool for designing CRISPR/Cas9-driven genetic modifications in diverse populations. G3 Genes Genomes Genet

Sun C, Huang Y, Lian S, Saleem M, Li B, Wang C (2021) Improving the biocontrol efficacy of *Meyerozyma guilliermondii* Y-1 with melatonin against postharvest gray mold in apple fruit. Postharvest Biol Technol 171:111351

Tang S-J, Sun K-H, Sun G-H, Chang T-Y, Wu W-L, Lee G-C (2003) A transformation system for the nonuniversal CUGSer codon usage species *Candida rugosa*. J Microbiol Methods 52(2):231–238

Tanner FW Jr, Vojnovich C, Van Lanen JM (1945) Riboflavin production by *Candida* species. Science 101(2616):180–1

Treu R, Falandysz J (2017) Mycoremediation of hydrocarbons with basidiomycetes—a review. J Environ Sci Health B 52(3):148–155

Tsoi R, Wu F, Zhang C, Bewick S, Karig D, You L (2018) Metabolic division of labor in microbial systems. PNAS 115(10):2526–2531

Tsui CKM, Daniel H-M, Robert V, Meyer W (2008) Re-examining the phylogeny of clinically relevant *Candida* species and allied genera based on multigene analyses. FEMS Yeast Res 8(4):651–659

Turner SA, Butler G (2014) The *Candida* pathogenic species complex. Cold Spring Harb Perspect Med 4(9):a019778

Uhl MA, Johnson AD (2001) Development of *Streptococcus thermophilus* lacZ as a reporter gene for *Candida albicans*. Microbiology 147(Pt 5):1189–1195

Unrean P, Jeennor S, Laoteng K (2016) Systematic development of biomass overproducing *Scheffersomyces stipitis* for high-cell-density fermentations. Synth Syst Biotechnol 1(1):47–55

Ur-Rehman S, Mushtaq Z, Zahoor T, Jamil A, Murtaza MA (2015) Xylitol: a review on bio-production, application, health benefits, and related safety issues. Crit Rev Food Sci Nutr 55(11):1514–1528

Varga E, Maraz A (2002) Yeast cells as sources of essential microelements and vitamins B1 and B2. Acta Alimentaria 31:393–405

Van Vleet JH, Jeffries TW (2009) Yeast metabolic engineering for hemicellulosic ethanol production. Curr Opin Biotechnol 20(3):300–306

Veras HCT, Campos CG, Nascimento IF, Abdelnur PV, Almeida JRM, Parachin NS (2019) Metabolic flux analysis for metabolome data validation of naturally xylose-fermenting yeasts. BMC Biotechnol 19(1):58

Vyas VK, Barrasa MI, Fink GR (2015) A *Candida albicans* CRISPR system permits genetic engineering of essential genes and gene families. Sci Adv 1(3):e1500248

Wang X, Li G, Deng Y, Yu X, Chen F (2006) A site-directed integration system for the nonuniversal CUG(Ser) codon usage species *Pichia farinosa* by electroporation. Arch Microbiol 184(6):419–424

Wang Y, Chu J, Zhuang Y, Wang Y, Xia J, Zhang S (2009) Industrial bioprocess control and optimization in the context of systems biotechnology. Biotechnol Adv 27(6):989–995

Wang H, La Russa M, Qi LS (2016a) CRISPR/Cas9 in genome editing and beyond. Annu Rev Biochem 85:227–264

Wang S, Cheng G, Joshua C, He Z, Sun X, Li R et al (2016b) Furfural tolerance and detoxification mechanism in *Candida tropicalis*. Biotechnol Biofuels 9:250

Wang J, Peng J, Fan H, Xiu X, Xue L, Wang L et al (2018) Development of mazF-based markerless genome editing system and metabolic pathway engineering in *Candida tropicalis* for producing long-chain dicarboxylic acids. J Ind Microbiol Biotechnol 45(11):971–981

Wang H, Roelants SL, To MH, Patria RD, Kaur G, Lau NS et al (2019) *Starmerella bombicola*: recent advances on sophorolipid production and prospects of waste stream utilization. J Chem Technol Biotechnol 94(4):999–1007

Wang J, Liu Y, Zhao G, Gao J, Liu J, Wu X et al (2020) Integrated proteomic and metabolomic analysis to study the effects of spaceflight on *Candida albicans*. BMC Genomics 21(1):57

Wei L, Liu J, Qi H, Wen J (2015) Engineering *Scheffersomyces stipitis* for fumaric acid production from xylose. Biores Technol 187:246–254

Werner N, Dreyer M, Wagner W, Papon N, Rupp S, Zibek S (2017) *Candida guilliermondii* as a potential biocatalyst for the production of long-chain α,ω-dicarboxylic acids. Biotechnol Lett 39

Wirsching S, Moran GP, Sullivan DJ, Coleman DC, Morschhäuser J (2001) MDR1-mediated drug resistance in *Candida dubliniensis*. Antimicrob Agents Chemother 45(12):3416–3421

Wohlbach DJ, Kuo A, Sato TK, Potts KM, Salamov AA, Labutti KM et al (2011) Comparative genomics of xylose-fermenting fungi for enhanced biofuel production. PNAS 108(32):13212–13217

Xiang Z, Chen X, Zhang L, Shen W, Fan Y, Lu M (2014) Development of a genetic transformation system for *Candida tropicalis* based on a reusable selection marker of URA3 gene. Hereditas (beijing) 36:1053–1061

Yan W, Qian X, Zhang W, Zhou J, Weiliang D, Xu B et al (2020) Enhanced 2-phenylethanol production by newly isolated *Meyerozyma* sp. strain YLG18 and characterization of its synthetic pathways 140:109629

Yan W, Gao H, Qian X, Jiang Y, Zhou J, Dong W et al (2021) Biotechnological applications of the non-conventional yeast *Meyerozyma guilliermondii*. Biotechnol Adv 46:107674

Yan et al (2021) Biotechnological applications of the non-conventional yeast *Meyerozyma guilliermondii*. https://doi.org/10.1016/j.biotechadv.2020.107674

Yehuda H, Droby S, Wisniewski M, Goldway M (2002) A transformation system for the biocontrol yeast, *Candida oleophila*, based on hygromycin B resistance. Curr Genet 40:282–287

Yoon G-S, Tae-Sik L, Chul K, Jin-Ho S, Yeon-Woo R (1996) Characterization of alcohol fermentation and segregation of protoplast fusant of *Saccharomyces cerevisiae* and *Pichia stipitis*. J Microbiol Biotechnol 6(4):286–291

Younes S, Bracharz F, Awad D, Qoura F, Mehlmer N, Brueck T (2020) Microbial lipid production by oleaginous yeasts grown on *Scenedesmus obtusiusculus* microalgae biomass hydrolysate. Bioprocess Biosyst Eng 43(9):1629–1638

Yu KO, Jung J, Kim SW, Park CH, Han SO (2012) Synthesis of FAEEs from glycerol in engineered *Saccharomyces cerevisiae* using endogenously produced ethanol by heterologous expression of an unspecific bacterial acyltransferase. Biotechnol Bioeng 109(1):110–115

Zhang C, Konopka JB (2010) A photostable green fluorescent protein variant for analysis of protein localization in *Candida albicans*. Eukaryot Cell 9(1):224–226

Zhang Y, Jia D, Sun W, Yang X, Zhang C, Zhao F et al (2018) Semicontinuous sophorolipid fermentation using a novel bioreactor with dual ventilation pipes and dual sieve-plates coupled with a novel separation system. Microb Biotechnol 11(3):455–464

Zhang X, Li B, Zhang Z, Chen Y, Tian S (2020) Antagonistic yeasts: a promising alternative to chemical fungicides for controlling postharvest decay of fruit. J Fungi 6(3):158

The manufacturer's authorised representative in the EU is Springer
Nature Customer Service Centre GmbH, Europaplatz 3, 69115 Heidelberg,
Germany. If you have any concerns regarding our products, please
contact ProductSafety@springernature.com

Printed and bound by CPI Group (UK) Ltd, Croydon, CR0 4YY
30/09/2026
02225462-0014